RADIO AND TELEVISION SERVICING

Editor
R. N. WAINWRIGHT
T. Eng. (CEI), F.S.E.R.T.

Radio and Television Servicing
1982–83 Models

MACDONALD & CO.
LONDON & SYDNEY

Copyright © Macdonald & Co. (Publishers) Ltd.
First published in 1983 by
Macdonald & Co. (Publishers) Ltd.
London & Sydney
Maxwell House
74 Worship Street
London EC2A 2EN

ISBN 0 356 09203 8

Filmset, printed and bound in Great Britain by
Hazell Watson & Viney Ltd, Aylesbury, Bucks

CONTENTS

PREFACE

At the present time, when the television industry is awaiting the introduction of cable and satellite links to extend viewing choice, circuit design will be generally found to be a development of refinement of earlier techniques. In certain instances however, a close inspection of some current models reviewed in this volume will reveal that facilities have already been included for additional modules to be fitted in anticipation of future trends. These become necessary to provide adaptation to the alternative colour systems and line standards in use around the world when received direct.

Perhaps the main innovation during the past year has been the release of the long-promised video disc player, delayed by difficulties in production of the discs themselves. Although offering an optimum quality of reproduction, their impact has been largely overlooked by a public which prefers the versatility of the now well-established video cassette-recorder at the expense of picture quality.

At the time of writing, the compact audio disc, using similar laser techniques, has been launched on the market with an initial list of titles to suit all tastes. Initial demonstrations have been most impressive and the system is worthy of enthusiastic support from the hi-fi buying public. Novelty in design, possibly showing the shape of things to come, is seen in a Sony clock radio where the digital read-out has been replaced by a time-telling voice synthesiser operated at the touch of a button.

In the now well-established format, the first part of this volume contains circuits and service information for the main chassis of a wide range of colour and monochrome television receivers, with cross-references to relevant information in earlier volumes being listed separately in the addendum. The second section presents a selection from the wide range of currently available audio equipment, including radios, record-players, cassette-recorders and combination audio systems. In view of the specialised knowledge and equipment required, no attempt is made in this volume to provide any guidance on the servicing of video cassette recorders. Most manufacturers offer short courses to familiarise engineers with the techniques of handling such equipment and it is stressed that no attempt should be made to interfere with any part of these machines without the prior study of the problems involved. A companion volume to *Radio and Television Servicing* dealing with the VCR is under consideration.

The valuable Supplementary Servicing Section to these volumes should not be overlooked as it provides a useful reference source of many modifications that have occurred in production. This section, which was first introduced in 1971–72 also suggests methods of dealing with unusual fault symptoms and may well repay a search for the solution to an elusive problem. Much other useful information, check lists and technical descriptions will also be found at the end of these earlier issues which emphasises the cumulative value of this collection of volumes.

In presenting this information, grateful thanks are offered to the manufacturers, distributors, and their representatives for the generous help and co-operation in providing the technical information used in the preparation of this volume. It is emphasised that modern electronic equipment is complex and sophisticated. Whilst every effort is made to ensure that the information given within these pages is correct, errors occurring in original service manuals may be reproduced here. The publishers and editor of this volume can therefore accept no responsibility for injury or damage to persons or property as a result of inaccuracies in circuits or text.

R.N.W.

TELEVISION SERVICING
(Colour and Monochrome)

ACKNOWLEDGEMENTS

Amstrad Consumer Electronics
Cap-10 Industries Ltd.
Doric Radio Ltd.
G.E.C. (Radio and Television) Ltd.
Grundig (G.B.) Ltd.
Hitachi Sales (U.K.) Ltd.
J.V.C. (U.K.) Ltd.
Luxor (U.K.) Ltd.
Mitsubishi Electric (U.K.) Ltd.
Philips Service
Photopia Ltd.
Sharp Electronics (U.K.) Ltd.
Sony (U.K.) Ltd.
Tatung (U.K.) Ltd.
Tech-Semco Ltd.
Teleton (U.K.) Ltd.
Thorn-E.M.I.-Ferguson
VEC (U.K.) Ltd.
Zanussi International (U.K.) Ltd.

AMSTRAD   Model CTV 1400

General Description: A portable colour television receiver comprising a main board with R.G.B. drives on the tube base. Integrated circuits are used in all small signal stages and a socket is provided for the connection of an earphone.

Note: A bridge rectifier is used directly across the mains rendering the chassis live irrespective of connections. An isolating transformer must be used when carrying out service on this chassis.

Mains Supplies: 220-240 volts, 50Hz.

Cathode Ray Tube: 370LHB22.

Loudspeaker: 8 ohms impedance.

Adjustments

Step	Function	Signal in	Signal out	Method	Remarks
1.	Black and White tracking	Tune to colour transmission	Monitor screen	Set Colour Control to minimum. Set Contrast Control to maximum	Ensure Colour Lock button is out
2	Black and White tracking continued	Off-tune set from transmission	Monitor screen	1. Set Normal/Service switch to SW601 to Service position (towards screen) 2. Set all Red, Blue and Green pre-sets VR801-805 to approx mid-way 3. Turn screen control on L.O.P.T. to min (fully anti-clockwise)	Check vertical frame of picture disappears
3	Black and White tracking continued	Off-tune set from transmission	Connect scope between TP12 (pin 7 IC604) and chassis	Adjust Brightness control so that square wave amplitude on scope is 2·5V peak to peak	
4	Black and White tracking continued	Off-tune set from transmission	Monitor screen	1. Adjust screen control on L.O.P.T. so that dim coloured line is seen on screen 2. Adjust upper-red and upper-blue drive pre-sets VR804 and 805 to achieve a near-white line; then adjust the green bias pre-set VR802 for the final white adjustment	Ensure scope is removed from circuit
5	Black and White tracking continued	Re-tune to transmission	Monitor screen	1. Set Normal/Service switch SW601 to Normal position (away from screen) 2. If necessary, make fine adjustments for correct Black and White using lower-red, lower-blue, and green bias pre-sets VR801, 803 and 802.	Check Black and White picture appears

9

Step	Function	Signal in	Signal out	Method	Remarks
6	Focus	1. Inject signal from T.V. pattern gen. via aerial socket (crosshatch + circle pattern) 2. Tune T.V. to pattern gen.	Monitor screen	Adjust Focus control on L.O.P.T. for finest pattern lines	Brightness, Contrast and Colour controls at normal operating settings
7	Vertical size	1. Inject signal from T.V. pattern gen. via aerial socket (crosshatch + circle pattern) 2. Tune T.V. to pattern gen	Monitor screen	1. Adjust Vertical hold control VR701 on rear of set for optimum picture lock 2. Adjust vertical size pre-set VR703 for perfect circle pattern	Brightness, Contrast and Colour controls at normal operating settings
8	Horizontal shift	1. Inject signal from T.V. pattern gen. via aerial socket (crosshatch + circle) pattern) 2. Tune T.V. to pattern gen	Monitor screen	1. Adjust Horizontal hold control VR702 for optimum picture lock 2. Adjust Horizontal shift pre-set VR616 to centralise circle between left and right side of screen	Brightness, Contrast and Colour controls at normal operating settings
9	R.F. A.G.C.	1. Inject signal from T.V. pattern gen. via aerial socket (crosshatch + circle pattern) 2. Tune T.V. to pattern gen. 3. Adjust pattern-gen. signal for 1mV O/P	Connect D.C. Voltmeter to A.G.C. pin on tuner (C404)	Adjust R.F. A.G.C. pre-set VR601 for 7·0V D.C.	Brightness, Contrast and Colour controls at normal operating settings
10	Sub-Brightness	1. Inject signal from T.V. pattern gen. via aerial socket (grey scale bars pattern) 2. Tune T.V. to pattern gen.	Monitor screen	Adjust Sub-Brightness pre-set VR704 so that 4 darkest bars in pattern appear black	Set Brightness control to min. Set Contrast control to max. Set Colour control to min. Ensure Colour Lock button is out

Step	Function	Signal in	Signal out	Method	Remarks
11	Sub-Contrast	1. Inject signal from T.V. pattern gen. via aerial socket (grey scale bars pattern) 2. Tune T.V. to pattern gen.	Monitor screen	Adjust Sub-Contrast pre-set VR607 so that 2 lightest bars in pattern appear white	Set Brightness control to max. Set Contrast control to min. Set Colour Control to min. Ensure Colour Lock button is out
12	Sub-Colour	1. Inject signal from T.V. pattern gen. via aerial socket (colour bar pattern) 2. Tune T.V. to pattern gen.	Monitor screen	Adjust Sub-Colour pre-set VR608 so that pattern appears black and white	Set Brightness control to max. Set Contrast control to max. Set Colour control to just above min. Ensure Colour Lock button is out
13	Auto-Colour	1. Inject signal from T.V. pattern gen. via aerial socket (colour bar pattern) 2. Tune T.V. to pattern gen.	Connect D.C. Voltmeter to pin 16IC 604 (wiper of VR615)	Adjust Auto-Colour pre-set VR615 for 5·0V D.C.	Set Brightness control to max. Set Contrast control to max. Set Colour control to mid position. Ensure Colour Lock button is in
14	Colour Delay Line	1. Inject signal from T.V. pattern gen. via aerial socket (Demodulator or Delay pattern) 2. Tune T.V. to pattern gen.	Monitor screen	1. Set coils L611 and L612 and pre-set VR611 to approx mid-way 2. Adjust trimmers TC601 and TC602 for nearest to correct demodulator or delay pattern 3. Make fine adjustments for final correct pattern using VR611 and, if necessary, L611 and L612 4. Repeat steps 2 and 3 until no further improvement is observed	Set Brightness control to max. Set Contrast control to max. Set Colour control to max. Ensure Colour Lock button is out

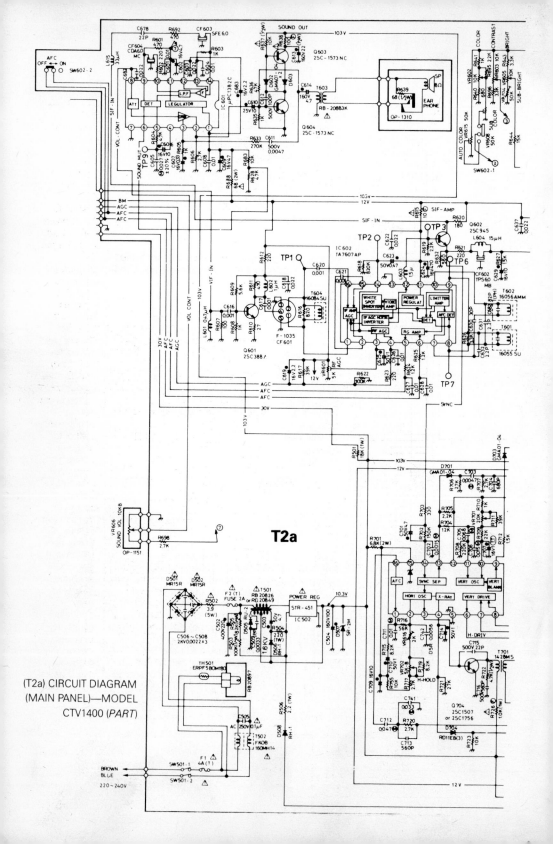

(T2a) CIRCUIT DIAGRAM
(MAIN PANEL)—MODEL
CTV1400 (*PART*)

T2a

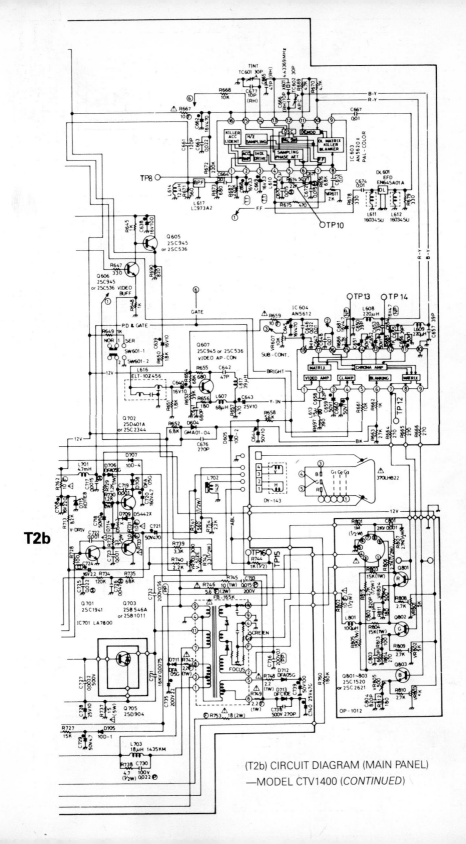

T2b

(T2b) CIRCUIT DIAGRAM (MAIN PANEL)
—MODEL CTV1400 *(CONTINUED)*

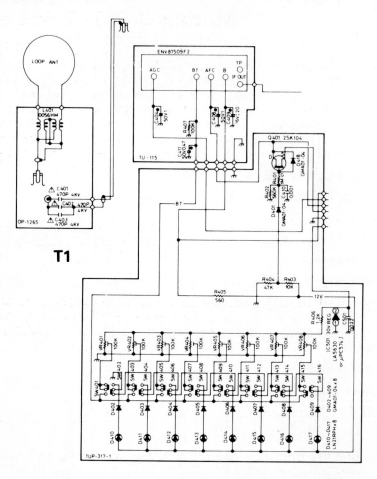

T1

(T1) CIRCUIT DIAGRAM (TUNER SECTION)—MODEL CTV 1400

BLAUPUNKT F.M. 120 Chassis, Models 7 660, 7 661 etc.

General Description: A mains-operated colour television chassis incorporating a 20, 22, or 26-in. pre-converged cathode ray tube and options for alternative audio systems (high power and stereo) and conversion to SECAM. The circuit of the basic PAL chassis is given here as a general guide. A switch-mode power supply is used which isolates the main chassis from the live supply but normal precautions should be observed when servicing this section of the receiver which is live irrespective of mains connections.

Mains Supply: 220 volts, 50Hz.

Cathode Ray Tubes: A51-420X, A56-701X, A67-701X.

Access for Service

After removing the rear cover, the main chassis may be hinged down to a horizontal position.

Adjustments (at normal operating temperature)

Adjustment Voltage U28/U34: U28 = +127V (A51-420X), U34 = +145V (A56-701X, A67-701X). Receive a transmission. Set contrast and brightness to minimum. V.T.V.M. to measuring point 534 and ground. With R420, adjust voltage.

Picture Height: R734.

Picture Width: R742.

East-West Equaliser: R737.

Centring (Vertical): R729.

Centring (Horizontal): Displacement by opening the diodes D778 (to the left) and D777 (to the right).

N.B.: Only one diode must be eliminated.

Horizontal Synchronisation: Shortcircuit MP800 to ground. With R697, adjust Horizontal frequency to beat. Remove short.

Definition (Focus): R785.

Integrated Circuit Functions

V310/TDA1035: 1. Stabilised Voltage. 2. Demodulator. 3. Electr. Volume Control.

V500/TD3300: 1. Brightness/Contrast Black Level Clamp. 2. U-Demodulator. 3. V-Demodulator. 4. Matrix. 5. R. 6. Driver and Chroma Control Stage.

7. ACC and Chroma Amp. 8. Burst Phase Detector. 9. 90° Phase Shifter. 10. H/2 Switch. 11. Beam Current Limiter. 12. G. 13. PAL Ident., Flip Flop, Killer. 14. 4.43MHz Oscillator. 15. 9V stab. 16. H+V Gating and Blanking Logic. 17. B.

W700: 1. Sync. Separator. 2. V-Pulse Amplifier. 3. Phase Correction. 4. Burst Detector and Blanking Stage. 5. Pulse Output Stage. 6. Trigger Pulse Stage. 7. Horizontal Oscillator. 8. Switchover. 9. Phase Comparator.

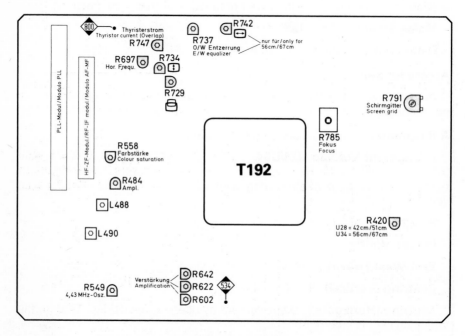

(T192) LOCATION OF PRE-SET CONTROLS—FM120 Chassis

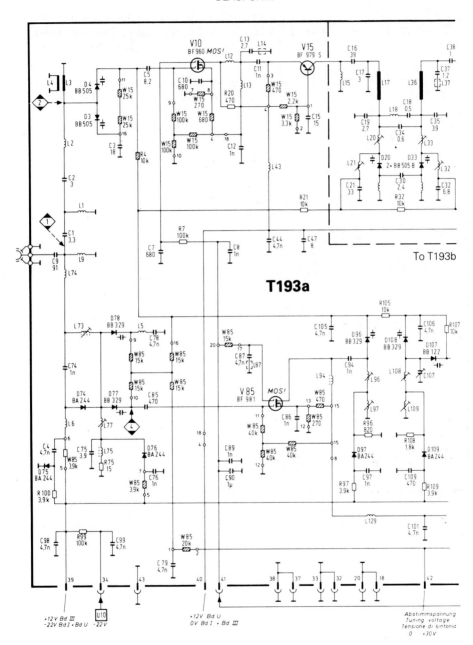

T193a

To T193b

(T193a) CIRCUIT DIAGRAM (R.F./I.F. MODULE)—FM120 CHASSIS (*PART*)

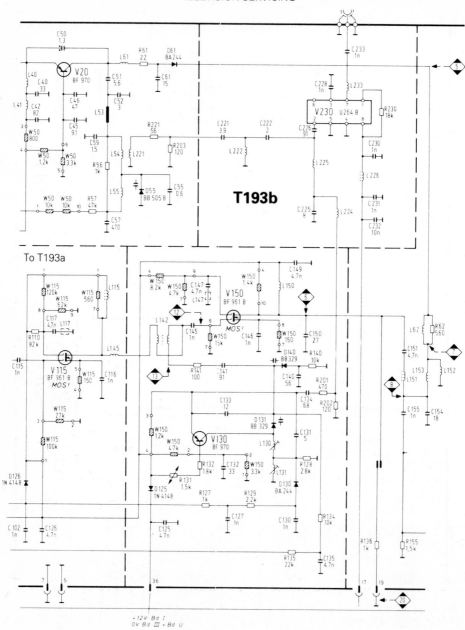

(T193b) CIRCUIT DIAGRAM (R.F./I.F. MODULE)—FM120 CHASSIS (*PART*)

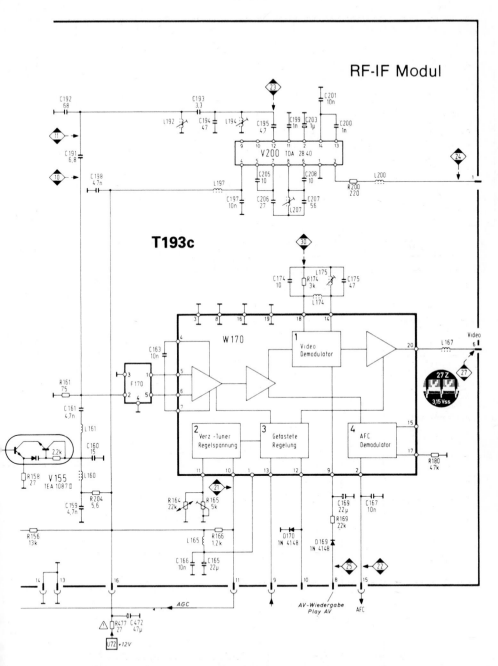

RF-IF Modul

T193c

(T193c) CIRCUIT DIAGRAM (R.F./I.F. MODULE)—FM120 CHASSIS (*CONTINUED*)

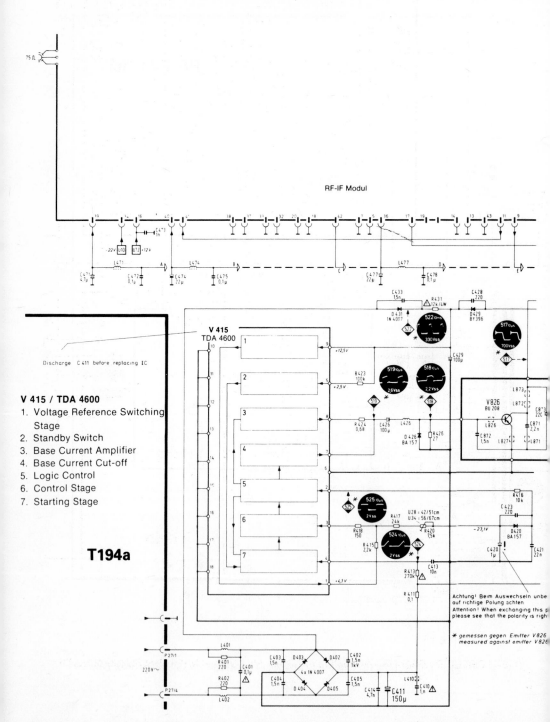

(T194a) CIRCUIT DIAGRAM (POWER SUPPLY AND A.F. MODULE)—FM120 CHASSIS (*PART*)

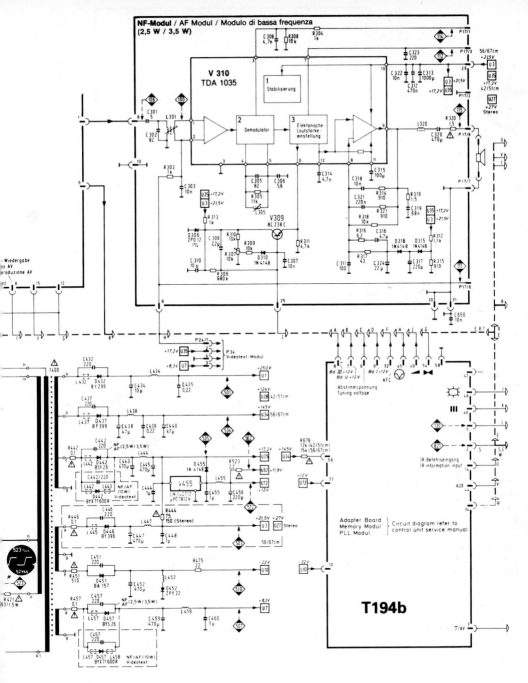

Chassis Board

(T194b) CIRCUIT DIAGRAM (POWER SUPPLY AND A.F. MODULE)—FM120 CHASSIS (*CONTINUED*)

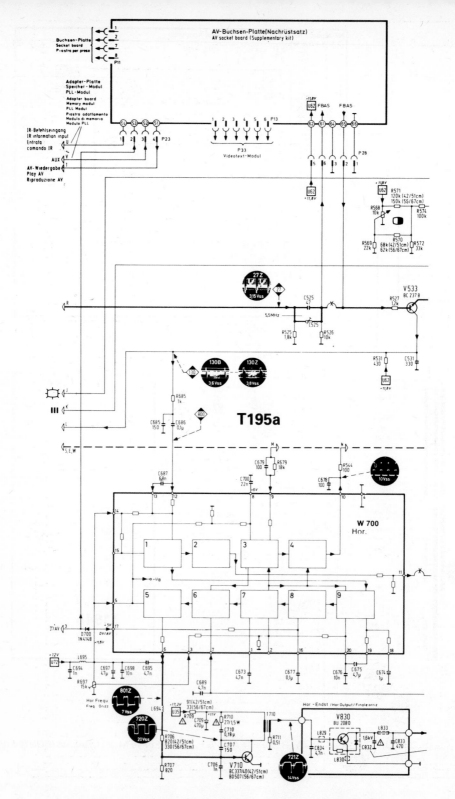

(T195a) CIRCUIT DIAGRAM (MAIN PANEL)—FM120 CHASSIS (*PART*)

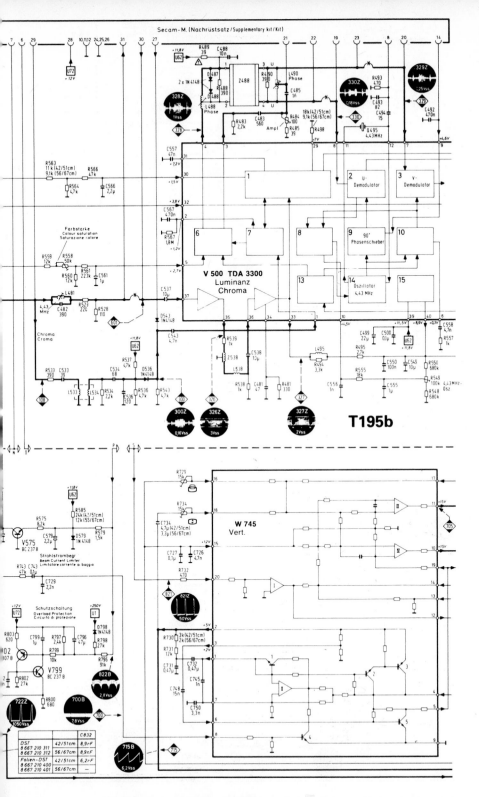

(T195b) CIRCUIT DIAGRAM (MAIN PANEL)—FM120 CHASSIS (*PART*)

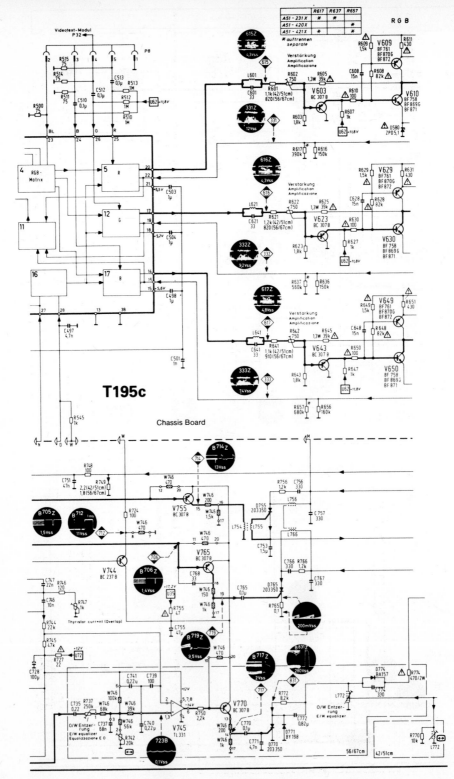

T195c

Chassis Board

(T195c) CIRCUIT DIAGRAM (MAIN PANEL)—FM120 CHASSIS (*PART*)

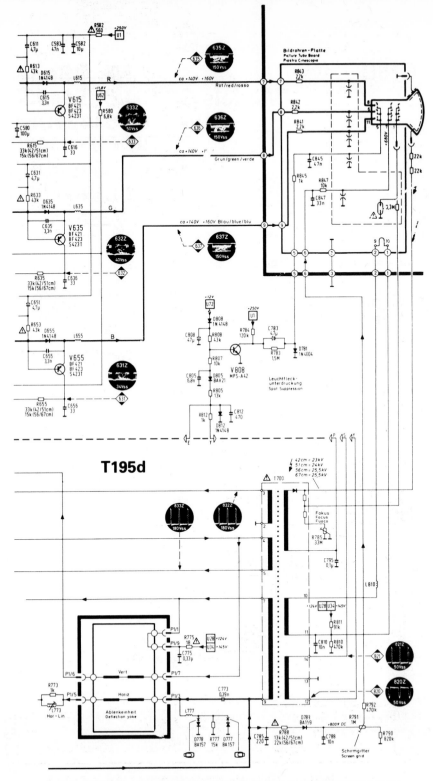

T195d

DORIC

Mk.4 Series, Models CD/CU/CV51402D, CD/CU/CV51404DRC, CD/CU/CV56402D, CD/CU/CV56404DRC, CD/CU/CV67402D, CD/CU/CV67404DRC

General Description: The Doric Mk.4 Series colour T.V. receivers are designed around the RCA S4 PIL range of picture tubes.

The 51cm receiver uses a 90° tube whilst the 56 and 67cm receivers use 110° tubes. The power supply and line scan panels differ between 90° and 110° versions.

All receivers use a switched mode power supply in which the S.M.P.S. transformer provides mains isolation of the receiver chassis. The receiver does not need an aerial isolator and direct audio and video connections are possible on all models.

Note: Approximately half the power supply circuitry is connected to the primary side of the S.M.P.S. transformer and is at mains potential. Care is necessary when servicing the receiver and the use of a separate isolation transformer is recommended.

The basic version with a conventional tuning system using a Preh 8 button tuner control unit and edge type rotary potentiometers for the customer controls. The de-luxe version has infra-red remote control and a frequency synthesis tuning system. On this model all the front panel controls are hidden behind a hinged flap located above the loudspeaker grille.

Adjustments

0RV1–TUNER A.G.C. TAKEOVER

Note: This adjustment is critical and should not be undertaken unless it is known that 0RV1 has been incorrectly set.

U.H.F. Signals Panel: Ensure that the receiver is switched off.

Connect the signal generator to the aerial input socket and adjust it for 1·5mV output at 600MHz.

Connect the D.C. voltmeter (2·5V range) across 0R17.

Set 0RV1 fully anti-clockwise.

Switch the receiver on and tune it to 600MHz (indicated by a minimum reading on the voltmeter).

Set 0RV1 fully clockwise, then turn it slowly anti-clockwise until the meter reading falls by 50mV.

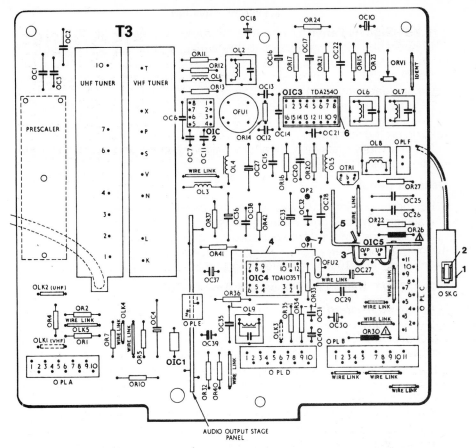

(T3) COMPONENT LAYOUT (SIGNALS PANEL—UNDERSIDE)—MK.4 SERIES

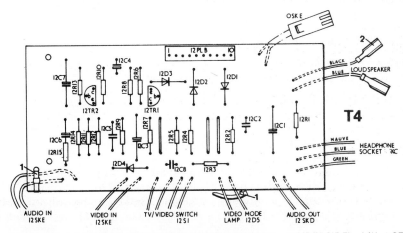

(T4) COMPONENT LAYOUT (AUDIO/VIDEO INTERFACE PANEL—UNDERSIDE)—MK. 4 SERIES

V.H.F. Signals Panel: Ensure that the receiver is switched off.

Connect the signal generator to the aerial input socket and adjust it for 1·5mV output at 200MHz.

Connect the D.C. voltmeter (10V range) between the junction of 0R17, 0R24 and earth (chassis).

Set 0RV1 fully anti-clockwise.

Switch the receiver on and tune it to 200MHz (indicated by a minimum reading on the voltmeter).

Set 0RV1 fully clockwise, then turn it slowly anti-clockwise until the meter reading just starts to fall.

Decoder panel

Equipment required: Pattern generator capable of providing phase, delay line adjustment, and grey scale patterns.

Oscilloscope and probe: bandwidth not less than 10MHz, sensitivity 100mV/cm or better.

1nF capacitor, 100nF capacitor, shorting link.

2L3, 2RV1. Delay Line Phase and Amplitude:

Preferred method:
Signal: None.

Connect the oscilloscope probe to link 2LK2 with the earth lead to the earthy end of 2C21. Disable the colour killer by connecting a link between the +12V line and pin 5 of 2IC1 (cathode of 2D1 to junction 2R2/2C3). Remove socket 2SKA and connect a 1nF capacitor between pins 1 and 11 of 2IC1. Switch the receiver on and adjust the oscilloscope for a line rate display. Adjust 2L3 and 2RV1 alternately for minimum sub-carrier output.

Alternative method:
Signal: Colour pattern generator's 'delay line adjustment' pattern.

Connect the pattern generator either to the aerial input or, if possible, to the video input; tune the receiver in or switch to video, as appropriate. Adjust 2L3 and 2RV1 for minimum delay line error.

2RV2A Oscillator Frequency:

Preferred method:
Signal: Colour pattern generator's 'phase' pattern.

Connect the pattern generator either to the aerial input or, if possible, to the video input; tune the receiver in or switch to video, as appropriate. Adjust 2RV2 for minimum static phase error.

Alternative method:
Signal: Any off-air picture.

Switch the receiver off. Disable the colour killer by connecting a link between the +12V line and pin 5 of 2IC1 (cathode of 2D1 to junction 2R2/2C3). Connect a 100nF capacitor between pins 8 and 13 of 2IC1, this will reduce the burst input to the phase detector to a very low value. Switch the receiver on

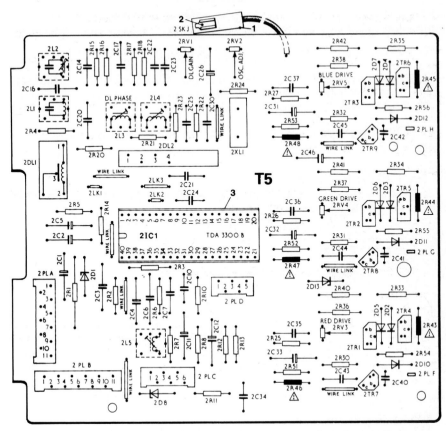

(T5) COMPONENT LAYOUT (DECODER PANEL—UNDERSIDE)—MK. 4 SERIES

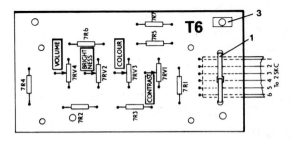

(T6) COMPONENT LAYOUT (POT. PANEL—UNDERSIDE)—MK. 4 SERIES

and adjust 2RV2 until the colour just runs through or is locked in the centre of the now much reduced locking range.

2RV3, 2RV4 and 2RV5. Red, Green, and Blue Drive Controls: Signal: Grey scale pattern.

Set 2RV3, 2RV4 and 2RV5 to maximum, i.e. fully clockwise. If necessary adjust any one or two controls to produce the correct peak white colour. Ensure at least one control remains at maximum.

Power Supply panel

Equipment required: D.C. voltmeter—accurate to $\pm 2\%$, preferably a digital meter.

4RV1. Set H.T.: Signal: Locked picture.

Set the Brightness, Colour, and Contrast controls to minimum and adjust 4RV1 for 110° receivers—150V; 90° receivers—124·6V, measured on the line stage supply rail. Connect the meter to the top end of 4L5 for this measurement.

Scan panel

3RV2. Line Hold: Signal: Off-air picture.

Switch off the receiver. Remove synchronisation by connecting a crocodile clip between pins marked 'OSC SET'. Switch on the receiver. Adjust 3RV2 until the picture is nearly stationary in the horizontal direction. Remove the crocodile clip.

3RV1. Line Shift: Signal: Test card.

Adjust 3RV1 to centre the picture in a horizontal direction.

3L1. Line Linearity: Signal: Test card or crosshatch.

Adjust 3L1 for best horizontal linearity.

3RV7 (110°), 3L3 (90°). Width: Signal: Test card.

110° receivers—Adjust 3RV7 for correct width. 90° receivers—Adjust 3L3 for correct width.

3RV6. E/W Amplitude (110° receivers only): Signal: Test card or crosshatch.

Adjust 3RV6 for straight vertical lines at the sides of the picture.

3RV3. Height: Signal: Test card.

Adjust 3RV3 for correct height.

3RV4. Field Linearity: Signal: Test card or crosshatch.

Adjust 3RV4 for correct vertical linearity.

3RV5. Field Shift: Signal: Test card.

Adjust 3RV5 to centre the picture in a vertical direction.

Focus: Signal: Test card or crosshatch.

Adjust the focus control (mounted on the tripler) for best focus at the picture centre.

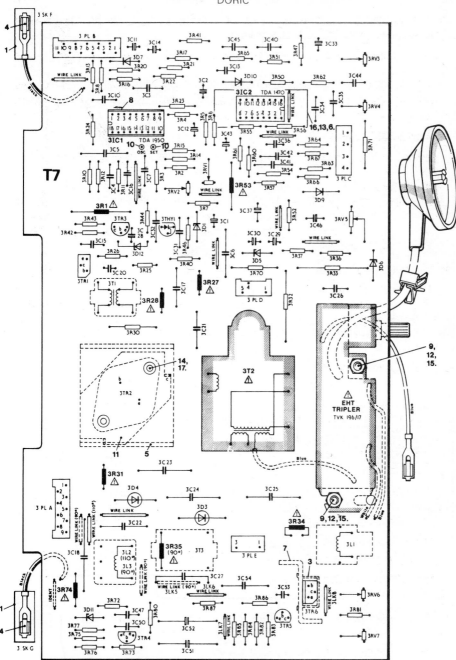

(T7) COMPONENT LAYOUT (SCAN PANEL—UNDERSIDE)—MK. 4 SERIES

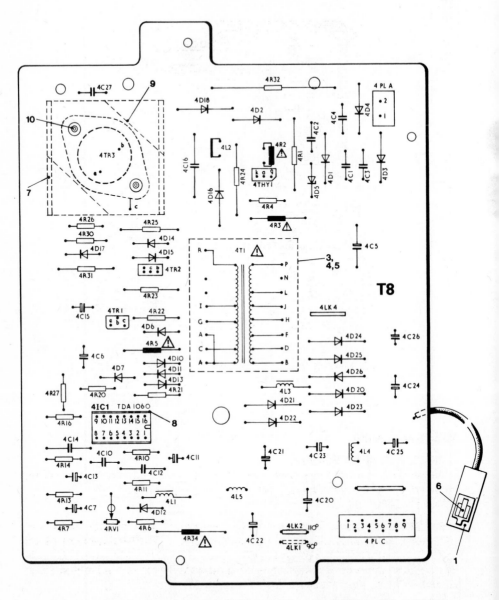

(T8) COMPONENT LAYOUT (POWER SUPPLY PANEL—UNDERSIDE)—MK. 4 SERIES

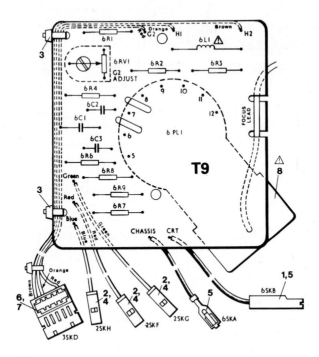

(T9) COMPONENT LAYOUT (TUBE BASE PANEL)—MK. 4 SERIES

Tube Base panel

6RV1. Adjust A1 Voltages:

Note: This control should not be touched unless known to have been incorrectly set or unless the picture tube has been changed.

Signal: Off-air picture.

Preferred method: Turn the Brightness, Colour, and Contrast controls to minimum. Using a carefully set-up oscilloscope look at the waveforms on the collectors of 2TR1, 2TR2 and 2TR3; at about 5ms/cm. Measure D.C. level of negative going tip for each collector and note which is the highest level. Adjust 6RV1 so that this level is 160V.

Alternative method (only to be used if the preferred method is not possible): Turn the Brightness, Colour, and Contrast controls to minimum. Using a D.C. voltmeter measure the collector voltages of 2TR1, 2TR2 and 2TR3, and note which voltage is greatest. Adjust 6RV1 so that this voltage is 185V.

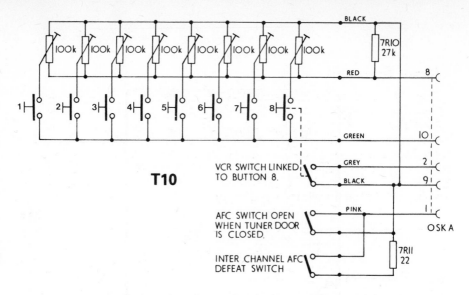

T10

VCR SWITCH LINKED TO BUTTON 8.

AFC SWITCH OPEN WHEN TUNER DOOR IS CLOSED.

INTER CHANNEL AFC DEFEAT SWITCH

8 button Tuner Control Unit without Band-switch

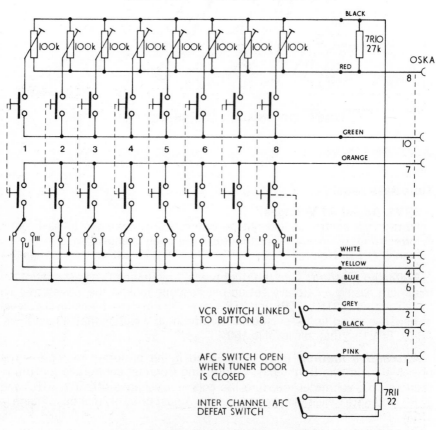

VCR SWITCH LINKED TO BUTTON 8

AFC SWITCH OPEN WHEN TUNER DOOR IS CLOSED

INTER CHANNEL AFC DEFEAT SWITCH

8 button Tuner Control Unit with Band-switch

(T10) CIRCUIT DIAGRAM (CONTROL UNITS)—MK. 4 SERIES

Circuit Waveforms

Waveforms were taken from a receiver operating from 240V mains. A Colour Bar signal was used and the Colour control was set for a normal saturation. A ÷10 passive low-capacitance probe was used for all the waveforms (note that the Y sensitivity figures represent the effective sensitivity and allow for the presence of the probe). Where appropriate the zero D.C. level with respect to chassis is indicated by E except for waveforms 31—38 where the zero D.C. level is with respect to Pin A of the power supply transformer 4T1 and is indicated by Ø.

T11a

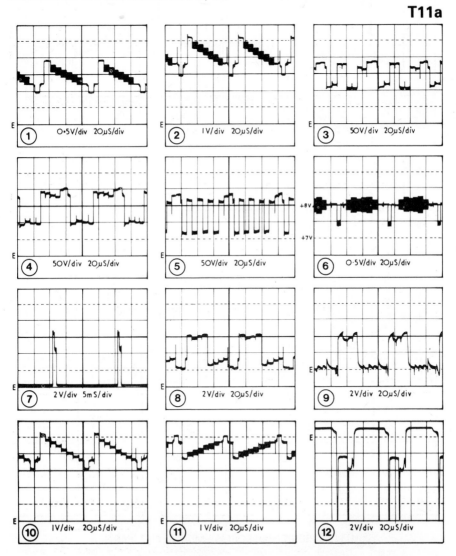

(T11a) CIRCUIT WAVEFORMS—MK. 4 SERIES (*PART*)

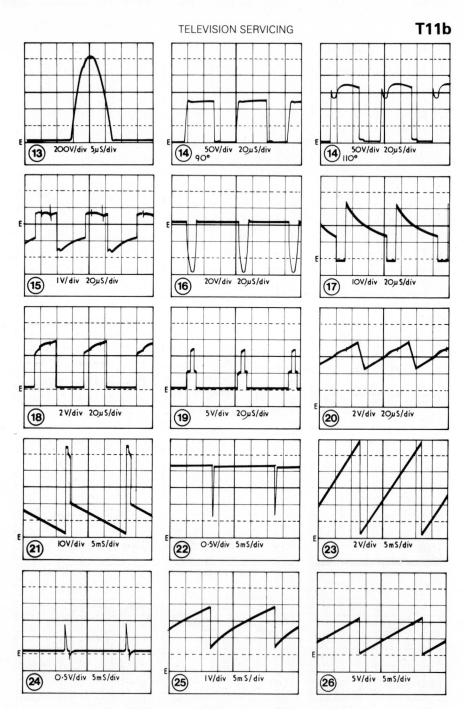

(T11b) CIRCUIT WAVEFORMS—MK. 4 SERIES (*PART*)

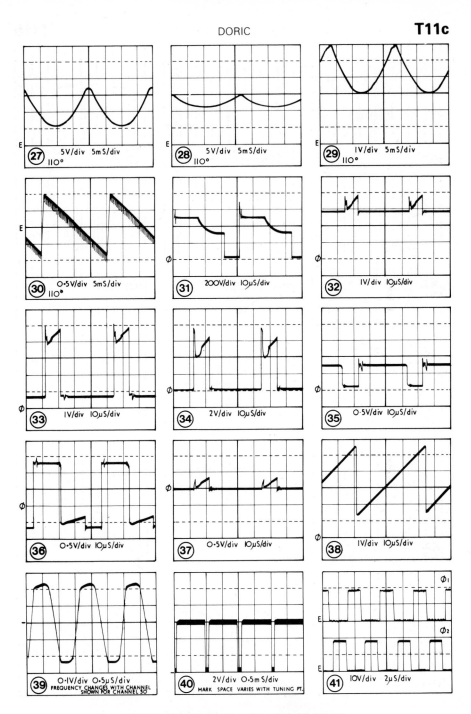

(T11c) CIRCUIT WAVEFORMS—MK. 4 SERIES (*PART*)

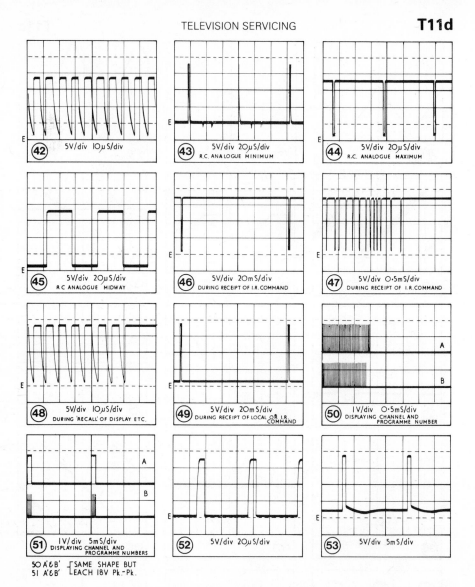

(T11d) CIRCUIT WAVEFORMS—MK. 4 SERIES (*CONTINUED*)

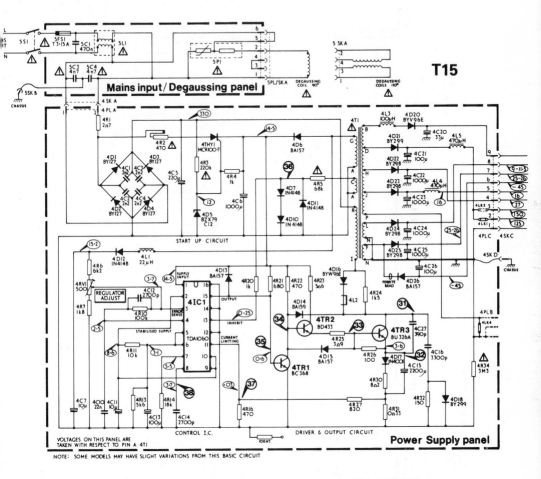

T15

(T15) CIRCUIT DIAGRAM (POWER SUPPLY PANEL)—MK. 4 SERIES

NOTE: SOME MODELS MAY HAVE SLIGHT VARIATIONS FROM THIS BASIC CIRCUIT

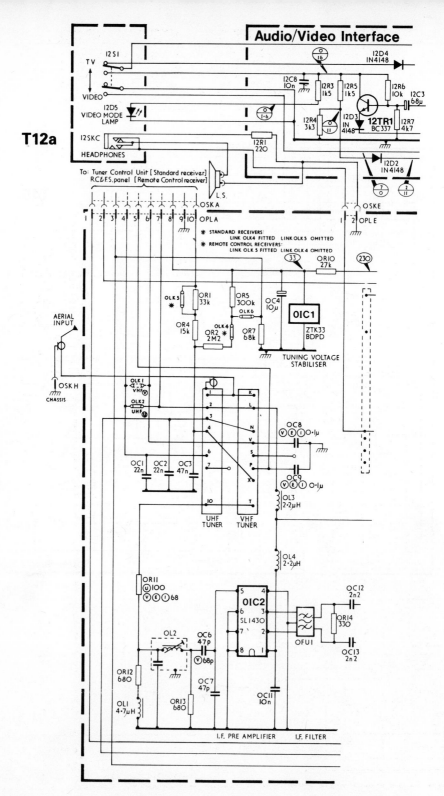

Audio/Video Interface

T12a

(T12a) CIRCUIT DIAGRAM (SIGNALS PANEL)—MK. 4 SERIES (*PART*)

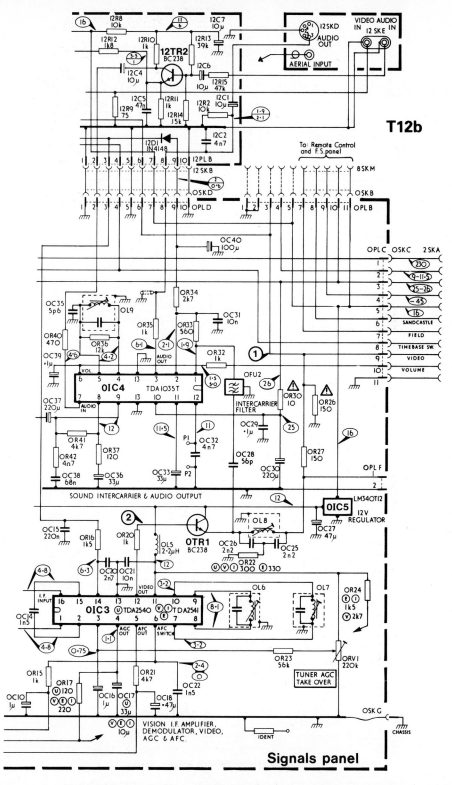

(T12b) CIRCUIT DIAGRAM (SIGNALS PANEL)—MK. 4 SERIES (*CONTINUED*)

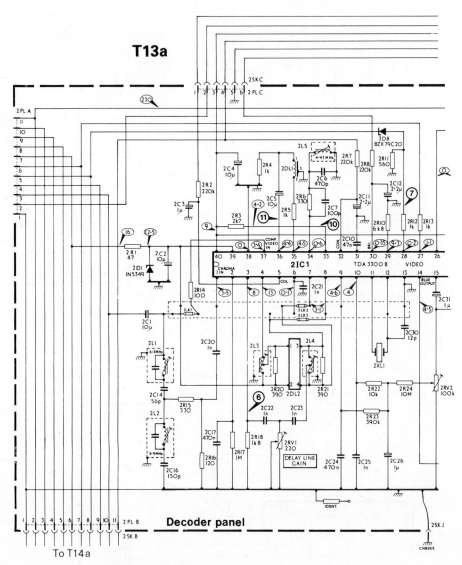

T13a

Decoder panel

To T14a

(T13a) CIRCUIT DIAGRAM (DECODER PANEL)—MK. 4 SERIES (*PART*)

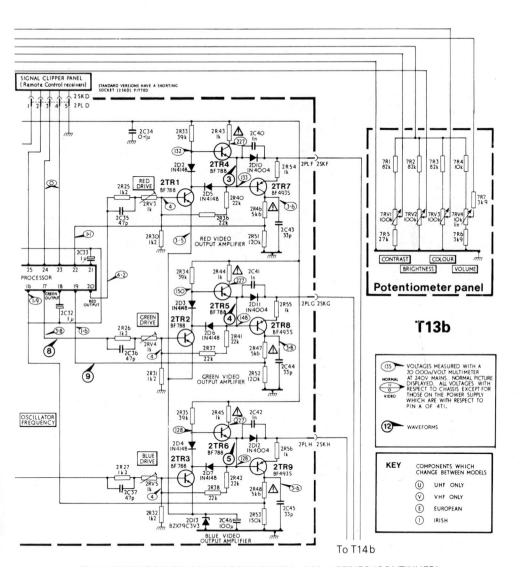

(T13b) CIRCUIT DIAGRAM (DECODER PANEL)—MK. 4 SERIES (*CONTINUED*)

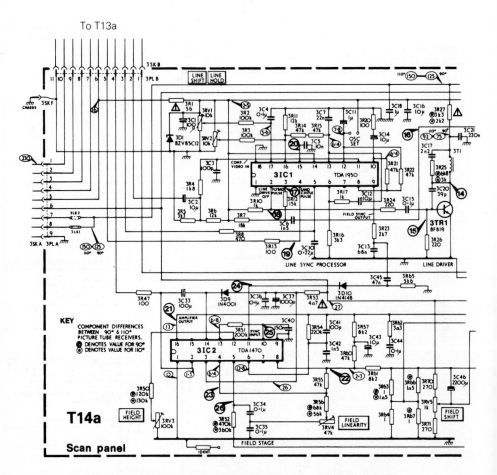

(T14a) CIRCUIT DIAGRAM (SCAN PANEL)—MK. 4 SERIES (*PART*)

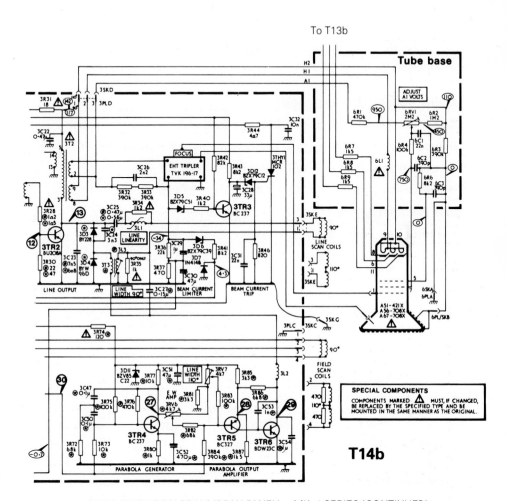

(T14b) CIRCUIT DIAGRAM (SCAN PANEL)— MK. 4 SERIES (*CONTINUED*)

FERGUSON   Model 3T14

General Description: A portable mains or battery operated entertainment centre comprising 3-in. monochrome television, A.M./F.M. radio and cassette tape-recorder with crystal controlled digital clock and alarm. A microphone is built in.

Mains Supply: 240 volts, 50Hz.

Batteries: 1·5 volts for clock (HP7); 9 volts (6×HP2) or 12 volt car battery.

(Note: Insertion of mains plug disconnects all batteries except HP7).

Consumption: Radio (65mA); T.V. or tape (650mA).

Wavebands: L.W. 150-250kHz; M.W. 540-1600kHz; F.M. 88-108MHz.

Bias Frequency: 36kHz plus 1·3kHz switchable.

Cathode Ray Tube: 85BV4.

Note: Clock battery must always remain fitted to ensure operation of receiver.

Adjustments

Height Adjustment: Adjust to partially exclude castellations at top and bottom of picture.
Inject 1mV from a pattern generator into external aerial socket and adjust RV702 for a well defined circle.

Vertical Linearity: Inject signal as for Height Adjustment but adjust RV704 to give a good circle shape.
To obtain a well defined circle it is advisable to repeat Height Adjustment and Vertical Linearity a few times to produce best results.

Vertical Hold: Inject a 1mV signal and adjust RV701 for a stationary image on the screen.

Horizontal Hold: Set RV705 to maximum. Connect a frequency counter to pins 1 and 2 of P2. Inject a 1mV signal and adjust T701 for a reading of 14·2kHz on counter.
With a voltmeter connected to TP9 adjust RV705 for a voltage reading of 0·85V at TP9.

R.F. A.G.C. Adjustment: Connect a voltmeter between TP7 and chassis.
Inject a 10mV signal from a pattern generator into the external aerial socket and adjust RV601 for 2·9V at TP7.

Tape Recorder Adjustments

Azimuth Adjustment: Playback a standard azimuth tape, and with an output meter connected in place of the loudspeaker, adjust the azimuth

46

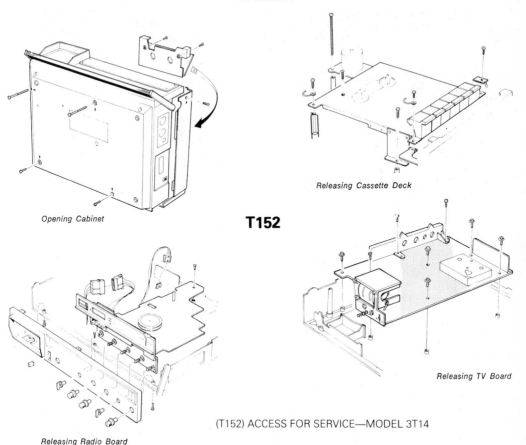

Opening Cabinet

T152

Releasing Cassette Deck

Releasing TV Board

Releasing Radio Board

(T152) ACCESS FOR SERVICE—MODEL 3T14

screw (nearest erase head) for a maximum meter reading whilst reducing the volume control to keep the output at a low level.

Record Bias Adjustment: Press 'Record' and 'Play' keys, and connect a sensitive voltmeter across R204. Adjust RV201 for a reading of 4·8mV. This corresponds approximately to a bias current of 480μA.

Radio Alignment Procedure

F.M. I.F. Alignment: Switch Selector to 'Radio' and Radio Band to 'V.H.F.'. Alignment can best be obtained using a V.H.F. wobbulator and a display unit. The wobbulator should be terminated with a 75 ohm resistor across the output and a capacitator of 10nF connected in series with the 'live' lead to the injection point. Connect the display unit to TP12; set the volume control at maximum and the tuning gang at minimum capacitance. Inject 10·7MHz signals to TP11. Adjust T101, T102 for maximum output and then adjust T103

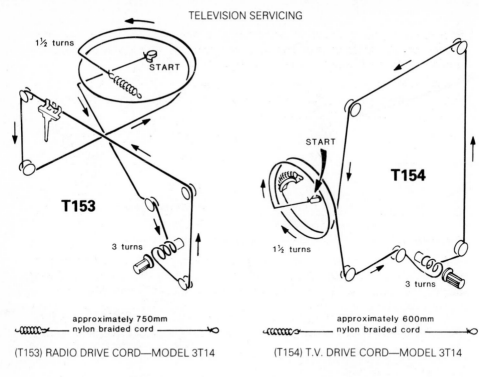

1½ turns

START

T153

3 turns

approximately 750mm nylon braided cord

(T153) RADIO DRIVE CORD—MODEL 3T14

START

T154

1½ turns

3 turns

approximately 600mm nylon braided cord

(T154) T.V. DRIVE CORD—MODEL 3T14

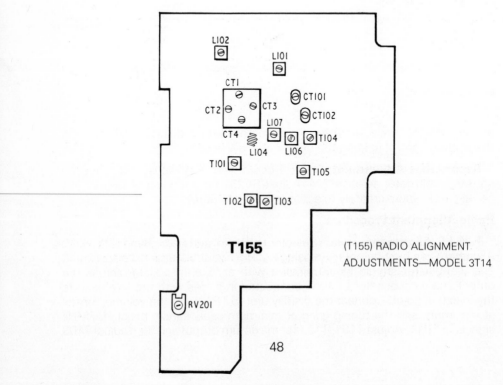

LI02
LI01
CTI
CT2 CT3
CT4
CT101
CT102
LI07
LI04 LI06 TI04
TI01
TI05
TI02 TI03

T155

(T155) RADIO ALIGNMENT ADJUSTMENTS—MODEL 3T14

RV201

to give an 'S' curve which is symmetrical above and below the base line. Switch on 10·7MHz marker and re-adjust T103, if necessary, to bring the marker to the centre of the straight portion of the 'S' curve.

F.M. R.F. Alignment: V.H.F./F.M./R.F. signals (22·5kHz deviation) are injected into TP11 and an output meter, set to 8 ohms impedance, is connected in place of the loudspeaker.

Range	Inject	Cursor Position	Adjust for Maximum
V.H.F.	86·5MHz	Extreme left	L104*
	109·5MHz	Extreme right	CT2
	90MHz	90MHz	L102
	106MHz	106MHz	CT1

Adjust by opening or closing coil turns

Repeat adjustments until no further improvement results.

A.M. I.F. Alignment: Connect an output meter, set to 8 ohms impedance, in place of the loudspeaker. Turn volume control to maximum. During alignment reduce the signal level as necessary to avoid exceeding 50mV.

Turn gang to maximum capacitance and select M.W.

Inject 468kHz (30% modulated) signals to TP13 and adjust T104 and T105 for maximum output. Repeat operations until no further improvement results.

A.M. R.F. Alignment: Signals are injected via a loop loosely coupled to the ferrite rod aerial.

Range	Inject	Cursor Position	Adjust for Maximum
M.W.	525kHz	Extreme left	L107
	1650kHz	Extreme right	CT4
	600kHz	600kHz	L105A*
	1400kHz	1400kHz	CT3
L.W.	142kHz	Extreme left	L106
	265kHz	Extreme right	CT102
	160kHz	160kHz	L105B*
	250kHz	250kHz	CT101

Adjust by sliding coils along ferrite rod

Repeat adjustments until no further improvement results.

Circuit Diagram Notes:

D.C. voltage measurements shown in rectangles were taken relative to the negative chassis line (except where otherwise indicated) under quiescent conditions with a 20,000 Ω/volt meter.

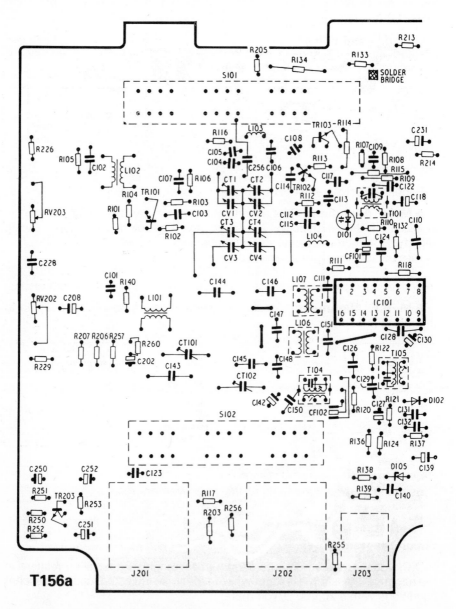

T156a

(T156a) COMPONENT LAYOUT (RADIO AND A.F. SECTIONS)—MODEL 3T14 (*PART*)

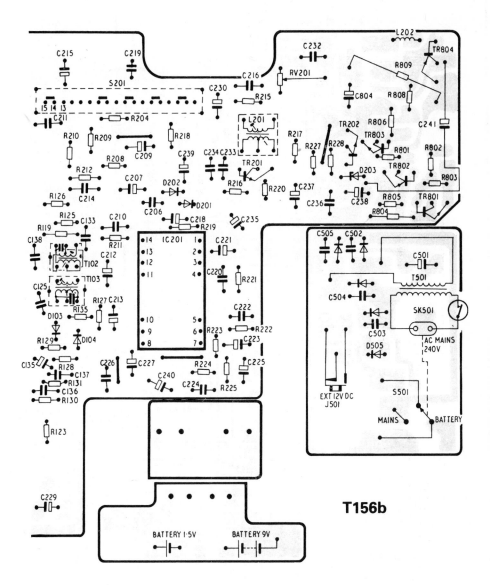

T156b

(T156b) COMPONENT LAYOUT (RADIO AND A.F. SECTIONS)—MODEL 3T14 (*CONTINUED*)

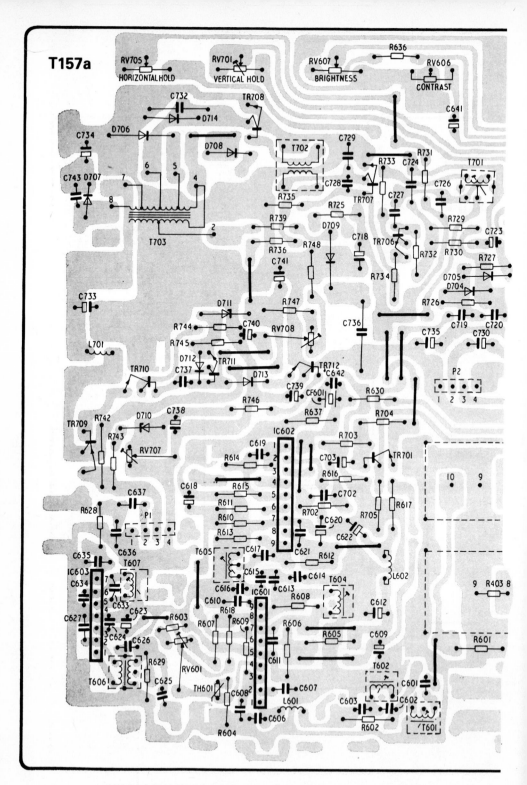

(T157a) COMPONENT LAYOUT (T.V. SECTION)—MODEL 3T14 (*PART*)

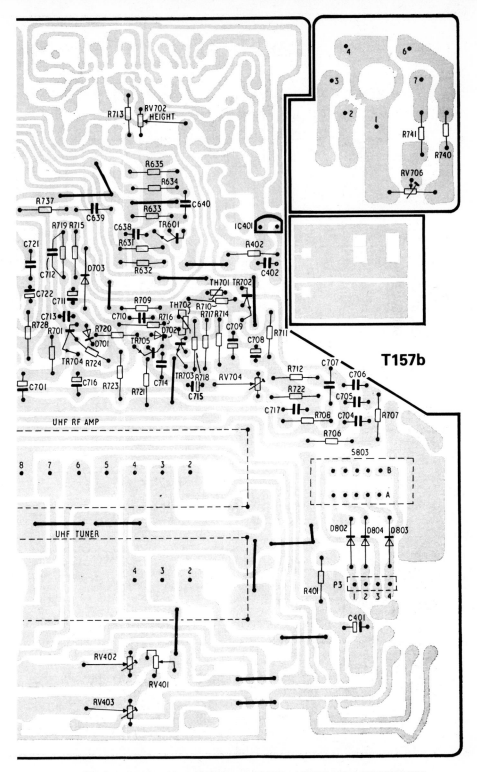

(T157b) COMPONENT LAYOUT (T.V. SECTION)—MODEL 3T14 (*CONTINUED*)

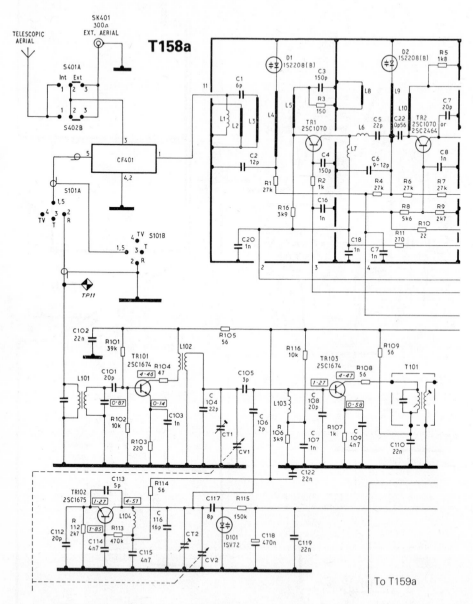

(T158a) CIRCUIT DIAGRAM (T.V. AND RADIO SECTION)—MODEL 3T14 (*PART*)

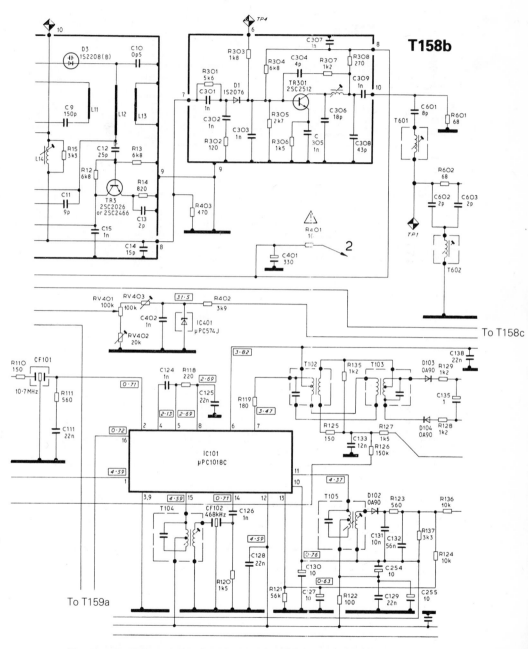

T158b

(T158b) CIRCUIT DIAGRAM (T.V. AND RADIO SECTION)—MODEL 3T14 (*PART*)

T158c

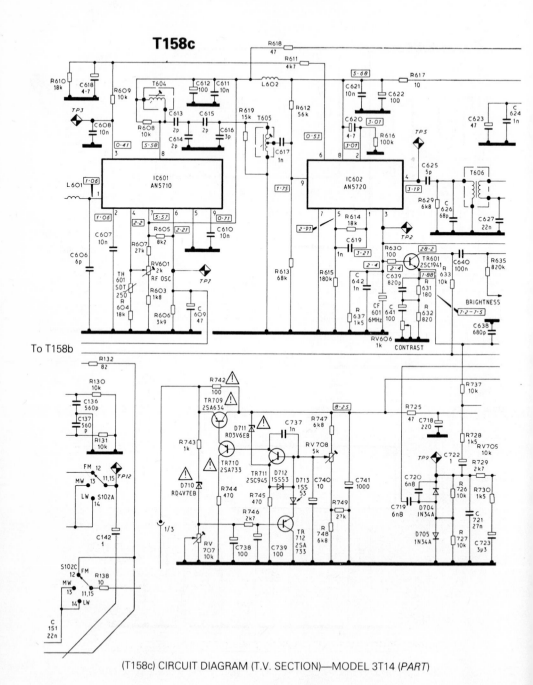

(T158c) CIRCUIT DIAGRAM (T.V. SECTION)—MODEL 3T14 (*PART*)

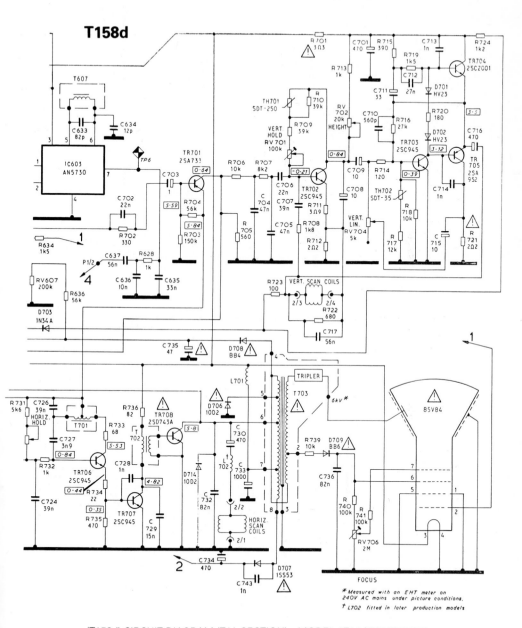

(T158d) CIRCUIT DIAGRAM (T.V. SECTION)—MODEL 3T14 (*CONTINUED*)

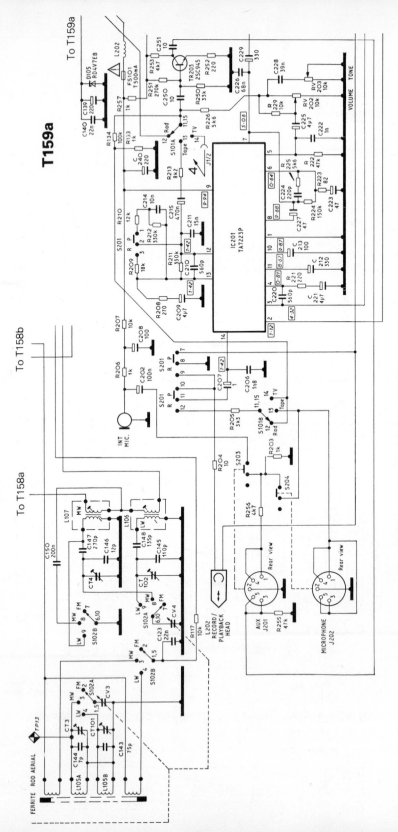

(T159a) CIRCUIT DIAGRAM (AUDIO SECTION)—MODEL 3T14 (*PART*)

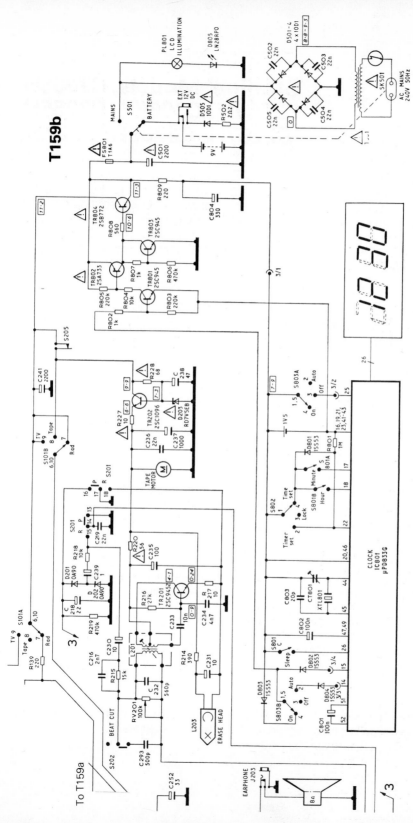

(T159b) CIRCUIT DIAGRAM (AUDIO SECTION)—MODEL 3T14 (CONTINUED)

G.E.C.

'Starline' Models, C2065H, C2067H, C2069H, C2265H, C2267H, C2269H, C2273H

General Description: A series of mains-operated colour television receivers incorporating a basic similar chassis. 20 or 22-in. versions are available with the option of manual or remote control with Teletext facilities in Models C2069, C2269. The basic chassis described here is that used in Model C2067 and may be used as a guide to all others in this group. The main differences occur in the R.G.B. drive panel and the teletext version of this is given separately.

Mains Supply: 240 volts, 50Hz.

Fuses: T2·5A (mains); T3·15A and thermal (C2067, C2267, C2273); T3·15A, 250mA, T500mA, T630mA (C2069, C2269).

Cathode Ray Tubes: 510VSB22 (20-in.); 569EGB22 (22-in.).

Access for Service (see Fig. T171)

Release leads from clips as required.
Release clips 'A' on both sides of chassis moulding.
Holding clips 'A' slide chassis to the rear of the cabinet and lift out.
Rotate chassis to a vertical position; slide leg 'B' into slot 'C' and hook strap over leg 'E'.
Care must be taken not to rotate chassis when parked as this could damage leg 'B'.
When chassis frame is restored to original position on completion of servicing, ensure that original lead dressing is followed.

Adjustments

Before attempting any adjustments, check H.T. line voltage:
Tune the receiver to a local channel.
Set Brightness, Contrast and Volume for best picture and sound.
Connect a voltmeter to cathode of D906 and ground.
H.T. reading should be +111V± 1·5V.
Line Oscillator: With receiver tuned to a local station, set Horizontal Hold control to its centre position.
Connect a 1µF/100V capacitor between TP701 and ground.
Adjust the line oscillator control to momentarily synchronise the picture.
Remove the capacitor.

Width Adjustment:
Note: The width should be adjusted under normal picture conditions.
To increase width, short-circuit the H.SIZE jumper lead.
To reduce width, cut the lead.

G.E.C.

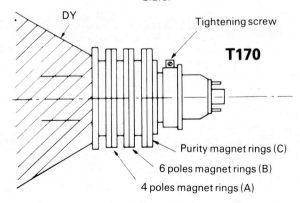

(T170) PURITY AND CONVERGENCE ASSEMBLY—MODELS C2065H ETC.

PARKING CHASSIS IN SERVICE POSITION

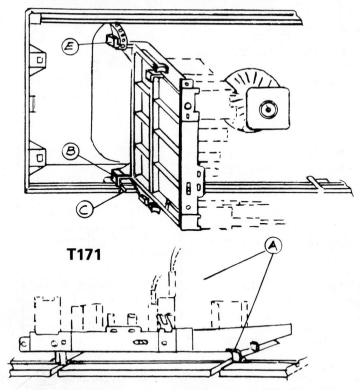

(T171) CHASSIS SERVICE POSITION—MODELS C2065H ETC.

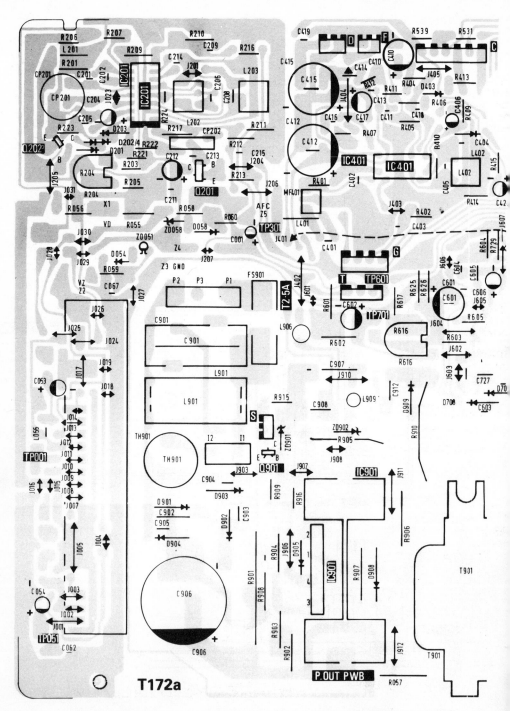

T172a

(T172a) COMPONENT LAYOUT (MAIN PANEL)—MODELS C2065H ETC. (*PART*)

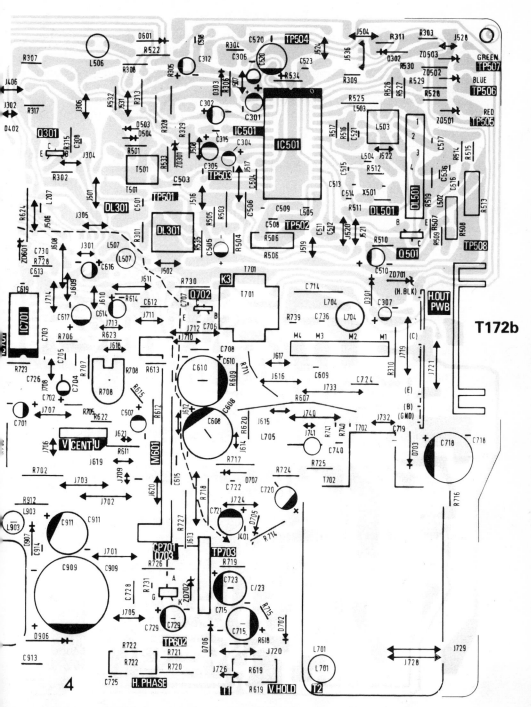

4

(T172b) COMPONENT LAYOUT (MAIN PANEL)—MODELS C2065H ETC. (*CONTINUED*)

Horizontal Shift: Display a test pattern and adjust R722 to position the left and right hand sides of the pattern, symmetrically.

Vertical Hold: To set the Vertical Hold-control to lock the field sync. under weak signal conditions:
Link TP601 and TP602 with a 240K resistor.
Adjust R619 to a point just before the picture begins to roll from top to bottom.
Remove the resistor.

Height Adjustment: Adjust R616 to obtain correct vertical amplitude.

Vertical Shift: With vertical hold properly set, connect the V.Cent link to N or U to centre the picture.

Focus: This is the upper control which projects from the line output transformer. Adjust as necessary.

Grey Scale (non-teletext models).

Set R.G.B. background controls R853, R863, R873 to midway position.
Set the Screen control (lower knob on line output transformer) to minimum anti-clockwise.
Set Red and Blue drive controls R856, R876 so that the screwdriver slots are parallel with the top of the P.C.B.
Turn the Brightness and Contrast controls to minimum.
Display a test pattern providing 100% white, e.g. a pulse and bar pattern.
Connect an oscilloscope to the green cathode.
Adjust the Brightness control to obtain 20V p-p.
Connect the oscilloscope to the red cathode and adjust R856 to obtain 20V p-p.
Connect the oscilloscope to the blue cathode and adjust R876 for 20V p-p.
Disconnect the oscilloscope and short-circuit the vertical scan coils by linking the extreme left and right hand pins nearest the screen, on the yoke assembly.
These pins are connected by green and yellow leads respectively, to pins 4 and 3 of plug M on the main circuit board.
Adjust the Screen control until a Red, Blue or Green line is just visible. Then adjust the background controls for the other colours to produce a white line.
Remove the short-circuit from the scan coils and adjust the Brightness control for a faint white raster.
Increase the contrast setting to check the high lights.
Adjust drive controls R856 and R876 if necessary.
Check grey scale low lights, correcting only with two background controls if necessary.

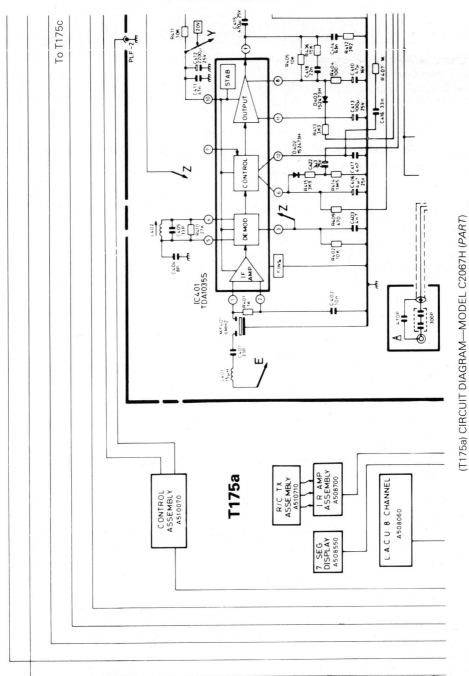

(T175a) CIRCUIT DIAGRAM—MODEL C2067H (PART)

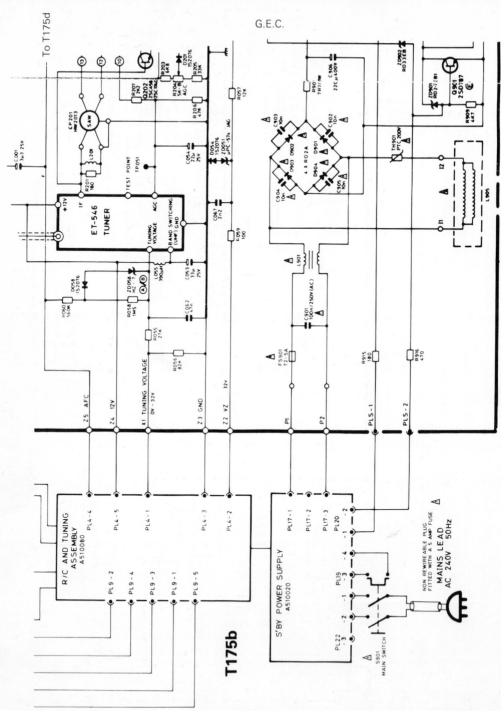

G.E.C.

T175b

T175EW CIRCUIT DIAGRAM. MODEL C990EW PART1

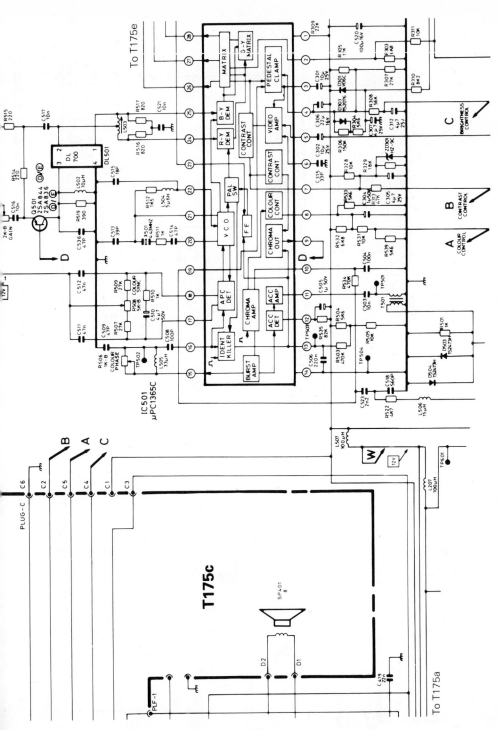

(T175c) CIRCUIT DIAGRAM—MODEL C2067H (*PART*)

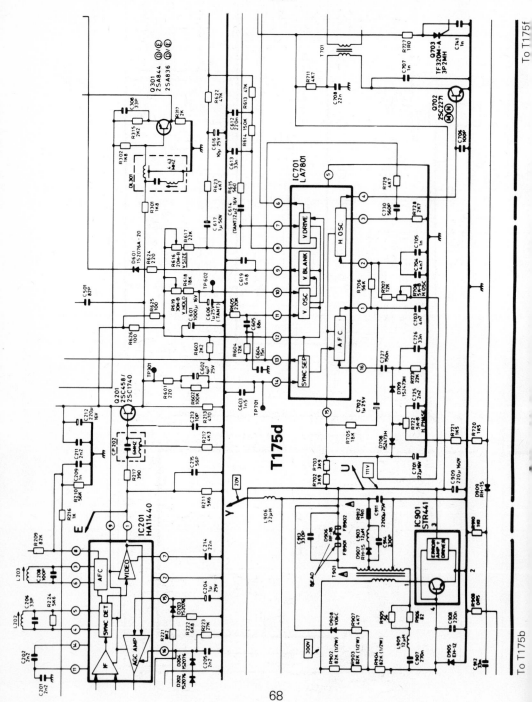

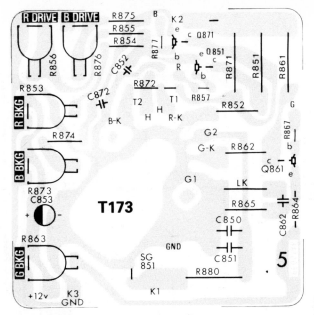

(T173) COMPONENT LAYOUT (C.R.T. BASE PC051)—MODELS C2065H ETC.

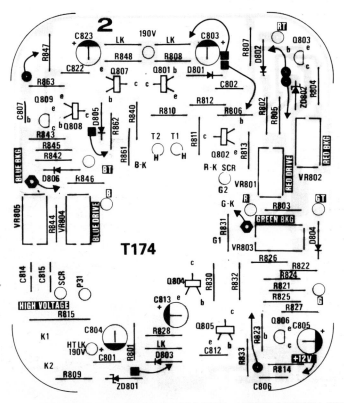

(T174) COMPONENT LAYOUT (C.R.T. BASE PC073)—MODELS C2069H, C2269H

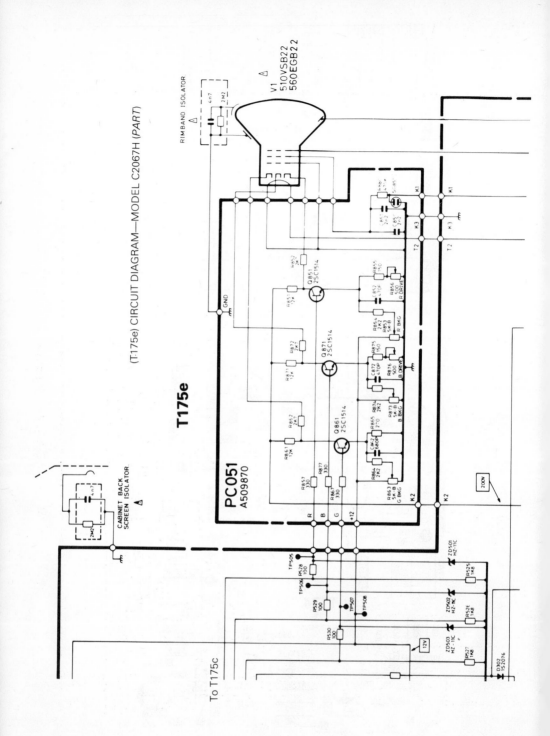

(T175e) CIRCUIT DIAGRAM—MODEL C2067H (PART)

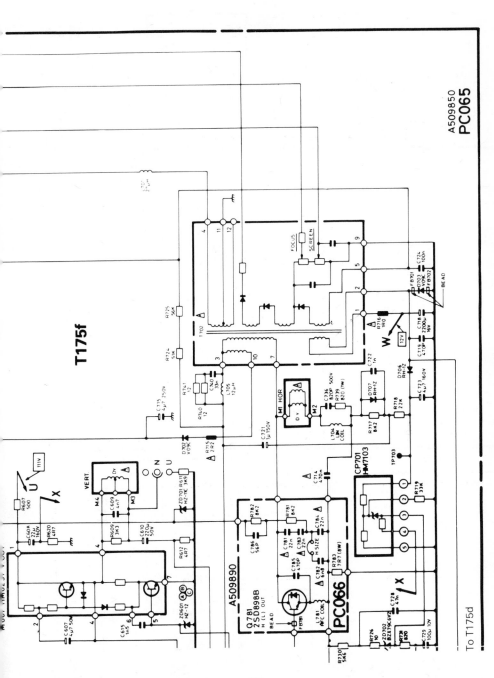

(T175f) CIRCUIT DIAGRAM—MODEL C2067H (CONTINUED)

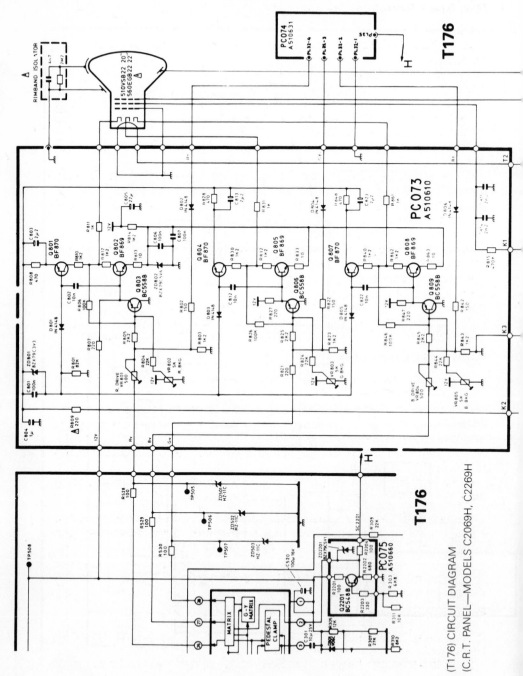

(T176) CIRCUIT DIAGRAM
(C.R.T. PANEL—MODELS C2069H, C2269H)

Grey Scale Procedure (Teletext models)

Initial Settings.

Turn the A₁ (screen) control on the line output transformer fully anti-clockwise.

Set the drive and background pots. midway.

Switch on and check that the receiver is working.

Adjust A₁ brightness and contrast as necessary.

On the interface board, disconnect the text drive plug PL32.

Put the receiver into text mode—no picture.

Set the background pots. to read 175V on a multimeter at the R.G.B. cathodes.

Short-circuit the vertical scan coils by linking the extreme left and right hand pins nearest the screen, on the yoke assembly. These pins are connected by green and yellow leads respectively, to pins 4 and 3 or plug M on the main circuit board.

Adjust the A₁ control to produce a very faint line.

Adjust the background pre-sets for neutral grey, but do not disturb the pot, corresponding to the most prominent colour displayed.

Restore vertical scan.

Replace PL32 and display a test card.

Adjust brightness and contrast for normal viewing.

Adjust the drive pots. for correct grey scale in the high lights.

If necessary reset the background pots. to correct the low light areas.

Purity and Convergence

The in-line type colour picture tube and saddle-toroidal scan coil provides convergence-free features. Since new tube assemblies are usually supplied complete with a sealed yoke assembly and the convergence quality is strictly controlled in the factory to secure its performance, reconvergence is not normally required. If the need does arise in the case of the replacement of a picture tube or a scan coil etc., follow the adjustment procedure step by step to obtain the best results.

Note: Appropriate safety precautions should be observed when replacing the C.R.T. and when adjusting the yoke assembly.

Preparation: Set the television receiver facing to the south and allow it to warm up for 20 minutes. Check the horizontal and vertical linearity, the picture width, and the focus. Display a crosshatch test card.

In case of the replacement of the picture tube, wind a layer of adhesive tape (bandage type is preferable) around the neck, to provide a suitable surface for the scan coil.

Purity Adjustments

The following preliminary procedure should be carried out before under-taking static convergence adjustments.

Push the scan coil fully up to the bell of the picture tube and disconnect the G-Y plug from the signal board. A broad magenta belt should appear on the screen.

Draw the scan coil gradually towards you and see that two oval areas (colours pale yellow and light blue) appear on each side of the screen.

Rotate the purity magnet (see static convergence) to make the pale yellow and blue areas equal.

Insert a rubber wedge between the C.R.T. and the scan coil at the top position, and tilt the scan coil backward at the top.

Slide the scan coil gradually towards you again until the two coloured areas disappear simultaneously.

Connect the G-Y plug and examine the white purity. If it is not satisfactory, move the scan coil until ideal white purity it obtained. Fasten the scan coil securely by tightening the screw.

Disconnect both G-Y and B-Y plugs to check red purity. If red purity cannot be obtained, repeat procedure.

Static Convergence (Fig. T170)

Display a crosshatch pattern. Remove the green O/P lead from TP507 on the main P.C.B. Adjust the customer controls to obtain an acceptable background level, then follow the procedure set out below for converging the centre of the screen.

Step	Convergence Lines	Magnet Adjustment
1	Vertical Blue and Red	Slide the 4 pole magnet ring tabs (A) toward or away from each other
2	Horizontal Blue and Red	Rotate both 4 pole magnet rings (A) together

Repeat steps 1 and 2 to achieve the best results then replace the green O/P lead. If the green vertical is out of position then open the 6 pole magnet (B) so that the red, blue and green converge at the centre. If the green horizontal is out of position then rotate the 6 pole magnet taking care not to alter the distance between the poles of the magnet rings.

Note: This convergence assembly is not fitted with a locking ring. When re-convergence is carried out, it will be necessary to carefully apply sufficient locking adhesive to the magnetic rings to maintain good purity and convergence after adjustment.

Dynamic Convergence

For convergence of red and blue vertical lines at top and bottom, disconnect the green O/P lead from TP507 and tilt scan coil up and down for best results. When the best result is obtained, insert a preliminary rubber wedge between the scan coil and C.R.T.

To converge horizontal lines top and bottom, tilt the sides of the scan coil to the right or left until convergence is obtained. Replace the green lead and check that complete convergence has been achieved. If satisfied, insert three fixing wedges to hold the scan coil in position using tape to stick them to the C.R.T.; then remove the preliminary rubber wedge.

I.C. and Transistor Voltages:

IC201

Pin	Volts	Pin	Volts
1	6·8V	9	12V
2	0V	10	9·2V
3	4·3V	11	5·2V
4	8V	12	5·2V
5	8V	13	5·2V
6	4·3V	14	5·1V
7	0V	15	6·2V
8	7·6V	16	6V

Pin 10 varies with A.G.C. control

IC401

Pin	Volts	Pin	Volts
1	2V	7	4·8V
2	2·1V	8	9·8V
3	5·5V	9	10V
4	4·5V	10	20V
5	4·5V	11	10V
6	Varies with Volume 0·6V-4·5V	12	9·5V

IC501

Pin	Volts	Pin	Volts
1	12·5V	15	8·2V
2	1V	16	4·5V
3	8·8V	17	8·2V
4	8·8V	18	8·2V
5	2·5V	19	−0·6V
6	2V	20	9·7V
7	8·4 Contrast	21	3·3V
8	6·7V Colour Control	22	3·3V
9	9·5V	23	−0·1V
10	2V	24	2·9V
11	1·3V	25	2·9V
12	8·0V	26	2·0V
13	6·2V	27	2·0V
14	Earth	28	2·0V

IC701

Pin	Volts	Pin	Volts
1	5·3V	9	0·6V
2	6·8V	10	6V
3	0·5V	11	−0·1V
4	0·3V	12	11·5V
5	0V	13	1V
6	0·5V	14	9·7V
7	3·3V	15	12·5V
8	0·3V	16	4·1V

M601

Pin	Volts	Pin	Volts
1	68V	5	0·6V
2	50V	6	34V
3	0V	7	0V
4	34V		

IC901

Pin	Volts
1	330V
2	Earth
3	111V
4	−0·3V

CP701

Leg	Volts
1	111V
2	9·7V
3	0V
4	0V
5	0V

Q703

	Volts
A Leg	110V
K Leg	0V
G Leg	0V

	Q201	Q202	Q301	Q501	Q702	Q781	Q901
E	5·5V	0V	9·5V	10·5V	0V	3·0V	−0·2V
B	6·0V	0V	8·8V	9·6V	0·3V	2·9V	−0·25V
C	12·5V	6V	2·5V	0V	34V	—	−0·3V

* Voltages taken with 20K Ω/V Meter AVO 8 with a colour bar displayed and Colour, Brill. and Contrast Controls at max.

Fault Finding

Under no H.T. conditions sets with standby power supply will go into a standby mode, otherwise go to *. To determine if the fault lies on the main chassis or on the standby power supply, disconnect PLS. If set now operates, the fault usually lies on the standby power supply board. However, an exception to this, is if the synchronising pulse, pin 2 flyback transformer to pin 4 IC901 is missing. So if set works with PLS disconnected, first check this pulse is reaching D909/C912 and if so, then proceed as shown below for 'Set Works' condition. Normally if this pulse is missing set will work but will have a whistle sound present with PLS disconnected.

No H.T.:

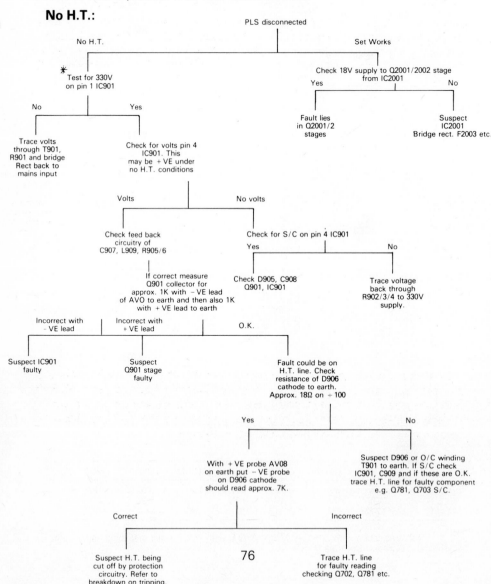

PLS disconnected

No H.T. — Set Works

* Test for 330V on pin 1 IC901

Check 18V supply to Q2001/2002 stage from IC2001 — Yes / No

No — Trace volts through T901, R901 and bridge Rect back to mains input

Yes — Check for volts pin 4 IC901. This may be + VE under no H.T. conditions

Fault lies in Q2001/2 stages

Suspect IC2001 Bridge rect. F2003 etc.

Volts — Check feed back circuitry of C907, L909, R905/6

No volts — Check for S/C on pin 4 IC901 — Yes / No

If correct measure Q901 collector for approx. 1K with − VE lead of AVO to earth and then also 1K with + VE lead to earth

Yes — Check D905, C908 Q901, IC901

No — Trace voltage back through R902/3/4 to 330V supply.

Incorrect with − VE lead — Suspect IC901 faulty

Incorrect with + VE lead — Suspect Q901 stage faulty

O.K. — Fault could be on H.T. line. Check resistance of D906 cathode to earth. Approx. 18Ω on ÷ 100

Yes — With + VE probe AV08 on earth put − VE probe on D906 cathode should read approx. 7K.

No — Suspect D906 or O/C winding T901 to earth. If S/C check IC901, C909 and if these are O.K. trace H.T. line for faulty component e.g. Q781, Q703 S/C.

Correct — Suspect H.T. being cut off by protection circuitry. Refer to breakdown on tripping.

Incorrect — Trace H.T. line for faulty reading checking Q702, Q781 etc.

76

Tripping H.T.: Under tripping conditions a surge is normally heard at switch on before H.T. is cut-off by protection circuitry. To ensure the fault is tripping H.T. disconnect PLS from the main chassis. If H.T. now available check for no E.H.T. If, however, set is still tripping, disconnecting R727 will enable the faulty stage to be detected. Before disconnecting R727, first check the regulation stage of IC901 and also make sure C781, C783, C784 are in circuit with each other and not dry jointed etc., otherwise component damage could occur with R727 disconnected. If these stages appear O.K. proceed as follows.

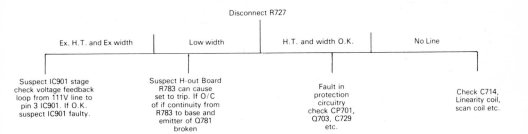

Disconnect R727

Ex. H.T. and Ex width	Low width	H.T. and width O.K.	No Line
Suspect IC901 stage check voltage feedback loop from 111V line to pin 3 IC901. If O.K. suspect IC901 faulty.	Suspect H-out Board R783 can cause set to trip. If O/C of if continuity from R783 to base and emitter of Q781 broken	Fault in protection circuitry check CP701, Q703, C729 etc.	Check C714, Linearity coil, scan coil etc.

On these models under no E.H.T. conditions or if the 12V supply is missing, the receiver will be put into a standby condition. However, if PLS is disconnected from the main chassis, the H.T. circuitry will function allowing voltage checks to be carried out, to determine where the fault lies. Once the presence of H.T. is established proceed as follows:

No E.H.T.:

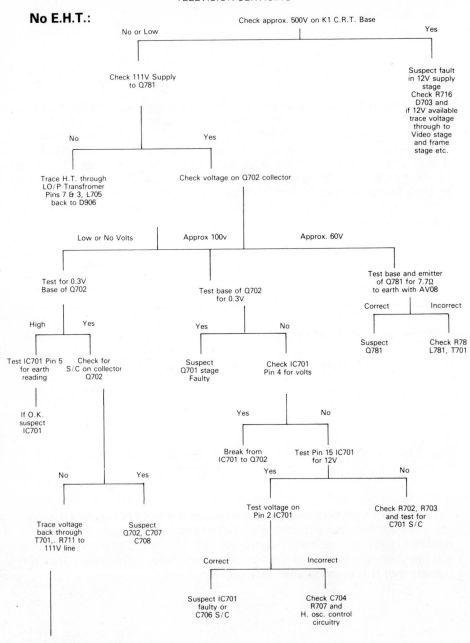

Check approx. 500V on K1 C.R.T. Base

No or Low — Yes

Check 111V Supply to Q781

Suspect fault in 12V supply stage Check R716 D703 and if 12V available trace voltage through to Video stage and frame stage etc.

No — Yes

Trace H.T. through LO/P Transfromer Pins 7 & 3, L705 back to D906

Check voltage on Q702 collector

Low or No Volts — Approx 100v — Approx. 60V

Test for 0.3V Base of Q702

Test base of Q702 for 0.3V

Test base and emitter of Q781 for 7.7Ω to earth with AV08

Correct — Incorrect

Suspect Q781

Check R78 L781, T701

High — Yes

Test IC701 Pin 5 for earth reading

Check for S/C on collector Q702

If O.K. suspect IC701

Yes — No

Suspect Q701 stage Faulty

Check IC701 Pin 4 for volts

Yes — No

Break from IC701 to Q702

Test Pin 15 IC701 for 12V

Yes — No

Test voltage on Pin 2 IC701

Check R702, R703 and test for C701 S/C

No — Yes

Trace voltage back through T701,. R711 to 111V line

Suspect Q702, C707 C708

Correct — Incorrect

Suspect IC701 faulty or C706 S/C

Check C704 R707 and H. osc. control circuitry

If 111V on R711, dry joint secondary of T701 supplying base Q781. If voltage Q702 collector now 80V approx. Suspect Q781 S/C Base - emitter

No Field Scan:

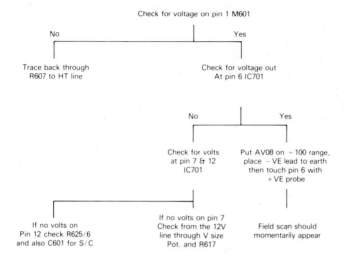

Check for voltage on pin 1 M601

No — Trace back through R607 to HT line

Yes — Check for voltage out At pin 6 IC701

No — Check for volts at pin 7 & 12 IC701

Yes — Put AV08 on ÷ 100 range, place − VE lead to earth then touch pin 6 with + VE probe

If no volts on Pin 12 check R625/6 and also C601 for S/C

If no volts on pin 7 Check from the 12V line through V size Pot. and R617

Field scan should momentarily appear

Some Field Fault Conditions:

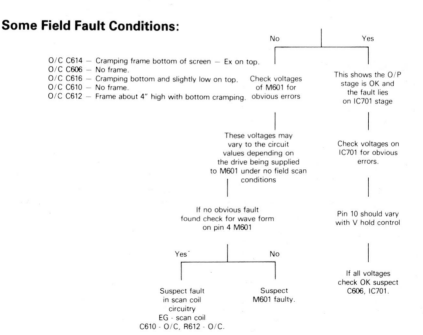

O/C C614 — Cramping frame bottom of screen — Ex on top.
O/C C606 — No frame.
O/C C616 — Cramping bottom and slightly low on top.
O/C C610 — No frame.
O/C C612 — Frame about 4" high with bottom cramping.

No — Check voltages of M601 for obvious errors

Yes — This shows the O/P stage is OK and the fault lies on IC701 stage

These voltages may vary to the circuit values depending on the drive being supplied to M601 under no field scan conditions

Check voltages on IC701 for obvious errors.

If no obvious fault found check for wave form on pin 4 M601

Pin 10 should vary with V hold control

Yes — Suspect fault in scan coil circuitry EG - scan coil C610 - O/C, R612 - O/C.

No — Suspect M601 faulty.

If all voltages check OK suspect C606, IC701.

No Colour: Connect a link between TP502 and TP508, then with an AV08 on the 10 volt range put the −VE lead to earth and the +VE lead to pin 13 of IC501 (TP503). Observe the result with colour control at maximum.

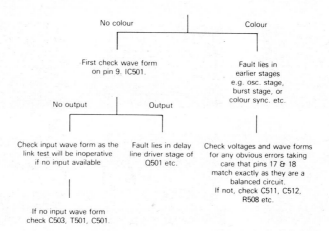

No Sound: First check 20V supply, if not available check R912, L903, D907 etc. Under normal conditions, if the input stage of IC401 is functioning correctly, an output appears at pin 3. If so, this can be used to determine the faulty stage in the following manner:

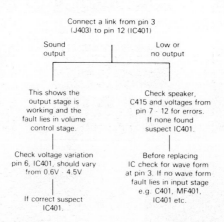

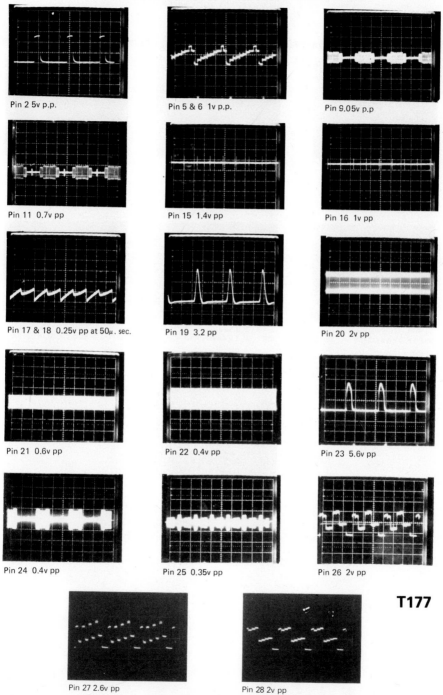

IC501 WAVEFORMS USING 10:1 PROBE AT 20µ SEC. (EXCEPT 17 & 18) ON COLOUR BAR SIGNAL. WITH COLOUR CONTROL. BRILL. & CONTRAST CONTROLS AT MAX.

Pin 2 5v p.p.

Pin 5 & 6 1v p.p.

Pin 9 .05v p.p

Pin 11 0.7v pp

Pin 15 1.4v pp

Pin 16 1v pp

Pin 17 & 18 0.25v pp at 50µ. sec.

Pin 19 3.2 pp

Pin 20 2v pp

Pin 21 0.6v pp

Pin 22 0.4v pp

Pin 23 5.6v pp

Pin 24 0.4v pp

Pin 25 0.35v pp

Pin 26 2v pp

Pin 27 2.6v pp

Pin 28 2v pp

T177

(T177) CIRCUIT WAVEFORMS (IC501)—MODELS C2065H ETC.

WAVEFORMS ON IC701 USING 10:1 PROBE

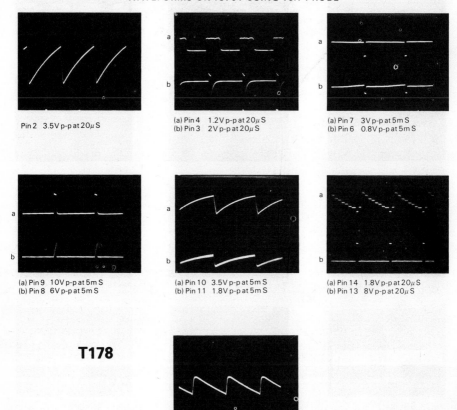

Pin 2   3.5 V p-p at 20 μ S

(a) Pin 4   1.2 V p-p at 20 μ S
(b) Pin 3   2 V p-p at 20 μ S

(a) Pin 7   3 V p-p at 5 m S
(b) Pin 6   0.8 V p-p at 5 m S

(a) Pin 9   10 V p-p at 5 m S
(b) Pin 8   6 V p-p at 5 m S

(a) Pin 10   3.5 V p-p at 5 m S
(b) Pin 11   1.8 V p-p at 5 m S

(a) Pin 14   1.8 V p-p at 20 μ S
(b) Pin 13   8 V p-p at 20 μ S

T178

Pin 16   0.8 V p-p at 20 μ S

(T178) CIRCUIT WAVEFORMS (IC701)—MODELS C2065H ETC.

82

GRUNDIG CUC120 Chassis
Models A3102/5, A4102/5, A4202/5, A5102/5, A6100/2/3/4, A6200/2/3/4, A7100/2/3/4, A7110/2/3, A7200/2/3, A7210/2/3

General Description: A colour television receiver chassis comprising 'mother' board with 'plug-in' modules and push-button selection. Variations in modules may occur and reference to the Grundig CUC220 and CUC720 Chassis, also described in this volume, may be of assistance in locating the correct circuit.

Mains Supplies: 160-260 volts, 50Hz.

Adjustments (using colour test bars)

Grey Scale. See Fig. T139: Disconnect C.R.T. cathode leads, set contrast to centre, brightness to centre, and saturation to minimum. Adjust R.G.B. output signals for equal amplitude with V.B., V.R. controls. Set line flyback levels to 180 volts with S.W.R., S.W.G., and S.W.B. controls. Reconnect cathode leads.

Set contrast to minimum and reduce brightness until peak white level is 160 volts. Advance S.G. control until peak white bar is just visible. Readjust S.W.R., S.W.G. and S.W.B. to remove any tinting. Set brightness to centre, contrast to maximum. Adjust V.R. and V.B. to remove any tinting from peak white. Check and repeat procedure if necessary.

Line Hold: Short circuit test point marked 'fo-Abgleich'. Adjust control-marked Z for stationary picture (as near as possible). Remove short circuit.

Service Hints for Switch-Mode Power Supply: Operate set with mains-isolating transformer (n.b. the primary side of the S.M. power supply is normally not isolated, if there is no secondary voltage or the power supply pulsates, disconnect the secondary circuits individually and check operation. Fault finding scheme if blocking function prevents oscillation.

Starting voltage (Pin 9/5) <8V, starting via Di616 and R616.

Reference voltage (Pin 1) approx. 6V.

Start pulse (Pin 4).

Base current drive (Pin 7).

Before replacement of IC631, C626 must be discharged. Power supply range 160 to 260V A.C.

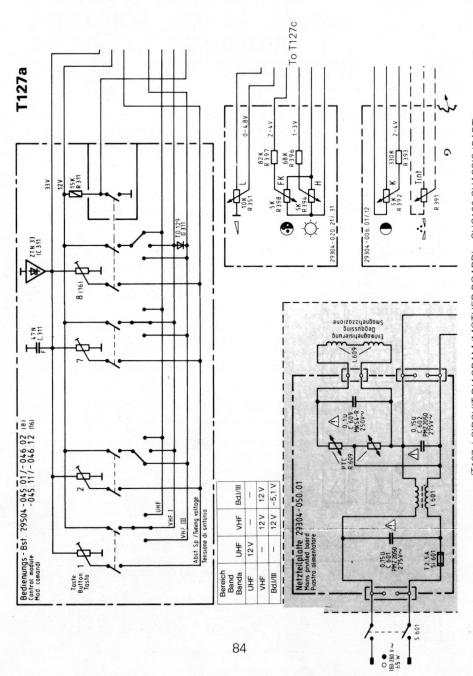

(T127a) CIRCUIT DIAGRAM (MOTHER BOARD)—CUC120 CHASSIS (*PART*)

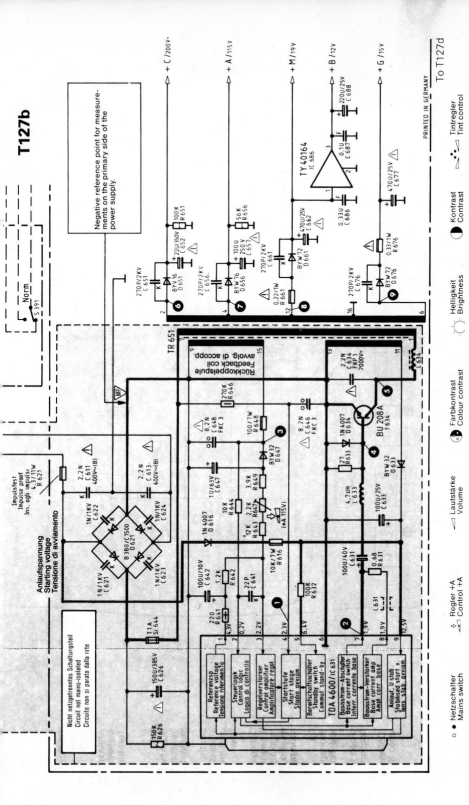

(T127b) CIRCUIT DIAGRAM (MOTHER BOARD)—CUC120 CHASSIS (PART)

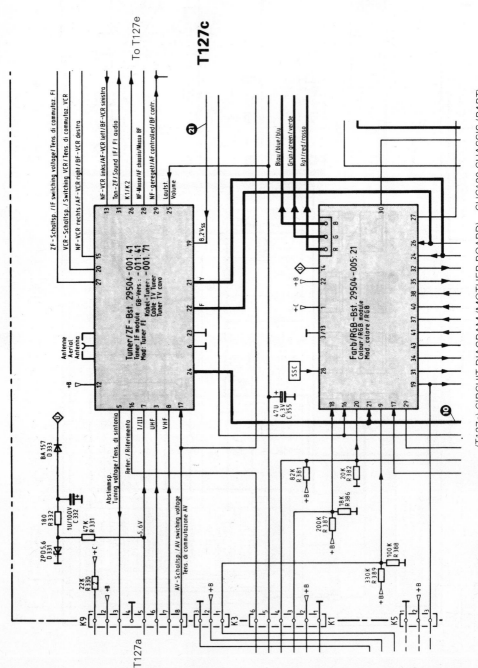

(T127c) CIRCUIT DIAGRAM (MOTHER BOARD)—CUC120 CHASSIS (PART)

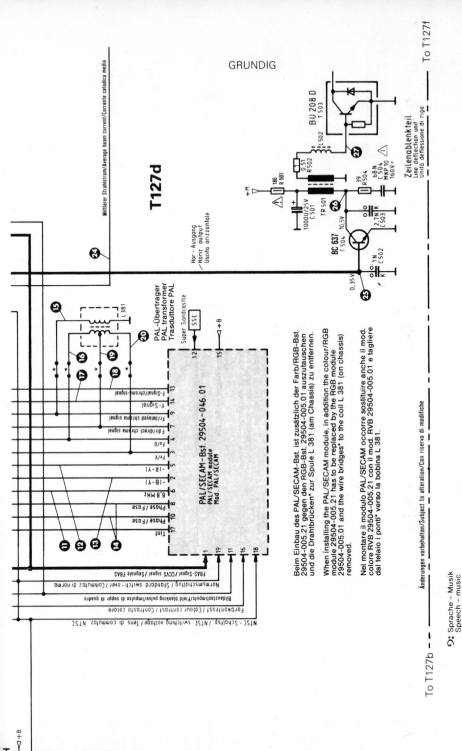

GRUNDIG

T127d

(T127d) CIRCUIT DIAGRAM (MOTHER BOARD)—CUC120 CHASSIS (PART)

Beim Einbau des PAL/SECAM-Bst. ist zusätzlich der Farb/RGB-Bst. 29504-005.21 gegen den RGB-Bst. 29504-005.01 auszutauschen und die Drahtbrücken* zur Spule L 381 (am Chassis) zu entfernen.

When installing the PAL/SECAM module, in addition the colour/RGB module 29504-005.21 has to be replaced by the RGB module 29504-005.01 and the wire bridges* to the coil L 381 (on chassis) removed.

Nel montare il modulo PAL/SECAM occorre sostituire anche il mod. colore RVB 29504-005.21 con il mod. RVB 29504-005.01 e tagliere dal telaio i ponti* verso la bobina L 381.

Änderungen vorbehalten/Subject to alteration/Con riserva di modifiche

To T127b

♪: Sprache – Musik
   Speech – music

To T127f

87

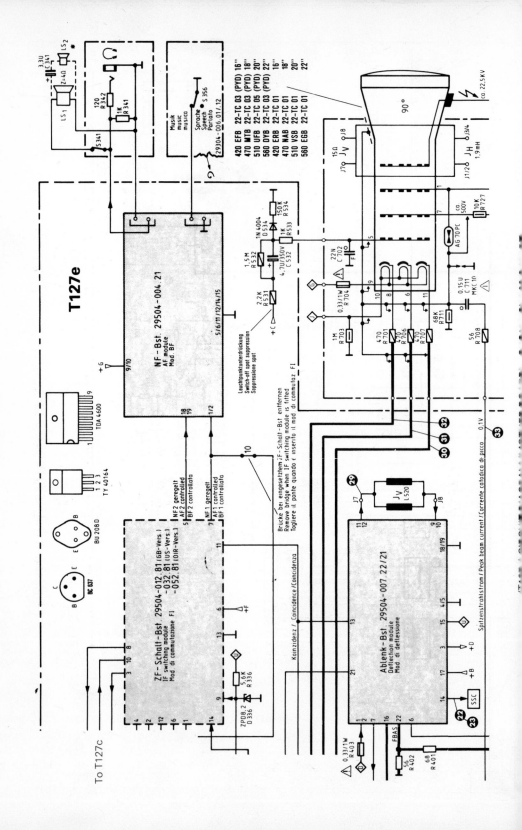

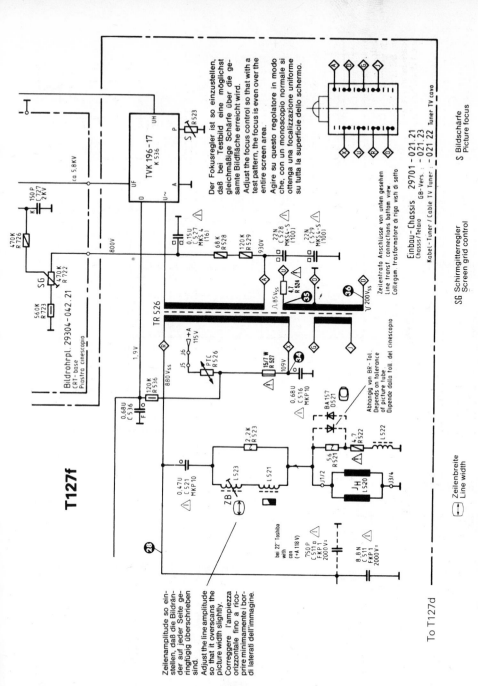

T127f

Der Fokusregler ist so einzustellen, daß bei Testbild eine möglichst gleichmäßige Schärfe über die gesamte Bildfläche erreicht wird.

Adjust the focus control so that with a test pattern, the focus is even over the entire screen area.

Agire su questo regolatore in modo che, con un monoscopio normale si ottenga una focalizzazione uniforme su tutta la superficie dello schermo.

Zeilentrafo Anschlüsse von unten gesehen
Line transf. connections bottom view
Collegam. trasformatore di riga visti di sotto

Einbau-Chassis   29701 - 021.21
Chassis/Telaio   GB-Vers.: - 021.23
Kabel-Tuner / Cable TV Tuner: - 021.22
Tuner TV cavo

⊕ Zeilenbreite
   Line width

SG Schirmgitterregler
   Screen grid control

S Bildschärfe
   Picture focus

Zeilenamplitude so einstellen, daß die Bildränder auf jeder Seite geringfügig überschrieben sind.

Adjust the line amplitude so that it overscans the picture width slightly.

Correggere l'ampiezza orizzontale fino a ricoprire minimamente i bordi laterati dell'immagine.

To T127d

(T127f) CIRCUIT DIAGRAM (MOTHER BOARD)—CUC120 CHASSIS (CONTINUED)

89

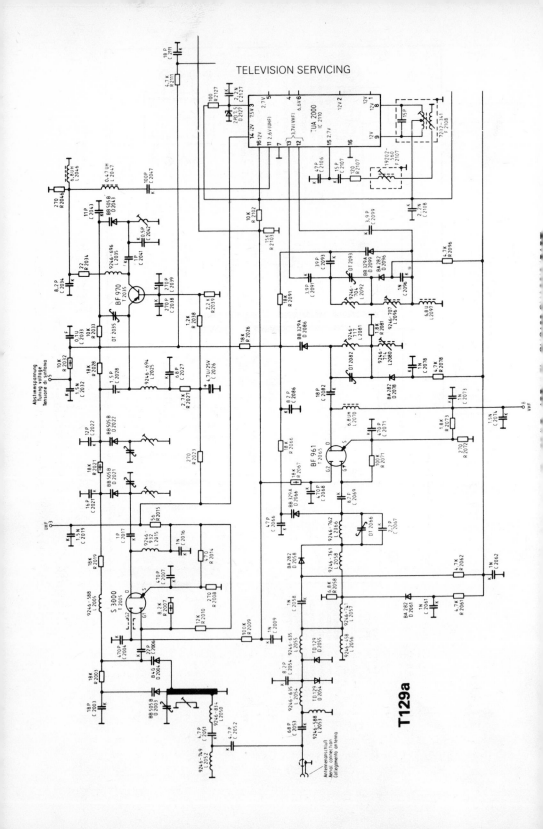

T129a

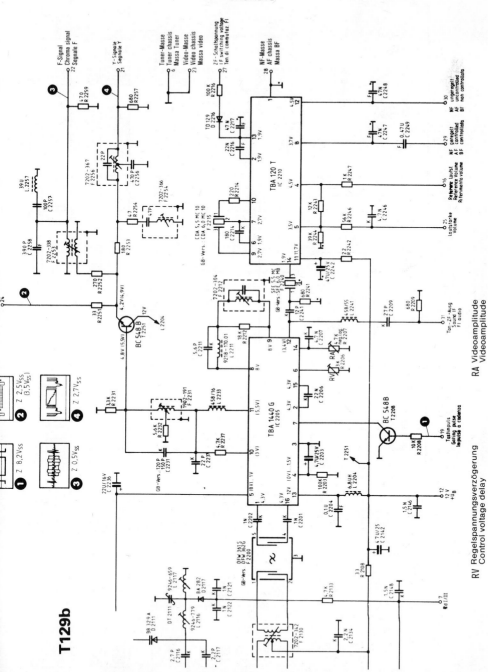

T129b

RV Regelspannungsverzögerung
Control voltage delay

RA Videoamplitude
Videoamplitude

(T129b) CIRCUIT DIAGRAM (TUNER/I.F. MODULE)—CUC120 CHASSIS (CONTINUED)

(T130) CIRCUIT DIAGRAM (DEFLECTION MODULE)—CUC120 CHASSIS.

GRUNDIG

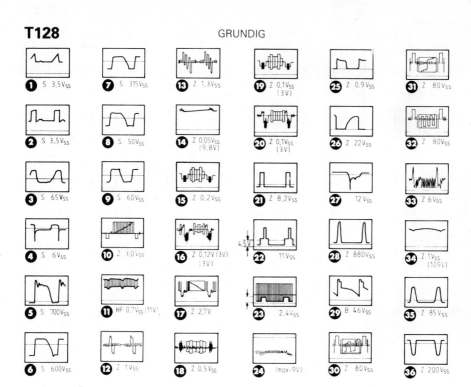

(T128) CIRCUIT WAVEFORMS (MOTHER BOARD)—CUC120 CHASSIS.

GRUNDIG CUC220 Chassis
Models A3402/5, A3412/5, A4402/5, A5402/5, A6400/2/3/4/7, A6410/2/3/4, A7400/2/3/7, A7401/2/3, A7410/2/3

General Description: A colour television receiver chassis fitted to a variety of receivers incorporating 16, 18, 20, and 22-ins. cathode ray tubes. The circuit comprises modules plugged into a 'mother' board. Circuits of individual modules are similar to many used in Grundig Chassis Types CUC120 and CUC720 which are detailed elsewhere in this volume, together with adjustment procedures.

Mains Supplies: 160-260 volts, 50Hz.

***Note*:** The chassis is 'live' irrespective of mains connection and an isolating transformer is recommended when carrying out service on this chassis.

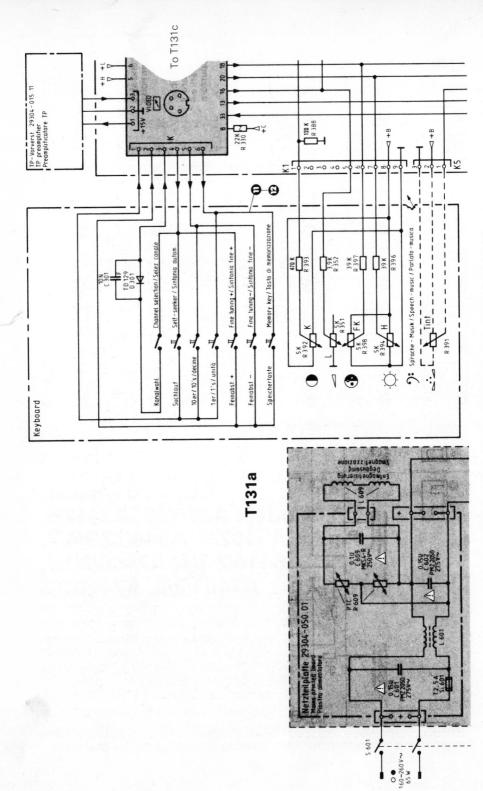

(T131a) CIRCUIT DIAGRAM (MOTHER BOARD)—CUC220 CHASSIS (*PART*)

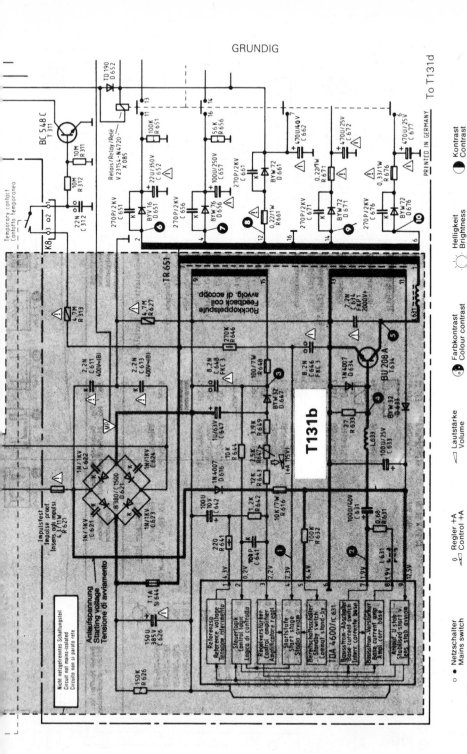

(T131b) CIRCUIT DIAGRAM (MOTHER BOARD)—CUC220 CHASSIS (*PART*)

To T131d

PRINTED IN GERMANY

o • Netzschalter
Mains switch

⊿ Lautstärke
Volume

δ Regler +A
⊿ Control +A

● Farbkontrast
Colour contrast

◐ Kontrast
Contrast

☼ Heiligkeit
Brightness

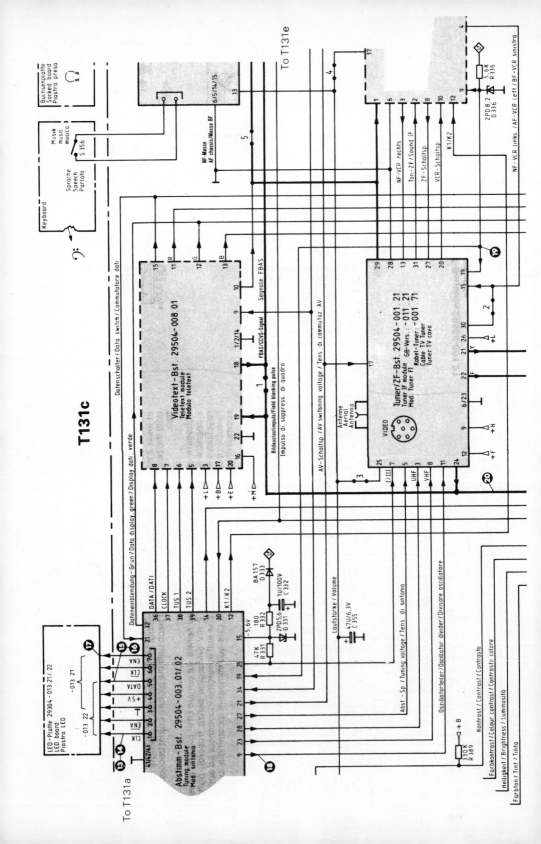

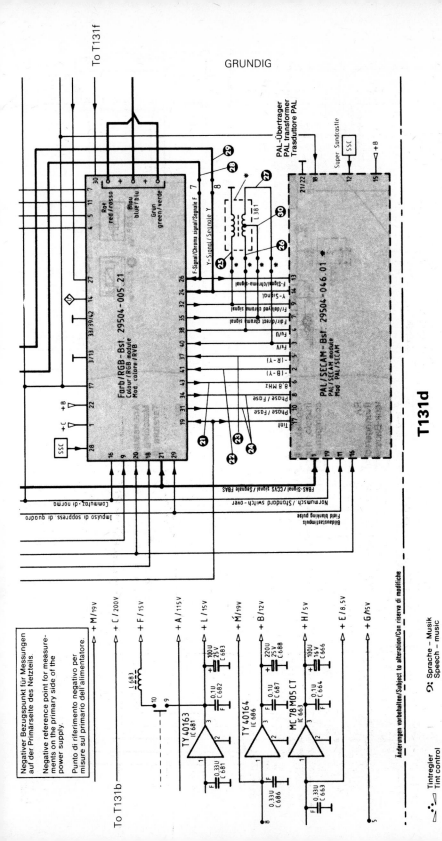

GRUNDIG

(T131d) CIRCUIT DIAGRAM (MOTHER BOARD)—CUC220 CHASSIS (PART)

T131d

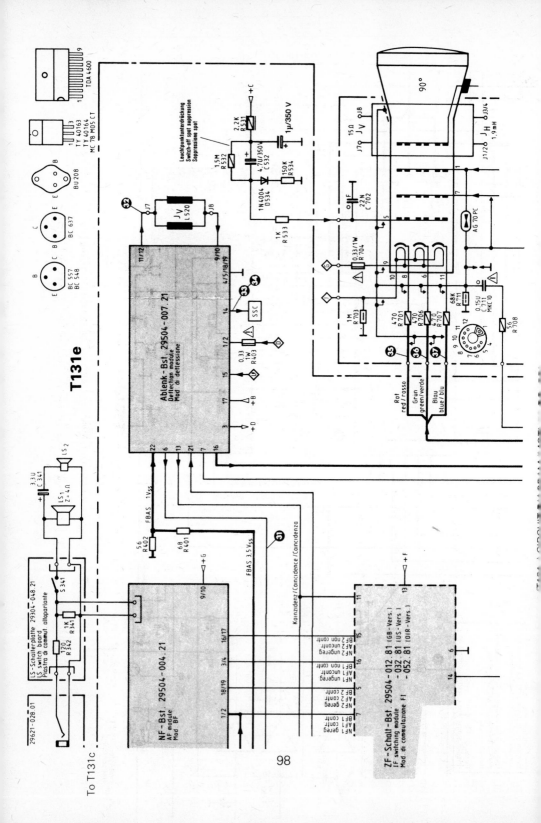

T131e

To T131c

98

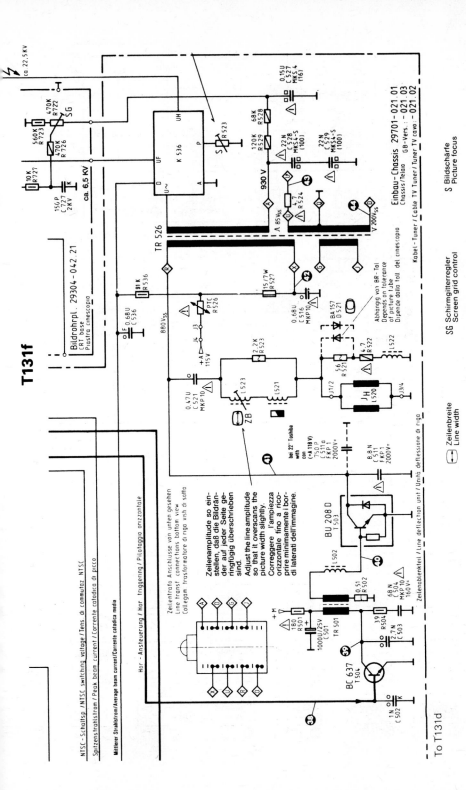

T131f

(T131f) CIRCUIT DIAGRAM (MOTHER BOARD)—CUC220 CHASSIS (*CONTINUED*)

To T131d

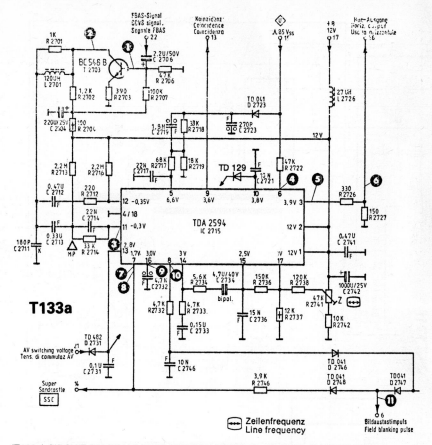

(T133a) CIRCUIT DIAGRAM (DEFLECTION MODULE)—CUC220 CHASSIS (*PART*)

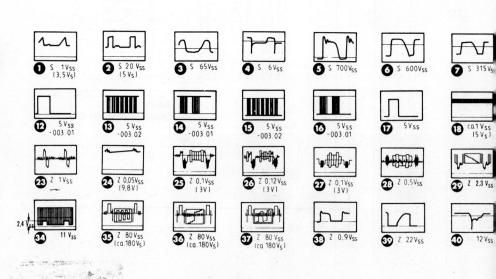

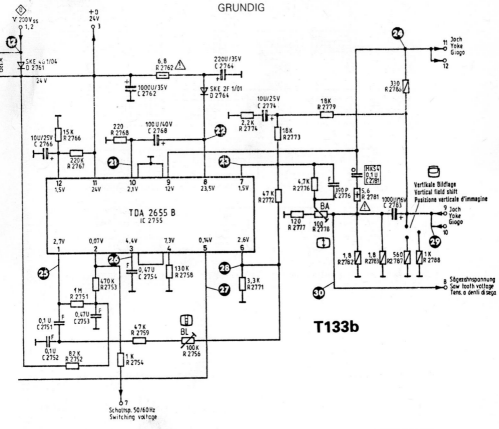

(T133b) CIRCUIT DIAGRAM (DEFLECTION MODULE)—CUC220 CHASSIS (*PART*)

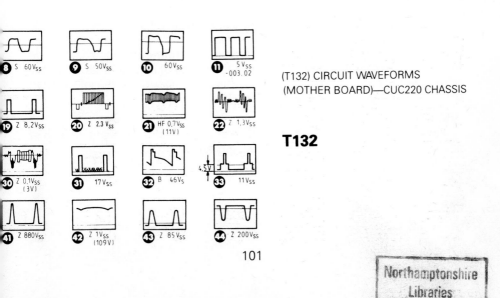

(T132) CIRCUIT WAVEFORMS
(MOTHER BOARD)—CUC220 CHASSIS

T132

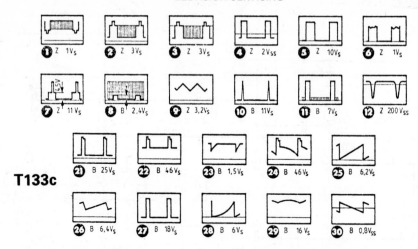

T133c

(T133c) CIRCUIT WAVEFORMS (DEFLECTION MODULE)—CUC220 CHASSIS (*CONTINUED*)

GRUNDIG

CUC720 Chassis
Models A8400/2/3/6/7,
A8410/2/3, A8420/2/3

General Description: A mains operated colour television chassis comprising a 'mother' board with plug-in modules. These modules may be common to a number of other chassis types and reference should be made to the Grundig CUC120 and 220 chassis for details of adjustments. A variety of remote tuning and control methods are also available, but the information given here relates only to the main chassis.

Mains Supplies: 160-260 volts, 50Hz.

***Note*:** The chassis is 'live' irrespective of mains connection and the use of an isolating transformer is strongly recommended.

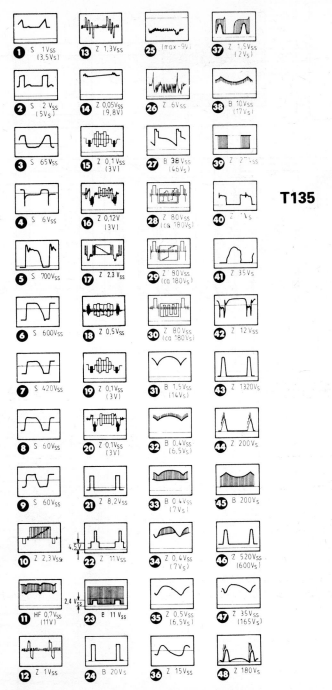

T135

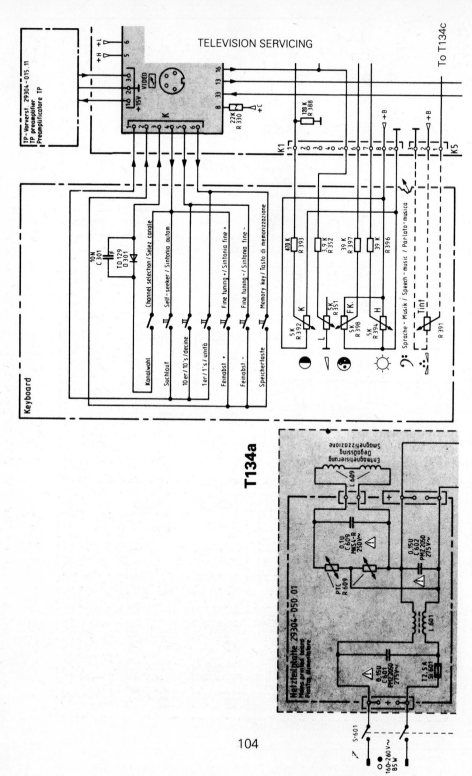

(T134a) CIRCUIT DIAGRAM (MOTHER BOARD)—CUC720 CHASSIS (PART)

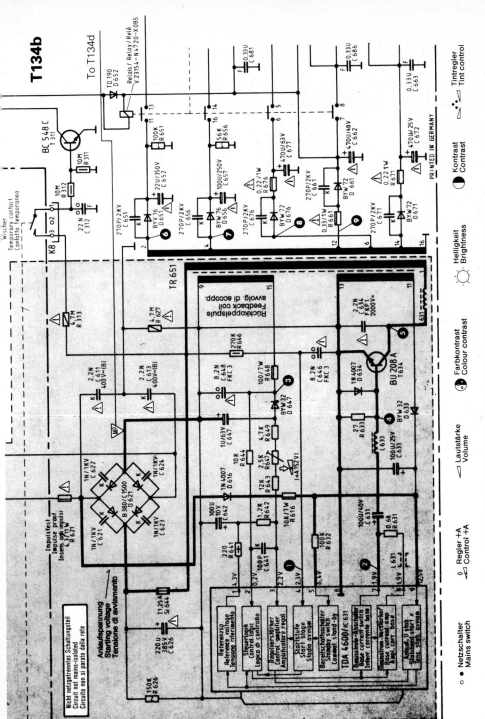

T134b

To T134d

(T134b) CIRCUIT DIAGRAM (MOTHER BOARD)—CUC720 CHASSIS (PART)

PRINTED IN GERMANY

— Netzschalter	— Lautstärke
o • Mains switch	Volume
	◉ Farbkontrast
	Colour contrast
⌀ Regler +A	— Helligkeit
⊲ Control +A	-☼- Brightness
	◖ Kontrast
	Contrast
	Tintregler
	Tint control

105

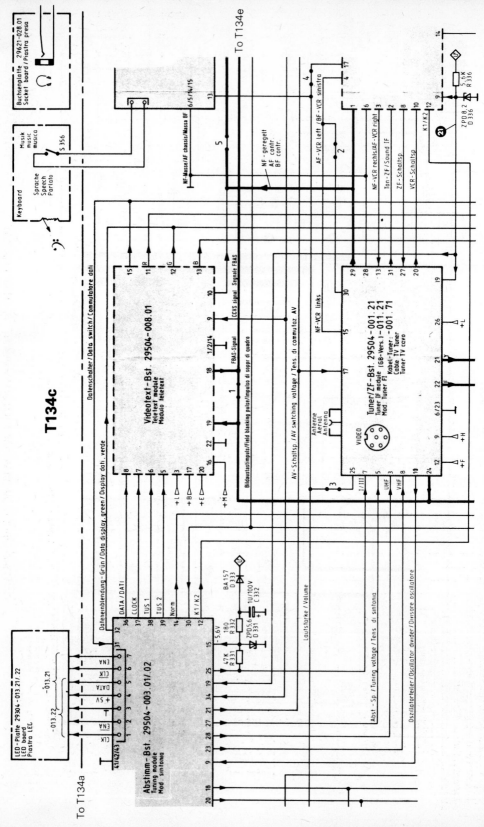

(T134c) CIRCUIT DIAGRAM (MOTHER BOARD)—CUC720 CHASSIS (PART)

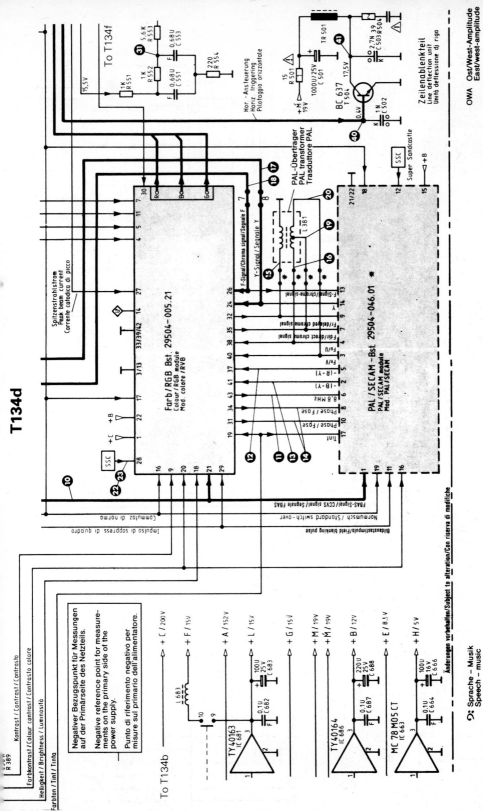

(T134d) CIRCUIT DIAGRAM (MOTHER BOARD)—CUC720 CHASSIS (*PART*)

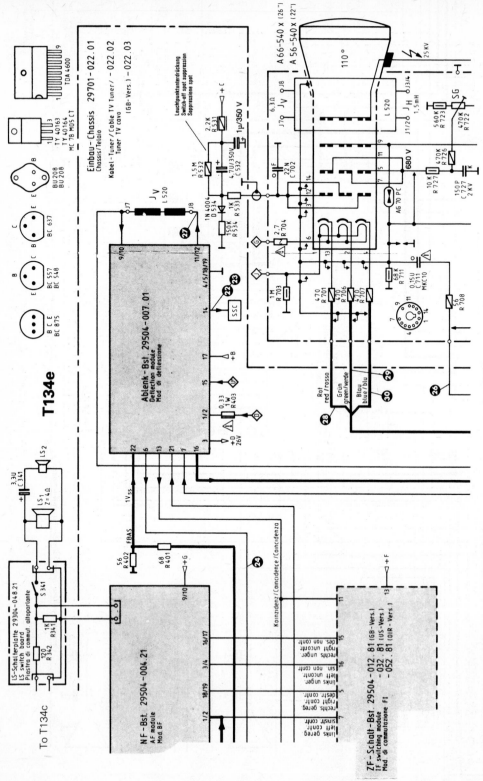

(T134e) CIRCUIT DIAGRAM (MOTHER BOARD)—CUC720 CHASSIS (PART)

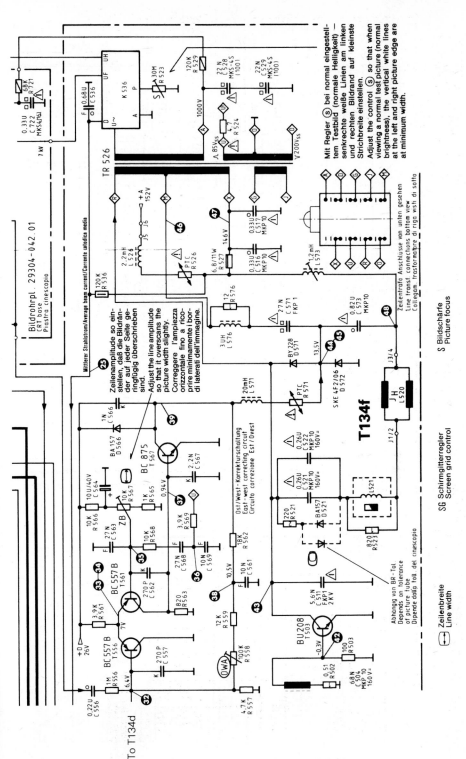

(T134f) CIRCUIT DIAGRAM (MOTHER BOARD)—CUC720 CHASSIS (CONTINUED)

(T136a) CIRCUIT DIAGRAM (TUNER/I.F. MODULE 001 21)—CUC720 CHASSIS (PART)

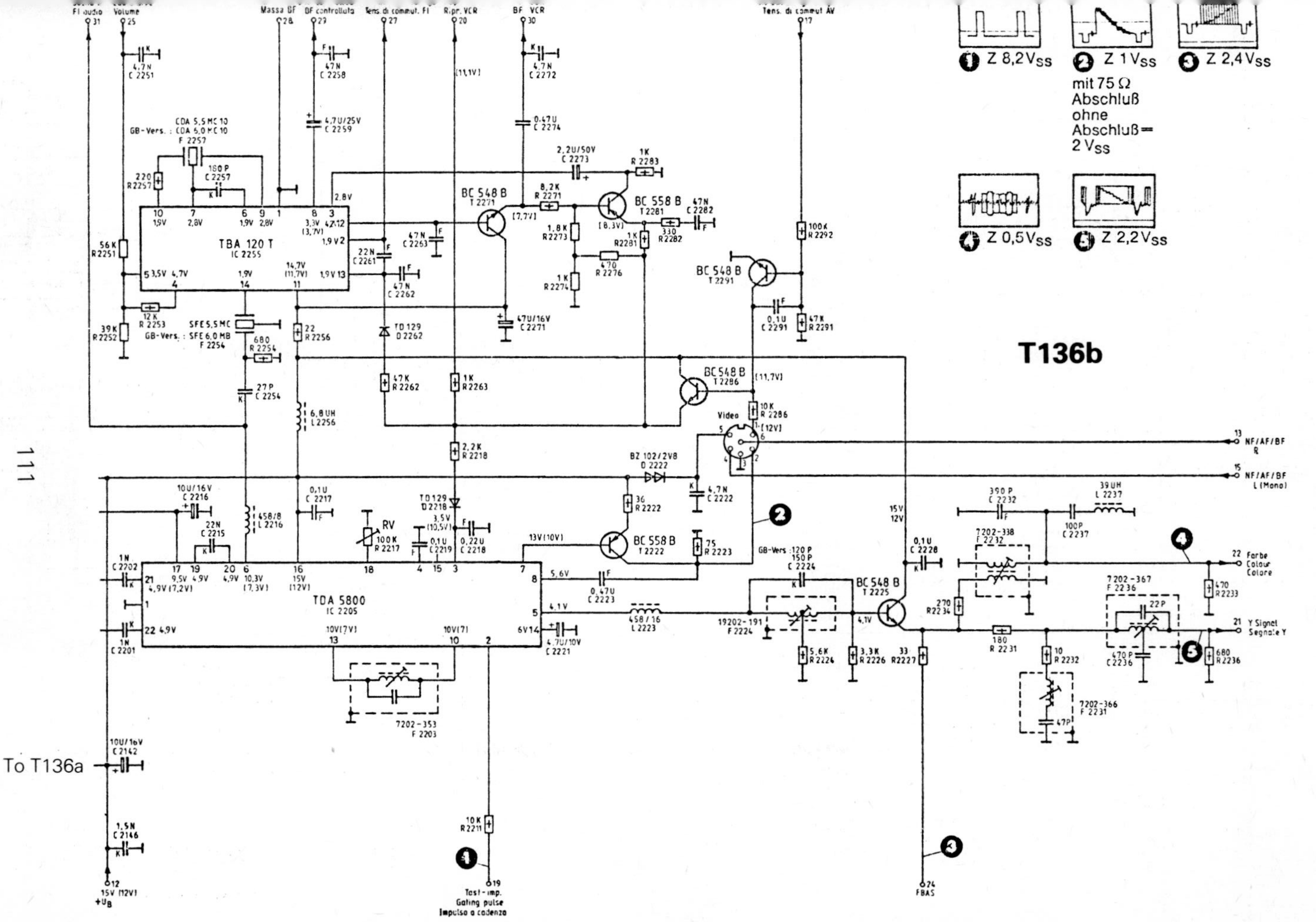

T136b

(T136b) CIRCUIT DIAGRAM (TUNER/I.F. MODULE 001:21)—CUC720 CHASSIS (CONTINUED)

111

To T136a

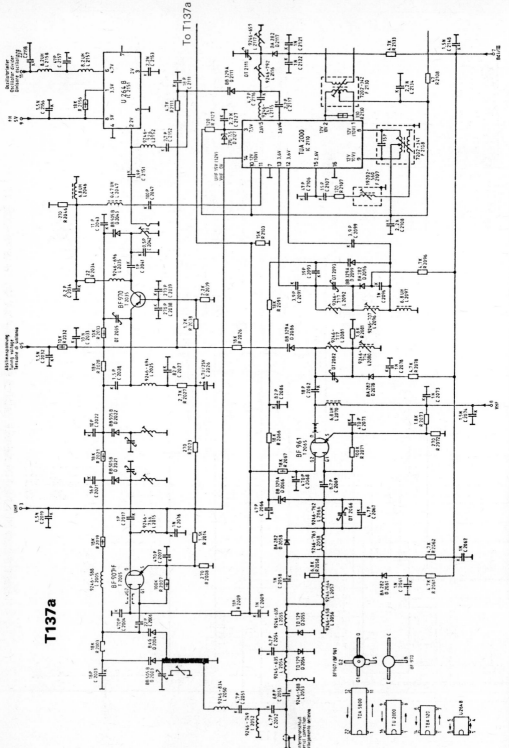

(T137a) CIRCUIT DIAGRAM (TUNER/I.F. MODULE 001.23 — CUC720 CHASSIS (PART)

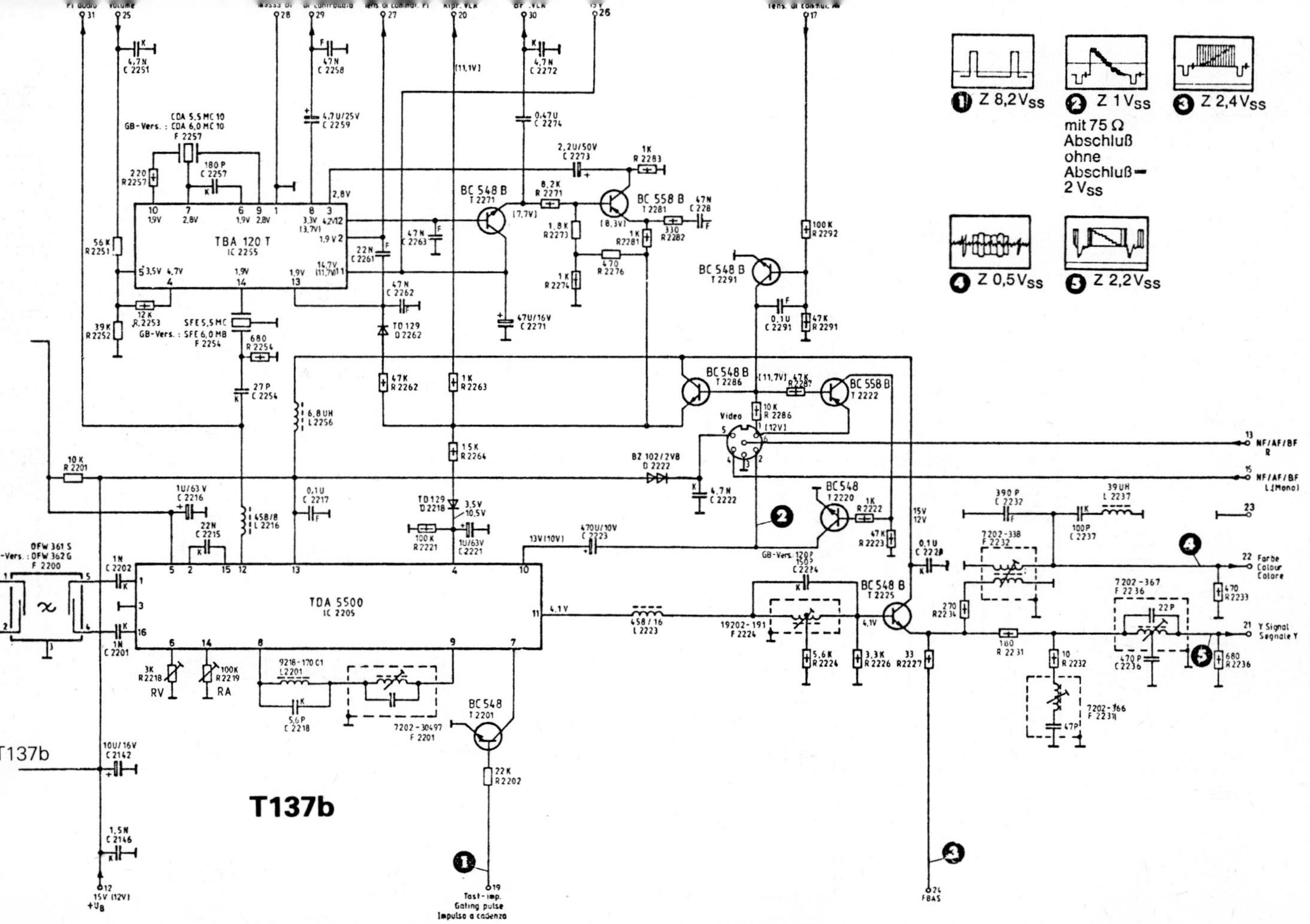

(T137b) CIRCUIT DIAGRAM (TUNER/I.F. MODULE 001.23)—CUC720 CHASSIS (CONTINUED)

T137b

To T137b

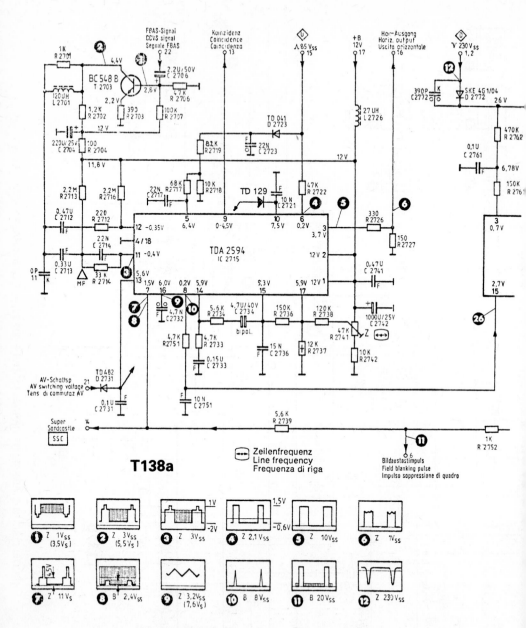

T138a

Zeilenfrequenz
Line frequency
Frequenza di riga

(T138a) CIRCUIT DIAGRAM (DEFLECTION MODULE)—CUC720 CHASSIS (*PART*)

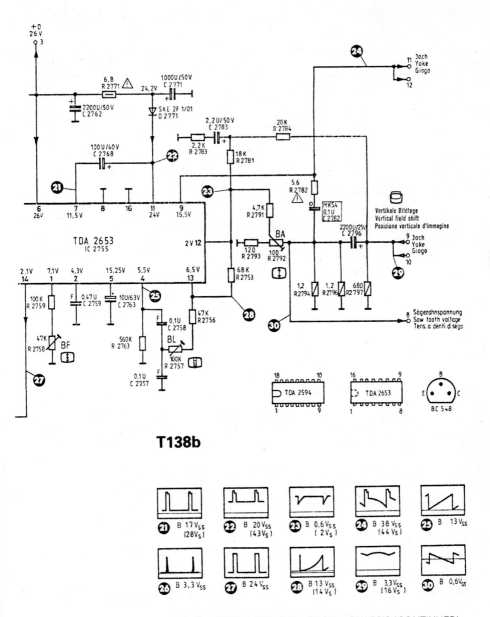

T138b

(T138b) CIRCUIT DIAGRAM (DEFLECTION MODULE)—CUC720 CHASSIS (*CONTINUED*)

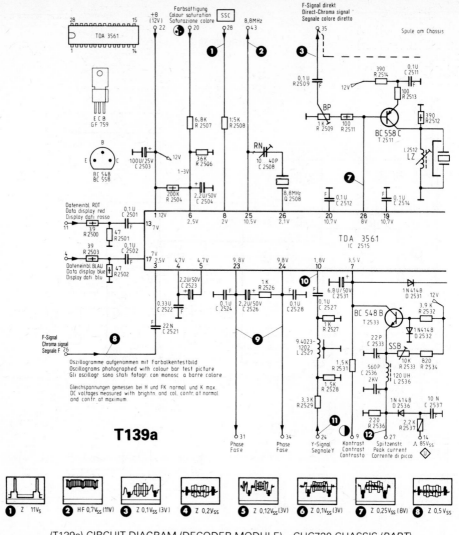

T139a

(T139a) CIRCUIT DIAGRAM (DECODER MODULE)—CUC720 CHASSIS (*PART*)

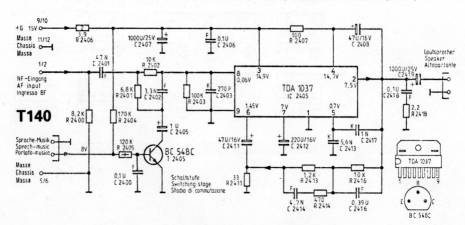

(T140) CIRCUIT DIAGRAM (A.F. MODULE)—CUC720 CHASSIS

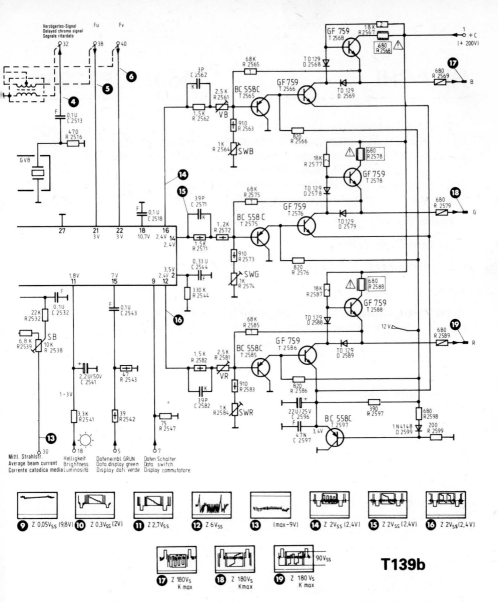

(T139b) CIRCUIT DIAGRAM (DECODER MODULE)—CUC720 CHASSIS (*CONTINUED*)

T139b

HITACHI   Model CPT2051

General Description: A mains operated colour television receiver with 20-in. in-line gun picture tube. Integrated circuits are used in all low level signal circuits on a main board with colour output drive amplifiers on the C.R.T. panel. The U.H.F. tuner used in this model is similar to that used in the G.E.C. model C1650 which is shown in the 1981–82 edition of Radio and Television Servicing. A socket is provided for the connection of an earphone.

Mains Supply: 240 volts, 50Hz.   **Fuse:** 2·5A.

Cathode Ray Tube: 510VSB22.

Loudspeakers: 12 and 5 ohms impedance.

Adjustments

+B Voltage Check: Tune the receiver to an active channel and synchronise picture. (Connect the receiver to an outlet of A.C. supply.)
Set Brightness and Contrast controls to obtain the best picture.
Connect a D.C. meter to 'Case of Q703' on main board and ground.
Check that +B voltage is 107±1·5 volts.

Horizontal Oscillator Adjustment: Tune the receiver to one of the T.V. stations in your area.
Set Horizontal Hold control to the mechanical centre of its range.
Connect capacitor 1μF (16V) between TP701 on main board and ground.
Adjust horizontal oscillator control (R708) on main board to momentarily synchronise picture.
Remove capacitor.

Horizontal Size Adjustment
Note: +220 volt power supply, A.G.C. and Grey Scale and Tracking should be adjusted before attempting Horizontal size adjustment.
Connect receiver to an outlet of A.C. supply and check picture width with test pattern or programme. If picture is too wide or narrow, short or cut leads (H. Size jumper) on Horizontal Outboard.
If lead is shorted, horizontal width increase.
If lead is cut, horizontal width decrease.

Horizontal Position Adjustment: Receive the test pattern.
Adjust so that the left and right sides of the test pattern are symmetrical using H. phase V.R. (R722).
Note: This V.R. can be varied by a screwdriver through a hole on the back cover.

Vertical Oscillator Adjustment:
Note: The vertical oscillator V.R. can be varied by a screwdriver through a hole on the back cover.

MAIN P.W. BOARD

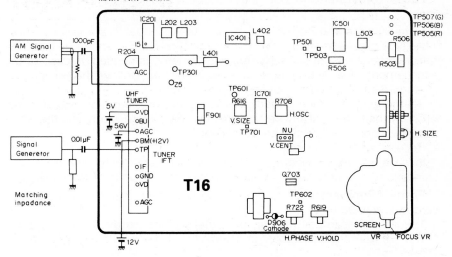

(T16) CIRCUIT ADJUSTMENTS (MAIN PANEL)—MODEL CPT2051

Tune the receiver to one of the T.V. stations.

Connect a 240K resistor between IC701, #12 and TP602, and adjust V. Hold (R619) just before the picture starts flowing from the top to the bottom. Remove the resistor.

Vertical Size Adjustment: Set vertical size (R616) control on main board. Adjust vertical size control for full scan.

Vertical Centring Adjustment: Before attempting vertical centring adjustment, set Vertical Hold control for properly synchronised picture. Vertical centring can be adjusted by connecting vertical tap to any one of the two terminals on main board.

C.R.T. Cut-off Adjustment (see Fig. T17):

Adjustment preparation: Set BGL-VR or R, B, G to Min, and the screen VR to minimum (FBT).

Set the R/B drive VR to TYP.

Receive the white pattern. 100% white.

Contrast: Min. Brightness: Min.

Connect the oscilloscope to the R, G, B cathodes (in sequence).

Adjustment: Increase the brightness until max. voltages of the R, G, and B cathodes are as shown.

Remove the oscilloscope to obtain a single lateral line.

Turn the screen VR clockwise to obtain a bright line.

Next, turn B.K.G. VR as the above, and perform the same operation.

Next, release the single lateral line, and adjust the high brightness/low brightness white balance.

Focus Adjustment: Adjust focus control on FBT for well defined scanning lines.

Convergence and Purity: This chassis television receiver incorporates an inline type tube, with saddle-toroidal scan coil and convergence-free features. Reconvergence is not normally required. When it is needed, in case of the replacement of a picture tube or a scan coil etc., follow the adjustment procedure step by step to obtain the best results.

Set the television receiver facing to the south, and soak it at least 10 mins.

Check the horizontal and vertical linearity, the picture width, and the focus as already described on the previous pages.

Receive a crosshatch test card.

In case of the replacement of the picture tube, wind adhesive tape (bandage type is preferable) around the neck, for securing of scan coil.

Purity Alignment: Before proceeding undertake preliminary static convergence adjustment.

1. Push the scan coil fully up to funnel part of the picture tube ground TP507 (G) from the main board and see a broad magenta belt appear on the screen.

2. Pull out the scan coil gradually towards you and see that two oval parts (coloured pale yellow and light blue) appear on each side of the screen.

3. Rotate the purity magnet (c) (Fig. T18), and set two coloured parts to become equal in area.

4. Insert preliminary rubber wedge between the funnel and the scan coil at the top position, and tilt the scan coil backward at the top.

5. Pull out the scan coil gradually towards you again until two coloured parts disappear simultaneously. See that the white purity is all right, if not, pull out the scan coil until white purity is obtained. Fasten the scan coil securely by tightening the screw.

6. Unless the red purity [Ground TP506 (B), TP507 (G)] is obtained, repeat the steps 2–5 again.

Static Convergence (See Fig. T18)

This procedure is the convergence alignment for the centre part of the screen.

Step	Convergence lines	Magnet	Movement
1	Vertical Blue and Red	4 poles (A)	sliding (a)
2	Horizontal Blue and Red	4 poles (A)	rotating (b)
3	Vertical B/R and Green	6 poles (B)	sliding (a)
4	Horizontal B/R and Green	6 poles (B)	rotating (b)
5	Repeat steps 1 to 4.		
6	If the correct convergence is not obtainable after carrying out the steps 1 to 5 above, displace the purity-convergence magnet assembly along the neck of the picture tube within the range of ±5mm and repeat steps 1 to 5.		

HITACHI

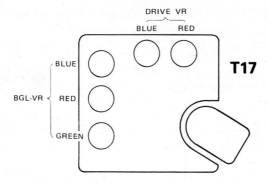

(T17) GREY SCALE ADJUSTMENTS—MODEL CPT2051

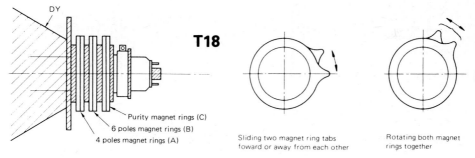

Purity magnet rings (C)
6 poles magnet rings (B)
4 poles magnet rings (A)

Sliding two magnet ring tabs
foward or away from each other

Rotating both magnet
rings together

(T18) STATIC CONVERGENCE—MODEL CPT2051

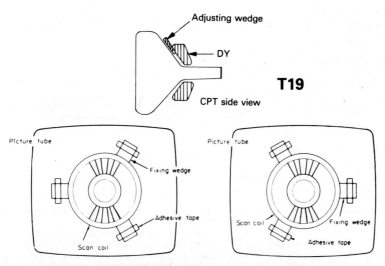

(T19) DYNAMIC CONVERGENCE—MODEL CPT2051

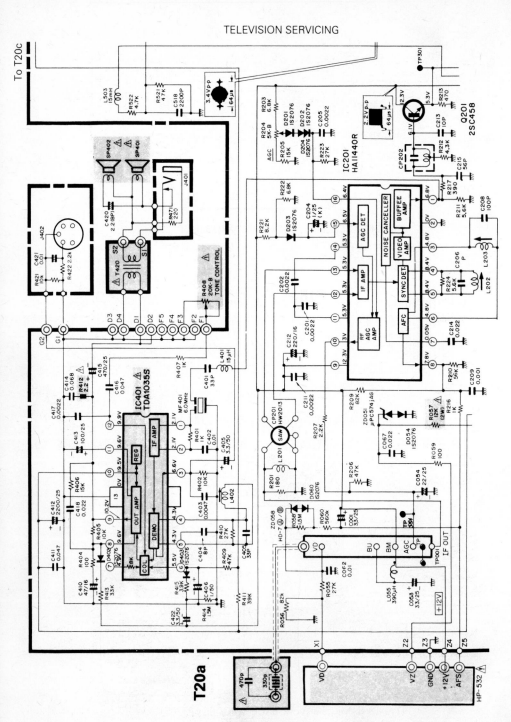

(T20a) CIRCUIT DIAGRAM—MODEL CPT2051 (*PART*)

122

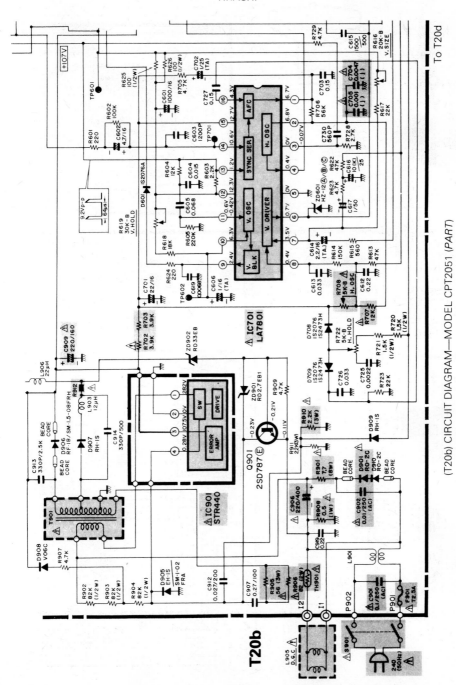

(T20b) CIRCUIT DIAGRAM—MODEL CPT2051 (PART)

T20b

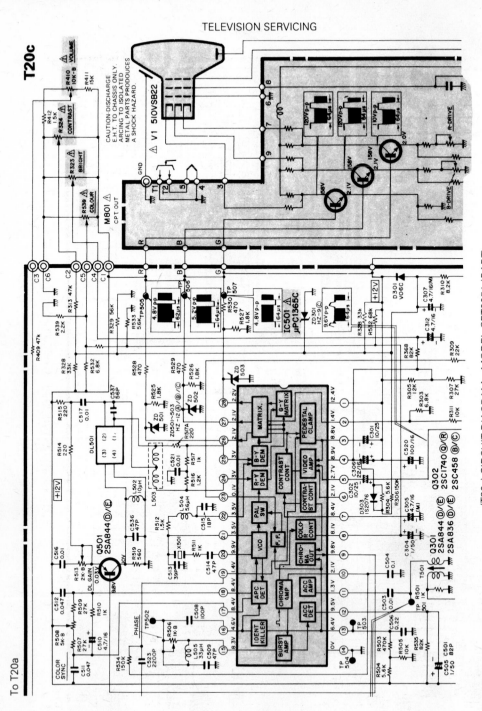

(T20c) CIRCUIT DIAGRAM—MODEL CPT2051 (*PART*)

124

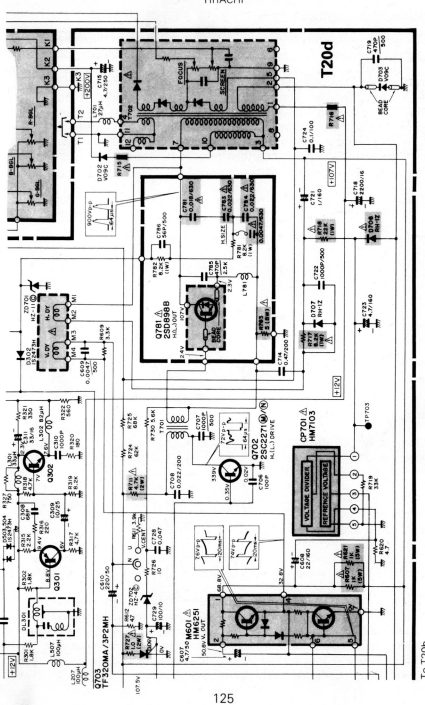

(T20d) CIRCUIT DIAGRAM—MODEL CPT2051 (CONTINUED)

To T20b

Dynamic Convergence (see Fig. T19)

For convergence of Red and Blue vertical lines at top and bottom, disconnect the Green O/P lead from DE/COD and tilt scan coil up and down for best results. Once the best result is obtained insert a rubber wedge between the scan coil and C.R.T. as shown.

To converge horizontal lines top and bottom, tilt the sides of the scan coil to the right or left until convergence is obtained. Replace the Green lead and check that complete convergence has been achieved. If satisfied, insert three fixing wedges to hold the scan coil in correct position using tape to stick them to the C.R.T. funnel; then remove rubber wedge. Two positions of wedging scan coil are shown.

I.T.T.   CVC801, CVC801/1, CVC803, CVC803/1 Series

General Description: The chassis consists of a power board with two plug in modules, one of which is the R.F./I.F./A.F. assembly CMR800 and the other the Decoder assembly CMD800. It can be used with 14-in., 16-in., and 20-in. 90° precision in-line picture tubes fitted with integral scan yoke and quick vision cathodes giving display of picture within approximately five seconds from switch on.

Either mechanical operated or infra-red remote control units may be used with this chassis.

CVC801 chassis with power board CMP801. This has beam current limiter set by R1077 for 14-in. and 16-in. tubes (i.e. link WL614 is removed disconnecting R1078).

CVC801/1 chassis with power board CMP801/1 for 20-in. tubes. This is identical to the CVC801 except that link WL614 is fitted connecting R1078 in parallel with R1077 thus delaying operating point of beam current limiter.

Later versions are designated CVC803 and CVC803/1 respectively.

The later chassis may be identified by the different printed circuit layout of the power board CMP803 or CMP803/1. In all respects other than the power board changes and the new C.R.T. base (CMB803) the details given for the CVC801 are applicable. The later CVC803 chassis may be used with the same types of C.R.T. as the earlier CVC801, but with a side contact adaptor to connect to the CMB803. The later CVC803/1 chassis may be used with either the same type of C.R.T. as the previous 20-in. model with a side contact adaptor to connect to the CMB803; or a new type of side contact C.R.T. (see 'Component Changes').

Mains Supply: 240V, 50Hz. The switched-mode power supply enables the receiver to operate satisfactorily with supply voltages down to approximately 180V A.C.

Cathode Ray Tubes: 14-in. (37 cm) 370-HUB-22-TC-01; 16-in. (42 cm) 420-ERB-22-TC-01; 20-in. (51 cm) 510-VSB-22-TC-01.

Chassis removal

The main chassis is secured with plastic lugs that slide into slots at the bottom of the cabinet moulding. Before commencing maintenance work, switch off the mains supply. Discharge the E.H.T. connector to the aquadag earthing braid before removing the chassis. The chassis can be pulled backwards and then lifted out from the cabinet taking care not to strain the connecting leads. To gain access to the back of the printed-circuit board, the chassis can be tilted through 90°. If necessary the entire chassis can be removed as follows:

Unplug all the inter-connecting plugs and sockets between the chassis and cabinet-mounted components. Withdraw the chassis away from the cabinet.

Module Removal

There are two plug-in modules only that can be removed from the chassis—the decoder module and the R.F./I.F. module. The decoder module may be plugged in from the rear of the main chassis P.C.B. for fault-finding purposes.

To take out the decoder, disconnect the two flying plugs, and remove the retaining screws and support bracket from the upper corner of the decoder board. Then unplug the decoder board from its socket.

To remove the R.F./I.F. module first disconnect the aerial socket and isolator from its moulded supporting bracket. This has a bayonet-type fitting and must be rotated through approximately 90°. Then remove the two retaining screws and washers from the underside of the chassis and the retaining screw and support bracket from the upper corner of the decoder board. The R.F./I.F. module can then be unplugged from the chassis.

Circuit Details

Several modules in these chassis are similar to units used in earlier models and descriptions may be found in previous editions of *Radio and Television Servicing*. Adjustment procedures will be found where these are relevant.

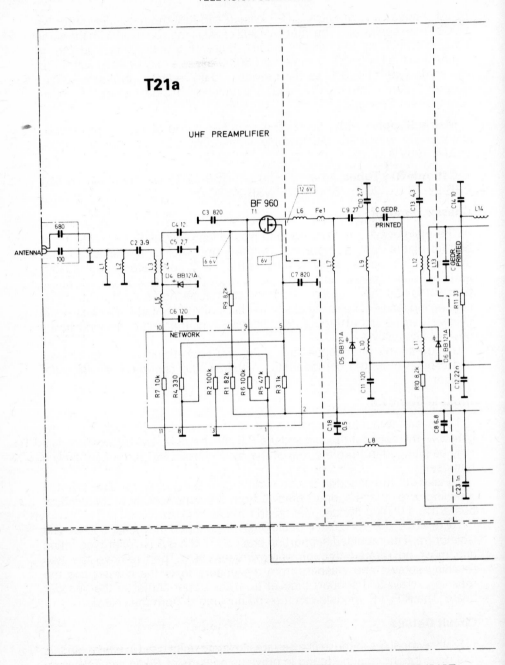

(T21a) CIRCUIT DIAGRAM (R.F./I.F. MODULE CMR800)—CVC800 SERIES (*PART*)

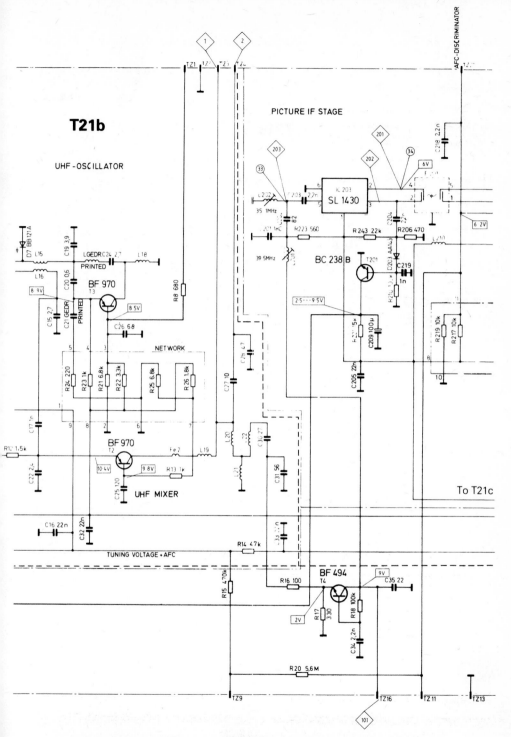

(T21b) CIRCUIT DIAGRAM (R.F./I.F. MODULE CMR800)—CVC800 SERIES (*PART*)

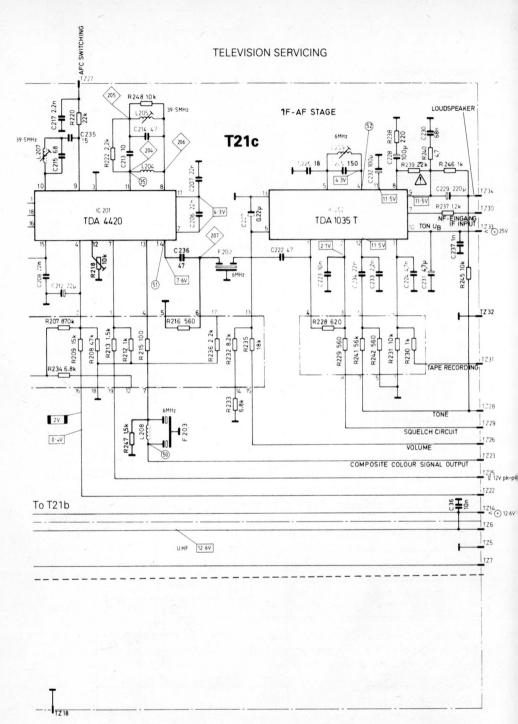

To T21b

(T21c) CIRCUIT DIAGRAM (R.F./I.F. MODULE CMR800)—CVC800 SERIES (*CONTINUED*)

Component Changes

The following components are different for the new Power Board:

Component	New value or type
C601	220 nF
C651	220 nF
C1001	330 pF
D700	BY438
D703	SE159
D704	SE159
D733 (added)	ZPD20
R512	Deleted
R605 (added)	10 kΩ
R607	4·7 kΩ
R702	Same (½ W)
R709	Same (½ W)
R719 (added)	2·2MΩ
R734	56 kΩ
R738	220 kΩ
R747	62 kΩ
R755	1·2MΩ
R1071	10MΩ
S2 (added)	Tape adaptor socket (selective fitting)
V1001 (20-in. models only)	A51-231X or 510-VSB-22-TC-01 (with side contact adaptor)

Adjustments

WARNING

This chassis is always live regardless of the mains lead connection and it is essential that a mains isolation transformer is used when carrying-out setting-up or fault finding.

Initial Settings: Switch on the receiver and adjust the Brightness and Contrast controls so that video noise can be seen. If necessary the A1 voltage may be adjusted for extra brightness.

Set the 110V supply by connecting a multimeter to any 110V test point and adjusting R746 to give a reading of 110V at zero C.R.T. beam current.

Connect a picture source with a castellated-edge pattern and adjust the tuner controls for a suitable picture.

Adjust the focus control R1001 for optimal focus at a point ⅓rd of the distance across the screen width.

Note: This setting-up procedure is written as a complete activity, and if discrete operations are executed separately, associated action described in previous operations may be necessary.

Picture Geometry

Horizontal Scan: Short-circuit TP601.

Adjust R613 to obtain minimum deviation from the nominal horizontal scan frequency.

Remove the short-circuit from TP601.

Check the 110V line by connecting a multimeter to any 110V test point, re-adjusting R746 if necessary to give 110V at zero C.R.T. beam current.

Vertical Scan: Short-circuit TP401, connect pin 8 of IC400 to chassis.

Adjust R422 so that the frame rolls from top to bottom of the screen once in each second.

Remove the short-circuit from TP401.

Adjust the horizontal amplitude control L504, vertical amplitude control R406 and picture shift controls R401 and R506 so that approximately ¼ of the castellations of the test pattern are visible on all sides of the picture. If the picture width is excessive, open circuit wire link 702 and re-adjust L504. (Switch off set first.)

Set the vertical linearity by adjusting R415 and re-adjust the vertical shift control R401 if necessary. Reset the focus control R1001 if necessary.

Grey Scale Tracking:

Preliminary Adjustments: Degauss the tube and chassis thoroughly, avoiding the neck components.

Set R708 fully clockwise, as viewed from the carbon track side of the potentiometer.

Set the black level potentiometers R352, R372, and R392 fully clockwise, as viewed from the copper side of the colour decoder P.C.B.

Adjust the three drive potentiometers R351, R371 and R391 to maximum, fully anti-clockwise as viewed from the copper side of the colour decoder P.C.B.

Low Lights: Connect a colour bar test pattern signal source.

Adjust the Brightness control to its mid-position.

Connect a shorting link to the blank raster pins (WL303) on the colour decoder to give a blank raster.

Adjust the appropriate black level potentiometers, R352 for TP303, R372 for TP304, and R392 for TP305 at the colour decoder to give 165V at each test point respectively, as measured with the multimeter.

Collapse the field by applying a shorting link across the two pins at TP402.

Reduce the ambient lighting.

Increase the A1 volts by adjusting R708 to obtain a coloured line.

Cbtain a white horizontal line, increasing those colours which are deficient, by adjusting the two appropriate black level potentiometers (selected from R352, R372 or R392) in an anti-clockwise direction, as viewed from the copper side of the colour decoder P.C.B.

Note: The third black level potentiometer (corresponding to the colour of the original line first appearing) must not be adjusted.

High-Lights: Remove the shorting link at TP402.

Remove the shorting link at the blank raster pins of the colour decoder.

Set the Colour control to its minimum level.

Set the Brightness control to its mid-position.

Adjust the Contrast control to obtain an acceptable picture.

Adjust the drive potentiometers R351, R371, and R391 on the colour decoder to obtain the correct grey scale, i.e. so that the picture screen shows no colour tint. After this adjustment at least one drive potentiometer must be at maximum, i.e. fully anti-clockwise as viewed from the copper side of the colour decoder P.C.B.

If necessary, re-adjust the appropriate black level potentiometers (R352, R372, or R392) to remove any low light colour errors.

T22a

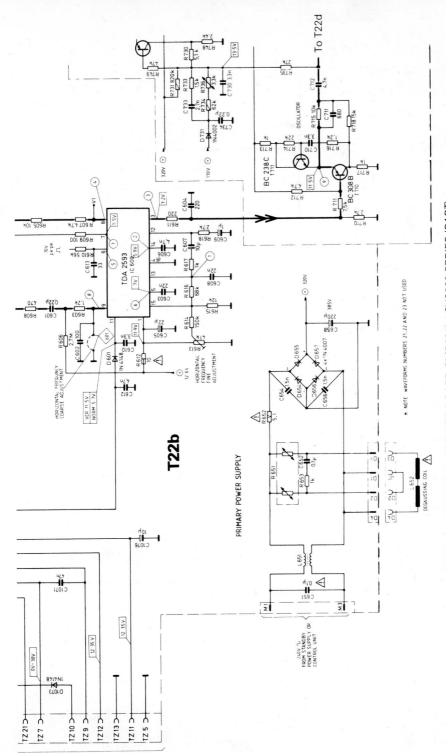

T22b

PRIMARY POWER SUPPLY

(T22b) CIRCUIT DIAGRAM (POWER BOARD CMP801)—CVC800 SERIES (*PART*)

* NOTE WAVEFORMS NUMBERS 21,22 AND 23 NOT USED

To T22d

(T22c) CIRCUIT DIAGRAM (POWER BOARD CMP801)—CVC800 SERIES (PART)

136

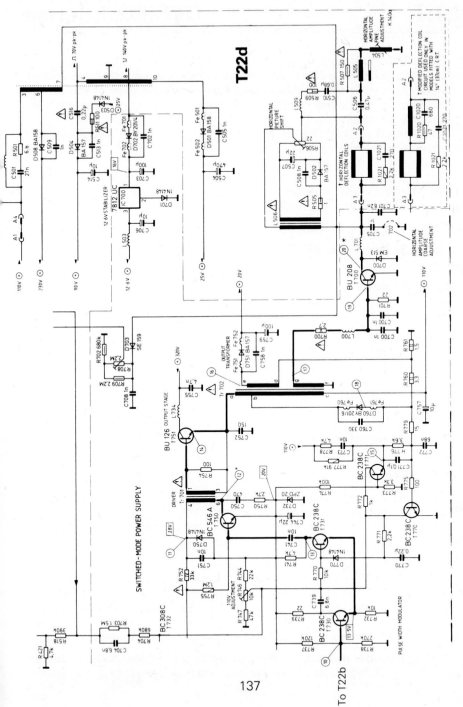

(T22d) CIRCUIT DIAGRAM (POWER BOARD CMP801)—CVC800 SERIES (CONTINUED)

I.T.T.   CVC802/1 Series

General Description: The CVC802/1 chassis is a remote control colour receiver, which uses a 20-in. 90° precision in-line picture tube fitted with an integral scan yoke and quick-vision cathodes giving display of the picture within approximately five seconds from switch-on. No convergence adjustments are necessary.

The chassis is made up of two plug-in modules: the colour decoder module and the H.F. module, and is easily removed from the cabinet by disengaging its plastic lugs from slits in the cabinet base.

The modules used here are basically similar to those used in the I.T.T. CVC800 and CVC801 chassis which is described in this and the 1981–82 volumes of *Radio and Television Servicing*.

Mains Supply: 240 volts, 50Hz.

Cathode Ray Tube: A51-231X or 510-VSB-22-TC-01 (with side contact adaptor).

Chassis Removal

The main chassis is secured with plastic lugs that slide into slots at the bottom of the cabinet moulding. Before commencing maintenance work, switch-off the mains supply. Discharge the E.H.T. connector to the aquadag earthing braid before removing the chassis. The chassis can be pulled backwards and then lifted out from the cabinet taking care not to strain the connecting leads. To gain access to the back of the printed circuit board, the chassis can be tilted through 90°. If necessary the entire chassis can be removed as follows:

Unplug all the inter-connecting plugs and sockets between the chassis and cabinet-mounted components. Withdraw the chassis away from the cabinet.

Module Removal

General: There are three modules that can be removed from the receiver. Two of these (the H.F. module and the colour decoder module) plug into the power board. The decoder module may also be plugged in from the rear of the power board P.C.B. for fault-finding purposes. The third module (the control unit) is secured to the receiver cabinet by screws.

H.F. module: To remove the H.F. module, first disconnect the aerial socket and isolator from its moulded supporting bracket. This has a bayonet-type fitting and must be rotated through approximately 90°. Then remove the two retaining screws and washers from the underside of the power board. The H.F. module can then be unplugged from the power board.

Colour decoder module: To remove the colour decoder module, disconnect the two flying sockets, remove the retaining screws and the support bracket from the upper corner of the decoder board, then unplug the decoder board from its sockets.

Control unit: To remove the control unit for fault-finding purposes remove the four securing screws and lift the unit out from the rear of the cabinet. If necessary the control unit can then be completely extracted from the cabinet by unplugging all the connections between it and the cabinet-mounted components.

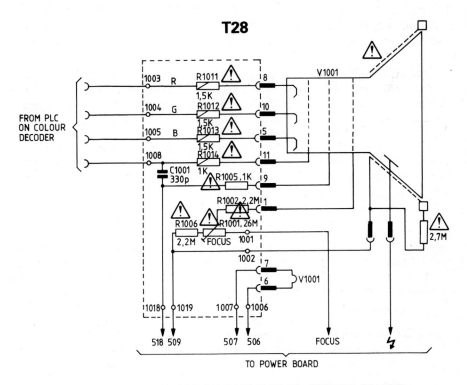

(T28) CIRCUIT DIAGRAM (C.R.T. BASE-CMB803)—CVC802/1 CHASSIS)

I.T.T.   IT T80-90° Chassis

General Description: The chassis comprises a horizontally mounted power board CVC823 and a vertically mounted signals board CVC824. It is intended for use with 20-in. and 22-in. precision in-line picture tubes fitted with an integral scan yoke and quick-vision cathodes giving display of picture within approximately five seconds from switch-on. There are significant circuit changes in the power supply module used in this chassis compared to the original ITT80 chassis. Other modules are similar to those used in the CVC800, CVC801 and CVC802, described elsewhere. Either mechanical operated or infra red remote control units may be used with this chassis.

Mains Supply: 240V, 50Hz. (The switched-mode power supply enables the receiver to operate satisfactorily with supply voltages down to approximately 180V A.C.)

Cathode Ray tubes: 90° P.I.L., quick-heat, 20-in. 510-VSB-22-TC-01; 90° P.I.L., quick-heat, 22-in. 560-EGB-22-TC-01. The chassis consists of two sections: a horizontally-mounted power board carrying the power supply, timebases and E.H.T. stages; and a vertically-mounted signals board carrying the H.F. module, interface module and colour decoder. Both boards slide into plastic channels fitted at the top and the base of the cabinet. The power board is held secure by two plastic latches and the signals board is secured by one posi-pan screw fitted at the bottom rear end of the board.

Power Board (CVC823)

To obtain access to the power board, press and hold the plastic latches at each front corner of the board and withdraw it carefully to the end of the channels, lifting upwards until the board rests at an angle of approximately 60° to the horizontal. If necessary, the power board can be removed as follows:

Switch-off the mains supply.

Discharge the E.H.T. connector to the aquadag earthing braid first.

Unplug all the inter-connecting plugs and sockets between the power board and cabinet-mounted components.

Press and hold the plastic latches at each front corner of the power board and withdraw it from the cabinet.

Signals Board (CVC824)

To obtain access to the signals board, remove the posi-pan screw fitted at the bottom rear of the board. Withdraw the board approximately 2½ inches until it locks into the withdrawn position. If necessary, the signals board can then be removed as follows:—

Switch-off the mains supply.

Discharge the E.H.T. connector to the equadag earthing braid first.

Unplug all the inter-connecting plugs and sockets between the signals board and cabinet-mounted components.

Withdraw the board approximately 1½ inches, then lift it upwards and out from the plastic channels.

Module Removal

Control Unit CMC77/2: To remove the control unit, remove the four posi-pan screws and then lift it up and out from the rear of the cabinet. If required, the control unit can be completely extracted from the cabinet by unplugging all connections between it and the cabinet-mounted components.

Control Unit CMC77/3: To remove the control unit, remove the four posi-pan screws and then lift it up and out from the rear of the cabinet. If required, the control unit can be completely extracted from the cabinet by unplugging all connections between it and the cabinet-mounted components.

Control Unit CMC90/3: To obtain access to the control unit, remove the rear-mounted plastic support bracket by turning the plastic fastener one quarter-turn anti-clockwise. Remove the two posi-pan screws and lift the unit up and out from the rear of the cabinet. If required, the control unit can be completely extracted from the cabinet by unplugging all connections between it and the cabinet-mounted components.

Control Unit CMC96/3: To obtain access to the control unit first remove the rear-mounted plastic support bracket by turning the plastic fastener one quarter-turn anti-clockwise and then remove the standby power supply from the cabinet base by unplugging the connecting leads to the control unit and pressing the four plastic securing clips, lifting it clear from the rear of the cabinet. To remove the control unit for fault-finding purposes remove the two posi-pan screws and lift the unit up and out from the rear of the cabinet. If it is required to operate the receiver at this stage reconnect the leads from the control unit to the standby power supply. If required, the control unit can be completely extracted from the cabinet by unplugging all connections between it and the cabinet-mounted components.

Control Unit CMC112/1: To remove the control unit, remove the five posi-pan screws and then lift it up and out from the rear of the cabinet. If required, the control unit can be completely extracted from the cabinet by unplugging all connections between it and the cabinet-mounted components.

Eight-Way Push-Button Unit: To remove the push-button unit, remove the four posi-pan screws and then lift it up and out from the rear of the cabinet. If required, the control unit can be completely extracted from the cabinet by unplugging all connections between it and the cabinet-mounted components.

Standby Power Supply: To remove the standby power supply, press the four plastic securing clips and then lift it clear of the cabinet. If required, the standby power supply can be disconnected and finally extracted from the cabinet by unplugging all the connectors between it and the cabinet-mounted components.

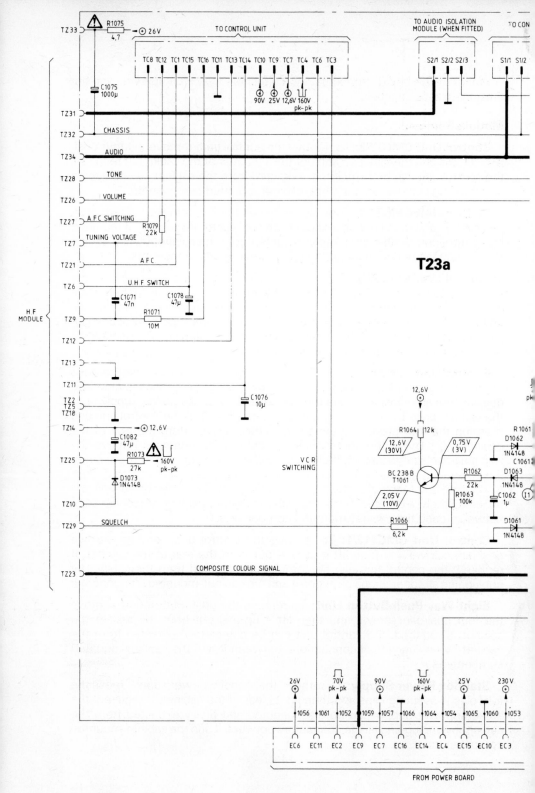

(T23a) CIRCUIT DIAGRAM (INTERFACE BOARD—CMN800)—ITT80–90° CHASSIS (*PART*)

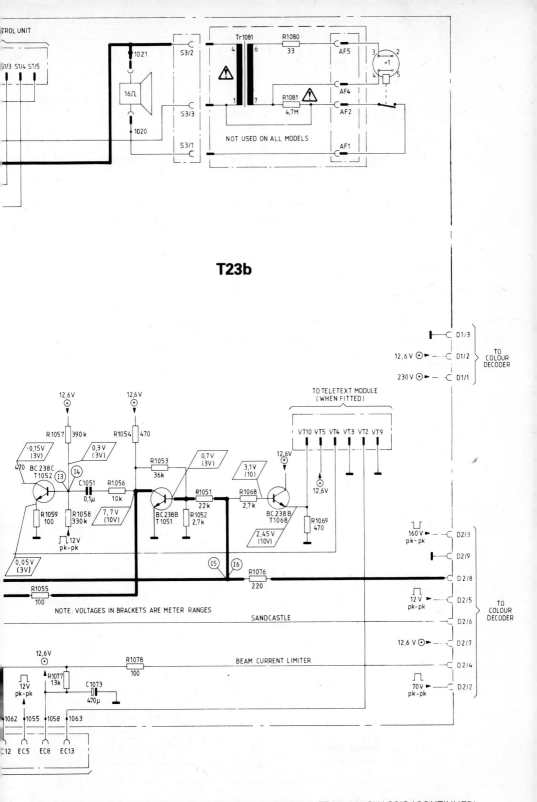

T23b

NOTE: VOLTAGES IN BRACKETS ARE METER RANGES

(T23b) CIRCUIT DIAGRAM (INTERFACE BOARD—CMN800)—ITT80–90° CHASSIS (*CONTINUED*)

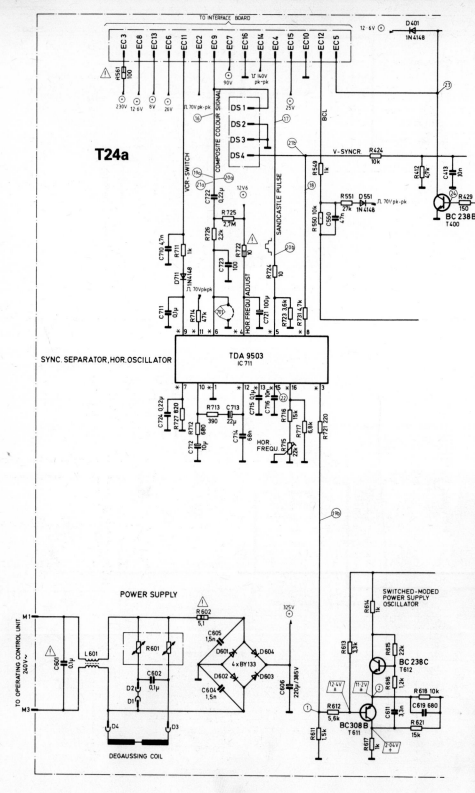

(T24a) CIRCUIT DIAGRAM (POWER BOARD—CVC823)—ITT80–90° CHASSIS (*PART*)

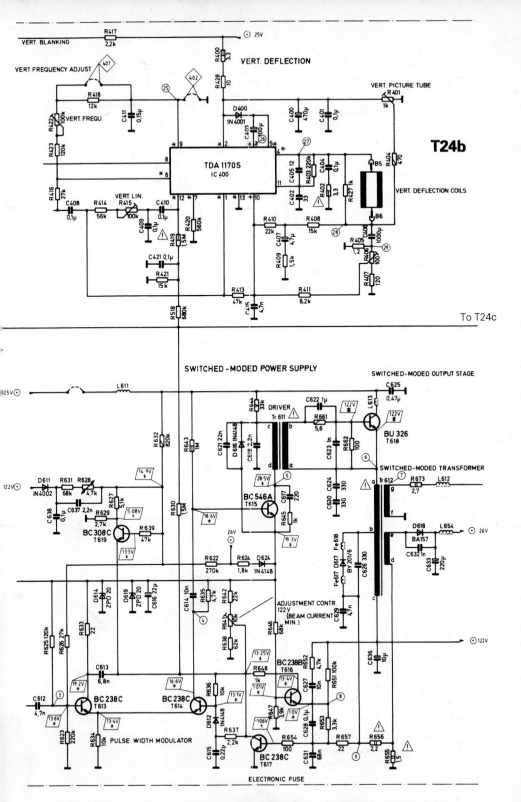

(T24b) CIRCUIT DIAGRAM (POWER BOARD—CVC823)—ITT80–90° CHASSIS (*PART*)

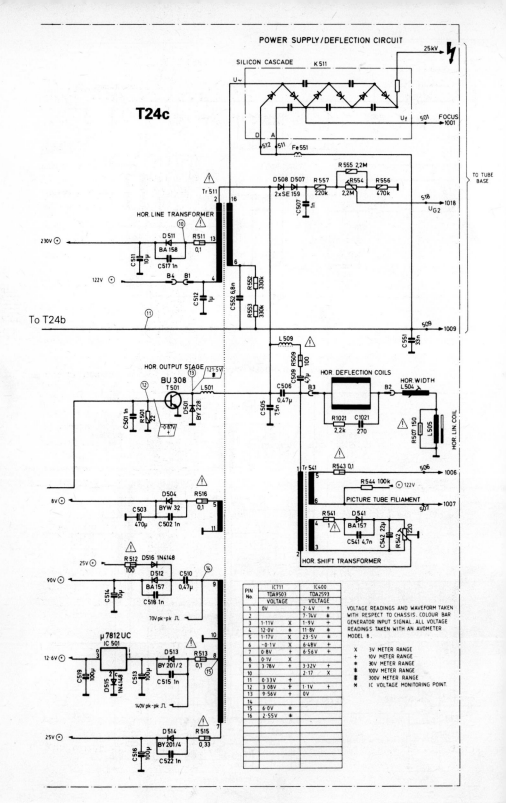

(T24c) CIRCUIT DIAGRAM (POWER BOARD—CVC823)—ITT80–90° CHASSIS (*PART*)

I.T.T.

ITT80-110° Chassis
Models 3432, 3702, 3732

General Description: This information covers the chassis and control units used in the 22-in. and 26-in. Trimline models.

The chassis use a common signal board CVC824. For 22-in. receivers the power board is type CVC830P and for 26-in. receivers type CVC825P. Both chassis are designed to drive 110° precision in-line picture tubes fitted with an integral scan yoke and quick-vision cathodes. These provide display of a picture within approximately five seconds from switch-on and convergence adjustments are not necessary.

Either mechanical operated or infra-red remote control units may be used with these chassis.

The remote control unit provides remote channel selection, control of brightness, colour, and volume; and selection of 'ideal' settings of volume, brightness, and colour, as determined by the pre-set controls in the receiver pre-set control compartment. On 26-in. Trimline receivers fitted with a toroidal wound isolation transformer the remote control system also provides remote control of an I.T.T. Video 2000 System V.C.R., connected via a special remote control interface to the appropriate socket at the rear of the receiver. The circuits and adjustments of the R.F., Decoder and interface modules are similar to those used in other I.T.T. chassis and are described elsewhere in this and previous volumes of *Radio and Television Servicing*. The new power boards (CVC825P and CVC830P) and C.R.T. Base circuits are given here.

Mains Supply: 240 volts, 50Hz.

Cathode Ray Tubes: A56-540X (22-in.); A67-701X (26-in.).

Chassis

The chassis consists of two sections: a horizontally-mounted power board carrying the power supply, timebases, and E.H.T. stages; and a vertically-mounted signals board carrying the H.F. module, interface module, and colour decoder. Both boards slide into plastic channels fitted at the top and the base of the cabinet. The power board is held secure by two plastic latches and the signals board is secured by one posi-pan screw fitted at the bottom rear-end of the board.

Power board

To obtain access to the power board, press and hold the plastic latches at each front corner of the board and withdraw it carefully to the end of the channels, lifting upwards until the board rests at an angle of approximately

147

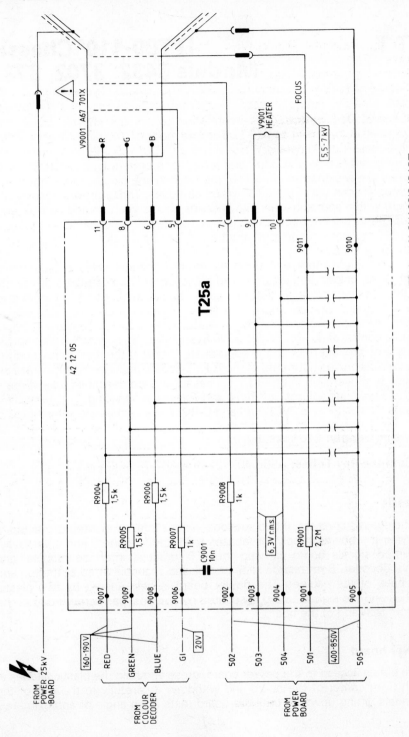

(T25a) CIRCUIT DIAGRAM (C.R.T. BASE—PART OF CVC825)—ITT80–110° CHASSIS (*PART*)

60° to the horizontal. If necessary, the power board can be removed as follows:

Switch-off the mains supply.

Discharge the E.H.T. connector to the aquadag earthing braid first.

Unplug all the inter-connecting plugs and sockets between the power board and cabinet-mounted components.

Press and hold the plastic latches at each front corner of the power board and withdraw it from the cabinet.

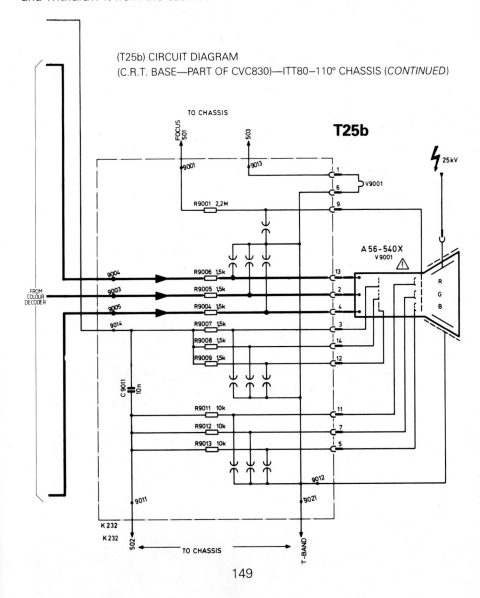

(T25b) CIRCUIT DIAGRAM

(C.R.T. BASE—PART OF CVC830)—ITT80–110° CHASSIS (*CONTINUED*)

149

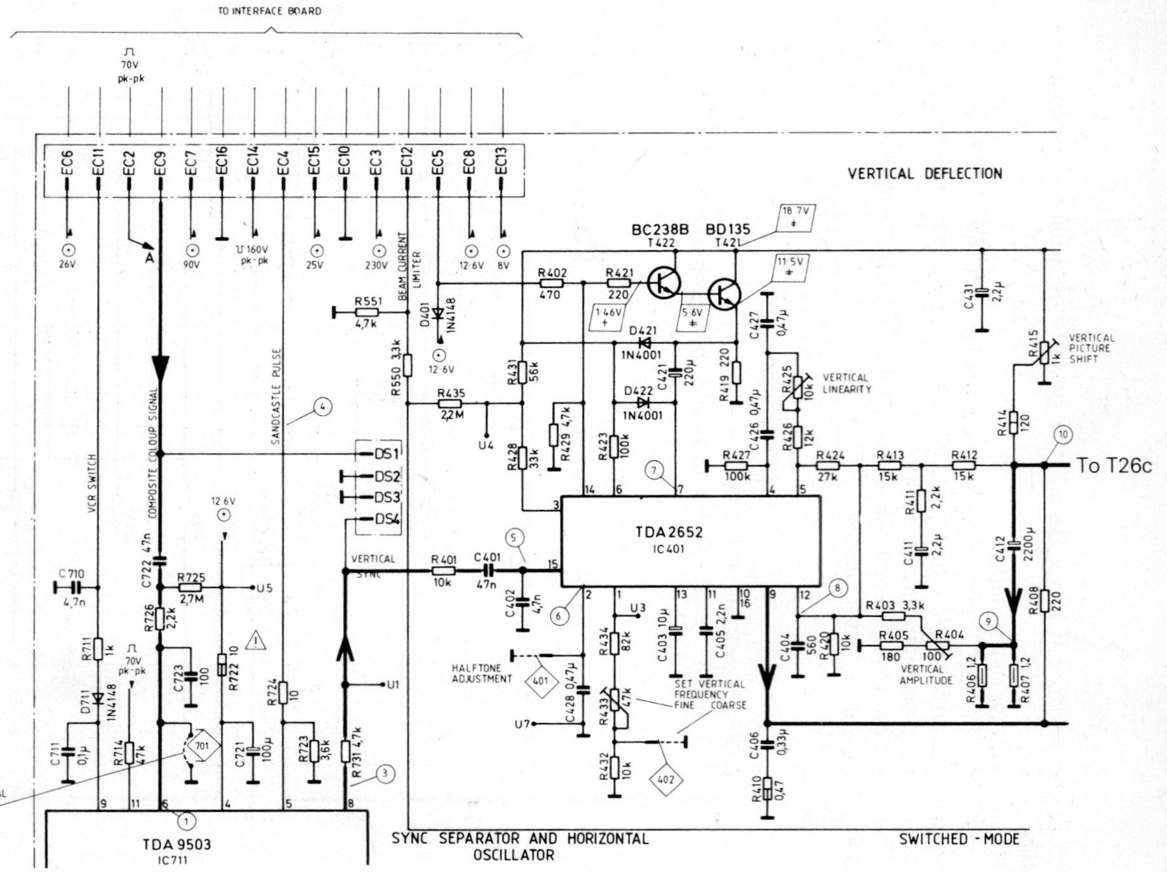

(T26a) CIRCUIT DIAGRAM (POWER BOARD—CVC825P)—ITT80–110° CHASSIS (PART)

T26a

PIN NO	IC711 TDA9503 VOLTAGE		PIN NO	IC401 TDA2652 VOLTAGE	
1	0V		1	8.4V	#
2	-		2	5.3V	+
3	1.17V	+	3	0.85V	X
4	12V	+	4	4.6V	+
5	12V	+	5	-	
6	0.2V	+	6	8.3V	+
7	1.22V	X	7	8.1V	+
8	-		8	18.1V	+
9	0.7V	+	9	-	
10	-		10	10.7	+
11	0.48V	+	11	0V	
12	1.4V	+	12	10.4V	+
13	9.6V	+	13	12V	+
14	-		14	12V	+
15	6.1V	+	15	1.5V	X
16	2.4V	+	16	0V	

ALL VOLTAGE READINGS TAKEN WITH
RESPECT TO CHASSIS AND WITH A
COLOUR BAR INPUT SIGNAL, USING AN
AVOMETER MODEL 8
WAVEFORMS WITH RESPECT TO CHASSIS

METER RANGES
X   3 VOLTS
*   10 VOLTS
+   30 VOLTS
#   100 VOLTS
300 VOLTS

150

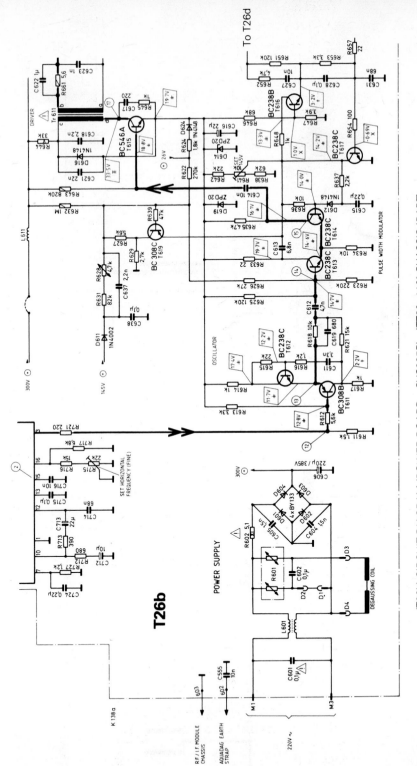

(T26b) CIRCUIT DIAGRAM (POWER BOARD—CVC825P)—ITT80–110° CHASSIS (PART)

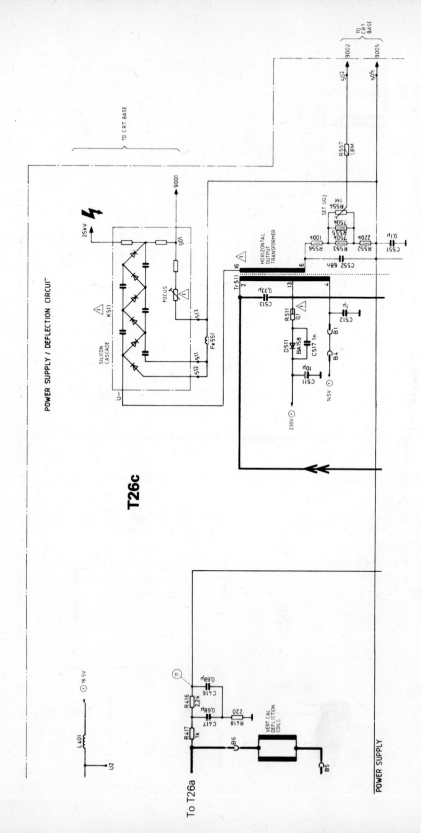

(T26c) CIRCUIT DIAGRAM (POWER BOARD—CVC825P)—ITT80–110° CHASSIS (*PART*)

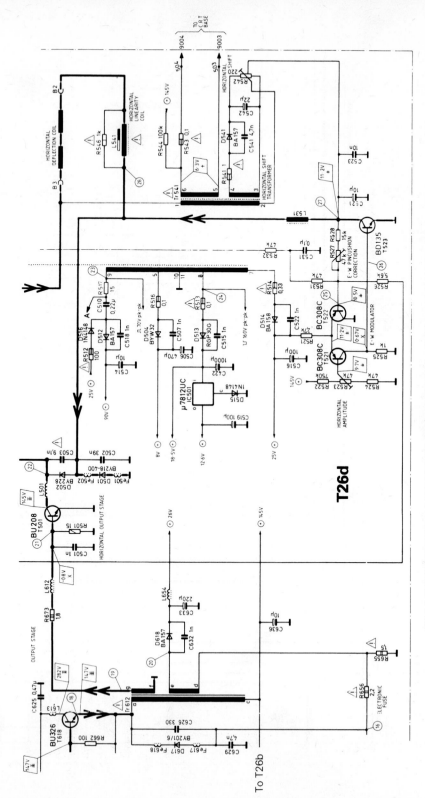

(T26d) CIRCUIT DIAGRAM (POWER BOARD—CVC825P)—ITT80–110° CHASSIS (CONTINUED)

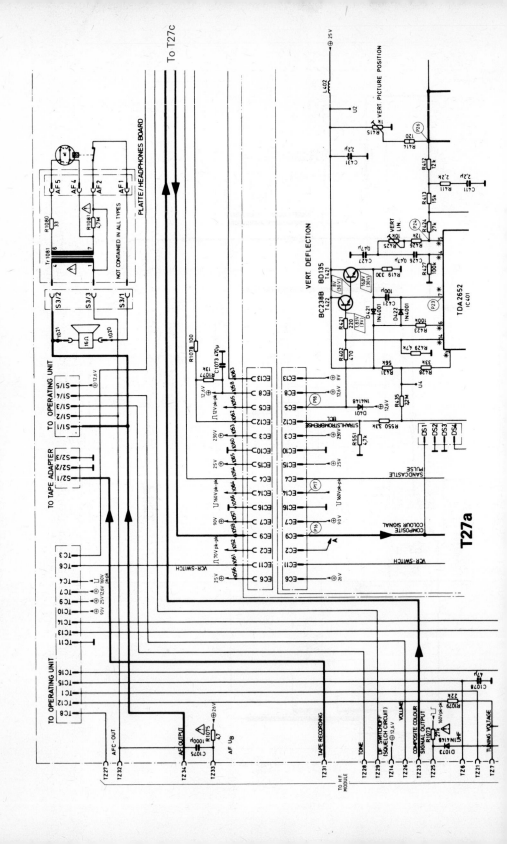

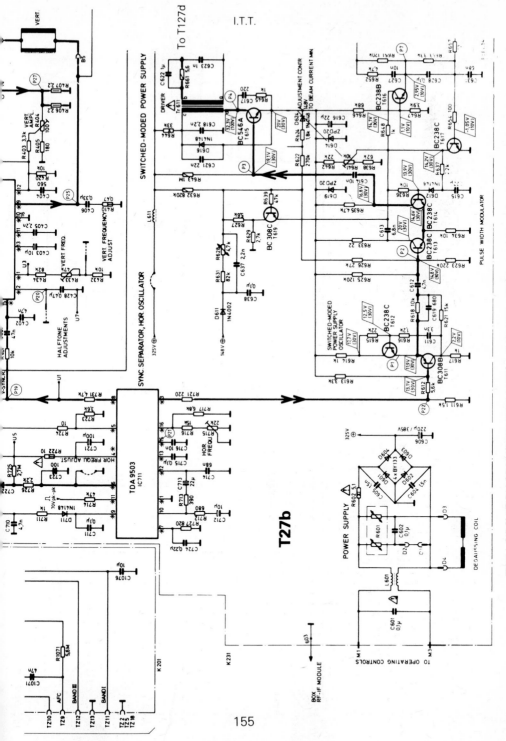

T27b

I.T.T.

To T127d

(T27b) CIRCUIT DIAGRAM (POWER BOARD—CVC830P)—ITT80—110° CHASSIS (PART)

155

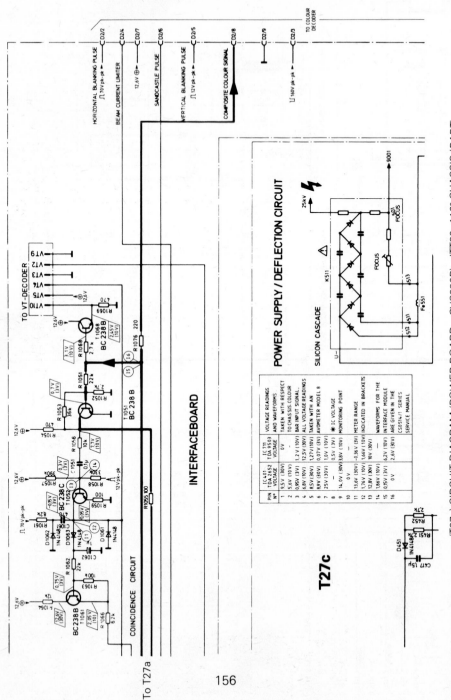

(T27c) CIRCUIT DIAGRAM (POWER BOARD—CVC830P)—ITT80–110° CHASSIS (*PART*)

T27c

POWER SUPPLY / DEFLECTION CIRCUIT

SILICON CASCADE

INTERFACEBOARD

COINCIDENCE CIRCUIT

TO VT-DECODER

To T27a

PIN N°	IC 401 TDA 2652 VOLTAGE	IC 711 TDA 9503 VOLTAGE	VOLTAGE READINGS AND WAVEFORMS
1	9.5V (30V)	0V	TAKEN WITH RESPECT TO CHASSIS. COLOUR BAR INPUT SIGNAL.
2	5.6V (10V)		ALL VOLTAGE READINGS TAKEN WITH AN AVOMETER MODEL 8
3	0.95V (3V)	1.2 V (10V)	
4	6.8V (10V)	12.5V (30V)	⚹ IC VOLTAGE
5	8.5V (30V)	1.27V(10V)	
6	8.9V (30V)	0.07V (3V)	10 MONITORING POINT
7	25V (30V)	1.0 V (10V)	
8		0.5V (3V)	METER RANGE INDICATED IN BRACKETS
9	14. 0V (30V)	3.8V (10V)	
10	0 V		WAVEFORMS FOR THE INTERFACE MODULE ARE GIVEN IN THE CS051A/1 SERIES SERVICE MANUAL
11	13.6V (30V)	-0.36V (3V)	
12	1.76V (10V)	1.66V (10V)	
13	12.8V (30V)	10V (30V)	
14	1.86V (10V)		
15	0.55V (3V)	6.2V (10V)	
16	0 V	2.6V (30V)	

156

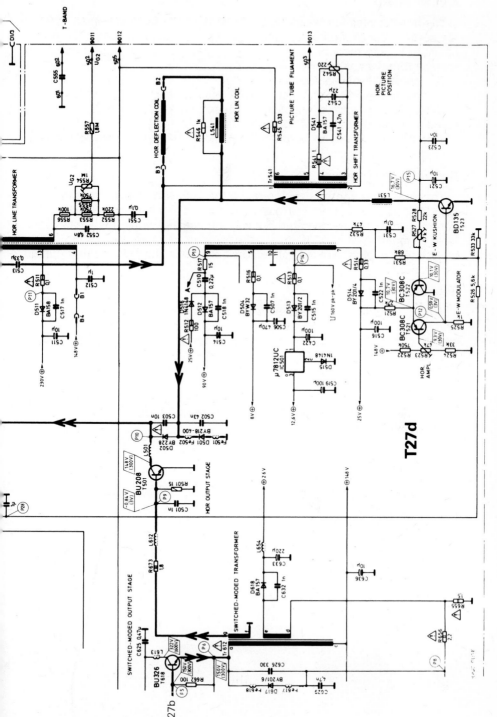

(T27d) CIRCUIT DIAGRAM (POWER BOARD—CVC830P)—ITT80–110° CHASSIS (CONTINUED)

Signals board

To obtain access to the signals board, remove the posi-pan screw fitted at the bottom rear of the board. Withdraw the board approximately 2½ inches until it locks into the withdrawn position. If necessary, the signals board can then be removed as follows:

Switch-off the mains supply.

Discharge the E.H.T. connector to the aquadag earthing braid first.

Unplug all the inter-connecting plugs and sockets between the signals board and cabinet-mounted components.

Withdraw the board approximately 1½ inches, then lift it upwards and out from the plastic channels.

Control panel removal

To release the control panel, remove the posi-pan screws from the rear, then lift it up and out from the rear of the cabinet. For servicing purposes, the control panel may be operated outside the cabinet within the limits of the inter-connecting wiring. If it is necessary to remove the control panel completely, unplug all of the connectors between it and the cabinet-mounted components.

Teletext module

Note: To remove the Teletext module, remove the four securing screws and then lift it clear of the mounting plate. If required, the Teletext module can be disconnected and finally extracted from the cabinet by unplugging all the connections between it and the cabinet-mounted components. It is not intended that the Teletext module should be repaired or adjusted in the field (due to the complexity of the logic circuits in it and the delicacy of integrated circuits used in it). In the event of failure of the module being established a replacement module must be fitted, obtainable from I.T.T. Consumer Products Services.

J.V.C.   Model CX-610GB

General Description: A portable colour television receiver with 6-in. cathode ray tube operating from mains or battery supplies. Integrated circuits are used in low level signal stages and a socket is provided for the connection of an earphone. The circuit is designed to accept various transmission standards which may be received from proposed satellite links.

Mains Supplies: 110–127, 220–240 volts, 50–60Hz.

E.H.T.: 14Kv.

Battery Supplies: 12–15 volts, 15 watts.

Adjustments

Power Supply: The regulated +B1 control has been factory adjusted and normally requires no adjustment. However, if any repairs have been made to the chassis it is recommended that this adjustment should be made.

Allow 5 minutes to warm up and tune in a station.

Connect an accurate D.C. voltmeter to TP-91.

Adjust the B1 Adj. Control for a reading of 10·7 volts D.C.

Note: Should +B1 control be set too high, it may cause possible component damage. Therefore, use an accurate voltmeter.

Sub Colour: While receiving a colour T.V. programme, set the Colour knob on the control panel to the central position. Then align the Sub Colour control on the Main P.B. assembly until the colour of skin looks natural.

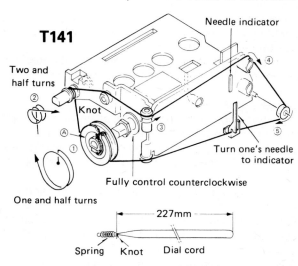

(T141) DRIVE CORD—MODEL CX-610GB

159

Sub Contrast and Sub Bright: While receiving a T.V. programme, set the Picture and the Bright knobs on the control panel to the central position. Then align both the Sub Contrast and the Sub Bright controls on the Main P.B. assembly until an ideal picture is obtained.

Horizontal Oscillator: Set the H. Freq. control to the mechanical centre position.

Connect the jumper clip between TP-33A and TP-33B.

Adjust the H. Freq. control until picture is in view and locks or drifts slowly back and forth.

Remove the jumper clip.

Make sure that the set maintains horizontal sync., when channels are switched.

C.R.T. Adjustments (Figs. T142, 143)

The picture tube is a precision in-line gun type. For this picture tube, dynamic convergence is carried out by a precision deflection yoke which eliminates the use of convergence yoke and convergence circuit. The deflection yoke and purity/convergency magnets assembly has been set at the factory and requires no field adjustments.

However, should the assembly be accidentally jarred or tampered with, some or all adjustments may be necessary.

Colour Purity and Vertical Centre: Loosen yoke retaining clamp. Remove wedges completely and clean-off dried adhesive from the picture tube. Paint is used to lock the tabs of the purity/convergence magnet assembly in place. The paint must be removed with the end of a screwdriver before any adjustments are attempted.

1. Select no signal channel.

2. Set the purity tabs in line horizontally. A long tab should be in the same direction as the other short tab.

3. Turn the Green cut-off control to maximum and the Red and Blue cut-off controls to minimum. Then adjust the screen control so that the green band can be seen.

4. Move the yoke slowly backward.

5. Rotate the two tabs with them kept at an angle, together in either direction so that the green band is centred on the picture tube.

6. Check the vertical centre position by displaying a horizontal line. Unless correct, bring it to the centre by rotating the two tabs, kept at an angle, together in either direction.

7. Repeat steps 5 and 6 alternately until the green band and the vertical centre come to the centre.

8. Move the yoke slowly towards the bell of the tube so that the whole surface of the picture tube is filled with a green pure raster.

9. Turning Red or Blue cut-off control to maximum and Green cut-off control to minimum, make sure of a red or blue pure raster.

10. Secure yoke retaining clamp.

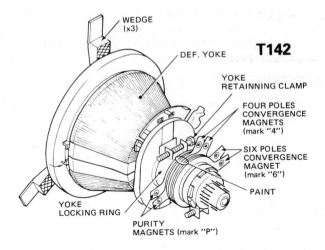

WEDGE (x3)

DEF. YOKE **T142**

YOKE RETAINNING CLAMP

FOUR POLES CONVERGENCE MAGNETS (mark "4")

SIX POLES CONVERGENCE MAGNET (mark "6")

PAINT

YOKE LOCKING RING

PURITY MAGNETS (mark "P")

(T142) SCANNING YOKE—MODEL CX-610GB, ETC.

[REAR VIEW]

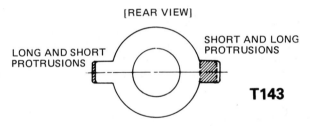

LONG AND SHORT PROTRUSIONS

SHORT AND LONG PROTRUSIONS

T143

Let the protrusions come in line

[FRONT VIEW]

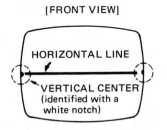

HORIZONTAL LINE

VERTICAL CENTER (identified with a white notch)

Bring the horizontal line nearest to the white notches shown in the dotted circles.

[FRONT VIEW]

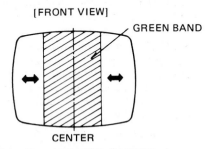

GREEN BAND

CENTER

Bring the green band to the center.

(T143) PURITY ADJUSTMENTS—MODEL CX-610GB, ETC.

Static Convergence (see Figs. T142, T143): Static convergence is achieved by four magnets located on the neck, nearest the base of the picture tube. The front pair of magnetic rings (closest to the purity tabs) are adjusted to converge the red and blue crosshatch lines.

The rear pair of convergence rings (closest to the base of the picture tube) are adjusted to converge the magenta (R/B) and green crosshatch lines.

Dynamic convergence is achieved by adjusting the static convergence.

Connect a crosshatch generator to the antenna terminals and adjust Brightness and Contrast control for a distinct pattern.

Rotate the front pair of tabs as a unit to minimise the separation of the red and blue lines around the centre of the screen. To adjust the convergence of red and blue, vary the angle between the tabs.

Rotate the rear pair of tabs as a unit to minimise the separation of the magenta (R/B) and green lines.

Adjust the spacing of the rear tabs to converge the magenta and green lines.

Apply paint to fix six magnets.

Fasten the magnet locking ring clockwise.

White Balance Adjustment (Black and White Tracking): Receive a black and white broadcast, or misadjust the Tuning control so that a colour picture becomes black and white.

Set the Red and Green drive controls for their mechanical centre.

Turn the Red, Green and Blue cut-off controls and the Screen control fully counter-clockwise.

Change a service tips as shown in Fig. T144.

Turn Screen control slowly clockwise until a very faint horizontal line appears.

Turn the Cut-off control of the colour which has appeared first, clockwise by about 10° and then adjust the Screen control again so that the colour may shine faintly.

Turn the other colour Cut-off controls slowly clockwise until a reasonable white line appears.

Return a service tips to normal position.

Adjust the Red and Green drive controls for best white highlights.

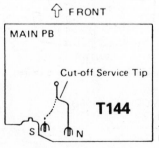

⇧ FRONT

MAIN PB

Cut-off Service Tip

T144

S   N

(T144) GREY SCALE ADJUSTMENT
—MODEL CX-610GB, ETC.

Reconnect the cut-off service tip from N to S, and a horizontal line will appear.

162

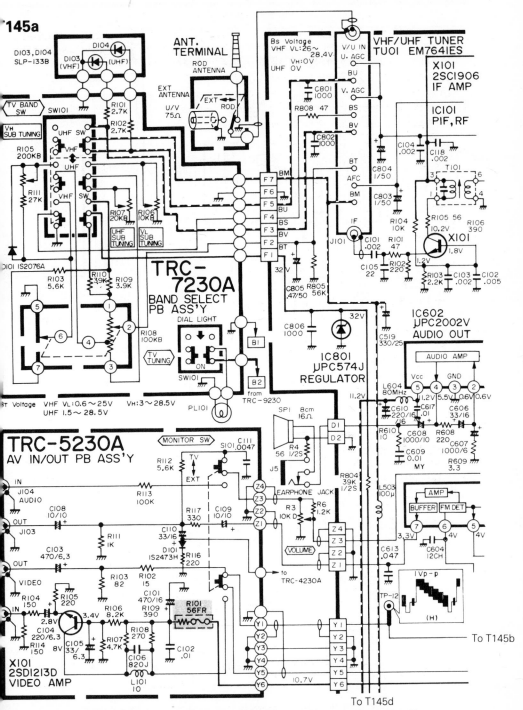

145a

(T145a) CIRCUIT DIAGRAM (MAIN BOARD)—MODEL CX-610GB (*PART*)

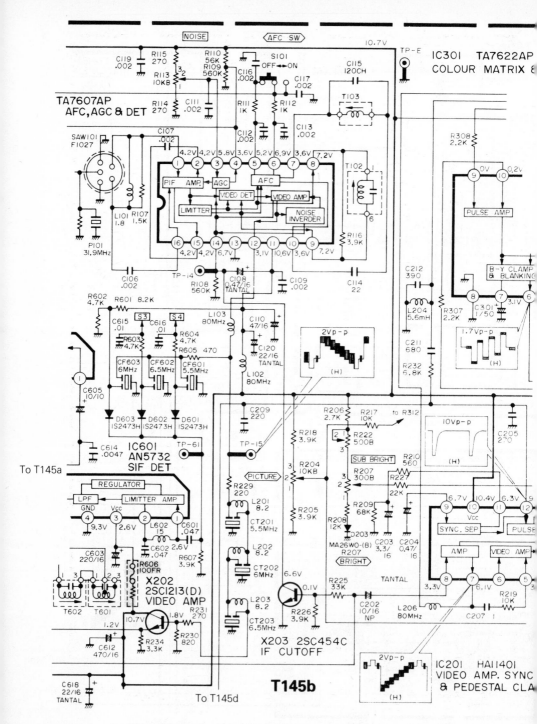

(T145b) CIRCUIT DIAGRAM (MAIN BOARD)—MODEL CX-610GB (*PART*)

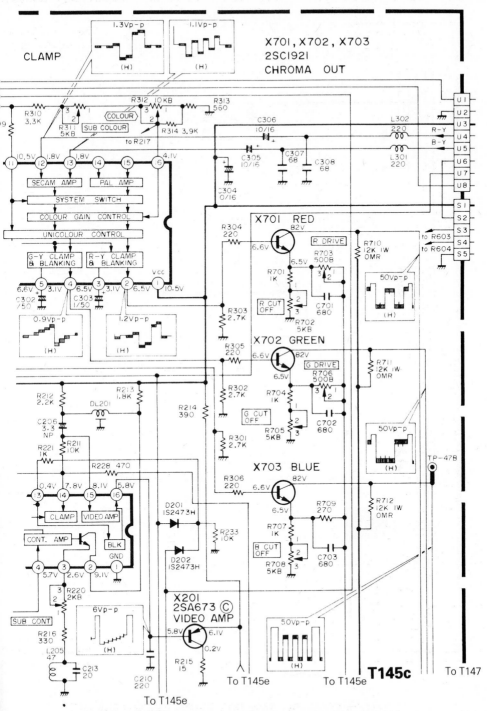

(T145c) CIRCUIT DIAGRAM (MAIN BOARD)—MODEL CX-610GB (*PART*)

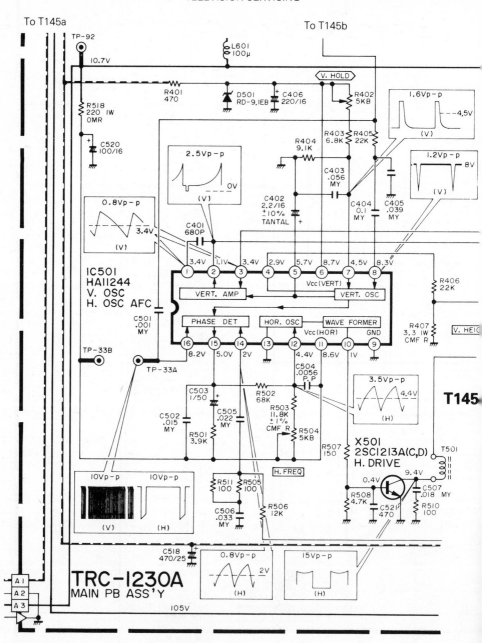

(T145d) CIRCUIT DIAGRAM (MAIN BOARD)—MODEL CX-610GB (*PART*)

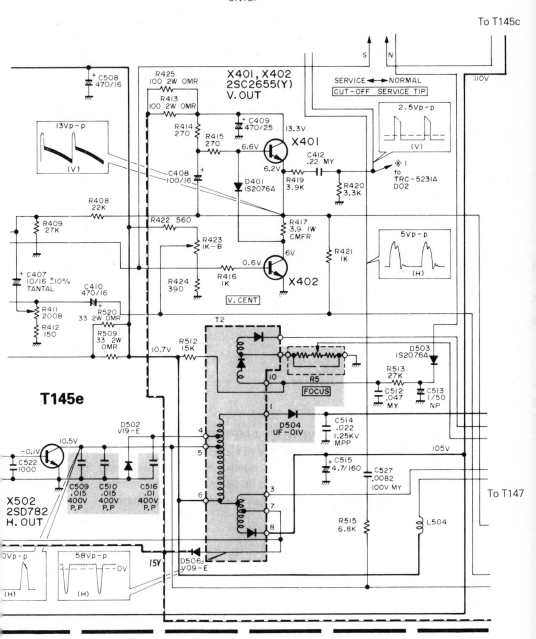

(T145e) CIRCUIT DIAGRAM (MAIN BOARD)—MODEL CX-610GB (*CONTINUED*)

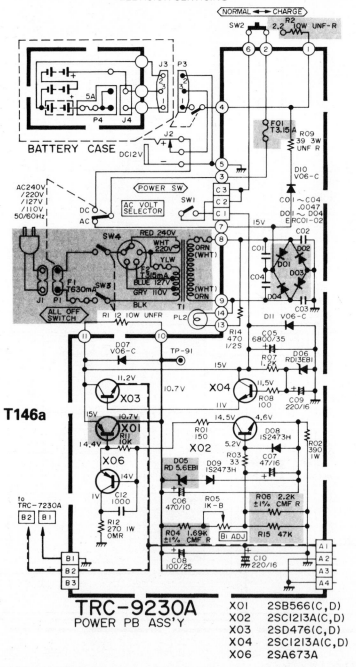

T146a

TRC-9230A
POWER PB ASS'Y

X01	2SB566(C,D)
X02	2SC1213A(C,D)
X03	2SD476(C,D)
X04	2SC1213A(C,D)
X06	2SA673A

(T146a) CIRCUIT DIAGRAM (POWER SUPPLY PANEL)—MODEL CX-610GB (*PART*)

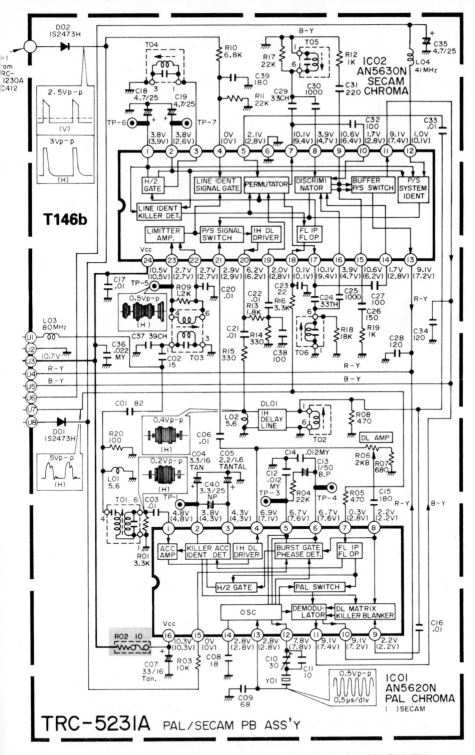

T146b

TRC-5231A PAL/SECAM PB ASS'Y

(T146b) CIRCUIT DIAGRAM (DECODER)—MODEL CX-610GB (*CONTINUED*)

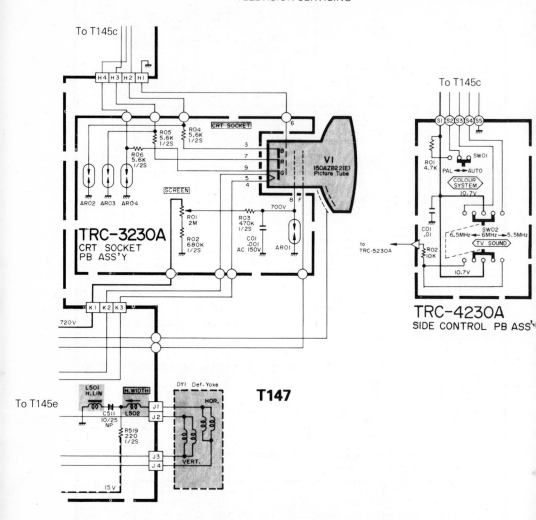

(T147) CIRCUIT DIAGRAM (C.R.T. AND CONTROL PANELS)—MODEL CX-610GB

J.V.C.   Model 7115GB

General Description: A 14-in. colour television receiver to P.A.L. specification operating on mains supplies. Integrated circuits are used in all low level signal stages and the chassis is isolated by mains transformer allowing the safe connection of earphone or tape recorder.

Mains Supplies: 200, 240 volts, 50Hz.

E.H.T.: 24Kv.

Adjustments

Colour Purity: The basic procedure is similar to that described for Model CX-610GB described previously in this volume.

Static Convergence and Dynamic Convergence (see Fig. T149): Connect a crosshatch generator to the antenna terminals and adjust Brightness and Contrast control for a distinct pattern.

Adjust the convergence around the edges of the picture tube by tilting the yoke, up-down and left-right, and temporarily install one wedge at the top of the yoke.

Rotate the four poles magnets as a unit to minimise the separation of the red and blue lines around the centre of the screen. To adjust the convergence of red and blue, vary the angle between the tabs.

Rotate the six poles magnets as a unit to minimise the separation of the magenta (R/B) and green lines.

Adjust the spacing of the six poles magnets to converge the magenta and green lines.

Apply paint to fix six magnets.

Remove the wedge installed temporarily on the yoke.

Tilting the angle of the yoke up, down, and sideways, and adjust the yoke so as to obtain the circumference convergence.

Insert three wedges to the position as shown in figure to obtain the best circumference convergence.

Wedge has a backing of double-sided adhesive tape, therefore tear off one side of adhesive tape, and fix the wedges.

White Balance: Receive a black and white signal.

Set the Red and Green drive controls for their mechanical centre.

Turn the Red, Green and Blue cut-off controls and the Screen control fully counter-clockwise.

Display a horizontal line. (Fig. T148.)

Turn Screen control slowly clockwise until a very faint horizontal line appears.

Turn the cut-off control of the colour which has appeared first, clockwise by about 10° and then adjust the screen control again so that the colour may shine faintly.

171

HOW TO OBTAIN A HORIZONTAL LINE

By reconnecting the cutoff service tip from "N" to "S", a horizontal line will appear.

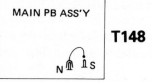

T148

(T148) PURITY SERVICE TAP—MODEL 7115GB

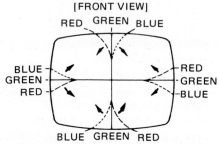

Tilting the yoke upward will move the lines as shown with the arrows.

T149

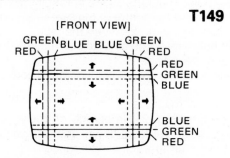

Tilting the yoke to the right will move the lines as shown with the arrows.

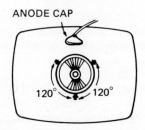

(T149) DYNAMIC CONVERGENCE—MODEL 7115GB

Turn the other colour cut-off controls slowly clockwise until a reasonable white line appears.

Return the black and white picture.

Adjust the Red and Green drive controls for best white highlights.

Service Adjustments

Sub Colour (20-in. only): Once, switch-off the main power switch, and once more switch it on.

Receive a colour T.V. programme.

Set the Tint knob to the central position.

Then adjust the Sub Colour control until the colour of human skin looks natural.

Sub Contrast and Sub Bright (20-in. only): Once, switch-off the main power switch, and once more switch it on.

Receive a T.V. programme.

Set the Contrast knob to the central position.

Then adjust both the Sub Contrast and the Sub Bright controls until an ideal picture is obtained.

Sub Colour (14-in. only): Receive a colour T.V. programme.

Set the Tint and the Colour knobs to the central position.

Then adjust the Sub Colour control until the colour of human skin looks natural.

Sub Contrast and Sub Bright (14-in. only): Receive a T.V. programme.

Set the Contrast and the Bright knobs to the central position respectively (where they click).

Then adjust both the Sub Contrast and the Sub Bright controls until an ideal picture is obtained.

Noise (R.F. A.G.C. Delay): This control is set at the factory and rarely requires any adjustment. If a snowy picture appears on a medium to weak station adjust the Noise control.

Turn control fully clockwise, maximum noise in picture.

Slowly turn the Noise control counter-clockwise until snow or noise in picture just disappears.

Note: Check operation on strong channels. If overloading occurs (bending, poor colour, loss of colour sync. etc.) make a compromise adjustment.

Horizontal Oscillator: Receiving a T.V. programme.

Connect the jumper clip between TP-33A and TP-33B.

Adjust the H. Freq. control until picture is in view and locks or drift slowly back and forth.

Remove the jumper clip.

Make sure that the set maintains horizontal sync., when channels are switched.

Horizontal Width: Adjust H. Width control coil (T504) by turning it with a hexagonal adjusting bar only if Right and Left sides of pictures cannot be seen.

Vertical Centre and Horizontal Centre: Centring is completed at the factory, although it may become distorted when C.R.T. is changed.

In such case, moving the grey wire tips (adj. point left, centre and right), move the picture up or down (left or right).

6MHz Trap: Tune in a local colour station preferably a programme with the least amount of movement and continuous audio. Turn core of trans- former T104 so that beating with sound signal disappears.

Search Sensitivity: Avoid attempting to conduct adjustment during normal operation. Should adjustment become necessary, adhere to the following procedures.

Select no signal channel.

Connect a tester between the ground terminal and the middle tap of the Search Sensitivity control.

Turn the Search Sensitivity V.R. fully clockwise and then slowly return it until the tester reading to falls within the range of $2 \cdot 0V \pm 0 \cdot 1V$ (20k Ω/V) tester value). (Confirm accuracy of the tester. Do not use a tester with instrumental error.)

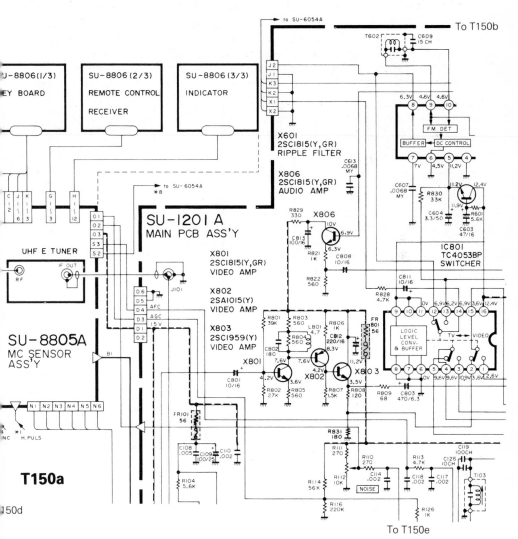

(T150a) CIRCUIT DIAGRAM—MODEL 7115GB (*PART*)

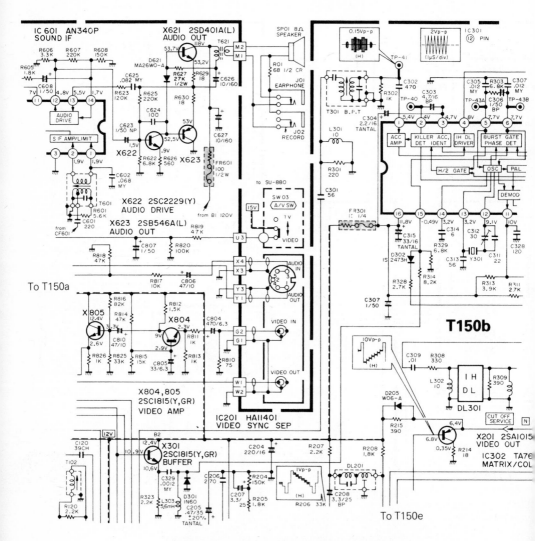

(T150b) CIRCUIT DIAGRAM—MODEL 7115GB (*PART*)

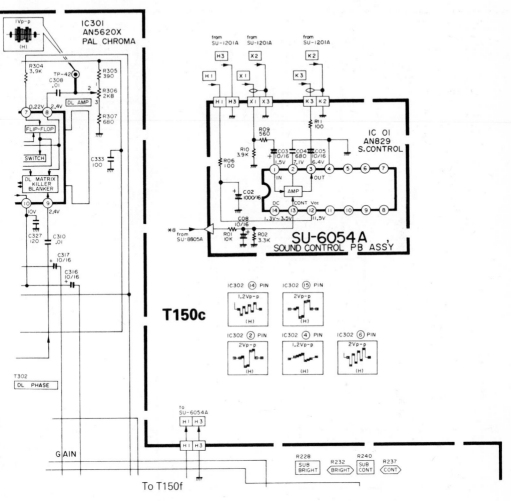

T150c

(T150c) CIRCUIT DIAGRAM—MODEL 7115GB (*PART*)

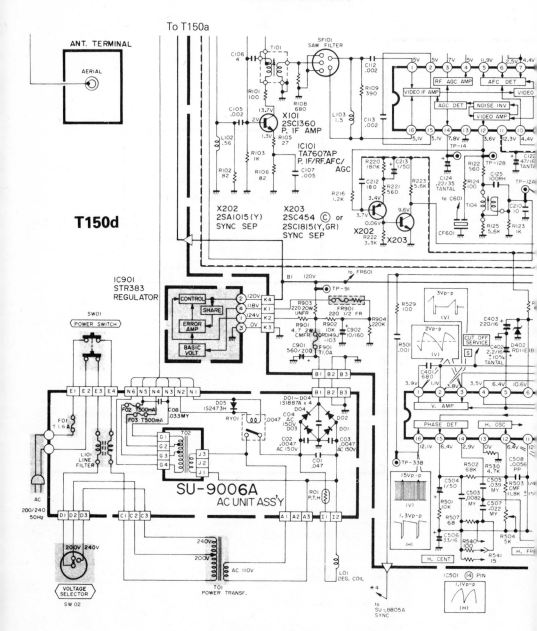

(T150d) CIRCUIT DIAGRAM—MODEL 7115GB (*PART*)

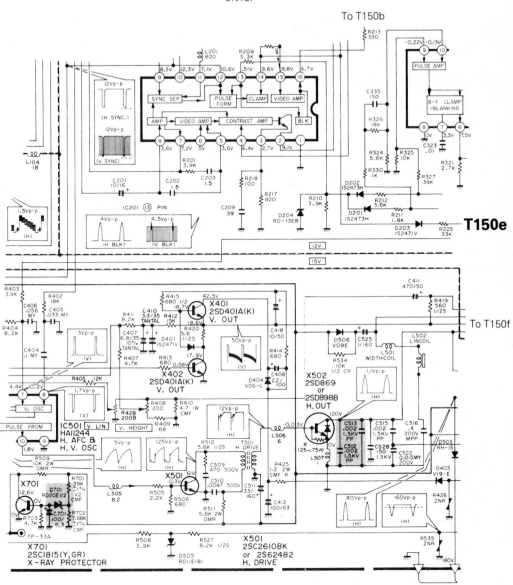

(T150e) CIRCUIT DIAGRAM—MODEL 7115GB (*PART*)

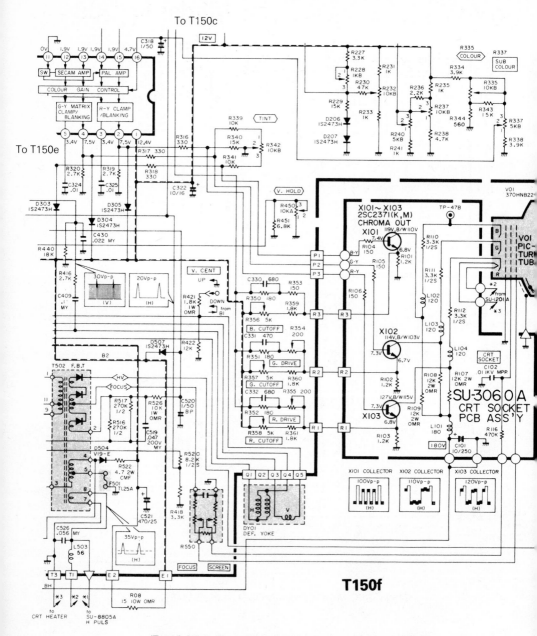

T150f

(T150f) CIRCUIT DIAGRAM—MODEL 7115GB (*CONTINUED*)

J.V.C.   # Model 7815GB

General Description: A 20-in. mains operated colour television receiver with remote control facilities. The chassis is isolated by mains transformer allowing the connection of earphone or tape-recorder. Integrated circuits are used in all low level signal stages.

Mains Supplies: 220, 240 volts, 50Hz.

E.H.T.: 27Kv.

Adjustments

In general, these follow the procedures outlined for J.V.C. models CX-610 and 7115 which are described elsewhere in this volume.

(T151a) CIRCUIT DIAGRAM—MODEL 7815GB (PART)

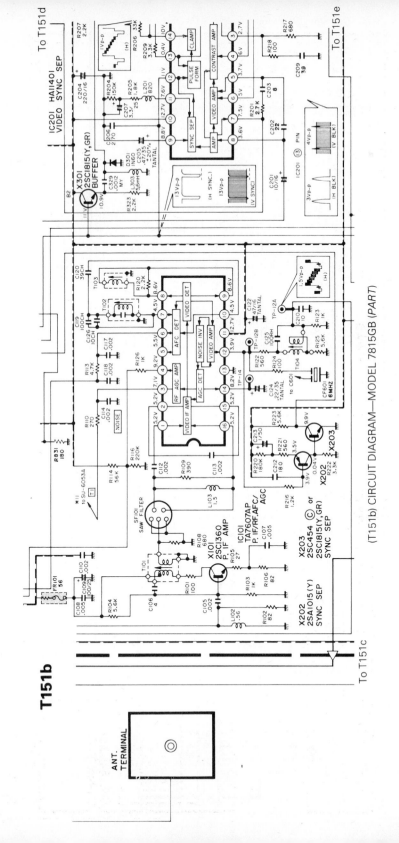

T151b

(T151b) CIRCUIT DIAGRAM—MODEL 7815GB (*PART*)

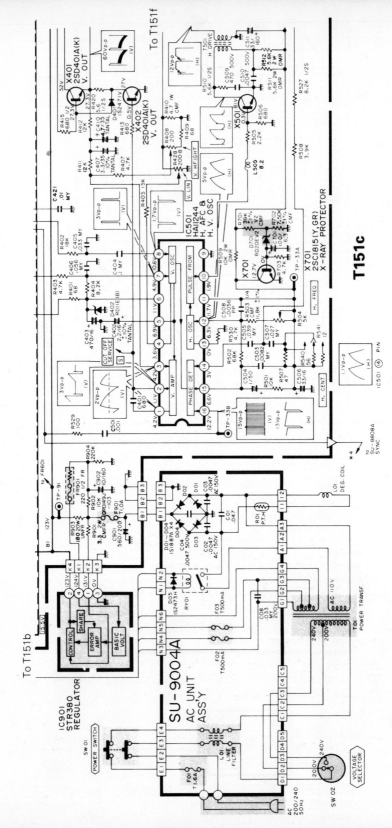

(T151c) CIRCUIT DIAGRAM—MODEL 7815GB (*PART*)

T151c

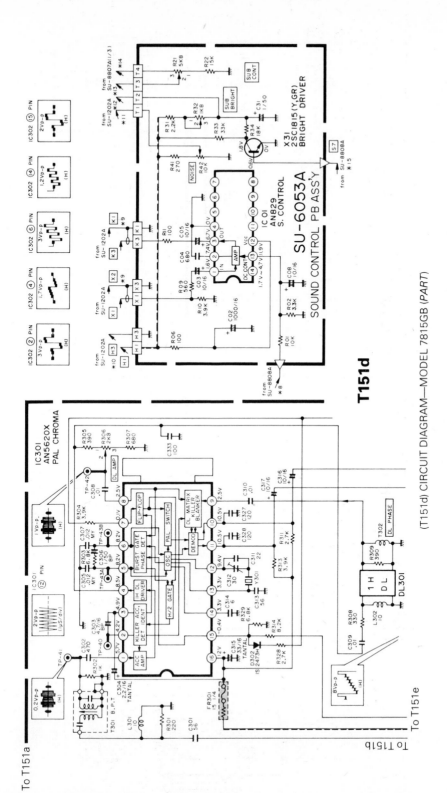

(T151d) CIRCUIT DIAGRAM—MODEL 7815GB (PART)

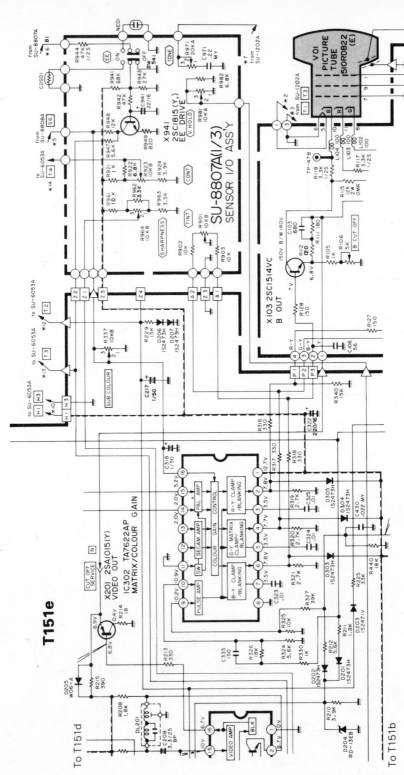

(T151e) CIRCUIT DIAGRAM—MODEL 7815GB (*PART*)

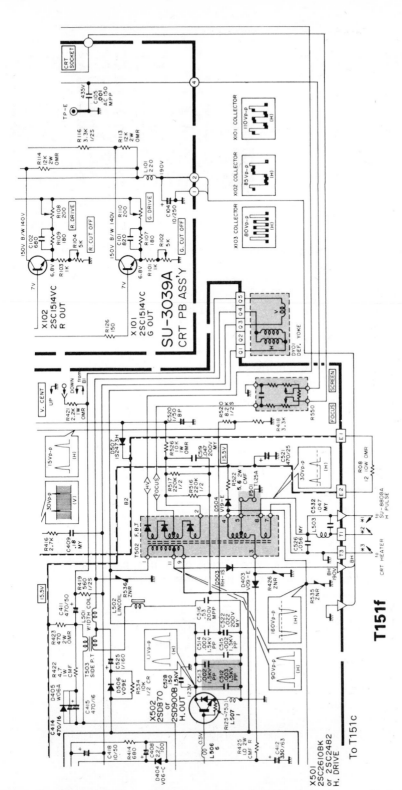

(T151f) CIRCUIT DIAGRAM—MODEL 7815GB (CONTINUED)

LUXOR

B1, B2 Chassis, Models 3781, 4211, 4214, 4781, 4791, 5111, 5114, 5191, 5601, 5608, 5609, 5681, 5691, 5694, 5698, 6711, 6718, 6791, 6794, 6798

General Description: A series of colour television receivers using a generally similar basic chassis containing a 'mother' board and plug-in modules for some sections. A number of different types of control panels are used incorporating an eight-channel mechanical tuner (B1 Chassis) or remote control facilities and digital read-out indication (B2 Chassis). Self-converging, in-line tubes with 90° or 100° deflection may be used and circuit variations associated with these options are given in these pages.

Note: The main chassis is isolated by the S.M.P.S. Power Supply, but care should be exercised when investigating conditions in the supply section which is 'Live' irrespective of mains connections.

Mains Supplies: 185–265V, 50Hz.

E.H.T.: 23/24KV.

H.T.: 129V (137V on larger receivers) at connection 3 on p.s.u.

Loudspeaker: 4 ohms impedance.

H.T. Adjustment: PN01.

Power Supply Fault Tracing (Fig. T207)

Note: Ensure that television receiver is connected to an isolated mains supply and fed via a variac. If a variac is not available use a tapped mains transformer to reduce mains input voltage by approximately half.

Warning: The primary side of the mains transformer is not isolated from the mains supply.

Place a voltmeter between 3 and chassis, switch on and gradually increase the mains input voltage. Does the unit start?—if not check:

(A). Voltage between emitter and collector of TN03, if it is zero, it could be due to the following:
Winding (7–11) LN03 open circuit.
RN07 open circuit.
RN32 open circuit.
FN02 open circuit.

188

RN23/RN08 open circuit.
Bridge rectifier defective.
FN01 open circuit.
Defective Mains relay in standby unit if fitted.
Other faults in mains standby unit including blown fuses.
Fault in mains input filter network.
LN02 Open circuit.

(B). If the voltage between emitter and collector of TN03 is correct, place an oscilloscope between the earth end of LN02 and base. Half wave pulses of around 30V pp should be obtained, if not check:
CN10 open circuit or dry joints.
RN12 open circuit or high resistance.
RN15 high resistance or open circuit.
DN07 open circuit.
TN03 open circuit.
TN02 or CN08 short-circuit.
DN06 short-circuit.

If the 500 mm/A fuse FN02 is blown or RN32 is open circuit check TN03 for short-circuits.

Should TN03 be short circuit replace TN02 also, reduce the setting of the variac to zero, monitor 3 and gradually increase mains input voltage. If the output voltage rises quickly above 129 (137) volts check the voltage across CN03—at the base of TN01 and at the collector of TN01. In the case of the latter, vary PN01 to ensure conduction of the transistor.

If no voltage exists across CN03 check:
Winding (1–5) open circuit.
DN03 open circuit.
RN11 open circuit or CN03 short-circuit.

If there is a voltage across CN03 but regulation is still not being obtained, check the following:
DN01 open circuit (no emitter voltage on TN01).
TN01 open circuit or defective.
RN03, RN05, RN06 or DN04 open circuit.
DN02 open circuit, CN08 defective.
RN01, PN01 or RN02, open circuit. (Low or no base voltage at TN01).

High-pitched buzzing noise from power supply:
In this case the trip has operated and the power supply is being heavily damped by a short-circuit across the 129V (137V) supply. However, overvoltage protection is provided by the trip circuit and the second part of note again applies (i.e. excessive output voltage due to poor regulation).

Low output voltages from power supply:
DN01 resistive.
TN02 defective.
CN08 defective.
CN04 open circuit.
TN01 resistive or defective in some other way.

189

Output voltages correct but begin to change after a period of time, trip may or may not operate:

DN01 defective (temperature sensitive)

TN01 defective (temperature sensitive)

CN03 becoming leaky.

Note: In the event of faults above, monitor the voltage at (3) and change the temperature of the components in the regulation circuit i.e. with freezer spray.

Shorted turns on transformer LN04 (unlikely).

DN05, DN06, DN19 Defective.

Summary

Use a variac and isolated mains source when working on power supply unit.

In the event of non starting, first check for 300 volt supply between earthy end of LN02 and collector of TN03.

If correct, check that the starting pulses are present and are of the correct amplitude.

Always measure conditions in the power supply with respect to the emitter of TN03 i.e. oscilloscope or meter negative or earth probe should be connected to the earthy end of LN02. Measuring from chassis will give incorrect readings.

If TN03 has to be replaced, replace TN02 also.

Carefully monitor (3) if it rises quickly above 129 (137V) when increasing mains input voltage, then regulation circuit is inoperative.

Do not operate power supply under open circuit load conditions. Disconnection of diodes DN12, DN13 and DN14 should be avoided if possible.

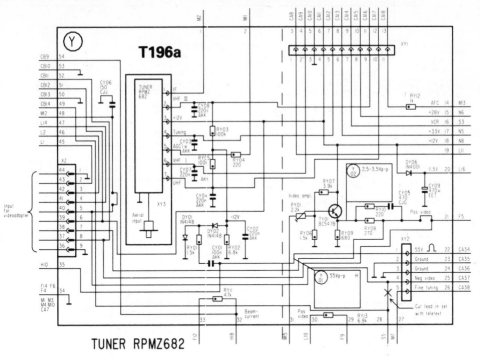

TUNER RPMZ682

(T196a) CIRCUIT DIAGRAM (TUNER PANEL RPMZ682)—B1, B2 CHASSIS (*PART*)

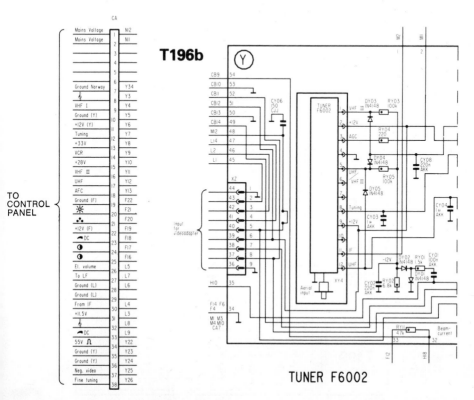

TUNER F6002

(T196b) CIRCUIT DIAGRAM (ALTERNATIVE FRONT END)—B1, B2 CHASSIS (*CONTINUED*)

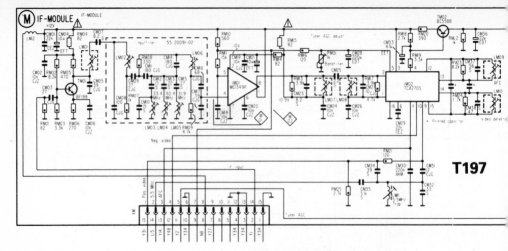

(T197) CIRCUIT DIAGRAM (I.F. MODULE)—B1, B2 CHASSIS

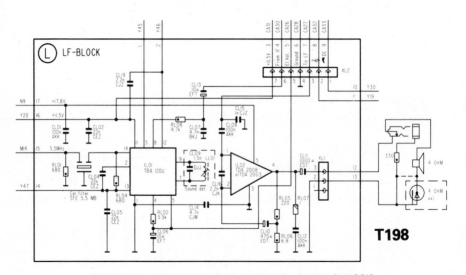

(T198) CIRCUIT DIAGRAM (A.F. SECTION)—B1, B2 CHASSIS

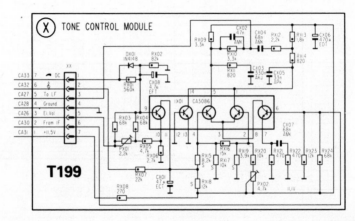

(T199) CIRCUIT DIAGRAM (TONE CONTROL MODULE)—B1, B2 CHASSIS

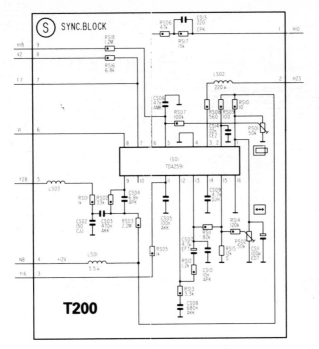

(T200) CIRCUIT DIAGRAM (SYNC. SEPARATOR)—B1, B2 CHASSIS

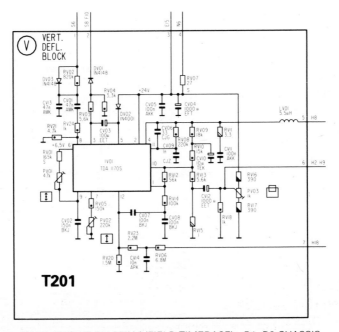

(T201) CIRCUIT DIAGRAM (FIELD TIMEBASE)—B1, B2 CHASSIS

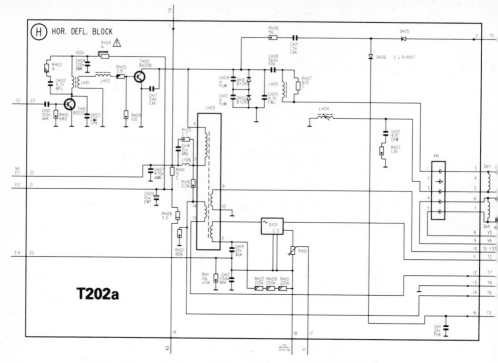

(T202a) CIRCUIT DIAGRAM (LINE OUTPUT STAGE)—B1, B2 CHASSIS (*PART*)

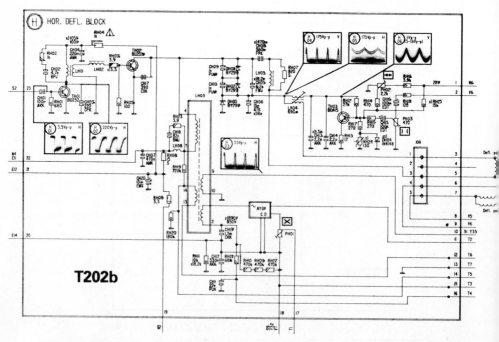

(T202b) CIRCUIT DIAGRAM (ALTERNATIVE LINE OUTPUT)—B1, B2 CHASSIS (*CONTINUED*)

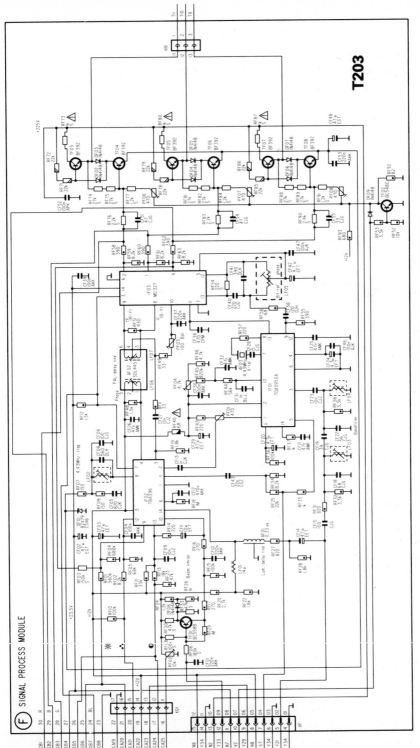

SIGNAL PROCESS MODULE

T203

(T203) CIRCUIT DIAGRAM (DECODER MODULE)—B1, B2 CHASSIS

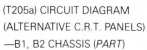

(T205a) CIRCUIT DIAGRAM
(ALTERNATIVE C.R.T. PANELS)
—B1, B2 CHASSIS (*PART*)

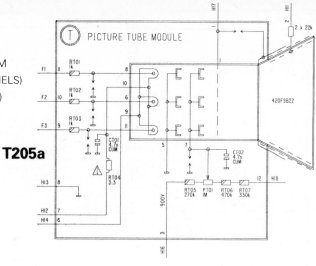

T205a

T205b

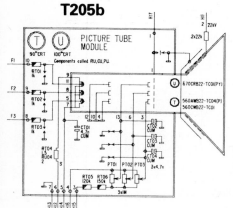

(T205b) CIRCUIT DIAGRAM (ALTERNATIVE C.R.T. PANELS)
—B1, B2 CHASSIS (*PART*)

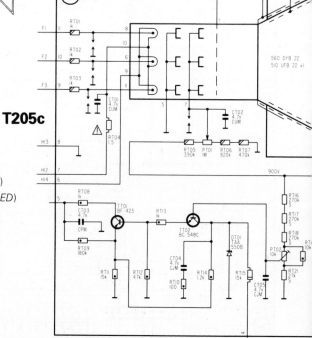

T205c

(T205c) CIRCUIT DIAGRAM
(ALTERNATIVE C.R.T. PANELS)
—B1, B2 CHASSIS (*CONTINUED*)

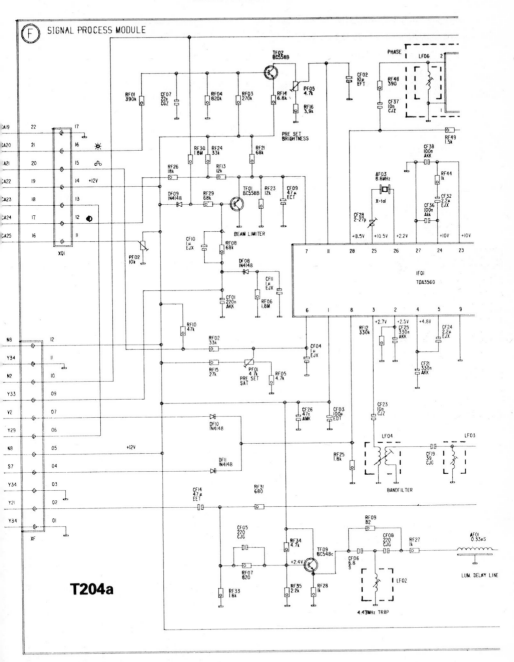

(T204a) CIRCUIT DIAGRAM (ALTERNATIVE DECODER MODULE)—B1, B2 CHASSIS (*PART*)

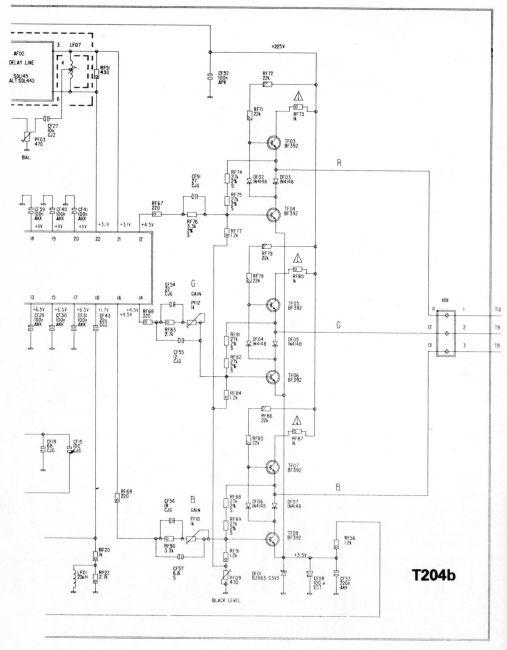

(T204b) CIRCUIT DIAGRAM (ALTERNATIVE DECODER MODULE)—B1, B2 CHASSIS (*CONTINUED*)

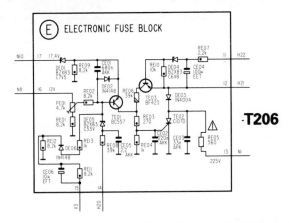

(T206) CIRCUIT DIAGRAM (AUTOMATIC CUT-OUT)—B1, B2 CHASSIS

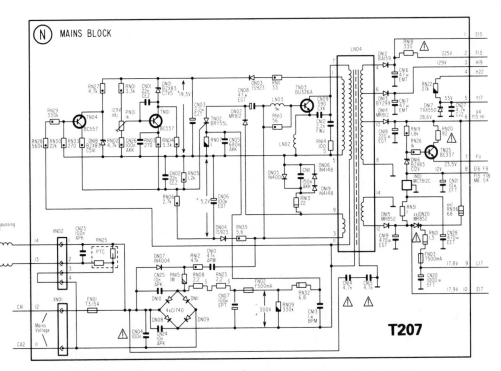

(T207) CIRCUIT DIAGRAM (S.M.P.S. POWER SUPPLY)—B1, B2 CHASSIS

MITSUBISHI

Models CT2206B, CT2207B, CT2606BM, CT2607BM

General Description: A series of mains-operated colour television receivers incorporating similar basic chassis and 110° in-line picture tube. B.M. models feature an infra-red remote control system for adjustment and channel selection. All models are isolated by mains transformer.

Mains Supply: 240 volts, 50Hz.

Cathode Ray Tubes: 560EVB22 (22-in.); 670BKB22 (26-in.).

Adjustments

R.F. A.G.C. (VR101): Turn A.F.T. on.

Set channel selector to a channel where cross-modulation or overload exists.

Turn R.F. A.G.C. control VR101 on P.C.B. Main slowly anti-clockwise until the noise disappears.

Receive all the channels available and make sure no noise or cross-modulation is observed.

Sub Cont. (VR201): Tune receiver into a standard colour bar signal of 65–90dB/μV.

Set Black level control VR271 at click stop position.

Set Picture control at the mechanical centre for model CT-2206B, or press 'Normalise' button' for model CT-2606BM.

Set Colour control to minimum.

Connect a D.C. ammeter with 3mA full scale between the test points TP91 (+) and TP1Z (−) on P.C.B.-Main.

Adjust Sub Cont. VR201 on P.C.B.-Main for beam current of 1100μA± 20μA for model CT-2206B, or 1300μA±20μA for model CT-2606BM on the meter.

SIF (L301): Tune receiver into a programme.

Set Volume control VR391 at a position where correct volume is obtained.

Adjust L301 on P.C.B.-Main for maximum volume.

If some buzz remains after completion of above procedure, adjust L138 to minimise the buzz.

Height and Linearity (VR401, VR464): Make sure the A.C. power supply voltage is at the specified value (240V).

Adjust Height control VR401 for approx. 90% vertical size of raster.

Adjust V. Lin. control VR464 on P.C.B. P.C.C. for symmetry of vertical linearity.

Set Black level control VR271 at the click-stop position and Picture control at the mechanical centre for model CT-2206B or press 'Normalise' button for model CT-2606BM.

Adjust Height control VR401 for normal vertical size.

Repeat steps above, if necessary.

P.C.C. (L452, VR461, VR462): Tune receive into a crosshatch signal.

Set Black level control at the click-stop position and Picture control at the mechanical centre for model CT-2206B, or press 'Normalise' button for model CT-2606BM.

Adjust P.C.C. Amp. VR461 on P.C.B. P.C.C. for the symmetry of vertical lines at both sides of raster.

Adjust Phase control VR462 on P.C.B. P.C.C. for straight vertical lines at both sides of the raster.

Repeat the steps above.

Adjust Top and Bottom Phase control coil L452 on P.C.B.-P.C.C. for straight horizontal lines at top and bottom of the screen.

Horizontal Freq. Control (VR501): If there is difficulty in maintaining horizontal sync. adjust VR501.

Short-circuit Pin 16 and Pin 11 of IC401 on P.C.B.-Main respectively.

Adjust VR501 for almost in sync. condition.

Remove the shorting lead across Pin 16 and Pin 11 of IC401.

Horizontal Centring (VR571): Make sure H. Freq. control VR501 has been adjusted.

Adjust H. Cent. VR571 on P.C.B.-Main to centre the raster.

Vertical Centring: Vertical position of picture can be slightly shifted upward by decreasing resistor R421 3·3K Ω on P.C.B.-Main in value.

Horizontal Width (VR463): Tune receiver to a programme.

Set Black Level control at the click-stop position and Picture control at the mechanical centre for model CT-2206B, or press 'Normalise' button for model CT-2606BM.

Make sure the Mains voltage is 240V.

Adjust VR463 on P.C.B.-P.C.C. for optimum horizontal size.

Focus (VR591): Tune receiver into a monochrome signal.

Set Black Level control at the click-stop position and Picture control at the mechanical centre for model CT-2206B, or press 'Normalise' button for model CT-2606BM.

Adjust Focus control VR591 on the rear right of P.C.B. Main for best overall definition and focus.

B4 Adjustment (VR901): Tune receiver into a programme.

Set Black Level control at the click-stop position and Picture control at the mechanical centre for model CT-2206B, or press 'Normalise' button for model CT-2606BM.

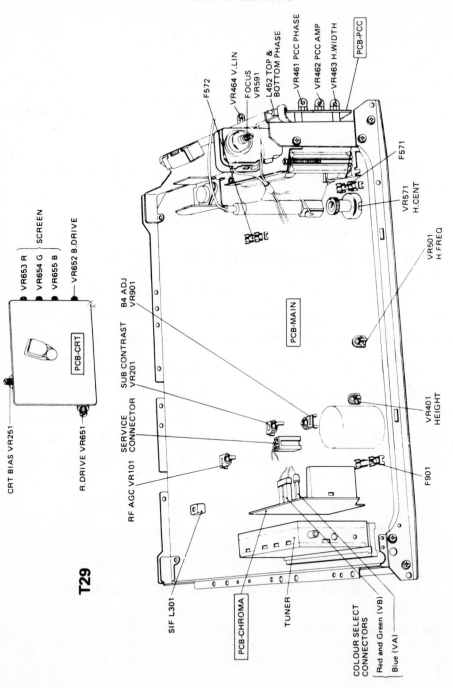

T29

CRT BIAS VR251

VR653 R
VR654 G } SCREEN
VR655 B
VR652 B.DRIVE

PCB-CRT

R.DRIVE VR651

SUB CONTRAST VR201

B4 ADJ VR901

SERVICE CONNECTOR

RF AGC VR101

PCB-MAIN

F572

VR464 V.LIN

FOCUS VR591

L452 TOP & BOTTOM PHASE

VR461 PCC PHASE

VR462 PCC AMP

VR463 H.WIDTH

PCB-PCC

F571

VR571 H.CENT

VR501 H.FREQ

VR401 HEIGHT

F901

SIF L301

PCB-CHROMA

TUNER

COLOUR SELECT CONNECTORS

Red and Green (VB)

Blue (VA)

(T29) LOCATION OF CONTROLS—MODEL CT2206B, ETC.

202

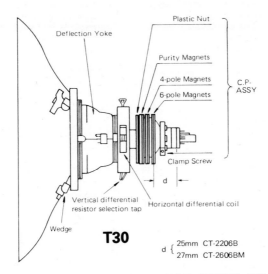

d $\left\{\begin{array}{l} 25mm \quad CT\text{-}2206B \\ 27mm \quad CT\text{-}2606BM \end{array}\right.$

(T30) CONVERGENCE YOKE—MODEL CT2206B, ETC.

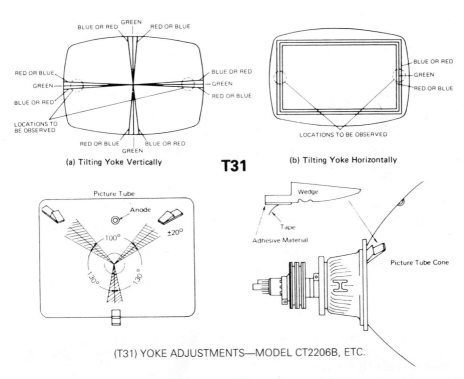

(a) Tilting Yoke Vertically

T31

(b) Tilting Yoke Horizontally

(T31) YOKE ADJUSTMENTS—MODEL CT2206B, ETC.

If some misconvergence still exists it is necessary to reposition one or other of the taps on the yoke housing. If misconvergence is of horizontal red and blue line crossover, the tap on top of the yoke should be moved in accordance with the diagram below.

If the error is of horizontal red and blue line misconvergence at the top and bottom, move the tap underneath the yoke in accordance with the diagram below.

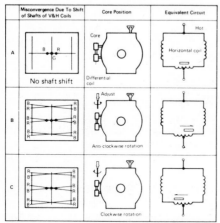

(a) Shift in vertical direction

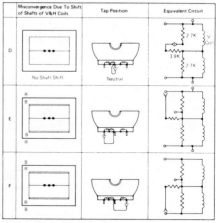

(b) Shift in horizontal direction

T32

Note: Set the tap positions at A or D where no adjustment is required.

(T32) CONVERGENCE ADJUSTMENTS—MODEL CT2206B, ETC.

Make sure the mains voltage is 240V.

Connect a D.C. voltmeter of 150V full scale between the test-point TP-91 (+) on P.C.B.-Main and the chassis ground (−).

Adjust VR901 on P.C.B. Main for 118V reading on the meter.

Convergence

The following procedure may be necessary when replacing picture tube, deflection yoke or P.C.B. C.R.T.

Preparation: Fit the deflection yoke on the neck of picture tube and push toward the cone.

Fit convergence and purity ring assembly to the neck of the picture tube and fasten the screw at the position where the distance between 6-pole magnet end and the base of picture tube is 25 mm for CT-2206B or 27 mm for CT-2606BM as shown in Fig. T30.

The tightening torque for C.P. Assy securing clamp screw must be 15kg.cm.

Before attempting I.T.C. adjustments, the set must be warmed up by running at normal setting for at least 30 minutes from cold, or if just from soak, at least 15 minutes.

Before adjusting purity, height, width, linearity, video bias, focus and grey scale should be accurately adjusted.

Demagnetise at the front and sides of the picture tube by means of degaussing coil.

Grey Scale:

1. Tune receiver in a black and white programme.

2. Set Black Level control at the click-stop position and Picture control at the midposition for model CT-2206B, or press 'Normalise' button for model CT-2606BM.

3. Turn Screen controls VR653, VR654 and VR655 on P.C.B. C.R.T. fully anti-clockwise, Drive controls VR651 and VR652 on P.C.B. C.R.T. at the mechanical centres.

4. Connect the Service Connector (Fig. T29) on P.C.B.-Main to obtain a horizontal line of low brightness at the centre of the screen.

5. Advance C.R.T. Bias VR251 on P.C.B.-C.R.T. until either the red, blue or green horizontal line appears on the screen.

6. Adjust the rest of the Screen controls to produce a white horizontal line.

7. Disconnect the Service Connector to produce raster.

8. Adjust red, and blue Drive controls VR651 and VR652 to make raster white over the entire screen.

9. Check overall black and white tracking throughout the normal brightness and contrast range.

If necessary, repeat steps (4) to (9). Accuracy of screen adjustment is important.

Beam Current (VR201): Make sure that the beam current is properly adjusted as described under 'Sub Cont.'

Preliminary Static Convergence: Tune receiver in a crosshatch signal.

Set Black Level control at the click-stop position and Picture control at the midposition.

Turn the plastic nut clockwise as observed from the back to loosen C.P. Assembly.

Adjust two 4-pole magnets to converge red and blue vertical and horizontal lines in the centre of the screen.

Adjust two 6-pole magnets to converge the red and blue lines on green lins.

If necessary, repeat the steps above.

Fasten the plastic nut by turning anti-clockwise (seen from back).

Note 1. Adjusment of 4-pole magnets affects red and blue beams to move them in the reverse directions with similar movement; relative angle of two tabs for amount of shift while rotating two 4-pole magnets placed together for rotating shift of beams.

2. Adjustment of 6-pole magnets affects red and bue beams to move them in the same direction with similar movement; relative angle of two tabs for amount of shift while rotating two 6-pole magnets placed together for rotating shift of beams. (Continued on p. 215).

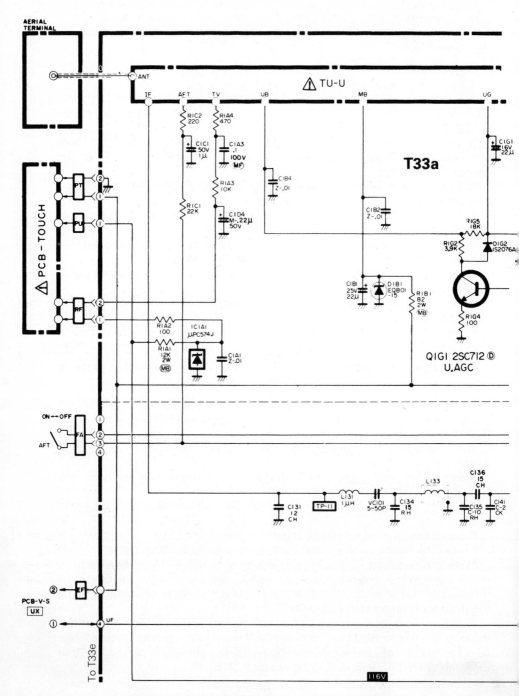

(T33a) CIRCUIT DIAGRAM (MAIN PANEL)—MODEL CT2206B, ETC. (*PART*)

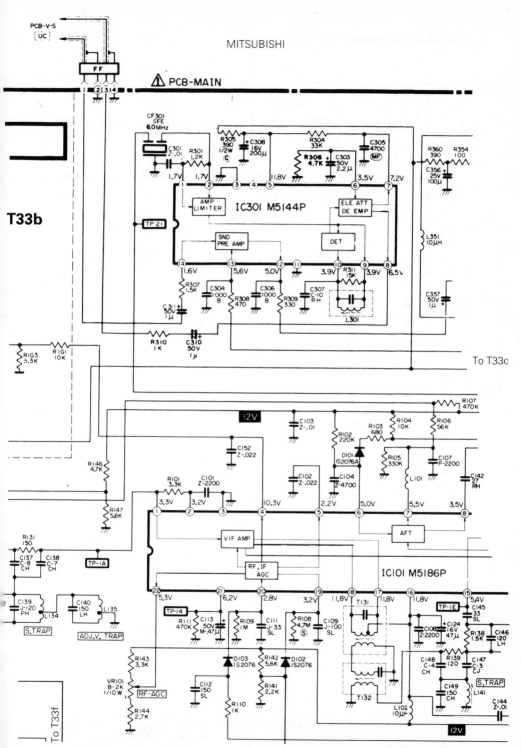

T33b

(T33b) CIRCUIT DIAGRAM (MAIN PANEL)—MODEL CT2206B, ETC. (*PART*)

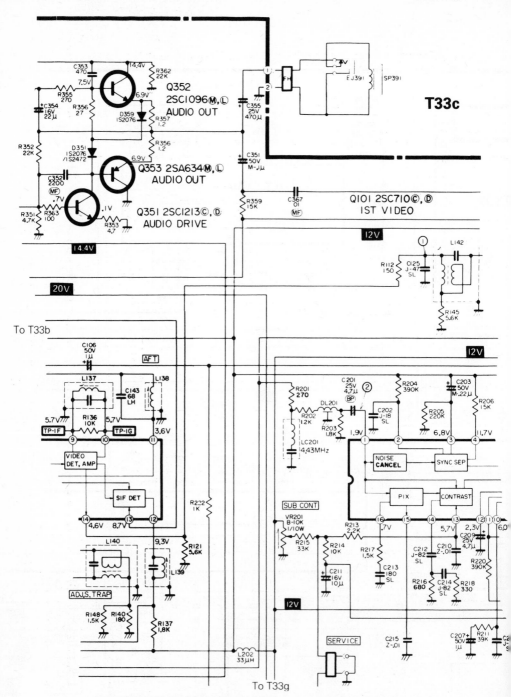

(T33c) CIRCUIT DIAGRAM (MAIN PANEL)—MODEL CT2206B, ETC. *(PART)*

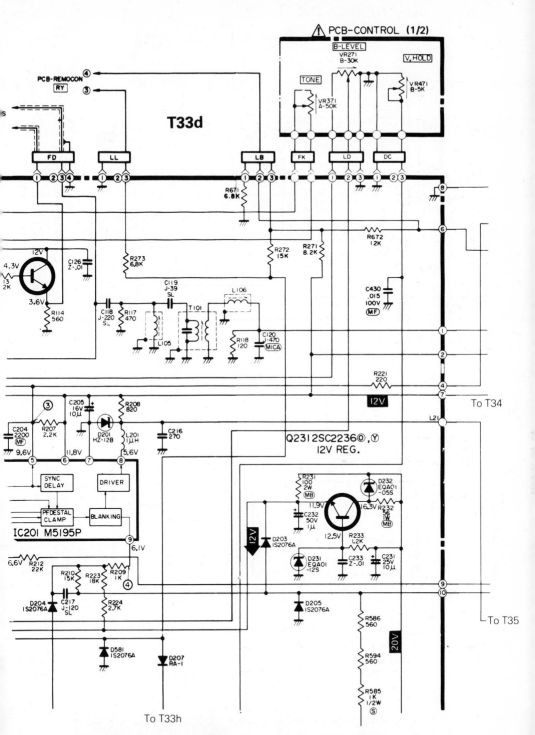

(T33d) CIRCUIT DIAGRAM (MAIN PANEL)—MODEL CT2206B, ETC. (*PART*)

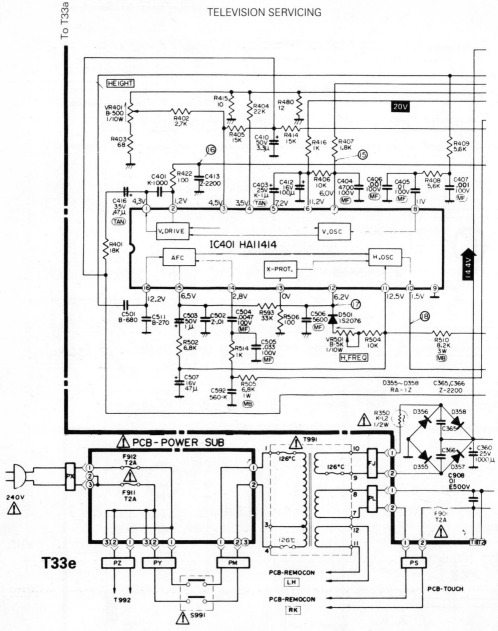

(T33e) CIRCUIT DIAGRAM (MAIN PANEL)—MODEL CT2206B, ETC. (*PART*)

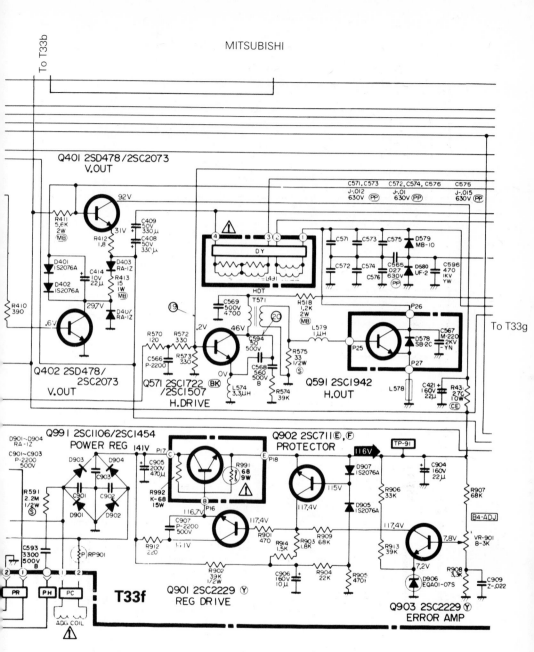

(T33f) CIRCUIT DIAGRAM (MAIN PANEL)—MODEL CT2206B, ETC. (*PART*)

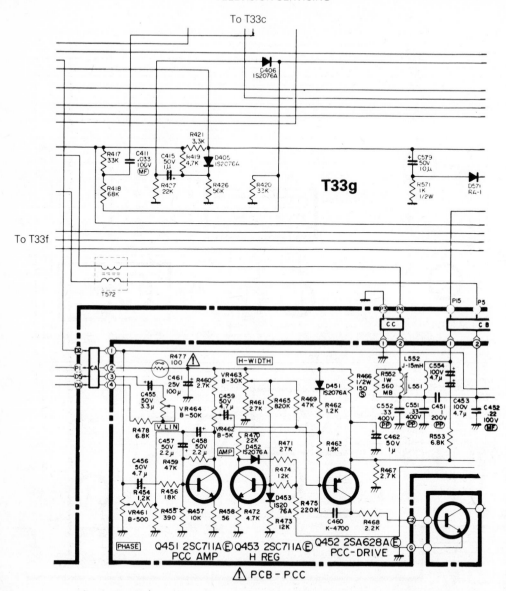

(T33g) CIRCUIT DIAGRAM (MAIN PANEL)—MODEL CT2206B, ETC. (*PART*)

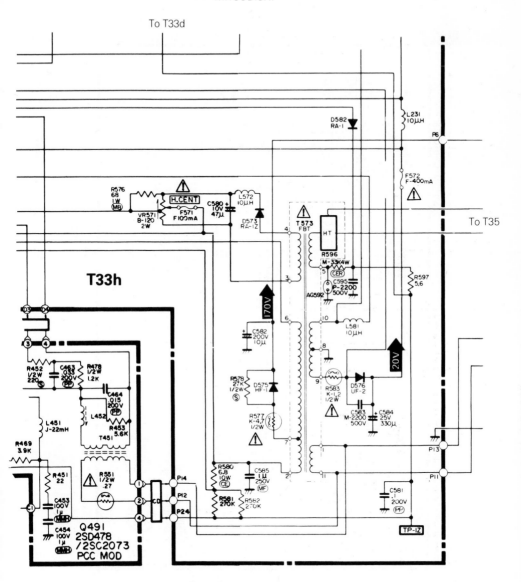

(T33h) CIRCUIT DIAGRAM (MAIN PANEL)—MODEL CT2206B, ETC. (*CONTINUED*)

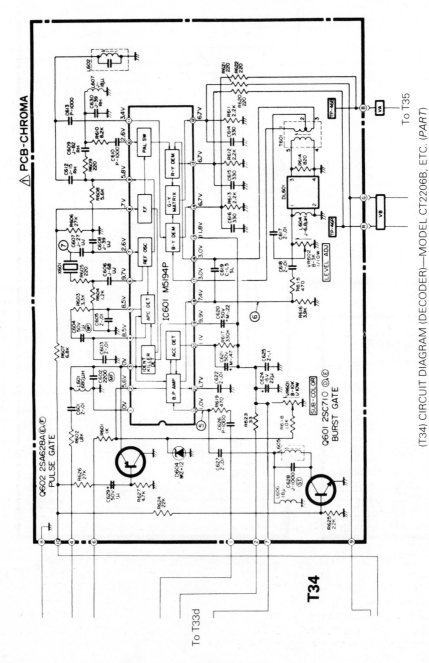

(T34) CIRCUIT DIAGRAM (DECODER)—MODEL CT2206B, ETC. *(PART)*

Purity: 1. Disconnect the connector V.A from P.C.B.-Chroma to produce yellow raster.

2. With the deflection yoke positioned fully forward, adjust purity magnet so that the yellow bar is at the centre of screen with normal horizontal centring.

3. Pull the yoke further back and position it in the centre of the available range for best purity. Confirm that the picture is straight without moving the yoke backwards or forwards and tighten the yoke locking screw. If necessary one wedge can be used to help fix the yoke position and should be stuck with tape only.

4. Check the single colour purity for each of the three colours and finally white uniformity.

5. If there is any problem encountered in (4) above, repeat the procedure from step (2).

6. Fasten the deflection yoke security screw temporarily.

Static Convergence: Black Level at the click-stop, Picture at the mid-position.

Adjust focus carefully.

Unlock the static rings and adjust the 4-pole pair for coincidence of red and blue beams in the centre of the C.R.T.

Adjust 6-pole rings to coincide the magenta lines with the green, in the centre to give white. If necessary repeat the procedure until static convergence is achieved.

Fasten the plastic nut by turning anti-clockwise seen from back, being careful not to move any of the pre-set positions.

Overall Convergence: By means of the colour select connectors disable the green gun to give magenta only.

Watching the top and bottom centre of the screen, move the yoke vertically up or down to coincide (if possible) red and blue vertical lines, in these positions. If not possible then the error should be balanced.

Watching the left and right hand side vertical lines, move the yoke horizontally to converge the red and blue at these points. If not possible then the error should be balanced.

Check the centre, top and bottom centre and left and right centre, for correct convergence. If not, repeat the necessary steps to obtain good convergence in these areas. The above adjustments should be carried out by using one rubber wedge temporarily, to help fix the yoke position, inserted at the top.

After completing the convergence adjustment three rubber wedges must be firmly fixed in position, approx. 130° apart with one of them directly underneath the yoke as shown in Fig. T32 and afterwards the temporary wedge must be removed.

Check overall convergence on a white grid pattern, confirming all the adjustments are correct and then tape the three rubber wedges in position with glass fibre tape or appropriate tape.

Note: After convergence has been completed the focus should not be re-adjusted. If this is necessary the convergence procedure may have to be done again.

Misconvergence of the three beams caused by gun rotation or incorrect yoke field distribution, cannot be corrected, but generally a balancing of errors will result in acceptable convergence.

The three rubber wedges should be fixed in position with the adhesive material at the end of the wedge or by injection of KE-48 RTV (silicon gum) between them and the C.R.T. funnel. If it is difficult to inject this, it may be necessary to use the R.T.V. before inserting the wedges.

If the purity, white uniformity, or overall convergence is not satisfactory, the above procedure should be done again.

After the adjustments are complete, all screws, plastic nut, etc. should be firmly fixed in place by appropriate use of locking paint.

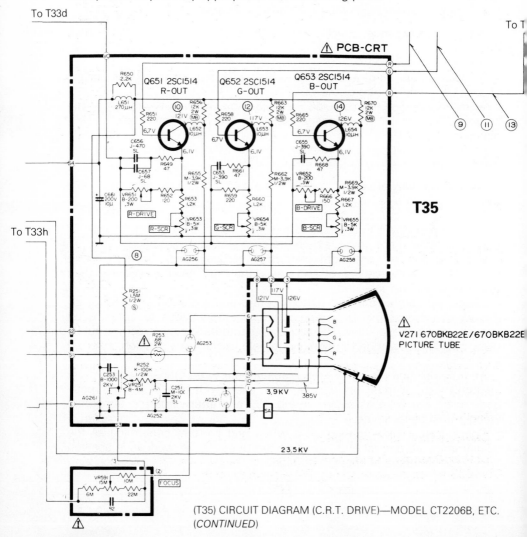

(T35) CIRCUIT DIAGRAM (C.R.T. DRIVE)—MODEL CT2206B, ETC.
(*CONTINUED*)

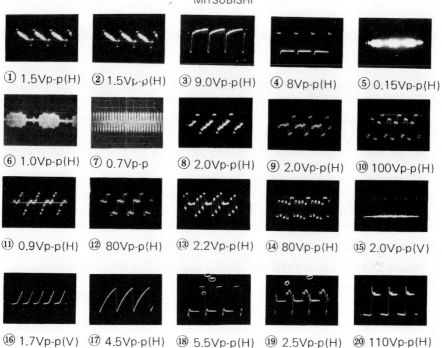

① 1.5Vp-p(H) ② 1.5Vp-p(H) ③ 9.0Vp-p(H) ④ 8Vp-p(H) ⑤ 0.15Vp-p(H)

⑥ 1.0Vp-p(H) ⑦ 0.7Vp-p ⑧ 2.0Vp-p(H) ⑨ 2.0Vp-p(H) ⑩ 100Vp-p(H)

⑪ 0.9Vp-p(H) ⑫ 80Vp-p(H) ⑬ 2.2Vp-p(H) ⑭ 80Vp-p(H) ⑮ 2.0Vp-p(V)

⑯ 1.7Vp-p(V) ⑰ 4.5Vp-p(H) ⑱ 5.5Vp-p(H) ⑲ 2.5Vp-p(H) ⑳ 110Vp-p(H)

T36

(T36) CIRCUIT WAVEFORMS—MODEL CT2206B, ETC.

MITSUBISHI Models CT2206BM, CT2206BX, CT2207BM, CT2207BX

General Description: A series of mains-operated colour television receivers incorporating a similar basic chassis and 110° in-line picture tube. A choice of touch-tuning or remote control is available and all chassis are isolated by mains input transformer.

Mains Supply: 240 volts, 50Hz.

Cathode Ray Tube: A56-540X.

Colour Decoder Panel: This is similar to that used in Mitsubishi Model CT2206B which is shown in Fig. T34 elsewhere in this volume.

Adjustments

Note: Most adjustments are similar to those required for Model CT2206B described previously and should be made with reference to that information. Variations from the procedures are as follows:

Horizontal Centring: Check that Line Frequency control is correctly set. Adjust VR501 on P.C.B.-Main.

B4 Adjustment (VR901): 1. Tune receiver into a programme.

2. Set Black Level control at the click-stop position.

3. Set Picture control at the mechanical centre for model CT-2206BX, or press 'Normalise' button 3 for model CT-2206BM.

4. Make sure the mains voltage is 240V.

5. Connect a D.C. voltmeter of 300V full scale between the testpoint TP-91 (+) in P.C.B.-Main and the chassis ground (−).

6. Adjust VR901 on P.C.B.-Main for 156V reading on the meter.

Video-Sound Circuit (For model CT-2206BM)

Video Output Level (VR2151): Tune receiver into a standard P.A.L. test signal (G card).

Connect a 75 Ω resistor to the video output terminal (Connector UT-1 of P.C.B.-V.S.), or connect across connectors UT-1 and UA-2.

Connect an oscilloscope across the connectors UT-1 and UT-2 (ground).

Adjust VR2151 of P.C.B.-V.S. for 1Vp-p (from the sync. tip to 80% white level).

Video Input Level (VR2101): The video output level VR2151 must have been adjusted prior to this adjustment.

Make sure that the connector XE remains connected on P.C.B.-V.S.

Apply a 400Hz sinewave signal to the connector UA-1 of P.C.B.-V.S. and set its amplitude to 1Vp-p.

Connect an oscilloscope to the connector UP-3.

Adjust VR2101 of P.C.B.-V.S. for 1·5p-p of 400Hz sinewave signal on the scope.

Sound-Sub (VR7N1): Tune receiver into a programme.

Press 'Normalise' button.

Adjust VR7N1 on P.C.B-Remocon for optimum volume.

Sound Input Level (VR7N1, VR3101): Prior to this adjustment, Sound Sub V7N1 on P.C.B.-Remocon must have been adjusted.

Tune receiver into a standard P.A.L. test signal (G card).

Apply a 400Hz sinewave signal of 1Vp-p to the connector UA-3 of P.C.B.-V.S. (Voltage at 600 Ω open end).

Connect a 'scope with 8 ohm load to the Din headphone terminal or to terminal F.H. of P.C.B.-Main with the connector F.H. disconnected. In this case, connect Din terminal so that the speaker of the set is disconnected.

Adjust VR3101 on P.C.B.-V.S. for 0·8±0·2Vp-p on the oscilloscope.

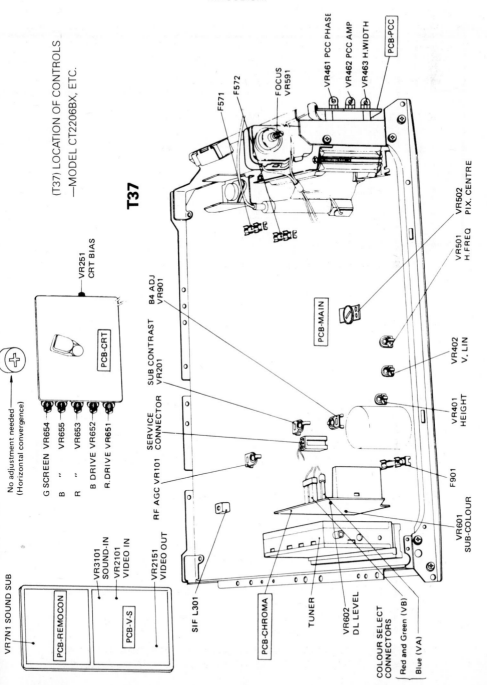

(T37) LOCATION OF CONTROLS
—MODEL CT2206BX, ETC.

T37

No adjustment needed
(Horizontal convergence)

VR251 CRT BIAS

PCB-CRT

G SCREEN VR654
B " VR655
R " VR653
B DRIVE VR652
R.DRIVE VR651

SUB CONTRAST VR201

B4 ADJ VR901

SERVICE CONNECTOR

RF AGC VR101

VR7N1 SOUND SUB

PCB-REMOCON

VR3101 SOUND-IN
VR2101 VIDEO IN

PCB-V-S

VR2151 VIDEO OUT

SIF L301

PCB-CHROMA

TUNER

VR602 DL LEVEL

COLOUR SELECT CONNECTORS

Red and Green (VB)
Blue (VA)

VR601 SUB-COLOUR

F901

VR401 HEIGHT

VR402 V. LIN

PCB-MAIN

VR501 H.FREQ

VR502 PIX. CENTRE

F571
F572

FOCUS VR591

VR461 PCC PHASE
VR462 PCC AMP
VR463 H.WIDTH

PCB-PCC

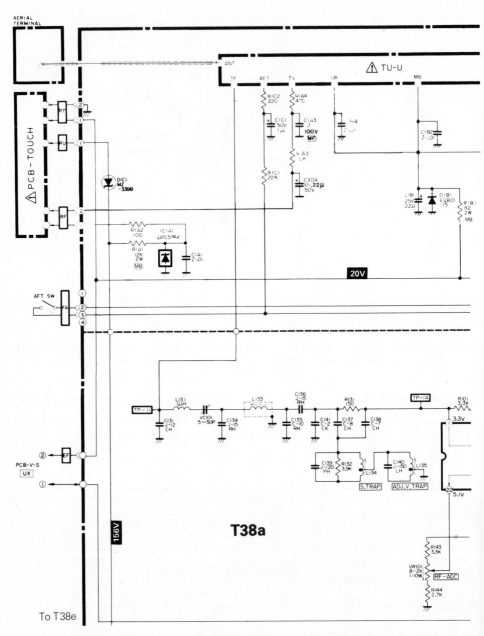

(T38a) CIRCUIT DIAGRAM (MAIN PANEL)—MODEL CT2206BX, ETC. (*PART*)

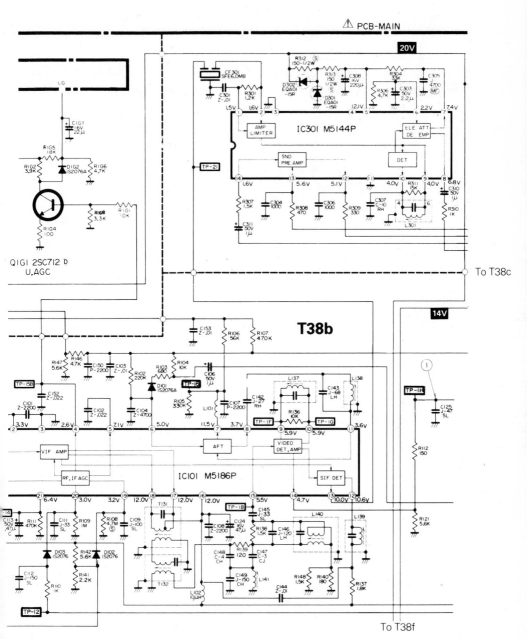

(T38b) CIRCUIT DIAGRAM (MAIN PANEL)—MODEL CT2206BX, ETC. (*PART*)

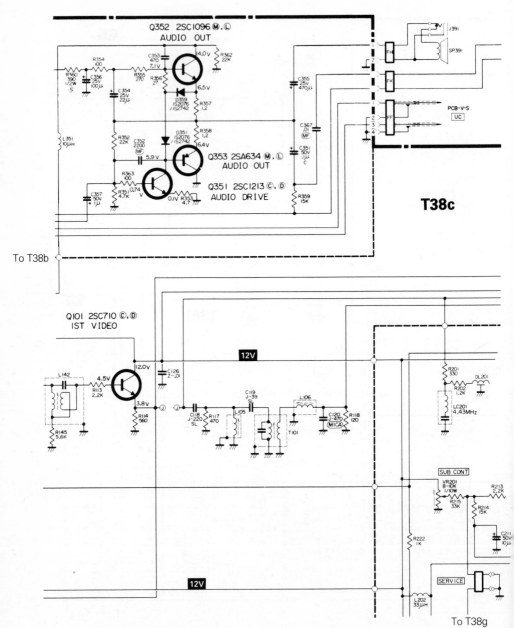

(T38c) CIRCUIT DIAGRAM (MAIN PANEL)—MODEL CT2206BX, ETC. (*PART*)

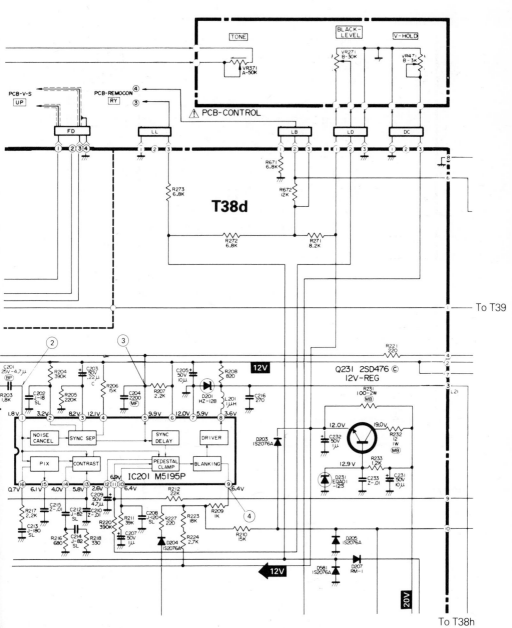

(T38d) CIRCUIT DIAGRAM (MAIN PANEL)—MODEL CT2206BX, ETC. (*PART*)

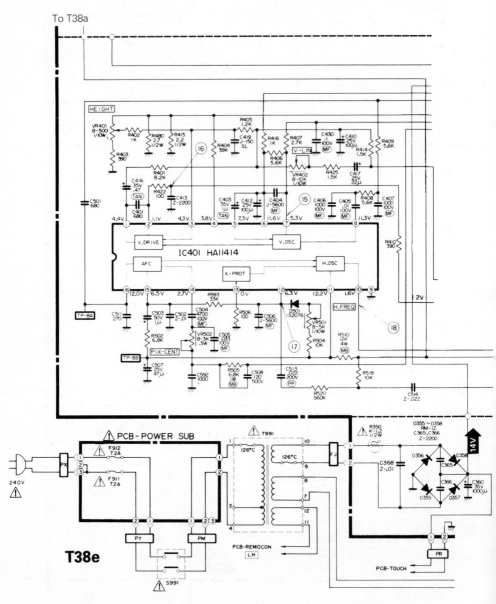

(T38e) CIRCUIT DIAGRAM (MAIN PANEL)—MODEL CT2206BX, ETC. (PART)

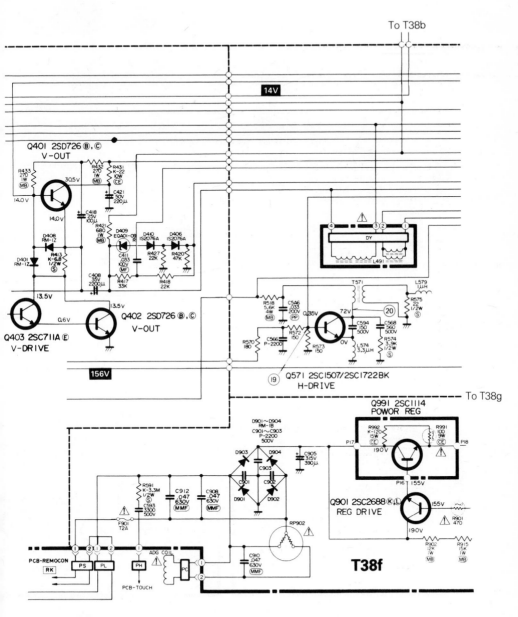

(T38f) CIRCUIT DIAGRAM (MAIN PANEL)—MODEL CT2206BX, ETC. (*PART*)

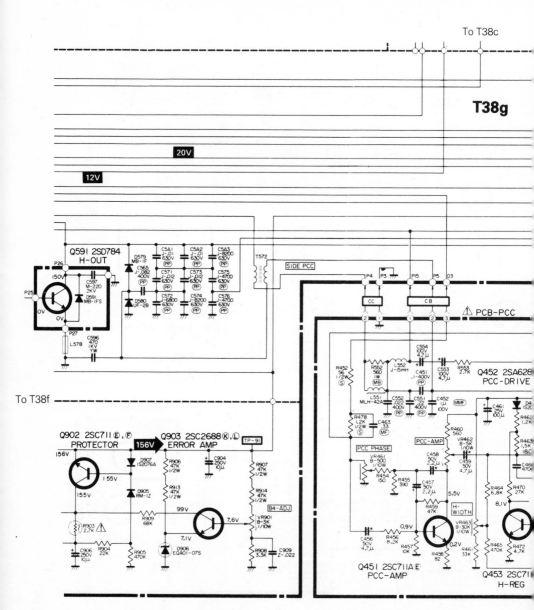

(T38g) CIRCUIT DIAGRAM (MAIN PANEL)—MODEL CT2206BX, ETC. (*PART*)

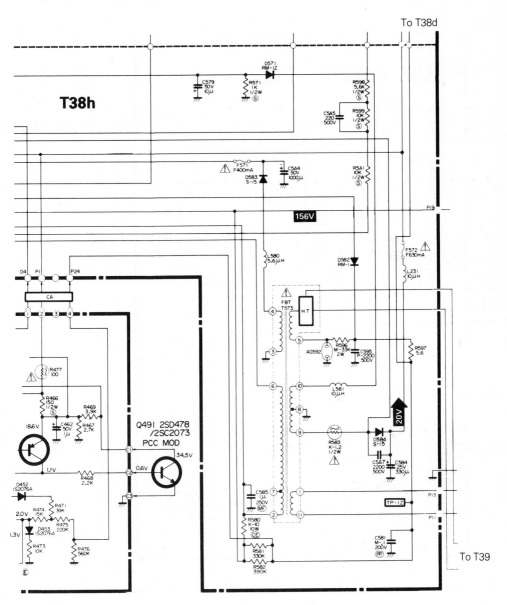

(T38h) CIRCUIT DIAGRAM (MAIN PANEL)—MODEL CT2206BX, ETC. (*PART*)

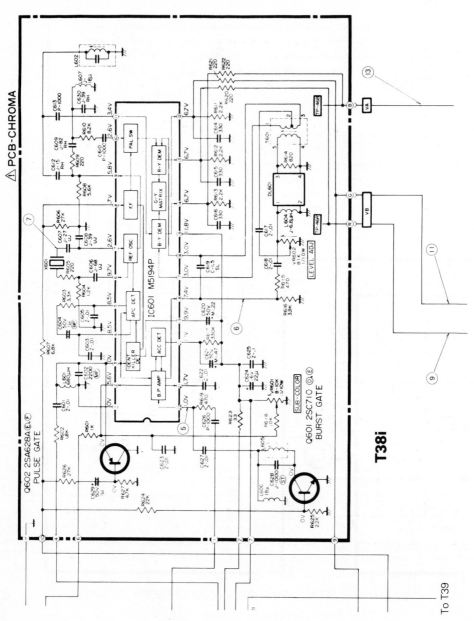

(T33i) CIRCUIT DIAGRAM (MAIN PANEL)—MODEL CT2206B, ETC. (CONTINUED)

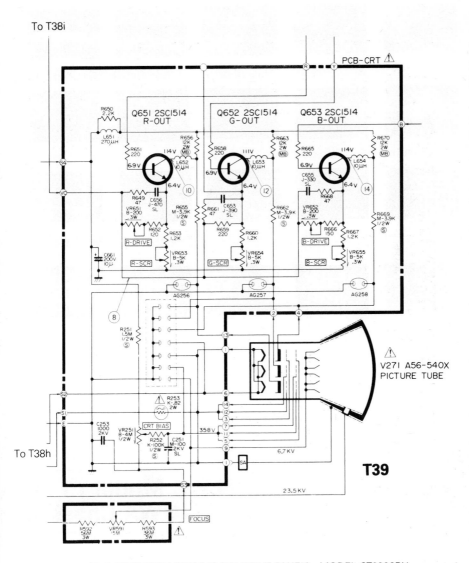

(T39) CIRCUIT DIAGRAM (C.R.T. DRIVE PANEL)—MODEL CT2206BX

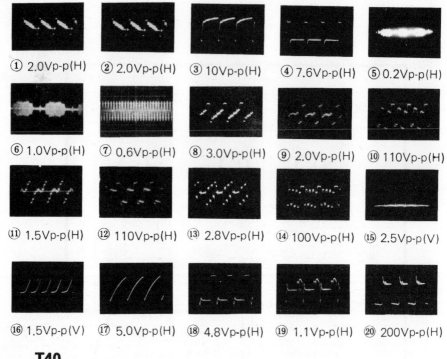

① 2.0Vp-p(H) ② 2.0Vp-p(H) ③ 10Vp-p(H) ④ 7.6Vp-p(H) ⑤ 0.2Vp-p(H)

⑥ 1.0Vp-p(H) ⑦ 0.6Vp-p(H) ⑧ 3.0Vp-p(H) ⑨ 2.0Vp-p(H) ⑩ 110Vp-p(H)

⑪ 1.5Vp-p(H) ⑫ 110Vp-p(H) ⑬ 2.8Vp-p(H) ⑭ 100Vp-p(H) ⑮ 2.5Vp-p(V)

⑯ 1.5Vp-p(V) ⑰ 5.0Vp-p(H) ⑱ 4.8Vp-p(H) ⑲ 1.1Vp-p(H) ⑳ 200Vp-p(H)

T40

(T40) CIRCUIT WAVEFORMS—MODEL CT2206BX, ETC.

N.E.C.

N.E.C. Model TT1/C (UWA Chassis)

General Description: A portable monochrome television chassis with 12- or 14-in. cathode ray tube. The receiver can operate from mains or battery supplies and the chassis is isolated by mains transformer allowing the use of an earphone.

Mains Supplies: 110/220 or 240 volts, 50Hz.

Cathode Ray Tubes: 310HEB4 (12-in.); 340BRB4 (14-in.).

Loudspeaker: 16 ohms impedance.

Adjustments

H.T.: Adjust VR701 for a reading of 11±0·3 volts at the fuse terminal (F701) on p.c.b.

Focus: The picture tube used in this receiver is electrostatically focused by means of a focus electrode connected to ground H7 (zero volt) through R814 located on the P.W.B.-C.R.T. socket. For 100 volt operation, cut the pattern of P.C.B.-C.R.T. socket and connect the electrode to H9 through R804 and R805. Select the operating voltage that produces the best focus.

PARTS   LOCATION

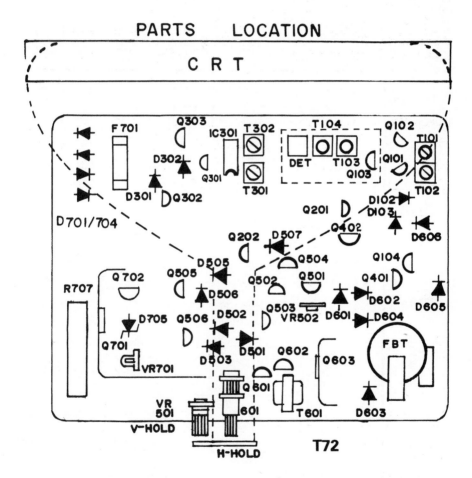

(T72) COMPONENT LAYOUT—MODEL TT1/C (UWA CHASSIS)

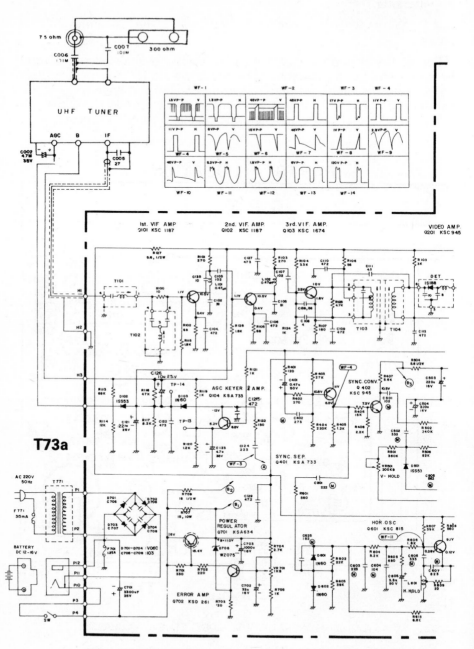

(T73a) CIRCUIT DIAGRAM—MODEL TT1/C (UWA CHASSIS) (*PART*)

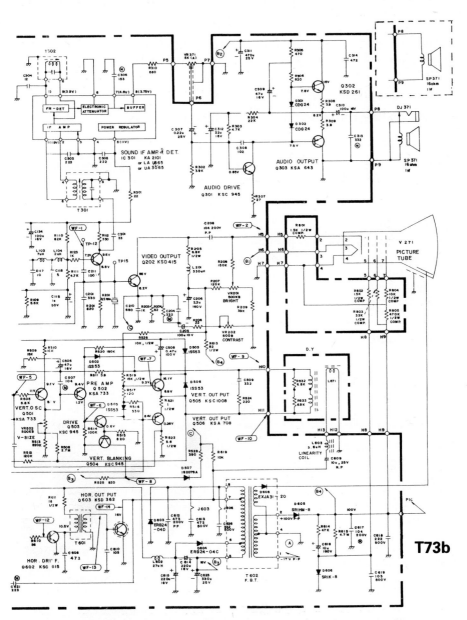

(T73b) CIRCUIT DIAGRAM—MODEL TT1/C (UWA CHASSIS) (*CONTINUED*)

General Description: A mains-operated colour television receiver with continuous channel tuning. The chassis is LIVE irrespective of mains connection and the use of an isolating transformer is recomended when carrying out service work on this chassis.

Mains Supply: 240 volts, 50Hz.

Cathode Ray Tubes: 320BXB22 or 320BMB22.

Dismantling

Remove 2 aerial socket mounting screws, 2 screws from under the carrying handle and 2 from the base to allow the back half of the case to be removed.

Adjustments

Power Supply: Pre-heat the receiver for at least 10 minutes.
Ensure mains input voltage is 240V±5V (A.C.).
Set the Brightness and Contrast controls to maximum.
Adjust VR1 for 115V (D.C.) at TP91 or J1 terminal.
Vary the Brightness and Contrast controls and ensure the voltage remains stable.

Sub-Brightness Adjustment: Turn fine tuning to obtain the best point then receive grey scale signal.
Set Brightness and Contrast to position 'Maximum' and Colour control to position 'Mid', C.T.C. to position 'Mechanical Centre'.
Connect the ammeter of 3mA D.C. range with limiter current check terminal.
Adjust limiter current to 0·65mA with Sub-Brightness control (VR701).

White Balance Adjustment:

a. Set Contrast 'Max' and R-Drive (VR903) B-Drive (VR904) to mechanical centre.

b. Set bias pots (VR905, VR906, VR907) to mechanical centre.

c. Set the service switch (SW701) to 'Service'.

d. Advance the C.R.T. screen pot (VR902) to provide a horizontal trace of the colour that appears first.

e. Blank out the trace that appeared with the appropriate bias pot (VR905, VR906 or VR907).

f. Advance the C.R.T. screen pot (VR902) further to illuminate the next colour, again blank out with the appropriate bias pot.

g. Adjust VR902 to just illuminate the last colour (see note below).

h. Re-adjust the former two bias pots (VR905, VR906 or VR907) to obtain a white trace.

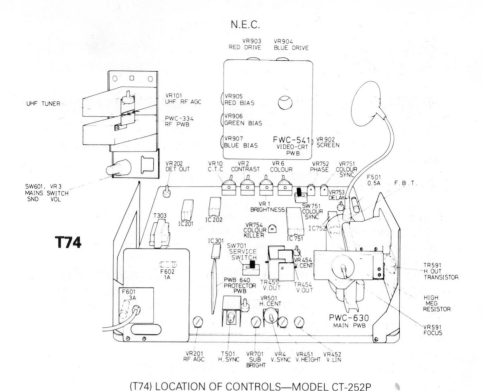

N.E.C.

(T74) LOCATION OF CONTROLS—MODEL CT-252P

T74

i. Re-set the service switch and on a grey scale pattern check the tracking by varying the Brightness and Contrast controls.

If necessary adjust the R and B drive controls (VR903 and VR904) for highlight correction.

Note: On some receivers it may not be possible to obtain a colour trace in step g. Should this be the case set the appropriate bias pot (VR905, VR906 or VR907) to 45° to the right of the mechanical centre and proceed with steps f to h. The best tracking will be obtained at the lowest brightness level of the trace.

After adjustment confirm that the indicated value of the ammeter is 0·65mA.

Brightness Adjustment: Set Brightness control to maximum position, Contrast control to minimum position, Colour control to Mid position and C.T.C. to Mechanical centre.

Adjust the Sub-Brightness control (VR701) as required.

Horizontal Sync. and Horizontal Centring Adjustment: Receive mono pattern signal, connect capacitor $220\mu F/16V$ between J13 and earth.

Adjust T501 for most stable picture.

Remove capacitor, confirm sync. by changing channel.

235

After adjustment horizontal sync., adjust pattern to centre position with VR501.

Confirm stable synchronism by changing channel.

Vertical Centring, Linearty and Vertical Height: Receive pattern signal, set service switch (SW701) to 'Service' position.

If horizontal line is not obtained on geometrical centre of screen, adjust vertical centring V.R. (VR454). (Horizontal line moves 3·5 m/m up and down side.)

Set service switch to position 'Normal', adjust height and linearity with VR451 and VR452.

Purity Adjustment: (Adjustment should be attempted immediately after 10 minutes pre-heating with full brightness and contrast.)

Degauss entire surface of picture tube, cabinet side, top, and back.

Tune the receiver to mono pattern.

Before attempting any adjustment, make sure static magnets are fixed in right position and tabs of (A) Purity Magnets, (B) 4-Pole Magnets, and (C) 6-Pole Magnets are at the position illustrated in Fig. T75.

Note: If each two tabs of 4-Pole Magnets and 6-Pole Magnets are lined up together correctly, it will provide zero magnetic field and R.G.B. static convergence error should be minimal.

1. Remove 'G-Y' tip from Video P.W.B. to provide magenta raster with R. & B.

2. Loosen Deflection Yoke clamp and slide the yoke backward.

3. Rotate the Purity Magnets for same angle in opposite directions from their original setting positions by both longer tabs lined up with short tabs at 9 o'clock and 3 o'clock position until the magenta vertical stripe is positioned in the centre of the screen.

4. Slide the yoke forward to obtain the best purity and tighten the clamp tentatively. If the yoke tends to bow down, insert an auxiliary wedge between picture tube funnel and the yoke gap from top side.

5. Connect 'G-Y' tip to Main P.W.B. and make sure the purity with white raster. Repeat the steps 1 to 5 if necessary.

6. Slide the Service Switch (SW701) to provide a horizontal line on the screen. If the horizontal line is tilted, rotate the deflection yoke until the horizontal line is positioned within ±5 mm from the geometrical centre of the screen.

If the horizontal line is not positioned within the above tolerance by the yoke rotation, rotate the Purity Magnets by maintaining the pre-set angle between both longer tabs until the horizontal line is positioned within the tolerance.

Note: Whilst this adjustment, V-Centre Tip must be set at the mechanical centre.

7. Slide the Service Switch (SW701) back to the normal position and check the purity.

N.E.C.

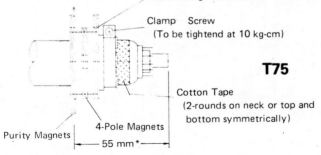

6-Pole Magnets
(Static Convergence Magnets
to converge Red/Blue to Green)

Clamp   Screw
(To be tightend at 10 kg-cm)

T75

Cotton Tape
(2-rounds on neck or top and
bottom symmetrically)

Purity Magnets

4-Pole Magnets

— 55 mm *—

* Distance between center of Purity Magnets and CRT stem end
should be 55 mm.
(Purity Magnets should be positioned at about 3 mm back from
edge of G3 electrode.)

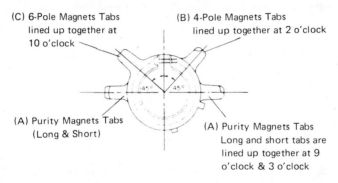

(C) 6-Pole Magnets Tabs
lined up together at
10 o'clock

(B) 4-Pole Magnets Tabs
lined up together at 2 o'clock

(A) Purity Magnets Tabs
(Long & Short)

(A) Purity Magnets Tabs
Long and short tabs are
lined up together at 9
o'clock & 3 o'clock

(T75) PURITY ADJUSTMENTS—MODEL CT-252P

If the Purity Magnets were rotated at step (6), make sure to touch up the
purity with magenta raster.
8. Repeat steps 1 to 7 if necessary.
9. Tighten the deflection yoke clamp.
After the purity adjustment, picture tube focus should be adjusted for the
best point.
Tune the T.V. to the monoscope pattern and adjust the focus with
maximum brightness and contrast.

Note: As the static convergence of the P.I.S. system C.R.T. varies with
focus voltage, it is necessary to adjust C.R.T. focus right after the Purity
adjustment and before attempting the static convergence adjustment.

237

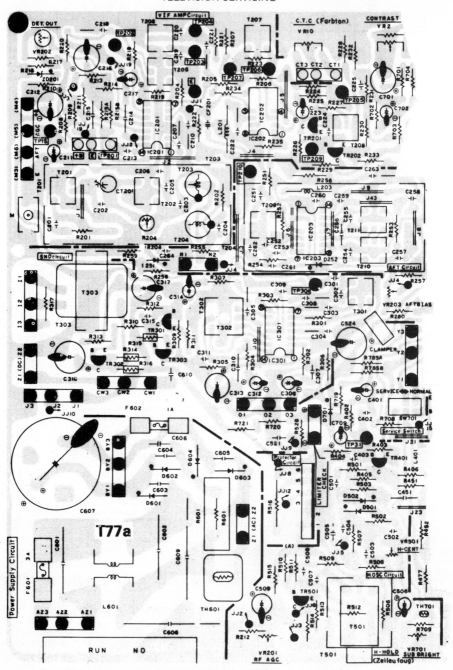

(T77a) COMPONENT LAYOUT—MODEL CT-252P (*PART*)

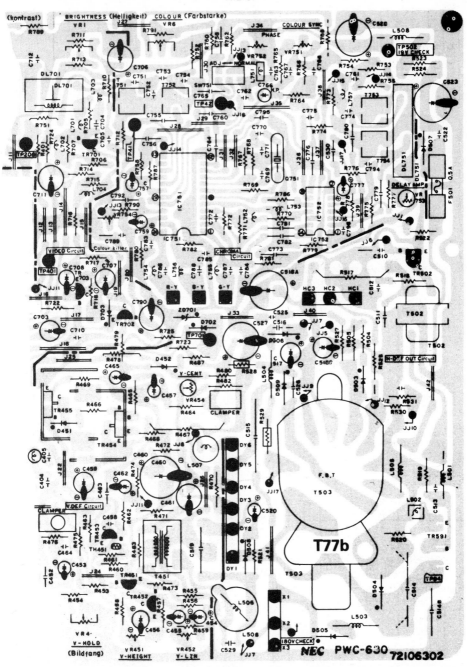

(T77b) COMPONENT LAYOUT—MODEL CT-252P (CONTINUED)

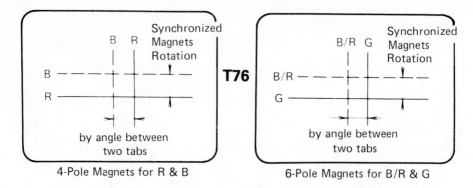

(T76) STATIC CONVERGENCE—MODEL CT-252P

Static Convergence Adjustment (Fig. T76):

1. Tune the receiver to a crosshatch pattern and adjust the brightness and the contrast just high enough to observe the pattern clearly.

2. Remove 'G-Y' tip from the Video P.W.B. to provide Red and Blue only on the screen.

3. Adjust 4-Pole Magnets by rotating two lined up tabs in opposite directions to converge Red and Blue vertical lines at the centre of the screen.

4. Rotate the 4-Pole Magnets together by maintaining preadjusted angle between two tabs to converge Red and Blue horizontal lines at the centre of the screen. If the operation misconverges vertical lines, touch up pre-adjusted angle between the two tabs to correct it.

5. Connect 'G-Y' tip to the Video P.W.B. to provide Red, Blue, and Green on the screen.

6. Adjust 6-Pole Magnets by rotating two lined up tabs in opposite directions to converge Red/Blue and Green vertical lines at the centre of the screen.

7. Rotate the 6-Pole Magnets together by maintaining pre-adjusted angle between two tabs to converge Red/Blue and Green horizontal lines at the centre of the screen.

If the operation misconverges vertical lines, touch up pre-adjusted angle between the two tabs to correct it.

8. With the steps 3 to 7, Red/Blue and Green should be converged at the centre of the screen.

9. Tighten the Lock Ring carefully without moving any combinations of the adjusted magnets.

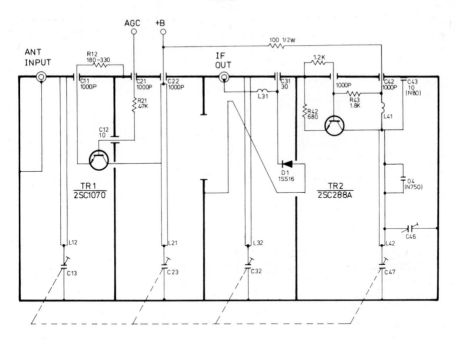

T78

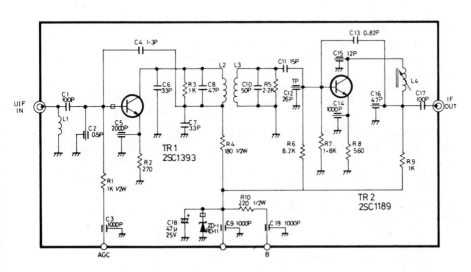

(T78) CIRCUIT DIAGRAM (U.H.F. TUNER AND I.F.)—MODEL CT-252P

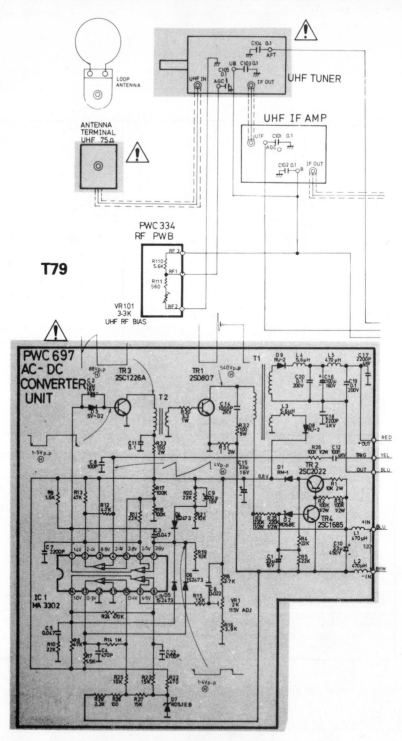

(T79) CIRCUIT DIAGRAM (POWER SUPPLY)—MODEL CT-252P

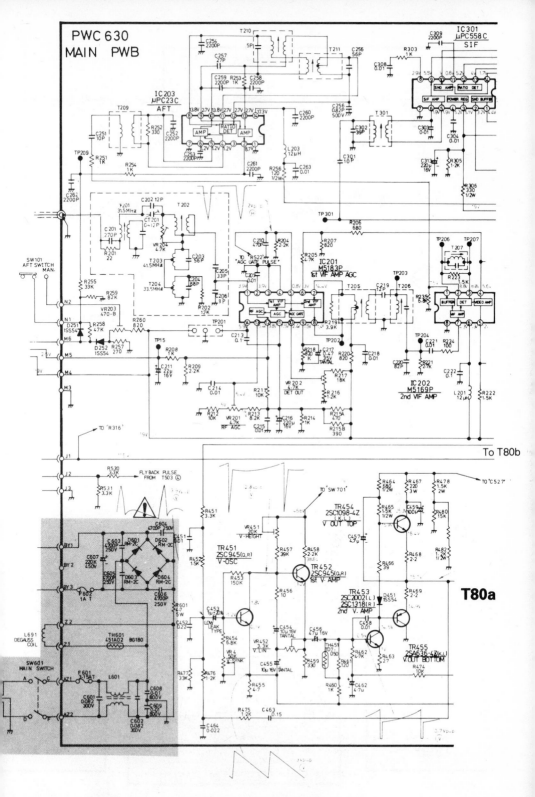

(T80a) CIRCUIT DIAGRAM (MAIN PANEL)—MODEL CT-252P (*PART*)

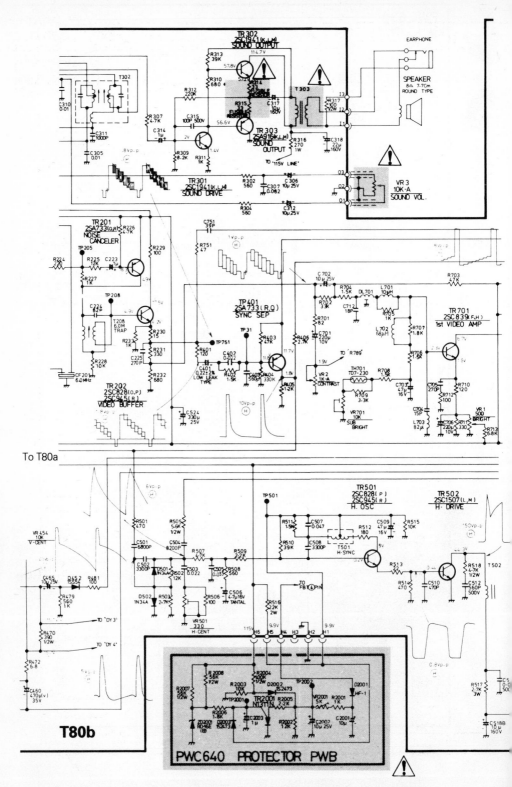

(T80b) CIRCUIT DIAGRAM (MAIN PANEL)—MODEL CT-252P (*PART*)

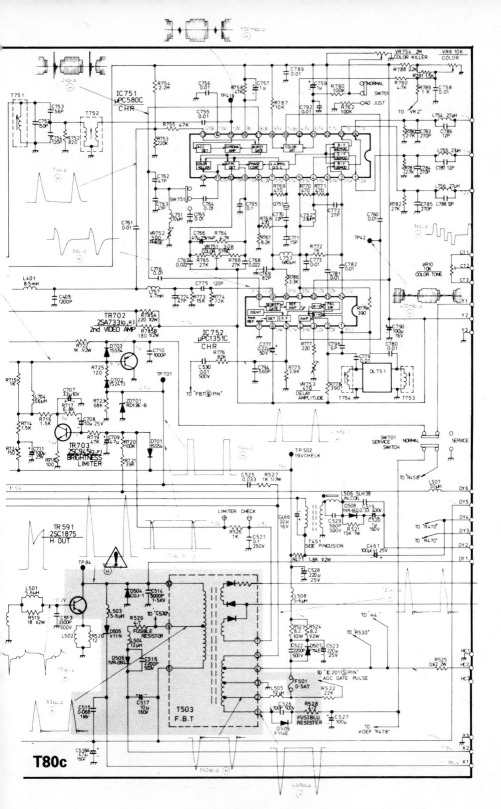

T80c

(T80c) CIRCUIT DIAGRAM (MAIN PANEL)—MODEL CT-252P (*CONTINUED*)

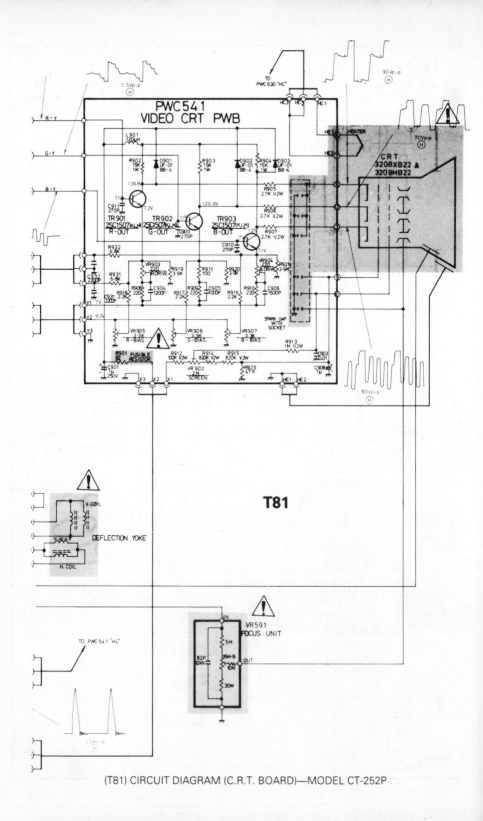

(T81) CIRCUIT DIAGRAM (C.R.T. BOARD)—MODEL CT-252P

PHILIPS   K30 Chassis (Ed. 2)

General Description: The K30 Chassis is described in the 1981–82 edition of *Radio and Television Servicing* and should be used in conjunction with the information given here. The introduction of design and production variations, mainly involved in the luminance/chrominance and power supply panels, has necessitated certain changes on the 'Mother' board. These changes are intended to provide increased compatibility of panels between chassis and individual models, as well as improvements in performance and reliability. The revised circuits are given here.

Because of the differences involved between the early and later-versions of the chassis, current production models are referred to as 'Edition 2'.

Service engineers should note that the same type and suffix numbers will be retained both for early and later-version models, and care will need to be taken to ensure that the correct items are indicated when spare parts require to be ordered.

Identification of Early and Later-Version Chassis

Panel type	Early chassis (Edition 1)	Later chassis (Edition 2)
Luminance/ chrominance panel	*Two* integrated circuit version IC3192-type TDA2560Q IC3223-type TDA2523/4Q	*One* integrated circuit version IC3035-type TDA3560
'Mother' board panel	Panel coding label: Non-Teletext—BA01 to BA05	Panel coding label: Non-Teletext—BA20 onwards Teletext—BA00 onwards

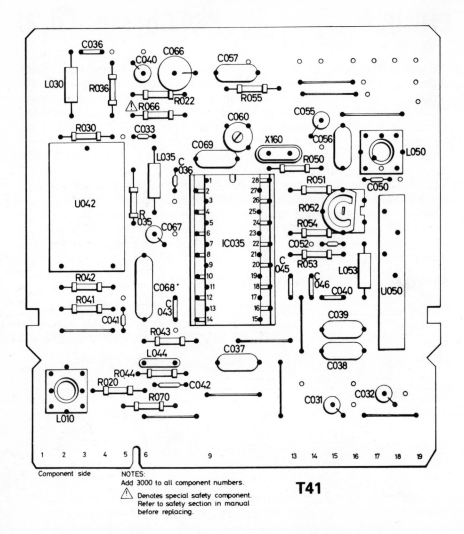

(T41) COMPONENT LAYOUT (LUMA/CHROMA PANEL)—K30 CHASSIS (ED. 2)

(T42) CIRCUIT DIAGRAM (LUMA/CHROMA PANEL)—K30 CHASSIS (ED. 2)

(T43) CIRCUIT DIAGRAM (SUPPLY CONTROL PANEL)—K30 CHASSIS (ED. 2)

PHILIPS   KT3 Chassis (Ed. 2)

General Description: The KT3 Chassis is described in the 1980–81 edition of *Radio and Television Servicing* and should be used in conjunction with the information given here. The introduction of design and production variations, mainly involved in the luminance/chrominance and power supply panels, has necessitated certain changes on the 'Mother' board. These changes are intended to provide increased compatibility of panels between chassis and individual models, as well as improvements in performance and reliability. The revised circuits are given here.

Because of the differences involved between the early and later-versions of the chassis, current production models are referred to as 'Edition 2'.

Service engineers should note that the same type and suffix numbers will be retained both for early and later-version models, and care will need to be taken to ensure that the correct items are indicated when spare parts require to be ordered.

Identification of Early and Later-Version Chassis

Panel type	Early chassis (Edition 1)	Late Chassis (Edition 2)
Luminance/ chrominance panel	*Two* integrated circuit version IC192-type TDA2560Q IC223-type TDA2523/4Q	*One* integrated circuit version IC035-type TDA3560
'Mother' board panel	Panel coding label: 8222 280 2077.1 or .4	Panel coding label: Teletext and Non-Teletext 8222 280 2077.5

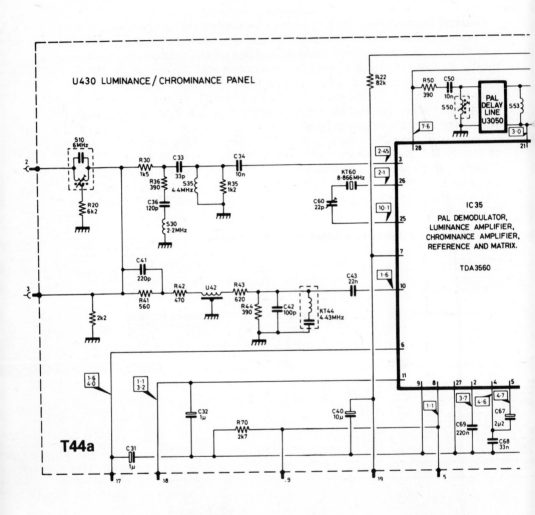

(T44a) CIRCUIT DIAGRAM (LUMA/CHROMA PANEL)—KT3 CHASSIS (ED. 2) (*PART*)

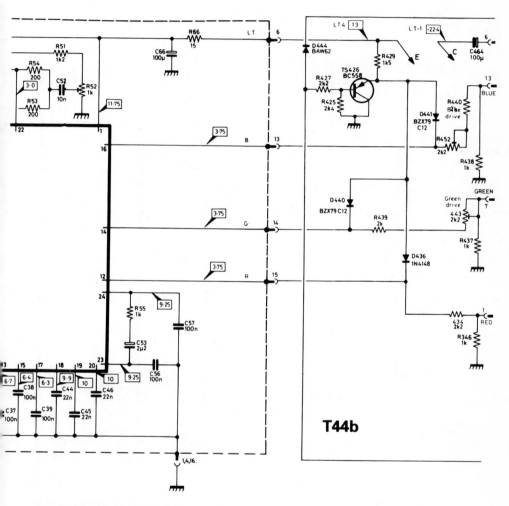

(T44b) CIRCUIT DIAGRAM (LUMA/CHROMA PANEL)—KT3 CHASSIS (ED. 2) (*CONTINUED*)

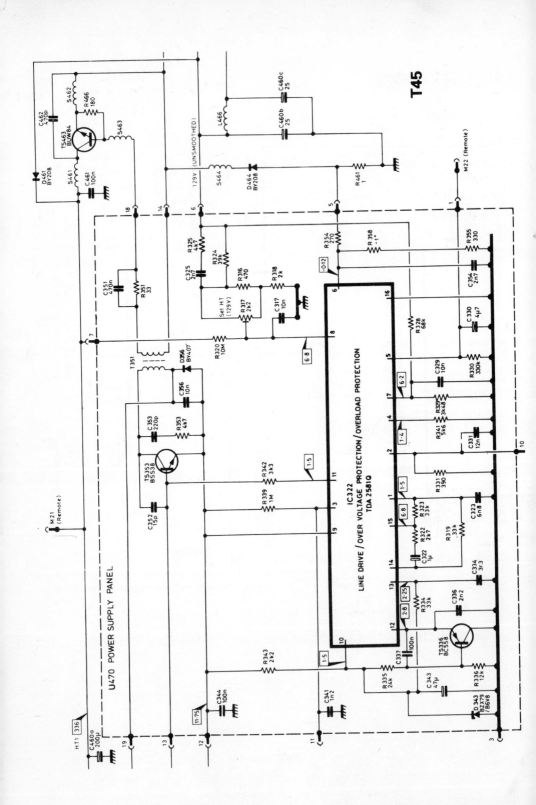

PLUSTRON Models CTV55, CTV55D

General Description: 5-in. portable colour television receivers covering V.H.F./U.H.F. bands and operating from mains, internal or external battery supplies. The chassis is isolated by mains transformer and sockets are provided for the connection of ancillaries.

Mains Supply: 240 volts, 50Hz.   **Batteries:** 10·8—15 volts.

Loudspeaker: 32 ohms impedance.

Access for Service

Remove a single screw from each side above the carrying handle and 3 screws from the base. The printed circuit boards are held by fixing screws.

Adjustments

H.T.: Adjust collector voltage of Q502 to 10·8 volts with VR502. Adjust collector voltage of Q501 to 9 volts with VR501.

Convergence (See Fig. T46): Slacken lock screw and set the magnet assembly approx. 18 mm along the tube neck.

Central Area: Inject a crosshatch signal and adjust 4-pole and 6-pole magnets to converge red and blue onto green as shown in Fig. T47.

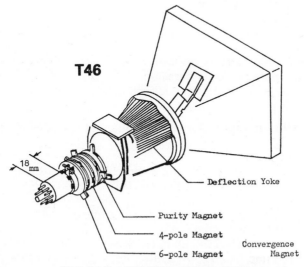

T46

18 mm

Deflection Yoke

Purity Magnet

4-pole Magnet

6-pole Magnet

Convergence Magnet

(T46) C.R.T. ASSEMBLY—MODEL CTV55

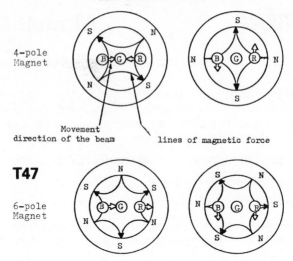

T47

6-pole
Magnet

(T47) STATIC CONVERGENCE ADJUSTMENTS—MODEL CTV55

Peripheral Area: Rotate and/or move deflection yoke backwards or forwards to obtain edge convergence.

Purity Adjustment: Turn the Low Light Amplitude (Red/VR614 and Blue/VR615) to minimum and adjust the raster on Green only.

Loosen screws on Deflection yoke and slide back the yoke fully.

Adjust the two Purity Magnets alternately so that green appears in the centre of the screen vertically.

Slide the yoke forward and adjust the yoke so that pure green covers screen.

If colour impurities shows on the screen, slide the yoke backwards or forwards until a pure picture is obtained.

Repeat convergence and purity adjustments until satisfactory. Retighten the fixing-screws.

High Light Adjustment: Allow 10 minutes at least before carrying out adjustments.

Inject colour bar signal.

Set the Remote Switch to ON and switch SW601 to manual.

Set the voltage of TP17 to 92–93V with VR613. (SW602 ON Service Switch.)

Set VR608 and VR605 to minimum.

Adjust the Screen Amplitude (VR618) under these conditions until you can clearly recognize the horizontal line.

Set the Service Switch (SW602) to NORMAL.

Adjust the red and blue High Light controls (VR616 and VR617) until all three colours are of equal intensity.

Check the voltage at TP17 after carrying out the above adjustments.

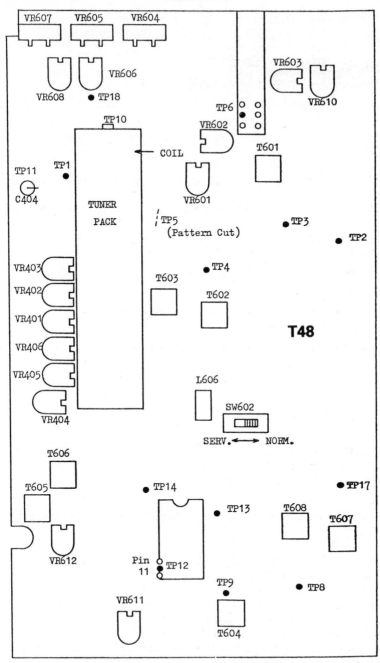

(T48) ADJUSTMENTS (PCB1 BOARD)—MODEL CTV55

Sub Bright Adjustment: Inject a Crosshatch Pattern.
Set the VR605 and VR607 to maximum (SW601 MANUAL).
Set the luminosity on the Picture screen with VR608.

Contrast Adjustment: Tune to test card 'F' (SW601 AUTO).
Set the voltage of TP18 with VR606 to 0·4V.

Low Light Adjustment: Set Contrast to minimum and Brightness to mid-position.

Adjust the Low Light amplitude controls (R-red and G-green) to the centre position.

Change the Service Switch (SW602) to 'Service'.

Set the collector voltage of Q607 to 100V by adjusting the green Low Light amplitude.

Turn the Screen Amp control gradually until the line of the collapsed field is just visible.

Adjust the Red and Blue Low Light controls until all colours are of equal intensity.

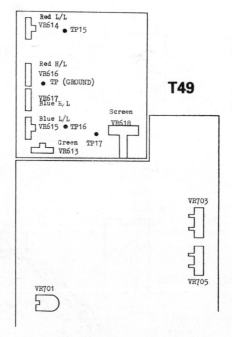

(T49) ADJUSTMENTS (PCB2 BOARD)—MODEL CTV55

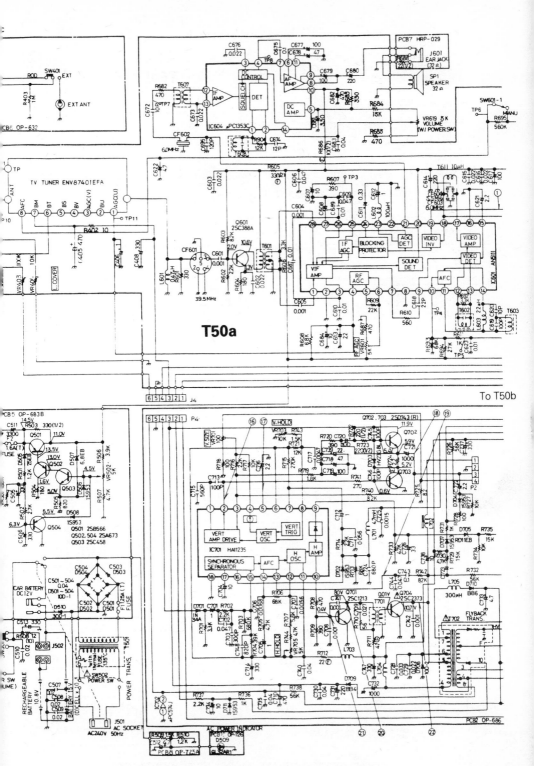

T50a

To T50b

(T50a) CIRCUIT DIAGRAM—MODEL CTV55 (*PART*)

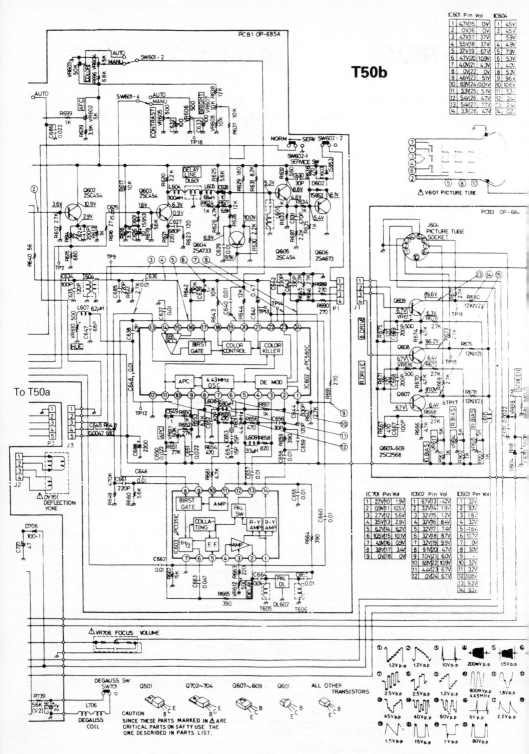

T50b

(T50b) CIRCUIT DIAGRAM—MODEL CTV55 (*CONTINUED*)

PLUSTRON

Models TV17, TV19

General Description: A mains operated monochrome television receiver chassis for 17- or 19-in. cathode ray tube. An integrated circuit is used for sound processing and the chassis is isolated by mains transformer.

Mains Supply: 240 volts, 50Hz.

Cathode Ray Tubes: 500CDB4 (19-in.); 440BGB4 (17-in.).

Loudspeaker: 16 ohms impedance.

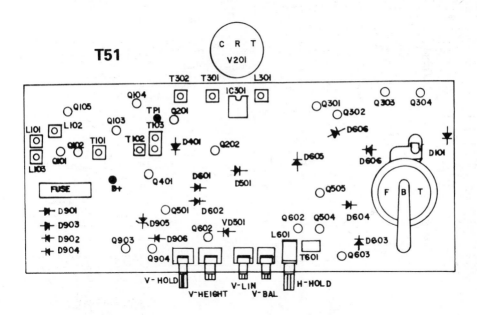

(T51) COMPONENT LAYOUT—MODEL TV17

(T53a) CIRCUIT DIAGRAM—MODEL TV17 (PART)

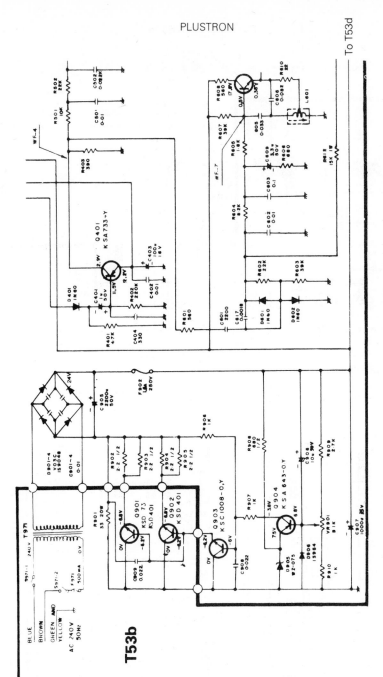

(T53b) CIRCUIT DIAGRAM—MODEL TV17 *(PART)*

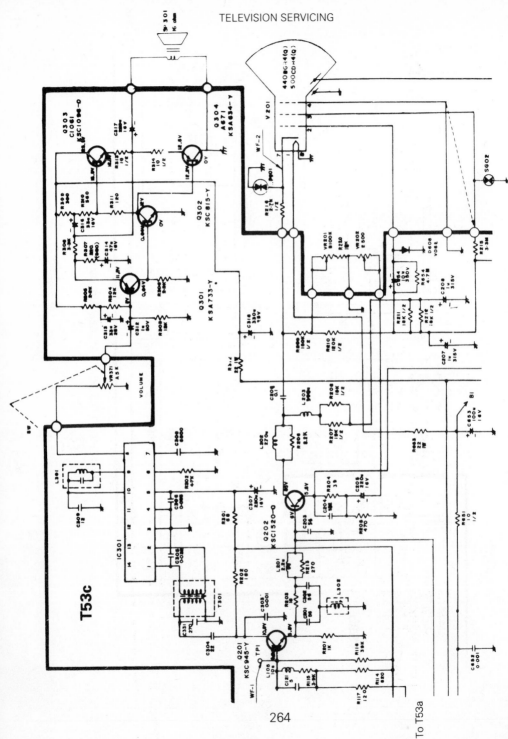

(T53c) CIRCUIT DIAGRAM—MODEL TV17 (*PART*)

To T53a

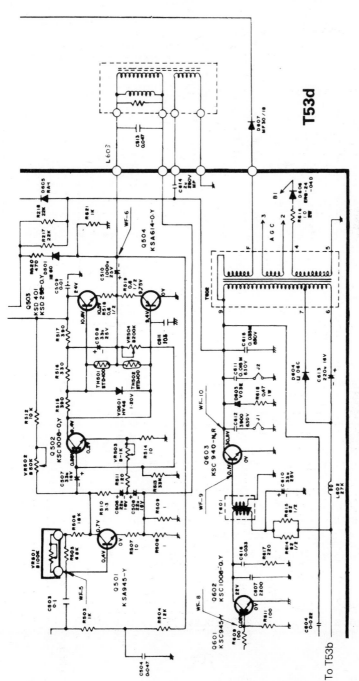

T53d

(T53d) CIRCUIT DIAGRAM—MODEL TV17 (CONTINUED)

To T53b

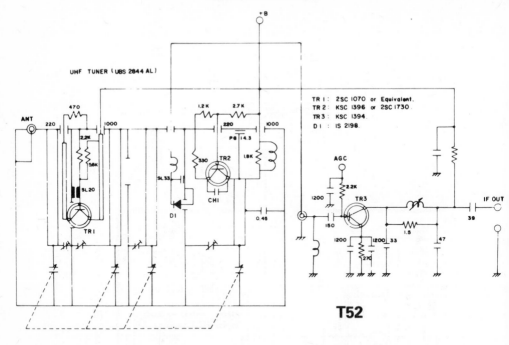

(T52) CIRCUIT DIAGRAM (U.H.F. TUNER)—MODEL TV17

SHARP   Model 5P-37H

General Description: A portable solid-state monochrome television receiver incorporating a 3-waveband radio and stereo cassette tape-recorder. A mains transformer provides chassis isolation and sockets are fitted for the connection of audio auxiliaries and external power sources.

Mains Supply: 240 volts, 50Hz.

Battery: 15 volts max. (22 watts).

Cathode Ray Tube: 140CNB4

Wavebands (Radio): L.W. 150–285kHz; M.W. 520–1620; F.M. 87·6–108MHz.

TV DIAL

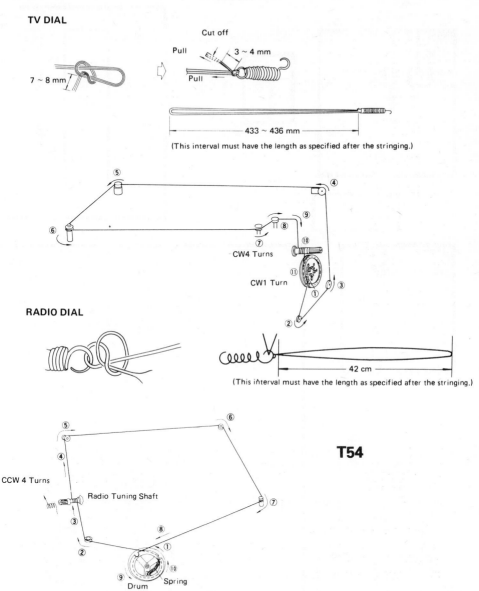

Cut off

Pull

3 ~ 4 mm

Pull

7 ~ 8 mm

433 ~ 436 mm

(This interval must have the length as specified after the stringing.)

⑤ ④ ⑨ ⑧ ⑥ ⑦ ⑩ ⑪ ① ③ ②

CW4 Turns

CW1 Turn

RADIO DIAL

42 cm

(This interval must have the length as specified after the stringing.)

⑤ ⑥ ④ ③ ⑧ ② ① ⑩ ⑨ ⑦

CCW 4 Turns

Radio Tuning Shaft

Drum

Spring

T54

(T54) DRIVE CORD ASSEMBLIES—MODEL 5P-37H

267

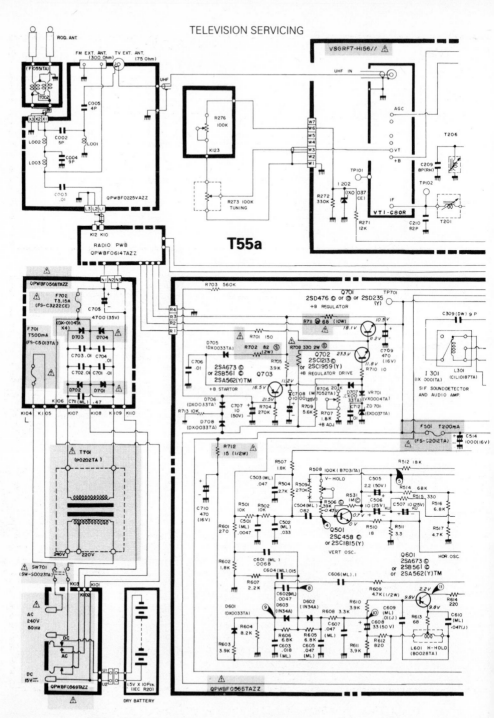

(T55a) CIRCUIT DIAGRAM (T.V. SECTION)—MODEL 5P-37H (*PART*)

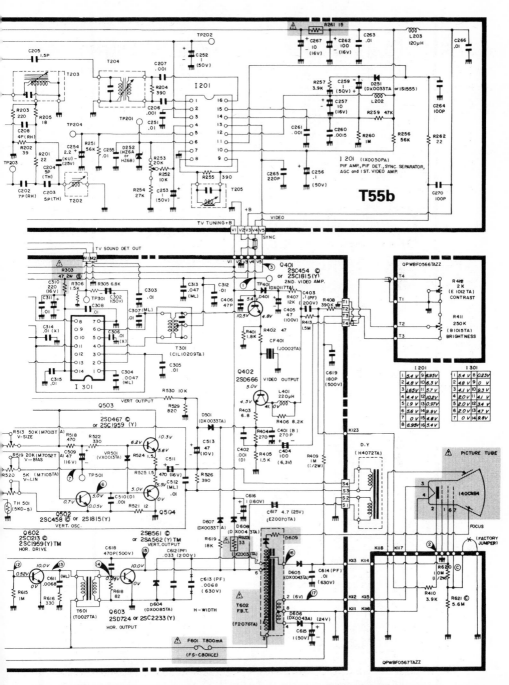

(T55b) CIRCUIT DIAGRAM (T.V. SECTION)—MODEL 5P-37H (CONTINUED)

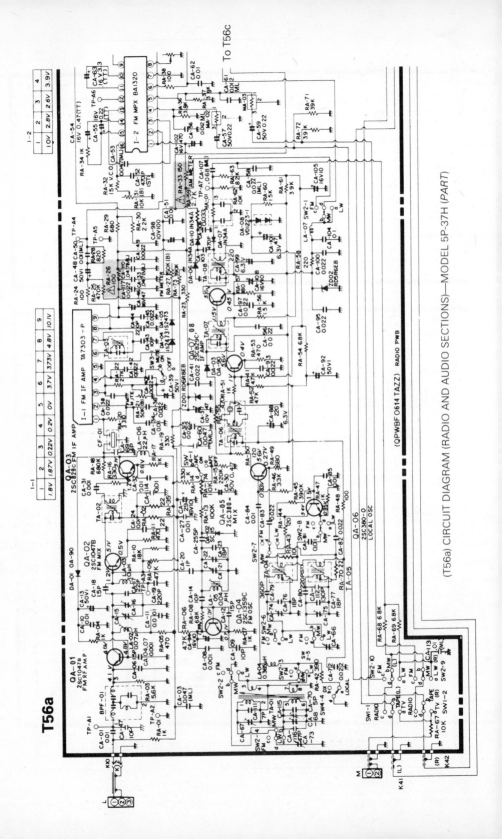

(T56a) CIRCUIT DIAGRAM (RADIO AND AUDIO SECTIONS)—MODEL 5P-37H (PART)

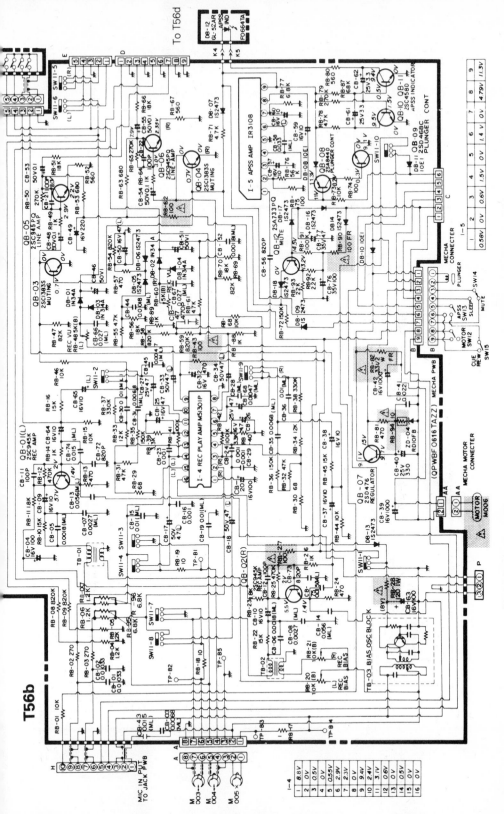

(T56b) CIRCUIT DIAGRAM (RADIO AND AUDIO SECTIONS)—MODEL 5P-37H (PART)

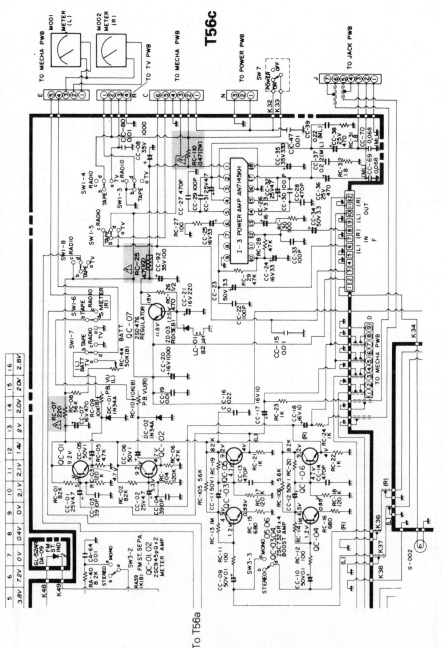

(T56c) CIRCUIT DIAGRAM (RADIO AND AUDIO SECTIONS)—MODEL 5P-37H (*PART*)

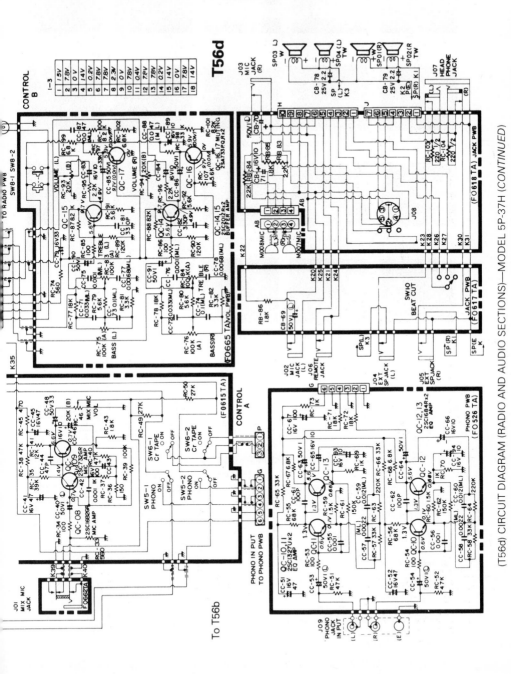

(T56d) CIRCUIT DIAGRAM (RADIO AND AUDIO SECTIONS)—MODEL 5P-37H (CONTINUED)

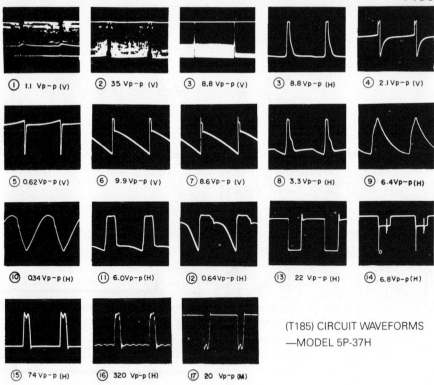

① 1.1 Vp-p (V) ② 35 Vp-p (V) ③ 8.8 Vp-p (V) ③ 8.8 Vp-p (H) ④ 2.1 Vp-p (V)

⑤ 0.62 Vp-p (V) ⑥ 9.9 Vp-p (V) ⑦ 8.6 Vp-p (V) ⑧ 3.3 Vp-p (H) ⑨ 6.4 Vp-p (H)

⑩ 0.34 Vp-p (H) ⑪ 6.0 Vp-p (H) ⑫ 0.64 Vp-p (H) ⑬ 22 Vp-p (H) ⑭ 6.8 Vp-p (H)

⑮ 74 Vp-p (H) ⑯ 320 Vp-p (H) ⑰ 20 Vp-p (M)

(T185) CIRCUIT WAVEFORMS
—MODEL 5P-37H

SHARP   Model 10P-18H

General Description: A portable monochrome television receiver with 3-waveband radio and monaural cassette tape recorder. The instrument operates from mains or battery supplies and is isolated by mains transformer allowing the safe connection of ancilliary signal sources. A microphone is built into the case.

Mains Supply: 240 volts, 50Hz.

Battery: 12 volts, 16 watts.

Wavebands (Radio): L.W. 150–285kHz; M.W. 520–1605kHz; F.M. 87·6–108·1MHz.

Bias Frequency (Tape): 61·7±5kHz.

Access for Service

Remove 7 screws from the back of the cabinet and pull out the U.H.F. channel selector and tuning knob to the main chassis to be withdrawn.

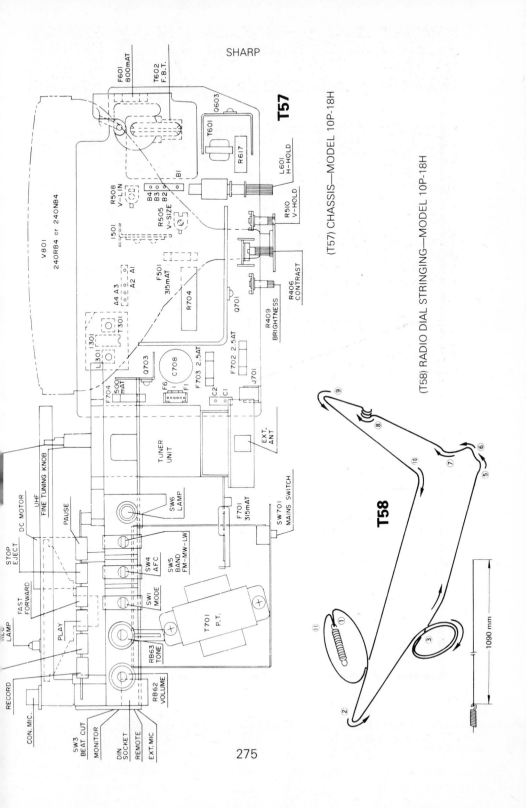

SHARP

T57

(T57) CHASSIS—MODEL 10P-18H

(T58) RADIO DIAL STRINGING—MODEL 10P-18H

T58

1090 mm

275

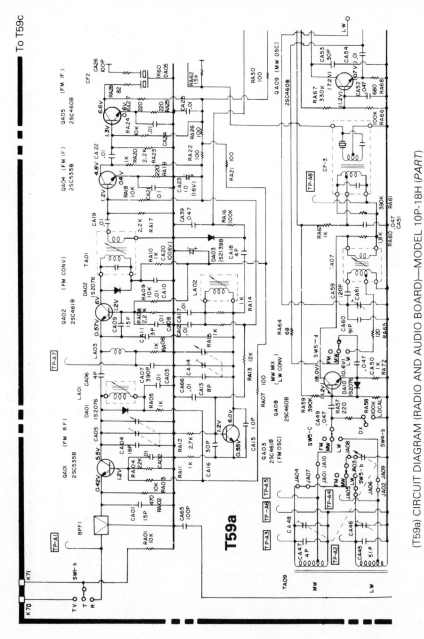

(T59a) CIRCUIT DIAGRAM (RADIO AND AUDIO BOARD)—MODEL 10P-18H (PART)

(T59b) CIRCUIT DIAGRAM (RADIO AND AUDIO BOARD)—MODEL 10P-18H (*PART*)

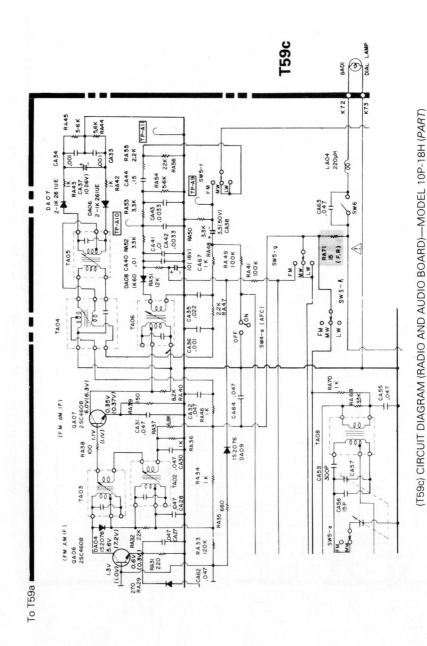

(T59c) CIRCUIT DIAGRAM (RADIO AND AUDIO BOARD)—MODEL 10P-18H (*PART*)

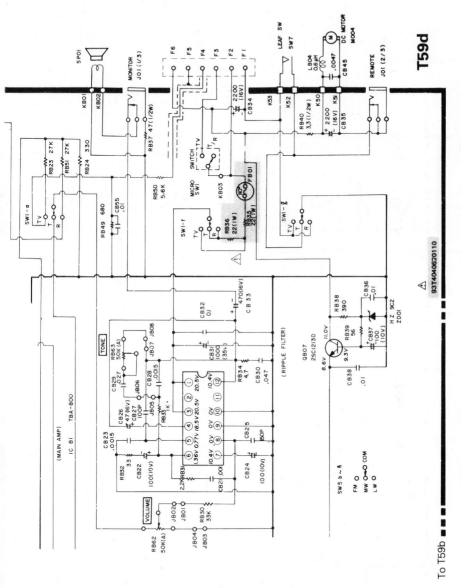

T59d

(T59d) CIRCUIT DIAGRAM (RADIO AND AUDIO BOARD)—MODEL 10P-18H (CONTINUED)

93T40620110

To T59b

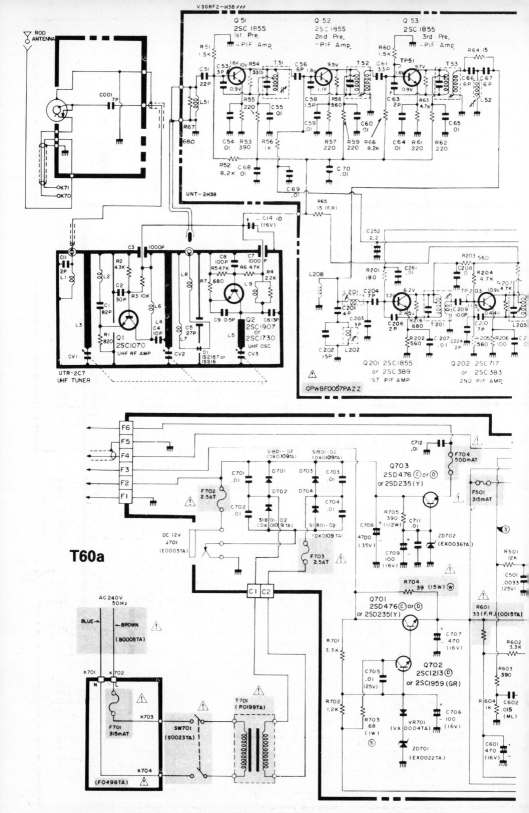

(T60a) CIRCUIT DIAGRAM (T.V. SECTION)—MODEL 10P-18H (*PART*)

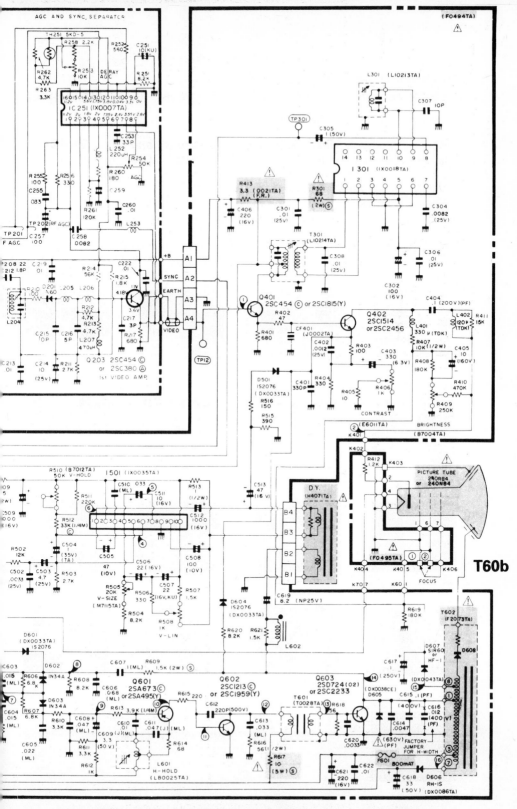

(T60b) CIRCUIT DIAGRAM (T.V. SECTION)—MODEL 10P-18H (CONTINUED)

Adjustments (Cassette)

Line Hold: L601. **Width:** Cut or connect C616. **Vertical Hold:** R510.

Height: R505. **Vertical Linearity:** R508.

Record/Playback Head A.C. Bias Adjustment and Bias Oscillation Frequency Check: Push the record and play buttons to get Record mode.

Connect V.T.V.M. and frequency counter between the test point TP-B1 and earth. (Use the shield plate to serve as earth.)

The bias oscillation frequency should be within 61·7±5kHz.

Next, adjust the bias-adjust control RB63 so that when the bias oscillation frequency is 61·7±5kHz, the bias current will be 400μA (say, corresponding to 4mV r.m.s. indicated by V.T.V.M. at the measuring point TP-B1).

Check that the bias oscillation frequency is within 61·7±5kHz.

T61

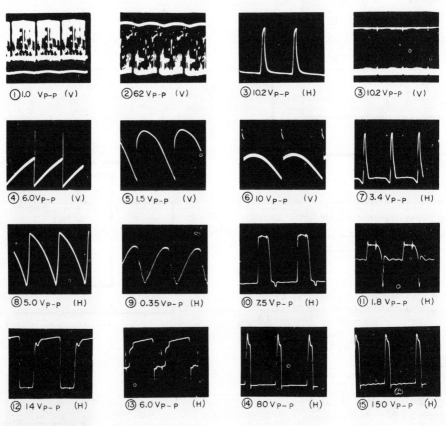

① 1.0 Vp–p (V)   ② 62 Vp–p (V)   ③ 10.2 Vp–p (H)   ③ 10.2 Vp–p (V)

④ 6.0 Vp–p (V)   ⑤ 1.5 Vp–p (V)   ⑥ 10 Vp–p (V)   ⑦ 3.4 Vp–p (H)

⑧ 5.0 Vp–p (H)   ⑨ 0.35 Vp–p (H)   ⑩ 7.5 Vp–p (H)   ⑪ 1.8 Vp–p (H)

⑫ 14 Vp–p (H)   ⑬ 6.0 Vp–p (H)   ⑭ 80 Vp–p (H)   ⑮ 150 Vp–p (H)

⑯ 45 Vp–p (H)   (T61) CIRCUIT WAVEFORMS—MODEL 10P-18H

SHARP   Model 10P-28H

General Description: A portable monochrome television receiver with 3-waveband A.M./F.M. radio and 2-track monaural cassette tape-recorder. The instrument operates from mains or battery supplies and sockets are provided for the connection of ancillaries.

Mains Supply: 240 volts, 50Hz.

Battery: 12 volts, 16 watts.

Wavebands: L.W. 150–285kHz; M.W. 520–1620kHz; F.M. 87·6–108MHz.

Bias Frequency: 61·7 ±5kHZ.

Erase: D.C.

H. Osc. Coil and H. Size Adjustment: Receive T.V. signals.
Rotate the shaft of the H. Osc. coil (L601) counter-clockwise to let the picture collapse, then rotate it clockwise until the picture again assumes its normal state: consider the pattern centre of the picture thus attained to be a point 'a'. Further rotate the coil clockwise to have the picture again break up, and after that, rotate it counter-clockwise so that the picture assumes the normal state once again: consider this pattern centre to be a point 'b'. Now, rotate the shaft of the H. Osc. coil so that the setting is in the middle between points 'a' and 'b'.

Chassis Removal

Whenever it becomes necessary to remove the chassis from the cabinet, proceed in the following manner.
Remove the seven screws from the back cabinet.
Pull out the U.H.F. channel Selector and Tuning knob from the front cabinet.
Pull out the main chassis.
Remove the two screws from the Tuner angle.
Remove the C.R.T. socket, coating earth lead and anode cap.
Loosen the deflection yoke clamping screw and pull the deflection yoke out of the C.R.T. neck.
Pull out the Radio Tuning knob, Tone, and Volume knob from the front cabinet.
Remove the five screws from the Radio and Cassette unit.

Cautions of Replacement of Telescopic Antenna: When replacing the telescopic antenna (QANTR0013TAZZ) with a new one, observe the following.
Clamp the nut with an appropriate torque.
After clamping the nut, apply adhesive agent to its top to prevent it from loosening.

Adjustment (Audio Section)

+B Voltage Checking: Set A.C. line voltage at 240V, 50Hz.

Set the mode switch to 'Radio' or 'Tape' position and be certain the voltage at the collector of QB03 (or at the emitter of QB04) is 12·2±0·5V. Here, the volume and tone controls must have been set to 'Max' position respectively (with a resistor of 16 ohm connected to the speaker terminal).

Adjustments (T.V. Section)

+B Voltage Checking: Connect the A.C. power cord plug to a wall outlet (A.C. 240V, 50Hz).

Set the mode switch to 'T.V.' position.

Adjust the unit to produce normal picture and sound, then be certain the +B voltage (at the emitter of Q701) is 11·5V. Only after that, proceed with the adjustment started below. More check that the +B voltage (at the emitter of QB04) for either of the radio and tape recorder circuits is 12·2V.

A.G.C. Adjustment: Set the R.F. A.G.C. control (R253) to 'MAX' position.

Receiving a test pattern, adjust the local oscillation to get normal picture.

Setting the input field intensity at 52±2dB, apply signal (50mV, 1kHz±50Hz, sine wave) generated by C.R. oscillator, to the tuner's test point TP202 through a resistor of 470 ohm and capacitor of 47μF connected in series.

Rotate the R.F. A.G.C. control (R253) counter-clockwise, until there are ripples on an oscilloscope (connected to the TP401).

Next, rotate the R.F. A.G.C. control (R253) gradually clockwise and stop where the amount of such ripples begins to reduce drastically.

Set the input field intensity to 80±2dB and check that there is no cross modulation, mis-sync. and too strong a contrast.

Video Trap (CF401) Checking: Let the test point (TP201) be grounded.

Set the Contrast and Brightness controls respectively to 'MAX' position.

Apply signal (6·0MHz, 100 dB, non-modulated) produced by a signal generator, to the TP401 (base of Q401) through a capacitor of 0·01μF.

Connect an oscilloscope, via a capacitor of 5 pF, to the cathode of C.R.T. and check that the amplitude of 6·0MHz signal is below 0·3Vp-p.

Note: The oscilloscope to be used must have an input impedance of 1M ohm, input capacitance; below 15 pF, band width; more than 10MHz. width; more than 10MHz.

Vertical and Horizontal Circuit Alignment

Rough Adjustment of Each Section: Set the A.C. line voltage to 240 volts.

Receive a test pattern in normal operating receiver condition.

Rotate the Brightness and Contrast controls to maximum clockwise.

Adjust the H-osc. coil (L601) to synchronize the picture horizontally.

Set the V. Hold control (R510) to synchronize the picture vertically.

Adjust the V. Linearity (R508) and V. Size (R505) controls to the best vertical linearity and picture size.

Both horizontal and vertical centring are accomplished by rotating the centring rings mounted on the rear of the deflection yoke assembly.

Vertical Circuit Adjustment: Adjust the V. Linearity control (R508) so that the picture is symmetrical in up-and-down direction having the best linearity.

Adjust the V. Size control (R505) to set the over-scanning to $9{\cdot}5\pm2\%$.

Rotate the V. Hold control (R510) clockwise and/or counter-clockwise to see that the picture is not deviated on either side of the screen.

Set the V. Hold control (R510) to its 'Centre' position.

Upon completion of the above adjustment, see there appears no crossover distortion (no white horizontal-bar at the centre of screen).

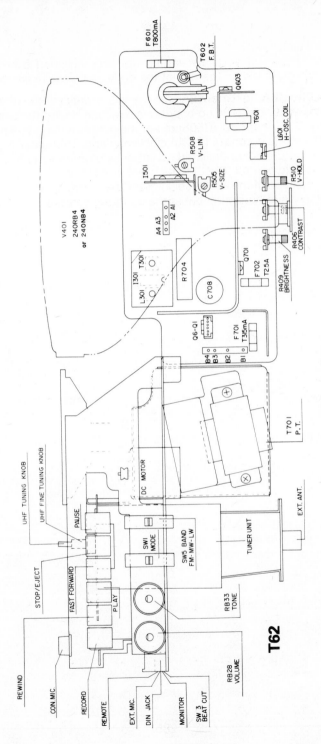

(T62) CHASSIS LAYOUT—MODEL 10P-28H

SHARP

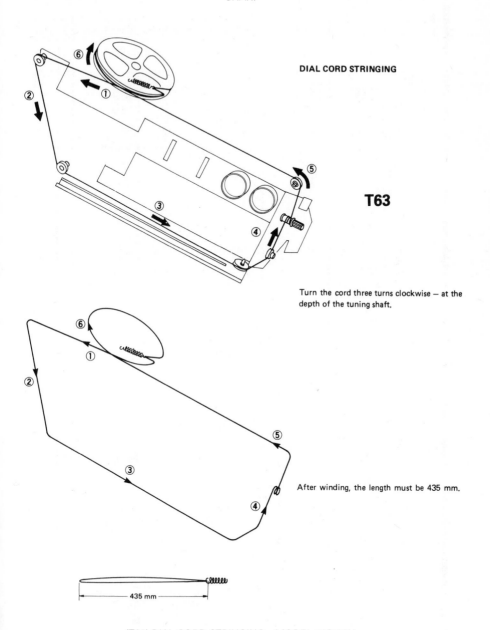

DIAL CORD STRINGING

T63

Turn the cord three turns clockwise — at the depth of the tuning shaft.

After winding, the length must be 435 mm.

435 mm

(T63) DIAL CORD STRINGING—MODEL 10P-28H

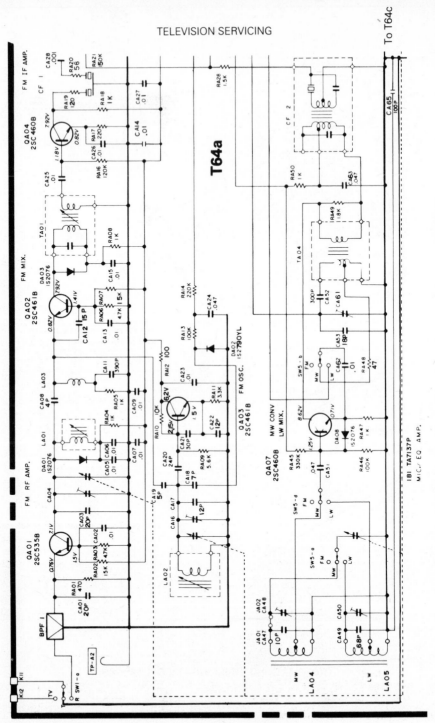

(T64a) CIRCUIT DIAGRAM (RADIO AND AUDIO PANEL)—MODEL 10P-28H (PART)

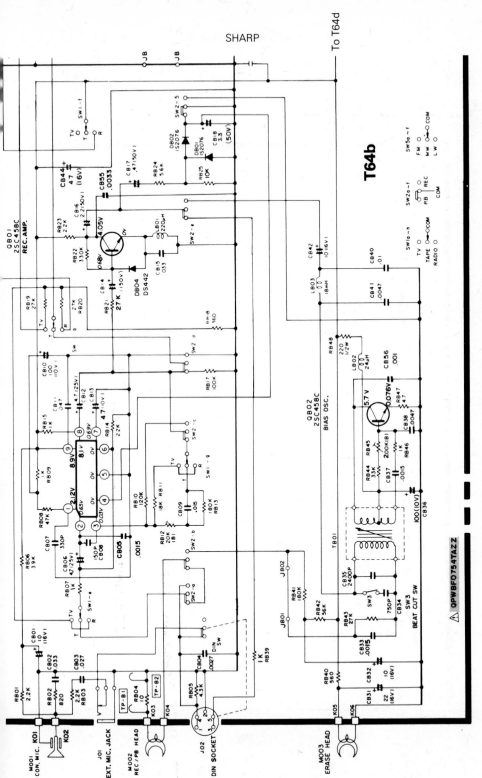

SHARP

T64b

To T64d

(T64b) CIRCUIT DIAGRAM (RADIO AND AUDIO PANEL)—MODEL 10P-28H (PART)

⚠ QPWBF0754TAZZ

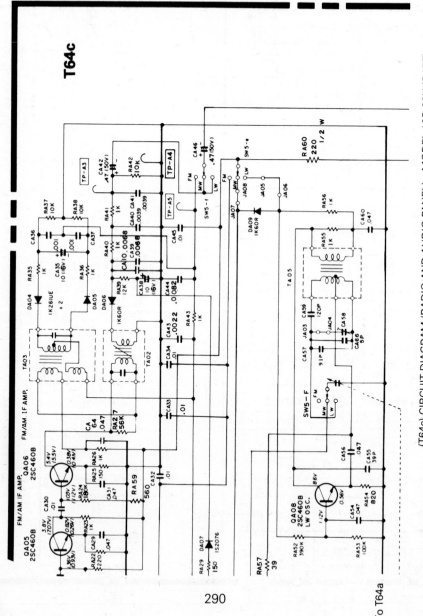

(T64c) CIRCUIT DIAGRAM (RADIO AND AUDIO PANEL)—MODEL 10P-28H (PART)

SHARP

(T64d) CIRCUIT DIAGRAM (RADIO AND AUDIO PANEL)—MODEL 10P-28H (CONTINUED)

291

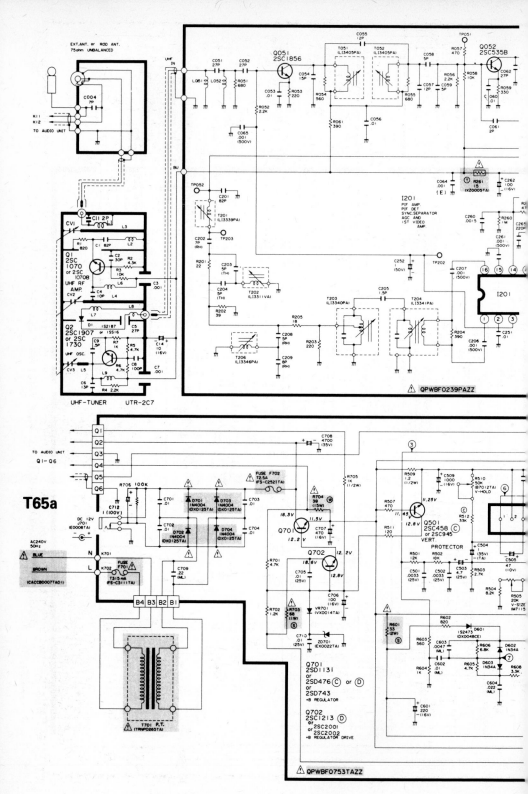

(T65a) CIRCUIT DIAGRAM (T.V. SECTION)—MODEL 10P-28H (*PART*)

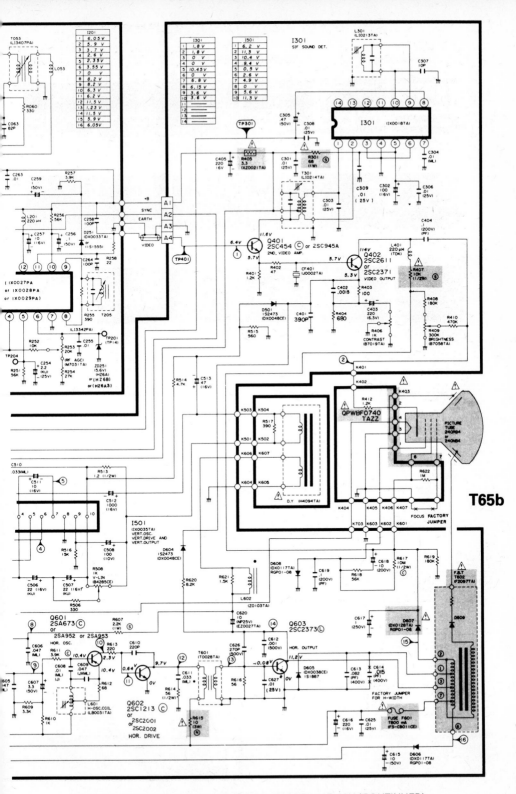

(T65b) CIRCUIT DIAGRAM (T.V. SECTION)—MODEL 10P-28H (*CONTINUED*)

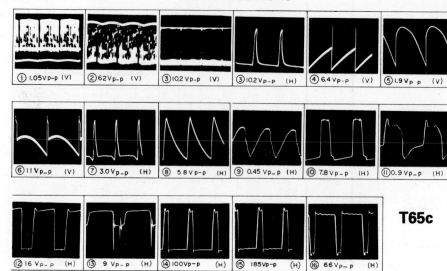

① 1.05Vp-p (V) ②62Vp-p (V) ③10.2 Vp-p (V) ③10.2Vp-p (H) ④6.4 Vp-p (V) ⑤1.9 Vp-p (V)

⑥11Vp-p (V) ⑦3.0Vp-p (H) ⑧5.8Vp-p (H) ⑨0.45 Vp-p (H) ⑩7.8Vp-p (H) ⑪0.9Vp-p (H)

⑫16 Vp-p (H) ⑬9 Vp-p (H) ⑭100Vp-p (H) ⑮185Vp-p (H) ⑯66Vp-p (H)

T65c

(T65c) CIRCUIT WAVEFORMS—MODEL 10P-28H

SHARP   Model 12P-41H

General Description: A portable 12-in. monochrome television receiver operating from mains or battery supplies. The chassis is isolated by mains transformer and sockets are provided for the connection of an external battery and earphones.

Mains Supply: 240 volts, 50Hz.

Battery: 12 volts, 16 watts.

Cathode Ray Tube: 210GNB4 or 310FZB4.

Loudspeaker: 32 ohms impedance.

Fuses: T315mA (mains); T2·5A (D.C.).

I.C. Voltages

IC201		IC301		IC1501	
Pin No.	*Volts*	*Pin No.*	*Volts*	*Pin No.*	*Volts*
1	6·05	1	1·83	1	5·7
2	5·9	2	1·83	2	11·45
3	3·7	3	0	3	10·3
4	2·6	4	0	4	9·4
5	2·35	5	10·6	5	0·53
6	3·55	6	0	6	2·45
7	0	7	6·37	7	4·83
8	8·2	8	5·73	8	0
9	8·2	9	3·57	9	5·07
10	6·3	10	3·58	10	11·45
11	6·2	11	—		
12	11·5	12	4·57		
13	1·25	13	—		
14	11·5	14	1·54		
15	5·9				
16	6·05				

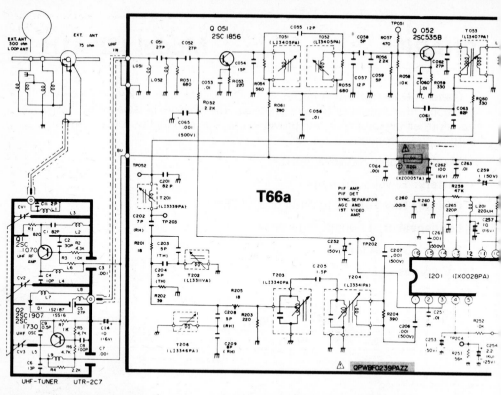

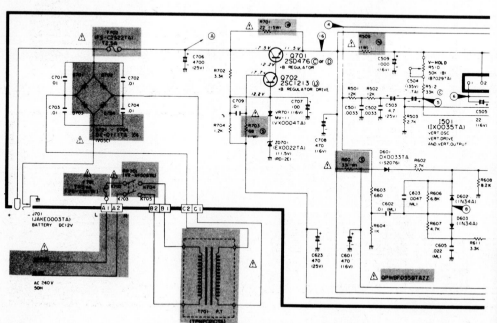

(T66a) CIRCUIT DIAGRAM—MODEL 12P-41H (*PART*)

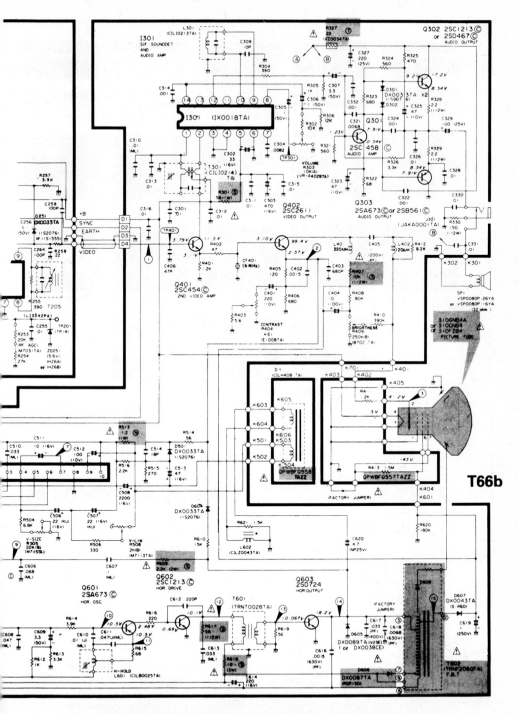

(T66b) CIRCUIT DIAGRAM—MODEL 12P-41H (*CONTINUED*)

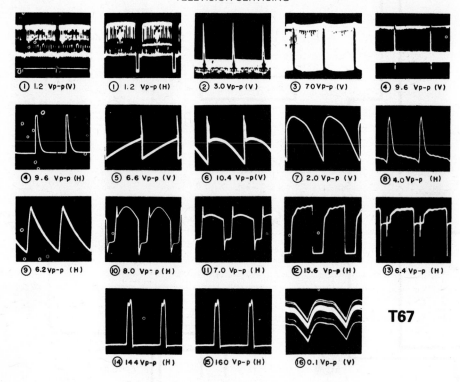

(T67) CIRCUIT WAVEFORMS—MODEL 12P-41H

SONY   Model KV-2040 UB

General Description: A 20-in. colour television receiver incorporating a Sony 'Trinitron' cathode ray tube. An eight-channel push-button selector is fitted and care should be taken when service is undertaken as the chassis is 'Live' irrespective of mains connection.

Mains Supply: 240 volts, 50Hz. **E.H.T.:** 23·7KV.

Loudspeaker: 8 ohms impedance.

Access for Service

The main chassis may be withdrawn after pulling out the claws engaged in each side of the bottom side rails.

Servicing Note: No meter connection should be attempted at the base of transistor Q651 as this will prevent H.T. supplies from operating.

Adjustments

H.T.: Set RV601 for a reading of 107 D.C. at the positive terminal of C606 with respect to chassis.

Line Hold: Connect a 1mfd/25v. capacitor across pins 2 and 3 of Socket D5 (Pin 2 earth). Adjust T501 for stability.

Purity (See Fig. T68b): (Preparation—degauss the screen and apply a white screen signal).
Loosen deflection yoke screw.
Adjust Purity control tabs to t.d.c.
Slide deflection yoke as far forward as it will go.
Disconnect leads G and B on the C board.
Adjust Purity control to centre a vertical red band on the screen.
Slide deflection yoke back for a uniform red screen.
Check green and blue rasters: to get a green screen, connect lead G on the C board and disconnect leads R and B; to get a blue screen, connect lead B on the C board and disconnect leads R and G).
After these checks, connect the leads R, G and B.
Tighten the deflection yoke screw.
Check if mislanding appears at corners. If mislanding is observed, correct it with rotatable disc magnets.

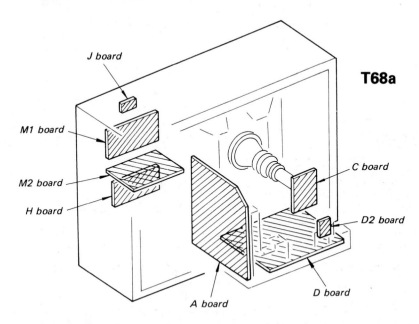

(T68a) CIRCUIT BOARD LOCATION—MODEL KV-2040UB

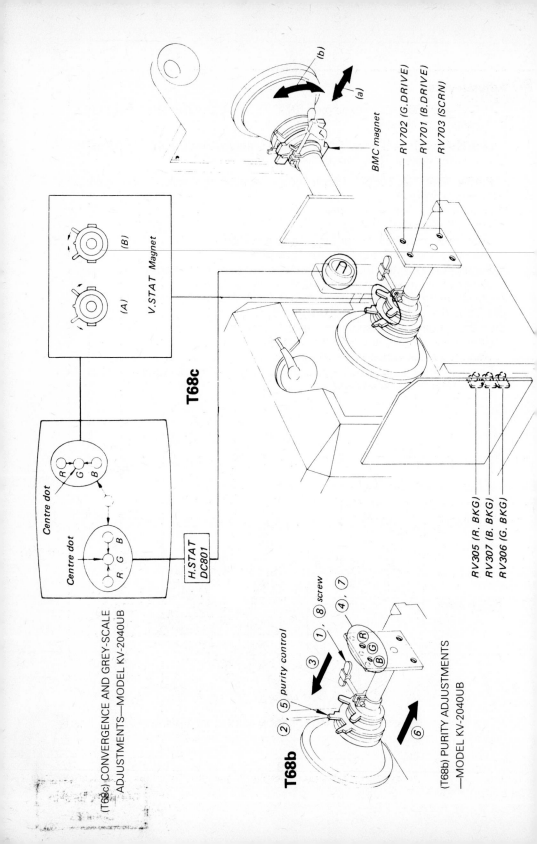

(T68c) CONVERGENCE AND GREY-SCALE
ADJUSTMENTS—MODEL KV-2040UB

Centre dot

Centre dot

R G B

R G B

H.STAT
DC801

T68c

V.STAT Magnet

(A)   (B)

BMC magnet

RV702 (G.DRIVE)
RV701 (B.DRIVE)
RV703 (SCRN)

(a)

(b)

RV305 (R. BKG)
RV307 (B. BKG)
RV306 (G. BKG)

purity control

① , ⑧ screw

③   ④   ⑦

② , ⑤

R
G
B

⑥

T68b

(T68b) PURITY ADJUSTMENTS
—MODEL KV-2040UB

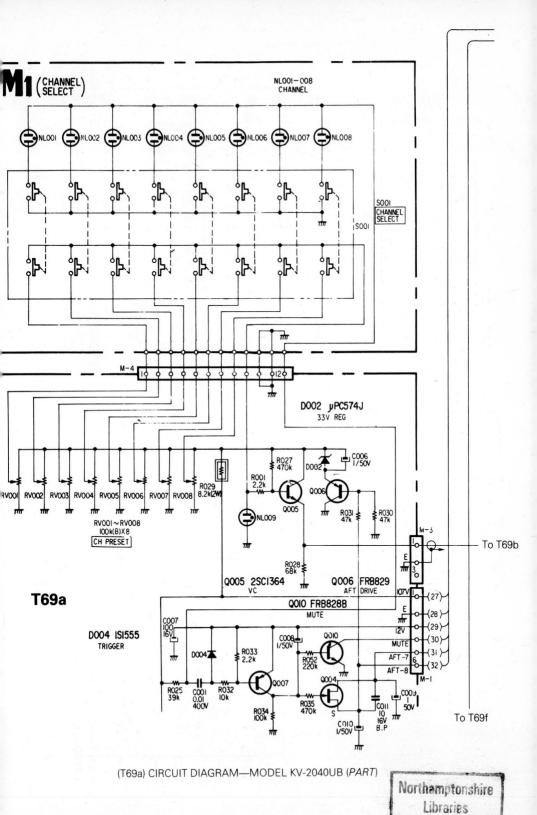

(T69a) CIRCUIT DIAGRAM—MODEL KV-2040UB (*PART*)

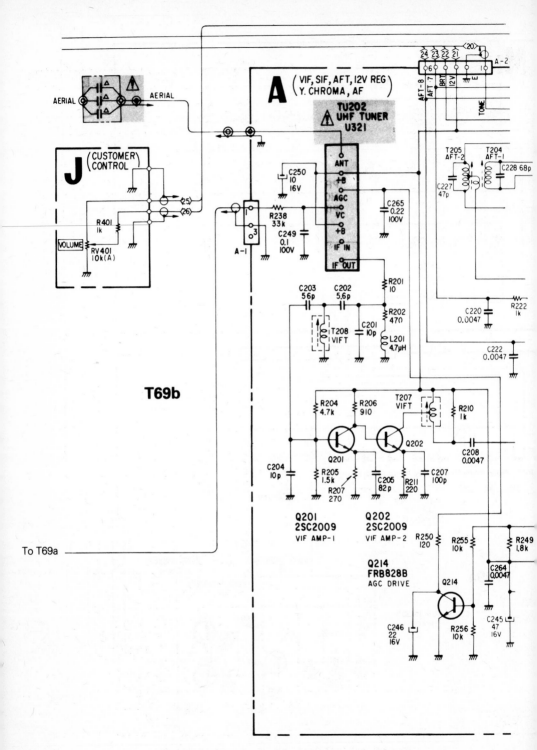

T69b

To T69a

(T69b) CIRCUIT DIAGRAM—MODEL KV-2040UB (*PART*)

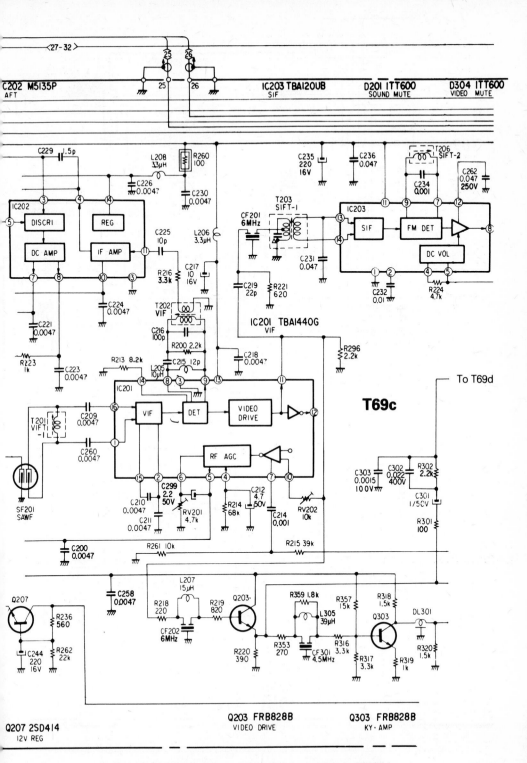

(T69c) CIRCUIT DIAGRAM—MODEL KV-2040UB (*PART*)

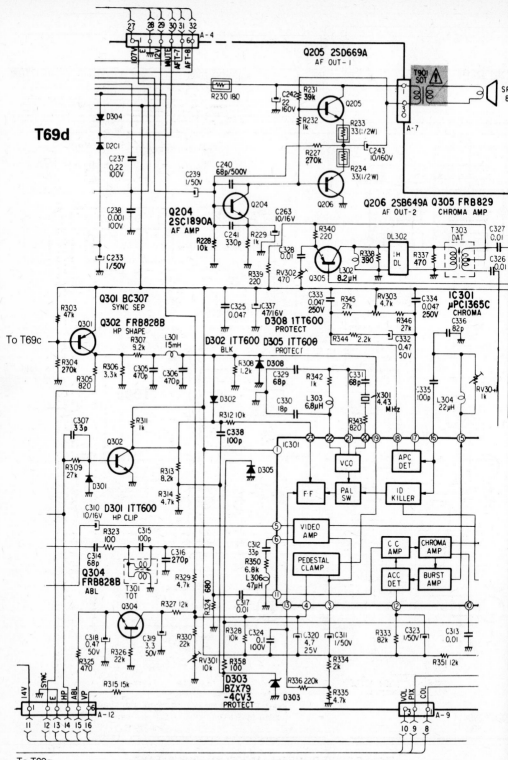

(T69d) CIRCUIT DIAGRAM—MODEL KV-2040UB (*PART*)

To T69g

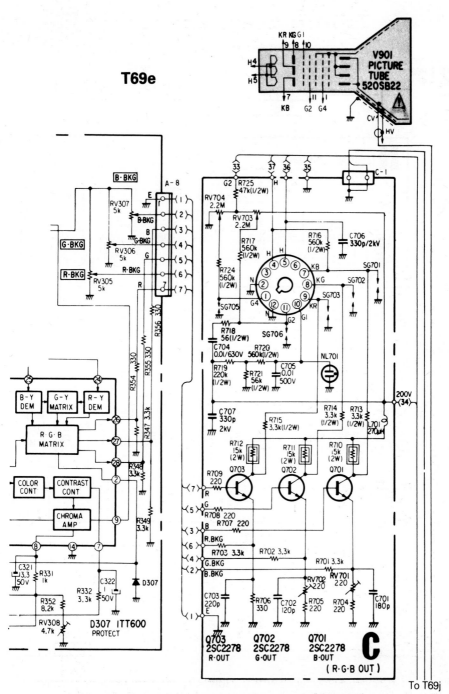

T69e

(T69e) CIRCUIT DIAGRAM—MODEL KV-2040UB (*PART*)

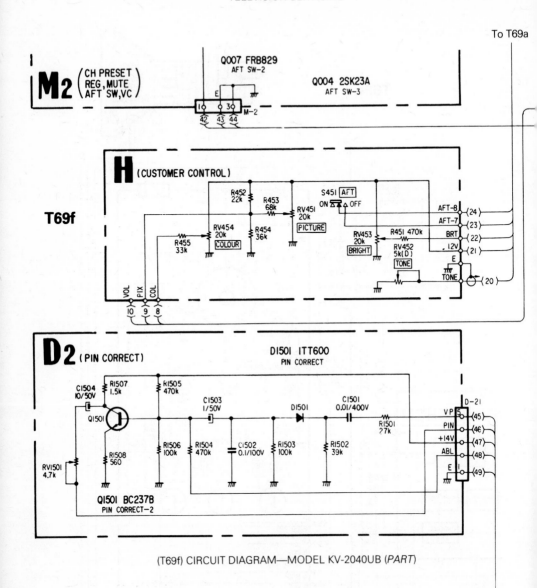

(T69f) CIRCUIT DIAGRAM—MODEL KV-2040UB (*PART*)

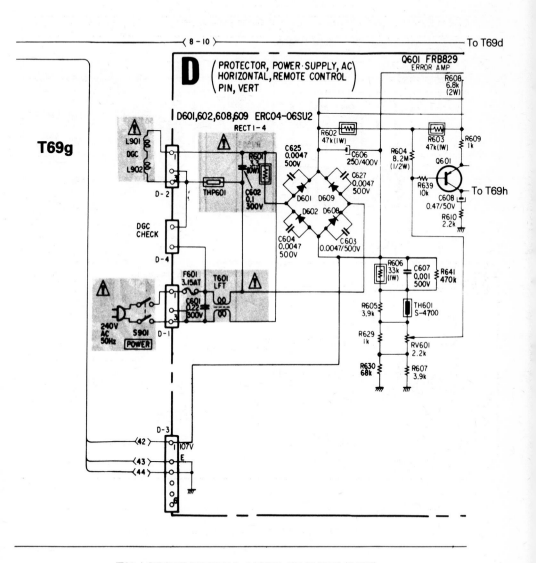

(T69g) CIRCUIT DIAGRAM—MODEL KV-2040UB (*PART*)

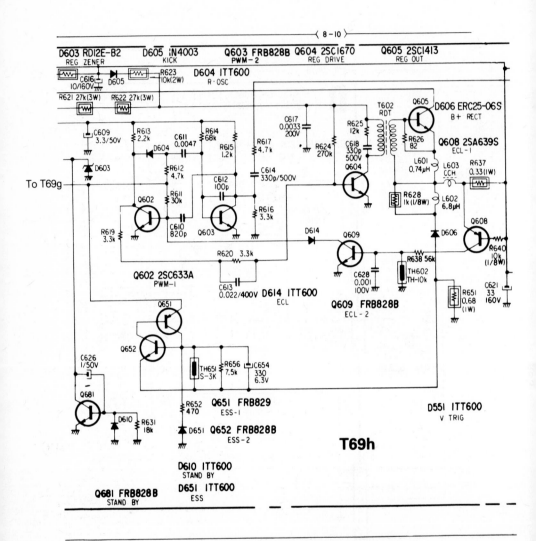

(T69h) CIRCUIT DIAGRAM—MODEL KV-2040UB (*PART*)

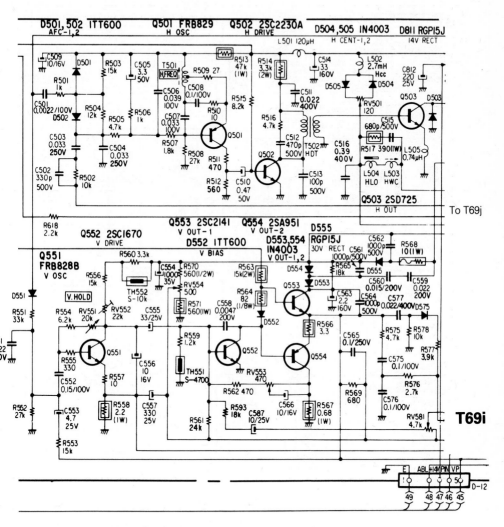

(T69i) CIRCUIT DIAGRAM—MODEL KV-2040UB (*PART*)

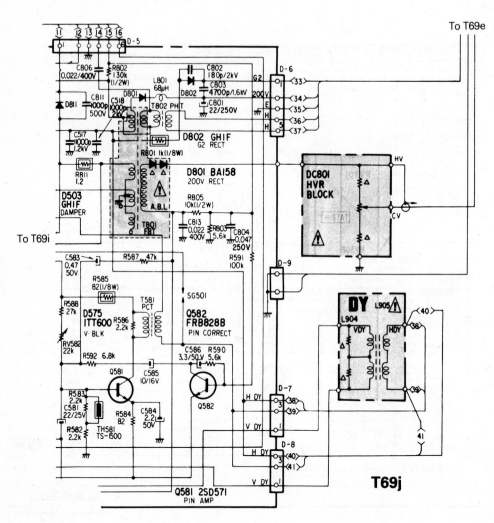

(T69j) CIRCUIT DIAGRAM—MODEL KV-2040UB (*CONTINUED*)

SONY

Model KV-2201
(Ye Chassis)

General Description: A mains-operated colour television receiver incorporating a 22-in. 'Trinitron' cathode ray tube and featuring an eight-channel manual push-button tuning system. Care should be exercised when servicing this chassis which is 'Live' irrespective of mains supply connections.

Mains Supplies: 200–240 volts, 50Hz.

Adjustments

H.T.: Set RV601 on F board to obtain a reading of 135 volts D.C. at the positive terminal of C606, for a mains input of 240 volts.

Line Hold: Connect a 10mfd/50V capacitor across Pins 5 and 6 of socket E-1 on the E board. Adjust T801 for stability.

Pincushion Adjustments (D1, D2 boards): Set P.I.C. and Brightness controls to centre. Adjust RV504 and RV511 to produce vertical side lines on a crosshatch picture. Turn the P.I.C. and Bright. fully anti-clockwise and adjust RV1091 to re-establish side verticals. Repeat the procedure several times. Adjust coil L508(PAC) to obtain horizontal crosshatch lines.

Purity and Grey Scale: These adjustments are similar to the methods adopted for Model KV-2040 described previously in this volume.

Dynamic Convergence (Fig. 70b): (Misconvergence at both sides of screen.)
Set RV509 and RV510 to mechanical centre.
Adjust H. STAT control so that green and blue dots coincide at centre of screen.
Adjust RV510 so that X1 is equal to X3.
Adjust RV509 so that X2 is equal to X3.
Repeat the procedure two or three times.

Top and Bottom Misconvergence: See Fig. T70c.

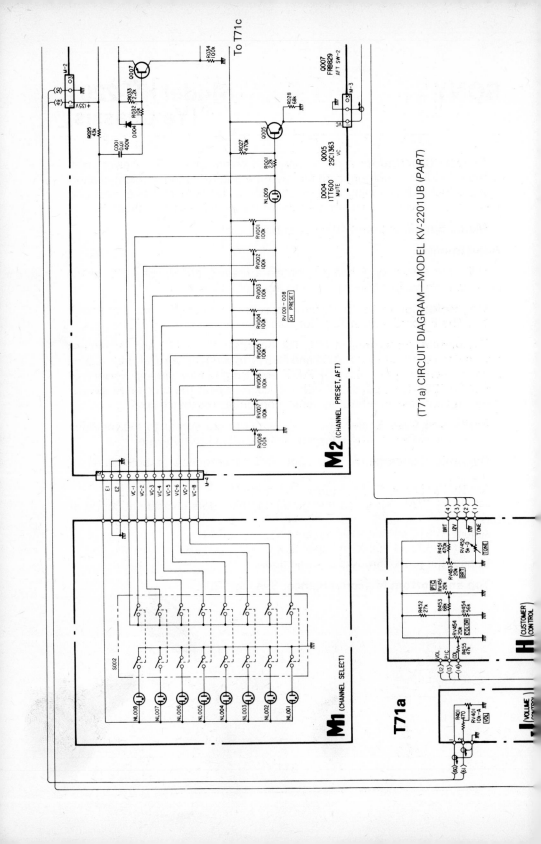

(T71a) CIRCUIT DIAGRAM—MODEL KV-2201UB (PART)

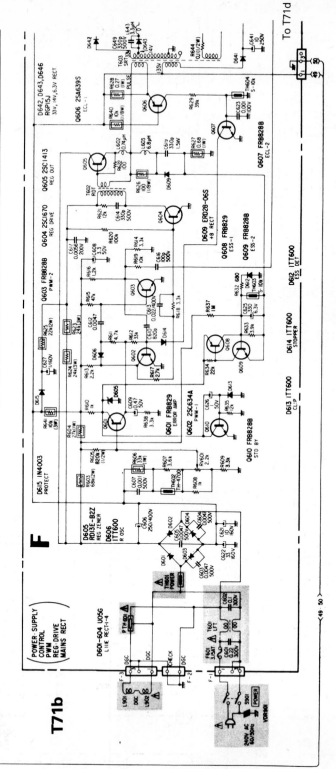

(T71b) CIRCUIT DIAGRAM—MODEL KV-2201UB (*PART*)

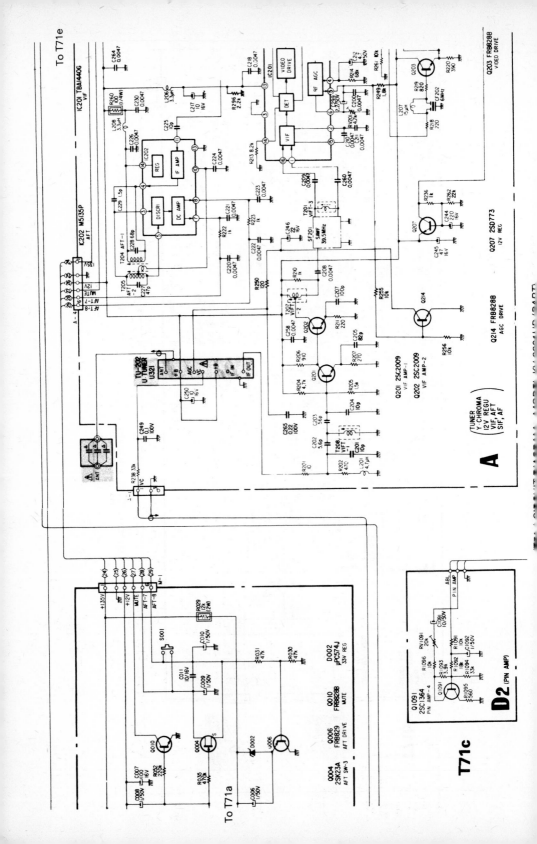

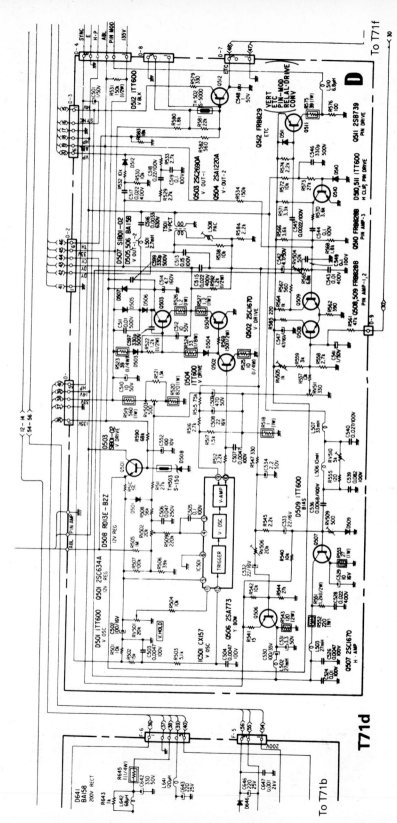

(T71d) CIRCUIT DIAGRAM—MODEL KV-2201UB (PART)

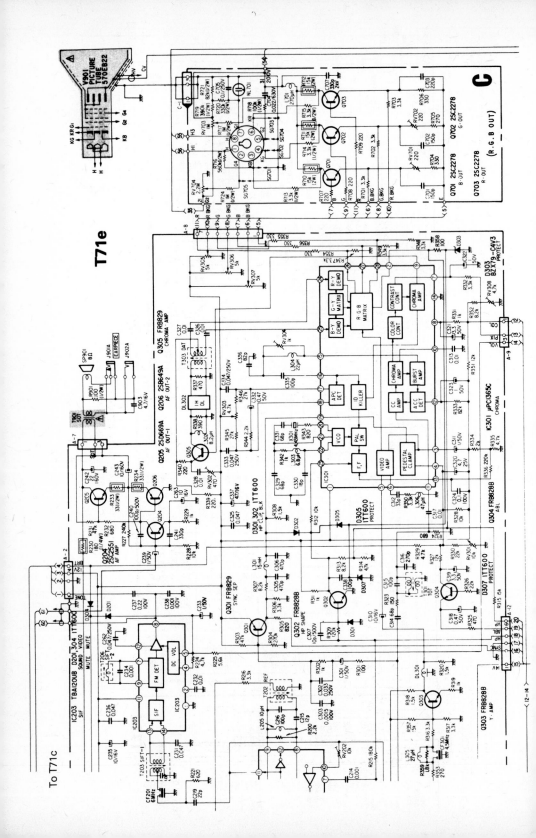

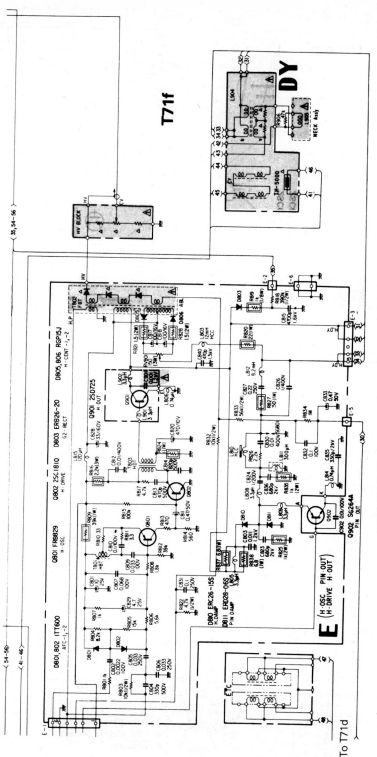

(T71f) CIRCUIT DIAGRAM—MODEL KV-2201UB *(CONTINUED)*

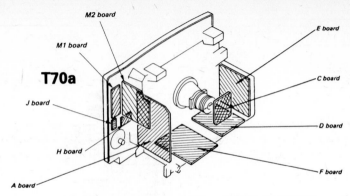

(T70a) CIRCUIT BOARDS LOCATION—MODEL KV-2201UB

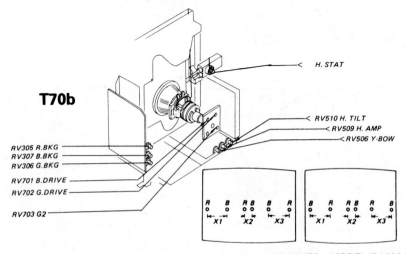

(T70b) GREY SCALE AND DYNAMIC CONVERGENCE ADJUSTMENTS—MODEL KV-2201UB

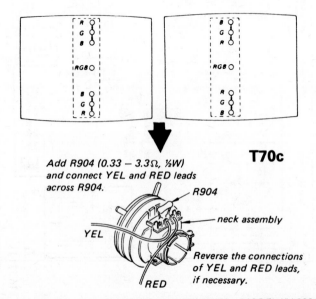

(T70c) DYNAMIC CONVERGENCE ADJUSTMENTS—MODEL KV-2201UB

SONY   Model TV-124 UB

General Description: A portable 12-in. monochrome television receiver operating from mains or battery supplies. The chassis is isolated by mains transformer and a socket is provided for the connection of an earphone.

Mains Supply: 240 volts, 50Hz.

Battery: 12 volts, 15 watts.

Cathode Ray Tube: 310GNB4A.

E.H.T.: 11KV.

Loudspeaker: 8 ohms impedance.

Access for Service

Remove 2 screws holding the aerial socket and 2 from the rear base of the cabinet. The main board can be drawn out and turned vertically after placing the lead from the EJ board on the outside of the mains transformer.

Adjustments

H.T.: Tune in a signal and adjust RV601 for 11·4±0·2 volts at the emitter of Q601 with respect to chassis.

Line Hold: T501 on chassis.

Line Width: Select a capacitor value between 15nF and o/c for C803 to obtain the correct width.

Height: RV552 on chassis.

T82

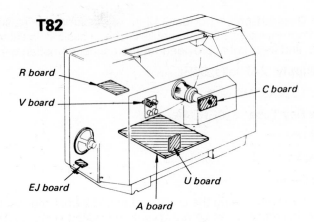

(T82) PANEL LOCATIONS—MODEL TV-124UB

T83

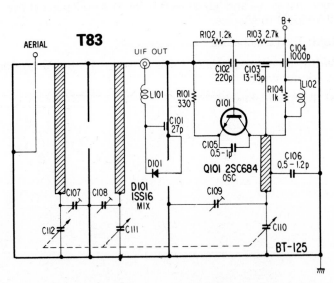

(T83) CIRCUIT DIAGRAM (U.H.F. TUNER)—MODEL TV-124UB

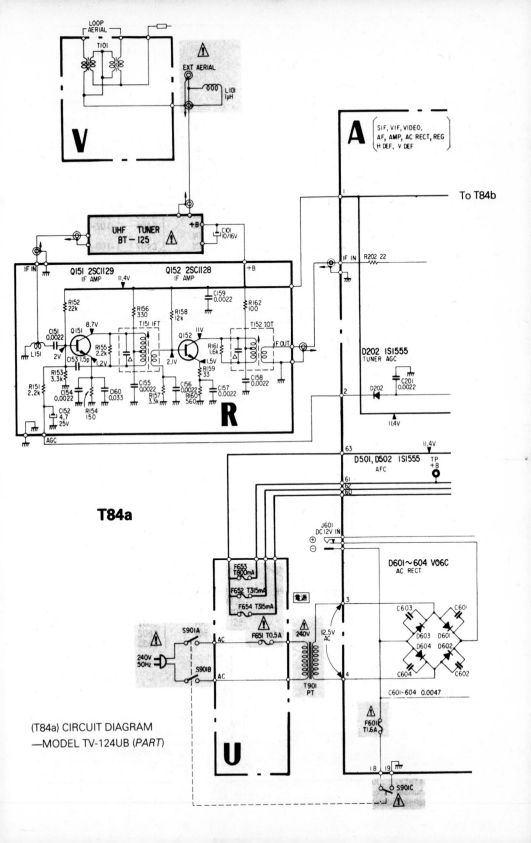

(T84a) CIRCUIT DIAGRAM
—MODEL TV-124UB (*PART*)

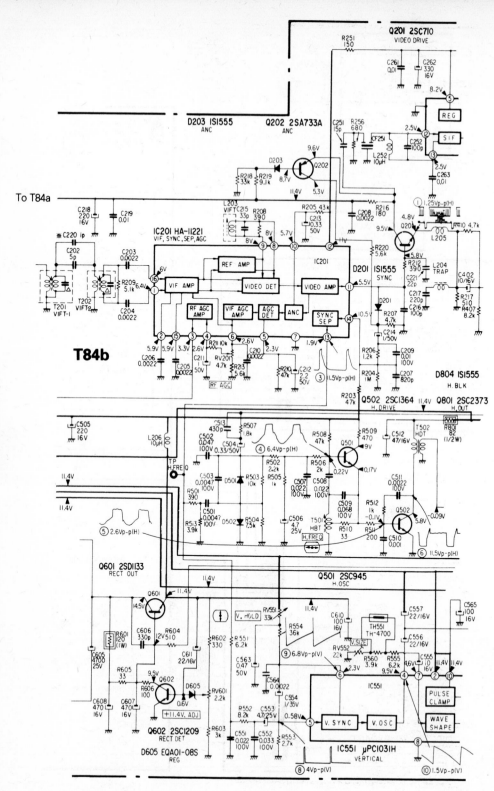

(T84b) CIRCUIT DIAGRAM—MODEL TV-124UB (*PART*)

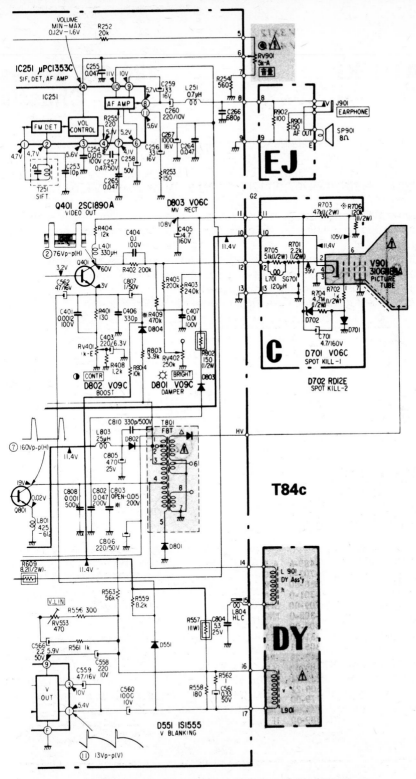

(T84c) CIRCUIT DIAGRAM—MODEL TV-124UB (*CONTINUED*)

TATUNG

120-125 Series
Chassis

The information here relates to the basic chassis used in the following Decca and Tatung models: CN1271, CN9276, DN1251, DN1256, DN1259, DN9256, DN9259, DP1257, DP9257, DT1253, DT9253, DV1254, DV9254, TN1201, TN9201, TP1207, TP9207, TT1202, TT9202, TV1203 and TV9203.

General Description: A mains-operated colour television receiver chassis incorporating an in-line gun 90° cathode ray tube and switched-mode power supply.

The main circuit panel is designed to be adaptable to a variety of television systems, frequency bands, and picture tube sizes. The specification of the chassis types is given in the table below.

Component differences are indicated on the circuit diagram.

Chassis Type	Television System	Frequency Bands	Picture Tube Sizes
120	PAL I	U.H.F. only	14″–20″
121	PAL I	U.H.F. only	22″
122	PAL B/G	V.H.F.+U.H.F.	14″–20″
123	PAL B/G	V.H.F.+U.H.F.	22″
124	PAL I	V.H.F.+U.H.F.	22″
125	PAL I	V.H.F.+U.H.F.	14″–20″
126	PAL/SECAM B/G	V.H.F.+U.H.F.	14″–20″

Mains Supplies: 200–240 volts, 50Hz.

E.H.T.: 24Kv.

Safety and Isolation

The chassis is always live regardless of the mains supply polarity. Hence, whenever servicing operations are being carried out, the receiver should be supplied through a mains isolation transformer of at least 200 watts rating.

The power supply section remains charged with respect to chassis for 30–60 seconds after switching off. Care should be taken when handling the chassis to avoid touching this area during this time.

Components marked with the Safety Symbol on the circuit diagram are safety approved types and should be replaced only with components supplied or approved by Tatung service department. It is also recommended that components not marked with the safety symbol should be replaced by parts of the type originally fitted, and this applies particularly to those resistors which are stood-off the printed circuit boards.

Waveforms

For the waveform diagrams the oscilloscope was fitted with a passive ÷ 10 probe and the receiver displaying colour bars. Waveforms 801–804 were taken with the oscilloscope set for 'single shot' operation to achieve clarity.

Access for Service

Cabinet Back: To remove the cabinet back, unplug the aerial, rotate the two plastic fasteners a quarter-turn anti-clockwise, and withdraw the back from the cabinet.

Chassis access position: Loosen the fixing screw from each of the two chassis support brackets. Note the warning to delay handling the chassis for approximately 60 seconds after switch-off. The main chassis may then be lifted slightly and partially withdrawn from the cabinet to provide access to preset adjustments.

The locations of all the pre-set adjustments on the main chassis and the picture tube base panel are shown in Figs. T113 and T114.

When access to the copper track side is required, the main chassis may be removed from its support brackets and mounted in a vertical position to the left side of the tube base. A U-shaped slot in the line scan transistor heat sink plate engages with the rear end of the left-hand chassis support bracket. A latching peg on the bracket fixed inside the cabinet top engages with a hole near the front edge of the circuit panel.

Note: On some models it may be necessary to unplug the degaussing coil in order to mount the chassis in its service position.

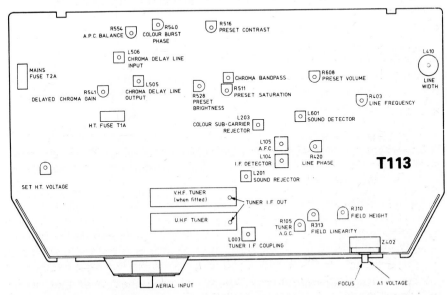

(T113) LOCATION OF CONTROLS (MAIN CHASSIS)—120 SERIES CHASSIS

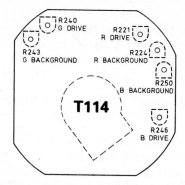

(T114) LOCATION OF CONTROLS (C.R.T. BASE)—120 SERIES CHASSIS

Replacement Procedures

Main Chassis Assembly: Discharge the final anode of the picture tube by connecting to the braid which contacts the outer graphite coating of the tube. Disconnect the E.H.T. lead from the anode button and connect momentarily to the chassis rail near the E.H.T. tripler to ensure discharge of the tripler unit.

Disconnect the earth lead tag from the picture tube earth braid and remove the tube base panel.

Unplug all leads connecting to the scan coils and the control assembly.

Loosen the two screws in the ends of the chassis support brackets. The chassis may then be lifted out and removed.

Removal of Picture Tube: Remove the main chassis assembly as described.

With the picture tube face down, position the cabinet on two padded blocks. Remove the four nuts and large washers and withdraw the tube assembly from the cabinet.

Transfer the degaussing coil and earth braids, etc., to the new picture tube and scan coil assembly.

Important: Do not disturb the tube neck components. These have been set for optimum performance during tube manufacture, and are an integral part of the tube system.

Adjustments after Replacement: Check that the earthing braid has been fitted correctly; replace the main chassis and all connections. Check that the earth lead from the tube base panel is connected to the tube earth braid.

Carry out adjustments of picture width and height, line phase, and focus as necessary. Set up A1 voltage and grey scale adjustments.

High Voltage Components: Before replacing the tripler unit, focus module or line output transformer, first remove the main chassis assembly as described above.

326

When fitting the new component, care should be taken to avoid sharp points on soldered high-voltage terminations and to maintain adequately spaced lead wire dressing.

Adjustments

Initial conditions: Turn down the A1 voltage control on Z402 to minimum, i.e., no beam current, or a low current.

Voltage Adjustment. Adjust the 'set 117V'. pre-set R813 to give 117V at TP43

Video Output

D.C. Adjustment: Set user Brightness control at centre of range.
Disable blanking by a short link from TP59 to chassis (across D501). Keep I.F. amplifier disabled with a shorting link from TP11 to TP12.
With voltmeter between pin 28 of IC501 and chassis, adjust pre-set Brightness control R528 to give 2·0V.
Check voltages on pins 26 and 27; to be also approximately 2·0V.
Set all three video drive pre-sets R221, R240, R246 to maximum.
With voltmeter connected between 'Red' video output at R906 and chassis, adjust R background pre-set R224 for 150V, then repeat for 'Green' and 'Blue' video outputs at R907 and R908, adjusting R243 and R250 respectively to give 150V.
Disconnect the voltmeter and remove the shorting links from the I.F. amplifier and blanking.

Pre-set Contrast and Subcarrier Rejector Adjustment: Set user contrast control to maximum.
Connect the oscilloscope probe to the emitter of Q201, across R206.
Tune the receiver to the colour bar signal, ensuring that the generator frequency is not close to any local off-air signal. The oscilloscope display should show waveform 201.
Transfer the oscilloscope probe to pin 28 of I501, using a local chassis track for the probe earth, to show waveform 505.
Turn user Colour Saturation control to minimum, so that the luminance signal only is seen on the oscilloscope.
Adjust pre-set contrast R516 to give 3·0V of video signal black to white (i.e., excluding the blanking pulse). If 3·0V cannot be achieved, set the control at a point where the amplitude just starts to decrease.
Adjust the subcarrier rejector coil L203 for minimum subcarrier content on the video waveform.

Video Output Gains Adjustment: Connect oscilloscope probe to 'Red' video output at R906, to display waveform 202 when receiver is tuned in to signal from colour bar generator.
Adjust R drive pre-set R221 to give 100V output black to white level. The blanking pulse amplitude should be approximately 45V.

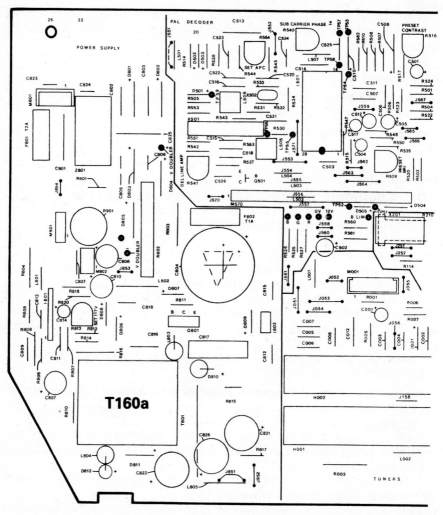

(T160a) COMPONENT LAYOUT—120 SERIES CHASSIS (*PART*)

Repeat above steps for the 'Green' and 'Blue' video outputs (waveforms 203 and 204), adjusting R240 and R246 respectively to give 100V black to white.

Disconnect oscilloscope probe.

Colour Saturation Adjustment: Set user colour control to give 7·5V on M501/6, i.e., approximately 1/3 from top of range.

Adjust pre-set saturation control R511 to give 100% saturation on the 'blue' waveform 506.

328

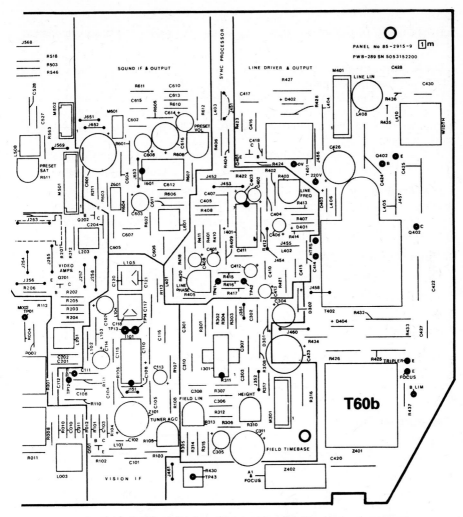

(T160b) COMPONENT LAYOUT—120 SERIES CHASSIS (*CONTINUED*)

Transfer the oscilloscope probe to observe waveforms 504 and 505 to check the 'red' and 'green' colour saturation is approximately 100%.

Grey Scale

Background Adjustment: Tune to colour bar test signal, set user Brightness and Contrast controls to mid-position and Colour control to minimum.

Turn up the A1 voltage control until the Black bar becomes just visible on the picture tube on one (or more) colours.

Adjust one or two only of the background pre-sets R224, R243, R250 corresponding to the colour(s) not illuminated to make them equally visible on the black bar.

Drive Adjustment: Adjust the user Contrast control as necessary to give a normally contrasted picture.

If the white bars show any colouration adjust one or two only of the drive pre-sets R221, R240, R246 as appropriate by turning *anti-clockwise* (viewed from the rear of the panel) to *reduce* the drive on the colours which predominate.

If after a number of adjustments the results are unsatisfactory, reset the initial conditions as in paragraphs relating to Video Output Adjustments.

Line Frequency: If it is found necessary to adjust the 31·25kHz oscillator, connect a shorting link across test points TP41 and TP42. Then adjust the line frequency pre-set potentiometer R403 until the picture 'drifts through' horizontally. Then remove the short-circuit link.

Voltage Measurements

Measured with a multimeter (20kOhm/V) on 10V D.C. range and with a normal picture displayed unless otherwise stated.

S.M.P.S.

Integrated Circuit: IC801 (TDA4600).

Heat sink tab to chassis: −315V (1000V range).

Important: *All other voltages on IC801 and Q801 are measured with respect to the heat sink tab of IC801 as a reference level for the power supply.*

Integrated Circuit: IC801.

Pin	V	Pin	V
1	4·4	6	0
2	0·1	7	1·8
3	2·2	8	1·8
4	2·2	9	16·5 (30V range)
5	6·3		

Transistor Q801 (BU426A):

Emitter V	Base V	Collector V
0	−0·2	300 (1000V range)
	(0·5V A.C.)	

Tuner and Vision I.F.

Voltage Regulator IC001 (TAA550): 32V (100V range).

Transistor Q101:

Emitter V	Base V	Collector V
1·9	2·7	11·6 (30V range)

Integrated Circuit IC101 (TDA2540):

Pin	V	Pin	V
1	5·0	9	8·0
2	5·0	10	3·2
3	0·7 (no signal 1·0V)	11	12·0 (30V range)
*4	0·1 to 10 (no signal 0·1V)	12	4·6
5	1·0 to 11 (on tune 6V)	13	0
6	11·6 (30V range)	14	7·3
7	3·2	15	5·0
8	8·0	16	5·0

* On U.H.F./V.H.F. receivers, when TDA2541 is used, pin 4 voltage becomes 2–10V (9·8V with no signal).

Video Pre-amplifier and Decoder

Transistors:

Ref.	Emitter V	Base V	Collector V
Q201	4·6	5·3	11·8 (30V range)
Q202	2·5	3·2	8·0
Q501	10·0	9·5	0

Integrated Circuit IC501 (UPC1365C):

Pin	V	Pin	V
1	12·0 (30V range)	15	8·0
2	0·7 (3V range)	16	4·5
3	8·6	17	8·2
4	8·4	18	8·4
5	2·5	19	0·2 (3V range)
6	1·8	20	9·6
7	6·3	21	3·2
8	5·3	22	3·2
9	9·4	23	0·1 (3V range)
10	2·0	24	2·8
11	1·3	25	2·8
12	7·0	26	2·2
13	6·0	27	2·2
14	0	28	2·2

Sync. and Line Panel

Integrated Circuit IC401 (TDA2576A):

Pin	V	Pin	V
1	1·35	9	0
2	0·1 (3V range)	10	1·6
3	4·6	11	4·1
4	4·5	12	2·7
5	6·2	13	12·0 (30V range)
6	4·4	14	0·8
7	2·4	15	5·2 (loss of line hold)
8	1·1	16	12·0 (30V range)

Line Scan and E.H.T.

Transistor	Emitter V	Base V	Collector V
Q401	0	−0·25	105 (300V range)
Q402	0	−1·0 (2·1V A.C.)	Do not measure

331

E.H.T. Measurement

Important Note: When measuring the E.H.T. voltage, the earth return from the meter must be connected to the C.R.T. Dag or braid earthing system and NOT to the chassis rail or heat sink.

Field Scan
Integrated Circuit IC301 (TDA1170):

Pin	V	Pin	V
1	4·8	7	No connection
2	23 (30V range)	8	0·05 (3V range)
3	2·3	9	Do not measure
4	15·2 (30V range)	10	Do not measure
5	23·5 (30V range)	11	0·65
6	6·7	12	3·0

Tube Base Panel
Picture Tube Socket:

Pin	Function	V D.C.	Meter Range
5	Grids	10	30V
6	Cathode, Green	136	300V
7	A1	600	1000V
8	Cathode, Red	136	300V
11	Cathode, Blue	136	300V

Heater voltage measured across pins 9 and 10 is typically 6·1V r.m.s. on a meter which reads true R.M.S. voltage up to 100kHz. On a multimeter (10V A.C. range), the reading is approximately 3·8V.

Video Amplifiers:
Measured on 10V D.C. range except where otherwise stated.

Transistors	Emitter V	Base V	Collector V
Q203, Q208, Q211	7·6	8·2	136 (300V range)
Q204, Q210, Q212	1·4	2·1	7·6
Q205	1·4	0·7	0·2
Q206	0	0·6	0·7
Q207	1·4	2·1	7·5

Sound I.F.
Integrated Circuit IC601 (TDA3190):

Pin	V	Pin	V
1	3·1	9	9·4
2	3·1	10	0·7
3	3·1	11	9·3
6	2·5	14	17·5 (30V range)
7	2·5	15	9·2
8	3·2	16	9·0

Switched Mode Power Supply (S.M.P.S.)

Circuit Description: The power supply is a self-oscillating switching type of converter, which, under normal conditions, operates at a free-running frequency of between 20 and 30kHz. All control functions are performed within IC801 (TDA4600).

The output transistor, Q801, switches the primary winding of the transformer T801 across the rectified mains voltage developed on the reservoir capacitor C804. The output voltages (152V and 18V) are generated at taps on the primary winding and rectified and smoothed by D812, D811, C822, C826 and C821.

The base driver for Q801 is generated on IC801, the current at pin 8 controlling the switch-on period and that at pin 7 controlling the turn-off conditions of Q801.

R808, R810 and C813 develop a sawtooth voltage at pin 4 which simulates the collector current of Q801. This is used to generate a sawtooth base drive at pin 8 to avoid over-saturation of Q801.

D808 and C814 rectify and smooth a voltage from the feedback winding on T801 which is proportional to the output voltages. This voltage is attenuated by R813, R812, R807 and R806 with respect to a reference voltage at pin 1 and applied to pin 3. Any changes in the output are transmitted via pin 3 to the control logic and the base current amplifier within the I.C. The frequency and duty cycle of the output pulses are therefore adjusted to correct for the changes.

R813 (set 148V) adjusts the proportion of voltage fed back and hence adjusts the output voltage. To complete the oscillator feedback loop, R814 feeds an attenuated voltage from the feedback winding on T801 to pin 2. This enables the I.C. to identify the points at which the output pulse crosses the zero voltage level and provide correctly timed base drive pulses to Q801.

The supply for the I.C. is developed by D806 and C808 from a third winding on T801. D805 and R802 provide a start-up supply for the I.C.

If an overload occurs on either of the outputs, the frequency and the duty cycle are reduced, thus limiting the power supplied to the load. If the overload becomes a short-circuit, the frequency reduces further to 1·4kHz and the power output is limited to a low level.

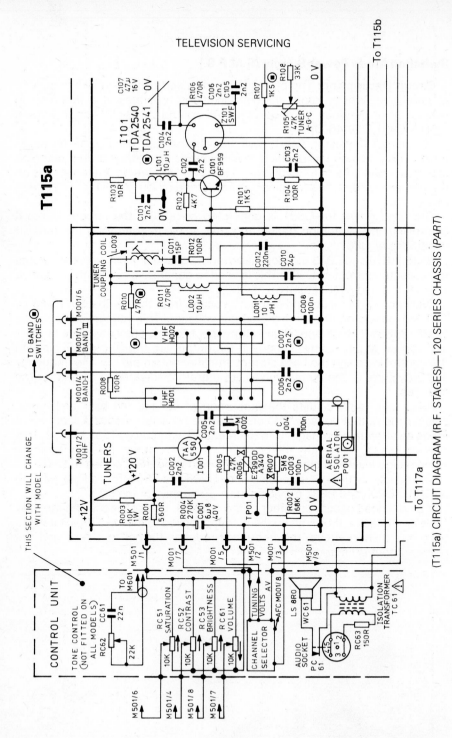

(T115a) CIRCUIT DIAGRAM (R.F. STAGES)—120 SERIES CHASSIS (*PART*)

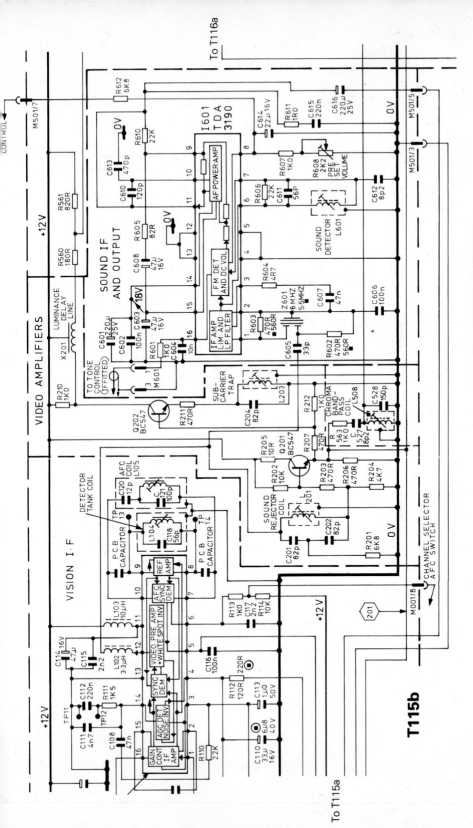

(T115b) CIRCUIT DIAGRAM (R.F. STAGES)—120 SERIES CHASSIS (CONTINUED)

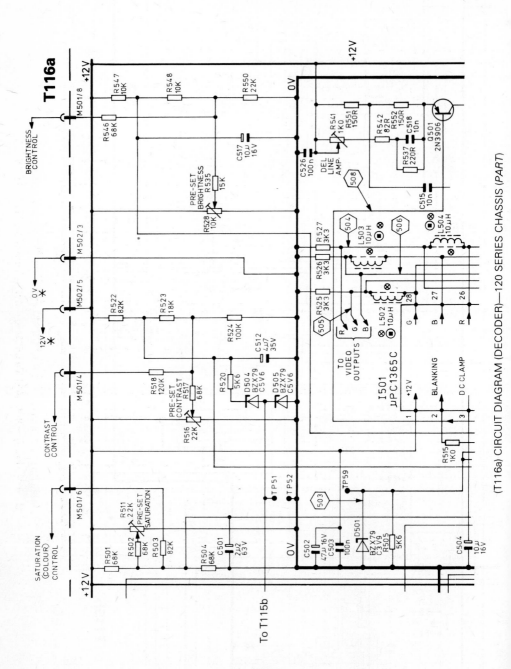

(T116a) CIRCUIT DIAGRAM (DECODER)—120 SERIES CHASSIS (PART)

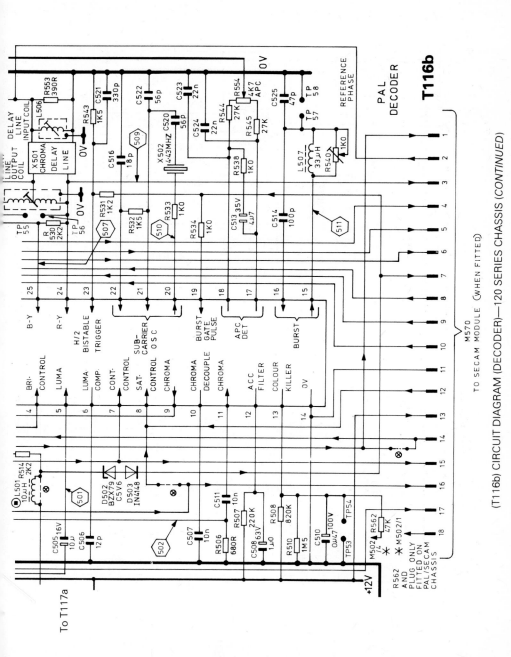

(T116b) CIRCUIT DIAGRAM (DECODER)—120 SERIES CHASSIS (CONTINUED)

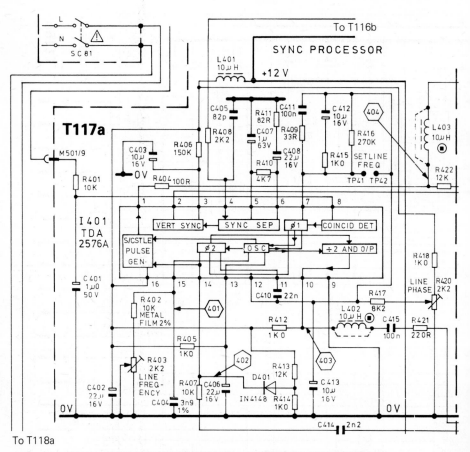

(T117a) CIRCUIT DIAGRAM (SYNC.)—120 SERIES CHASSIS (*PART*)

FIELD TIME BASE

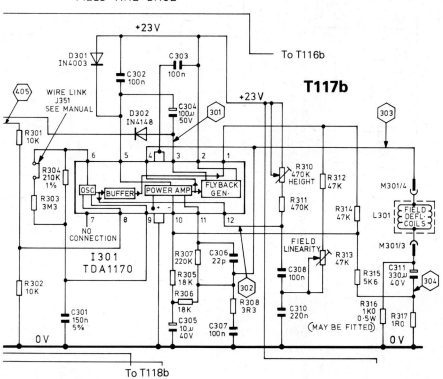

T117b

(T117b) CIRCUIT DIAGRAM (FIELD TIMEBASE)—120 SERIES CHASSIS (*CONTINUED*)

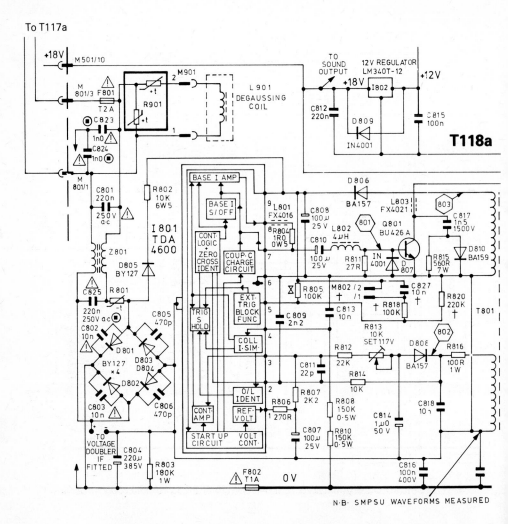

(T118a) CIRCUIT DIAGRAM (S.M.P.S.)—120 SERIES CHASSIS (*PART*)

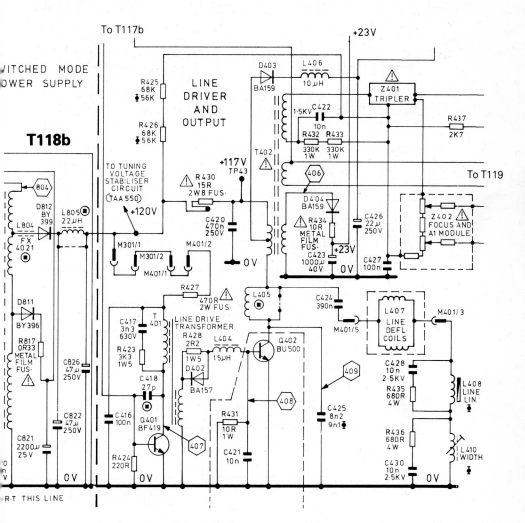

(T118b) CIRCUIT DIAGRAM (LINE OUTPUT STAGE)—120 SERIES CHASSIS (*CONTINUED*)

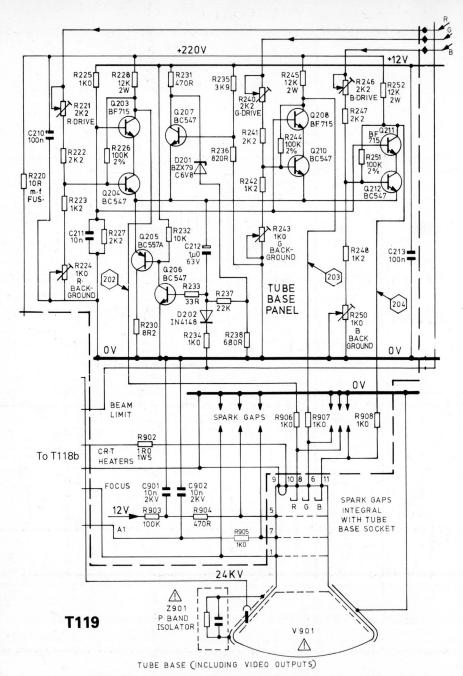

TUBE BASE (INCLUDING VIDEO OUTPUTS)

SYMBOL	DESCRIPTION
⊗	LINK OR CHOKE NOT FITTED ON SECAM MODELS
⊙	UHF/VHF MODELS
⎔	CHANGED FOR 121 CHASSIS
■	ALL SYSTEMS USING 5·5 MHZ SOUND
✳	TO PAL/SECAM SWITCH (IF FITTED)
†	COMPONENTS ADDED FOR FREQUENCY SYNTHESIS MODELS
✗	COMPONENTS REMOVED FOR FREQUENCY SYNTHESIS MODELS

(T119) CIRCUIT DIAGRAM (TUBE BASE PANEL)—120 SERIES CHASSIS

T120a

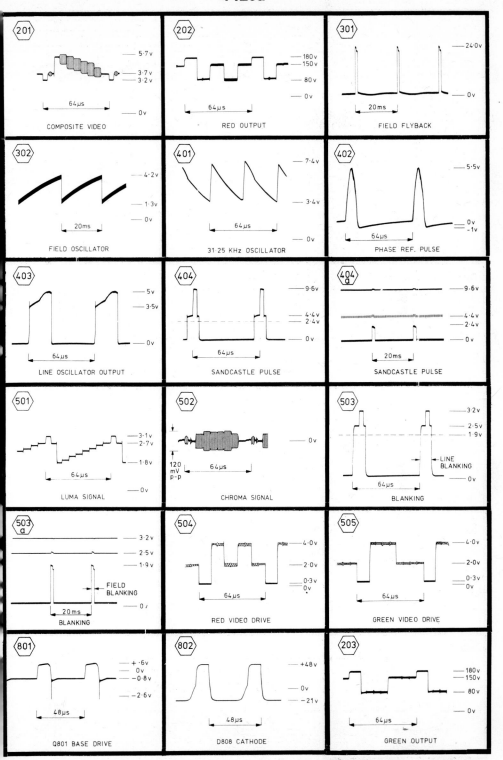

(T120a) CIRCUIT WAVEFORMS—120 SERIES CHASSIS (*PART*)

T120b

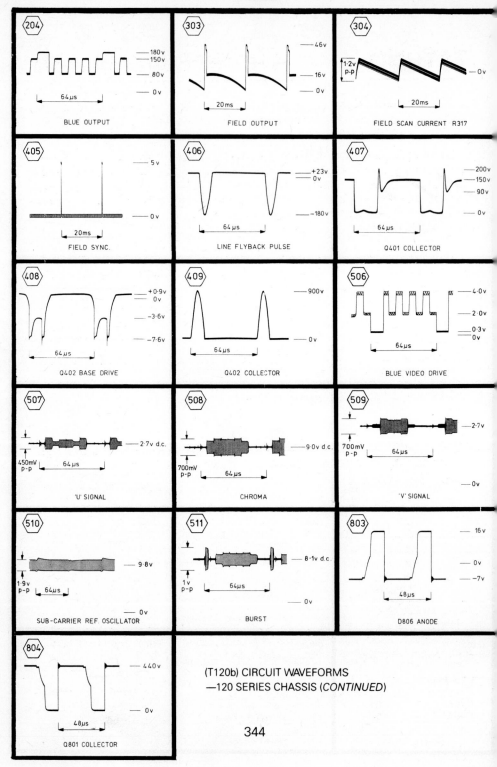

204	303	304
— 180v — 150v — 80v — 0v	— 46v — 16v — 0v	1·2v p-p — 0v
64µs	20ms	20ms
BLUE OUTPUT	FIELD OUTPUT	FIELD SCAN CURRENT R317

405	406	407
— 5v — 0v	+23v 0v — −180v	— 200v — 150v — 90v — 0v
20ms	64µs	64µs
FIELD SYNC.	LINE FLYBACK PULSE	Q401 COLLECTOR

408	409	506
+0·9v 0v −3·6v −7·6v	— 900v — 0v	— 4·0v — 2·0v 0·3v 0v
64µs	64µs	64µs
Q402 BASE DRIVE	Q402 COLLECTOR	BLUE VIDEO DRIVE

507	508	509
2·7v d.c. 450mV p-p 64µs	9·0v d.c. 700mV p-p 64µs	2·7v 700mV p-p 64µs — 0v
'U' SIGNAL	CHROMA	'V' SIGNAL

510	511	803
9·8v 1·9v p-p 64µs — 0v	8·1v d.c. 1v p-p 64µs — 0v	— 16v — 0v — −7v 48µs
SUB-CARRIER REF. OSCILLATOR	BURST	D806 ANODE

804	
— 440v — 0v 48µs	**(T120b) CIRCUIT WAVEFORMS —120 SERIES CHASSIS (CONTINUED)**
Q801 COLLECTOR	344

The information given here relates to the basic chassis used in the following Decca and Tatung models: DV1351, DV9351, DZ1352, DZ9352, TV1301, TV9301, TZ1302 and TZ9302.

General Description: A mains-operated colour television receiver chassis incorporating a 110° 22-in. or 26-in. cathode ray tube. Most of the circuitry used in this chassis is similar to that used in the 120 Series chassis described previously in this volume to which reference should be made. Variations are mainly concerned with the greater power requirements necessary for 110° scanning and the revised sections are detailed here.

Note:: The safety precautions necessary for the 120 Series apply equally to the 130 Chassis.

Chassis Type	Television System	Frequency Bands
130	PAL 1	U.H.F. only
132	PAL B/G	V.H.F.+U.H.F.
135	PAL 1	V.H.F.+U.H.F.
136	PAL/SECAM	V.H.F.+U.H.F.

Note: The H.T. setting of these chassis is higher than for the 120 Series. R813 should be adjusted to obtain 148V at TP43.

Voltage Measurements

Measured with a multimeter (20kOhm/V) with respect to chassis, with picture displaying colour bars at average viewing conditions.

Picture Tube Socket

Pin	Function	V d.c.	Meter Range
2	Cathode, Green	153	300
3	G1, Green	8·7	30
4	Cathode, Blue	145	300
5	A1, Blue	680	1000
7	A1, Green	680	1000
11	A1, Red	680	1000
12	G1, Red	8·7	30
13	Cathode, Red	145	300
14	G1, Green	8·7	30

Heater voltage measured across pins 1 and 6 is typically 6·2V R.M.S. on a meter which reads true R.M.S. voltage up to 100kHz. On a multimeter (10V A.C. range), the reading is approximately 4·65V.

Live Scan

Transistor	Emitter V	Base V	Collector V
Q401	0	−0·6	100V (300V range)
Q402	0	−0·5 (2·3V A.C.)	Do not measure

Transistor	Emitter V	Base V	Collector V
Q301	5·1	4·4	0
Q302	5·1	4·5	0·6
Q303	0	0·6	15·5 (30V range)

Field Circuit

Integrated Circuit I301 (TDA1670)

Pin	V
1	15·7 (30V range)
2	24 (30V range)
3	2·8
4	0·3
5	0
6	0·3
7	n.c.
8	0
9	4·0
10	7·1
11	n.c.
12	3·7
13	0·4
14	23·8 (30V range)
15	0

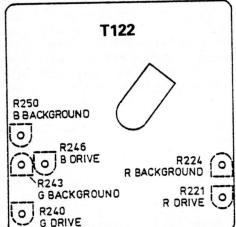

(T122) LOCATION OF ADJUSTMENTS
(TUBE BASE PANEL)—130 SERIES CHASSIS

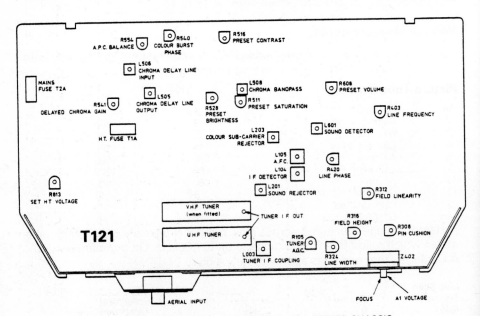

(T121) LOCATION OF ADJUSTMENTS—130 SERIES CHASSIS

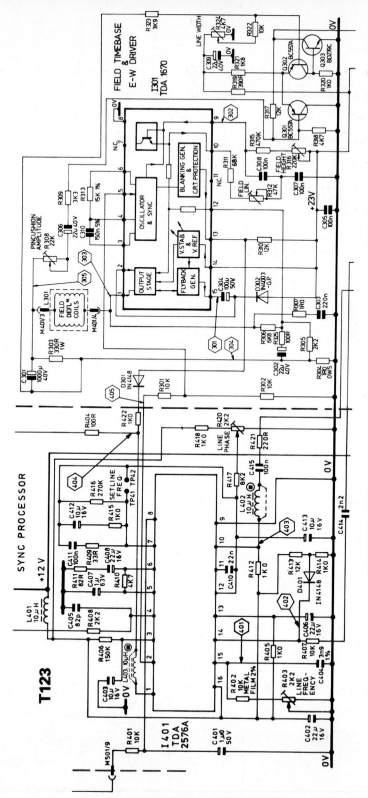

(T123) CIRCUIT DIAGRAM (SYNC. PROCESSOR)—130 SERIES CHASSIS

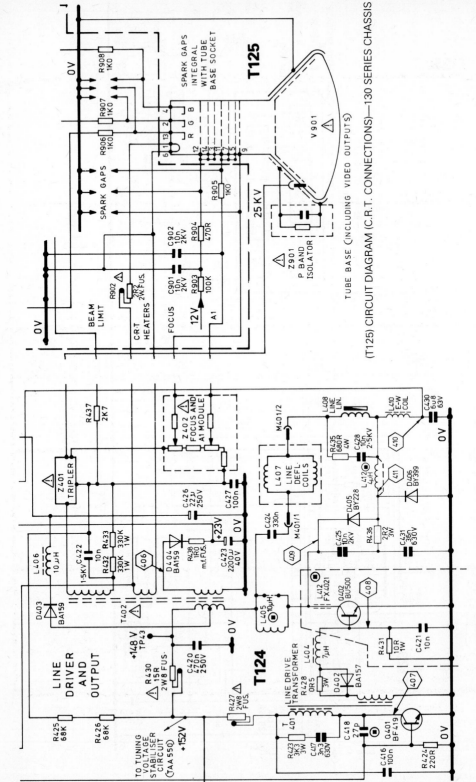

TUBE BASE (INCLUDING VIDEO OUTPUTS)

(T125) CIRCUIT DIAGRAM (C.R.T. CONNECTIONS)—130 SERIES CHASSIS

(T124) CIRCUIT DIAGRAM (LINE OUTPUT STAGE)—130 SERIES CHASSIS

T126a

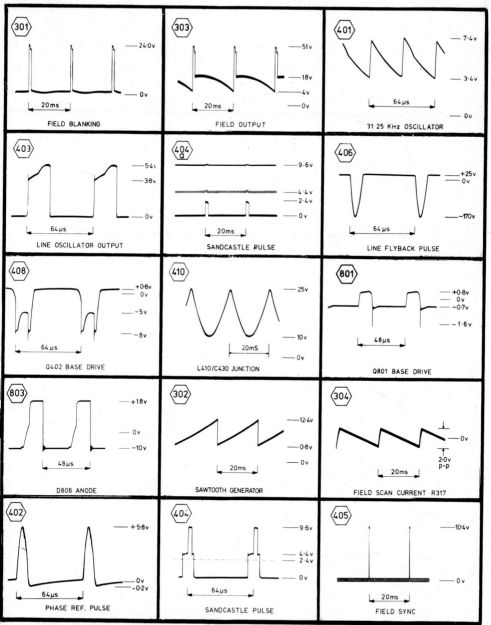

(T126a) CIRCUIT WAVEFORMS—130 SERIES CHASSIS (SEE ALSO 120 SERIES) (*PART*)

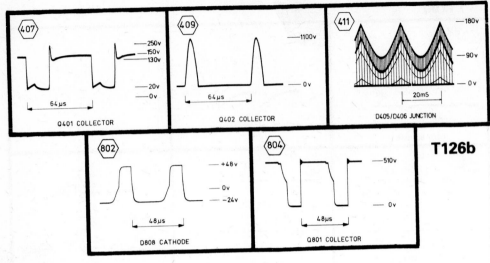

(T126b) CIRCUIT WAVEFORMS—130 SERIES CHASSIS (SEE ALSO 120 SERIES) (*CONTINUED*)

T.C.E.

1696, 1697
Series Chassis

General Description: A portable monochrome television receiver incorporating a 12-in. or 14-in. cathode ray tube and operating from mains or battery supplies. The chassis is isolated by mains transformer and a socket is provided for the connection of an earphone.

Mains Supply: 240 volts, 50Hz.

Fuses: The mains input is protected by a 250mA time-lag fuse and the low voltage supply line is protected by a 2A fuse.

Battery: 12 volts, 18 watts.

Connection to 12V Battery (Consumption 1·4A approx.): The receiver can be operated from a 12V battery by plugging a suitable connecting lead into the socket provided at the rear of the cabinet. A suitable plug and lead is provided with the receiver. Insertion of the plug automatically switches

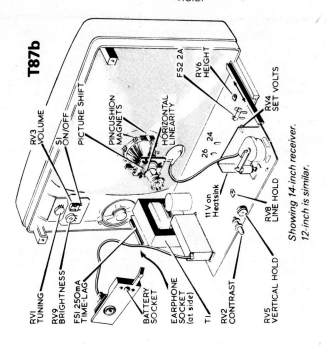

T87b

RV3 VOLUME

S1 ON/OFF

PICTURE SHIFT

PINCUSHION MAGNETS

HORIZONTAL LINEARITY

FS2 2A

RV6 HEIGHT

RV4 SET VOLTS

26 24

11 V on Heatsink

RV8 LINE HOLD

RVI TUNING

RV9 BRIGHTNESS

FS1 250mA TIME-LAG

BATTERY SOCKET

EARPHONE SOCKET (at side)

T1

RV2 CONTRAST

RV5 VERTICAL HOLD

Showing 14-inch receiver. 12-inch is similar.

(T87b) GENERAL CHASSIS LAYOUT
—1696/7 SERIES CHASSIS (CONTINUED)

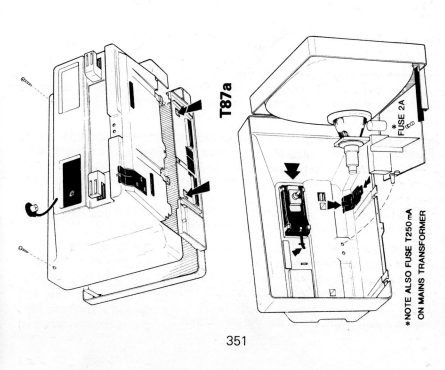

T87a

FUSE 2A *

*NOTE ALSO FUSE T250mA
ON MAINS TRANSFORMER

(T87a) ACCESS FOR SERVICE—1696/7 SERIES CHASSIS (PART)

over the low voltage supply line from the rectifier to the battery input. The receiver circuits are protected against incorrect polarity connections.

This protection does *not* involve blowing either of the two fuses fitted. Additional connecting leads are available, equipped with a plug at one end to fit the receiver socket.

These are terminated as follows:

Type TA83: Fitted with a plug to fit a dashboard cigar lighter, for car or boat with *negative earth.*

Type TA82: Similar to TA83 but wired for a *positive* earth system.

Type TA81: Terminated with 'crocodile' clips for direct connection to the battery terminals.

E.H.T.: 11·5Kv.

Adjustments

Vertical Hold (RV5): Where more than one channel is available, make the adjustment on the channel with the weakest signal. The adjustment is externally accessible using a screwdriver through a hole in the casing.

Picture Shift: The picture can be moved in any direction by rotating both control rings together around the tube neck. The position of the rings relative to one another alters the extent of the picture movement.

Width and Horizontal Linearity: The correct width and horizontal linearity is obtained by adjusting the linearity sleeve (in the scan coil) to obtain the correct width with the optimum linearity.

Height (RV6): Adjust to partially exclude the castellations at top and bottom of picture.

Contrast and Brightness (RV2 and RV9): Adjust the Contrast control in conjunction with the Brightness control to produce a correctly balanced picture in which all the grey tones as well as the highlights and shadows are present in correct proportions.

Horizontal Hold (RV8): Normally this coil should not require adjustment on installation but if the need for adjustment arises proceed as follows: With the chassis in its normal position, short-circuit Tag 24 to Tag 26 and adjust RV8 for a floating but resolved picture. On removal of the short-circuit the picture should lock.

After adjustment check for correct setting by interrupting the signal, i.e. withdraw the aerial connector or select another channel.

Set H.T. Volts (RV4): This should not need readjustment unless the factory setting has been disturbed or an associated component replaced. Connect an Avometer Model 8 between the large heat sink and chassis, and with the receiver connected to mains supply adjust RV4 for a reading of 11V D.C.

Vertical linearity and focus are not adjustable.

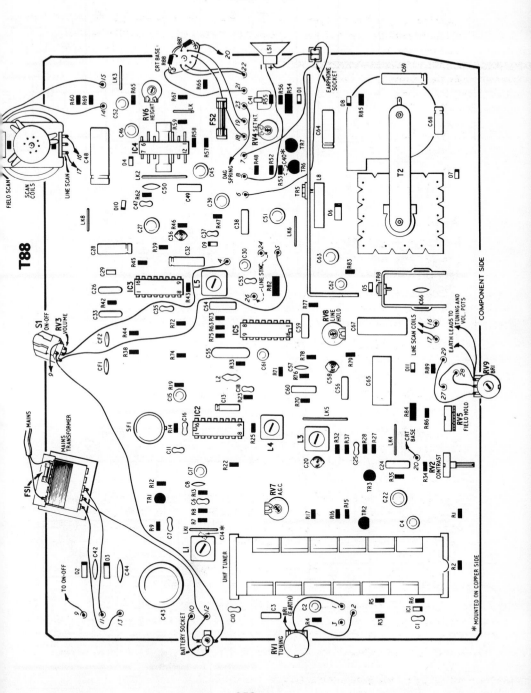

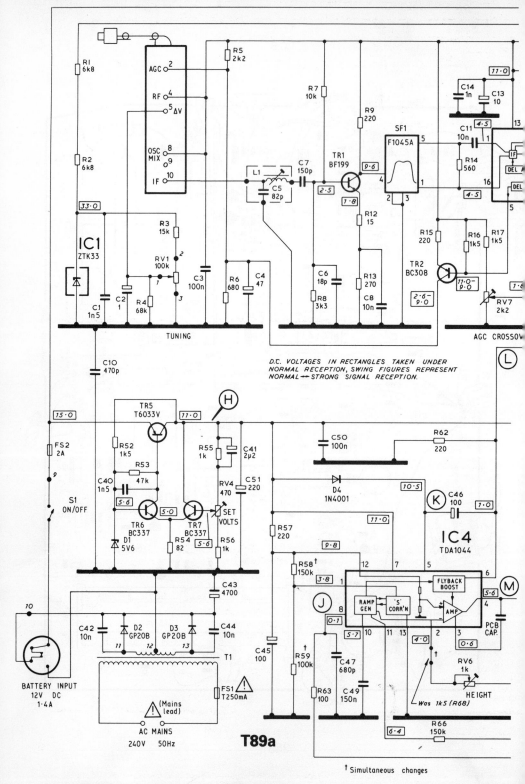

D.C. VOLTAGES IN RECTANGLES TAKEN UNDER NORMAL RECEPTION, SWING FIGURES REPRESENT NORMAL ⟶ STRONG SIGNAL RECEPTION.

† Simultaneous changes

(T89a) CIRCUIT DIAGRAM—1696/7 SERIES CHASSIS (*PART*)

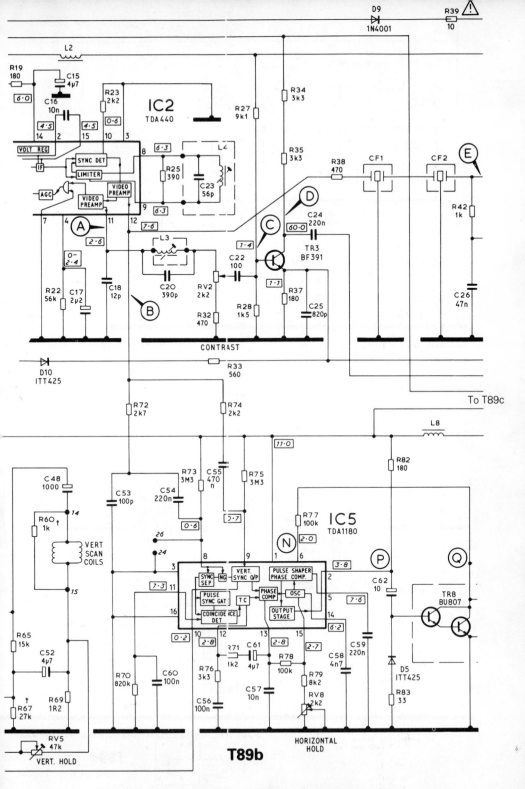

(T89b) CIRCUIT DIAGRAM—1696/7 SERIES CHASSIS (*PART*)

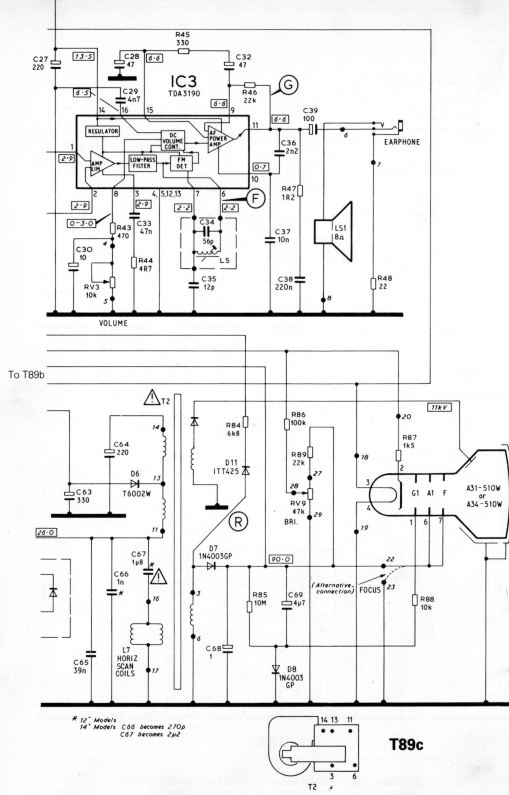

(T89c) CIRCUIT DIAGRAM—1696/7 SERIES CHASSIS (*CONTINUED*)

Circuit Waveforms (Fig. T90)

These were taken from a typical receiver at the points indicated by corresponding letters in the circuit diagram. The voltage and time figures refer to the sensitivity per division of the graticule. The receiver was set up for normal reception (test card with tone on sound) and the oscillograms were taken via a ÷10 probe having an input capacitance of 12pF in parallel with 10M Ω. The mixed mode timebase facility was used for J, K, and L.

Modifications

The following modifications have been introduced during the course of production.

Ref.	Early Chassis	Later Chassis	Remarks
TR2	BC308A	BC308	Improved availability
C14	10nF	1nF	Improved stability at maximum contrast
R3	15k Ω	10k Ω	To adjust tuning scale for SC4 Mk3 tuner
R39△	18 Ω or 10 Ω	10 Ω fusible	Protection in event of failure of C27
R55	1·8 Ω	1k Ω	To assist the set up of 11V line
R56	1·8 Ω	1k Ω	
R67	33k Ω	27k Ω	To overcome field cramping
'D' coils	6D02-014-005	06D2-016-007	—
R59	33k Ω	100k Ω	Component changes linked with changes of 'D' coils from 06D2-014-005 (3 Ω vert. scan) to 06D2-014-007 (8 Ω vert. scan). Appropriate values must be retained with associated coils
R60	220 Ω	11k Ω	
R58	100k Ω	150k Ω	
R68	1·5k Ω	Replaced by link	

T90

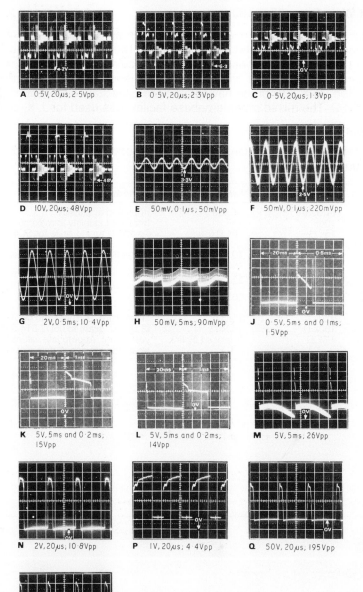

A   0·5V, 20μs; 2·5Vpp

B   0·5V, 20μs; 2·3Vpp

C   0·5V, 20μs; 1·3Vpp

D   10V, 20μs; 48Vpp

E   50mV, 0·1μs; 50mVpp

F   50mV, 0·1μs; 220mVpp

G   2V, 0·5ms; 10·4Vpp

H   50mV, 5ms; 90mVpp

J   0·5V, 5ms and 0·1ms; 1·5Vpp

K   5V, 5ms and 0·2ms; 15Vpp

L   5V, 5ms and 0·2ms; 14Vpp

M   5V, 5ms, 26Vpp

N   2V, 20μs; 10·8Vpp

P   1V, 20μs; 4·4Vpp

Q   50V, 20μs; 195Vpp

R   20V, 20μs; 120Vpp

(T90) CIRCUIT WAVEFORMS—1696/7 SERIES CHASSIS

358

T.C.E.

TX9 Chassis Units TA92, TA92A and TA126

General Description: Details of the TX9 Chassis were given in the 1981–82 edition of *Radio and Television Servicing*. Subsequent development has led to the introduction of a new type Power Supply section which described here together with information regarding the Battery Converter Units available for each version of the power supply.

Series 1044 Main Panel

Chopper Power Supply: PC1044 represents a further development of the TX9 Chassis with the introduction of a chopper power supply, resulting in reduced weight, lower component count and simplified operation.

The new panel can be substituted for the earlier panels (PC1001 and PC1040) but for battery operation must be used in conjunction with add-on Battery Converter Type TA126 which is specifically designed for use with PC1044.

A quick means of identifying PC1044 is to look for the fuse which is positioned behind the tuner (towards the tube), as viewed from the rear of the receiver. Receivers initially fitted with PC1044 will have serial number prefixed by GB.

The signal and timebase circuits remain basically as described previously (i.e. PC1040). An exception is the S.A.W.F. filter panel which now incorporates automatic A.G.C. crossover adjustment, dispensing with the need for a pre-set pot. and also giving improved performance in the presence of strong signal adjacent channel interference. The new panel carries the number PC1531.

Installation, Pre-set Adjustments and Alignment: These are the same as for PC1040 except that the A.G.C. adjustment (RV36) has been deleted.

Note also that the Set 115V adjustment is now referenced RV171 and the voltage can be checked at pin 5 of PL17 (see accompanying diagram Fig. T85b). Measurements are made with respect to the line output, heat sink, or tuner case, etc. (**not** chassis frame which is isolated).

Power Supply Fault Tracing

In contrast to earlier versions (PC1001 and PC1040) where overload faults can cause the fuse to blow, PC1044 automatically reduces the 115V line in the event of excessive current drain (approx. 1A limit).

In the event of a continuous overload on the 115V rail, the Power Supply will repeatedly attempt to restart giving rise to an audible oscillation from the Chopper transformer.

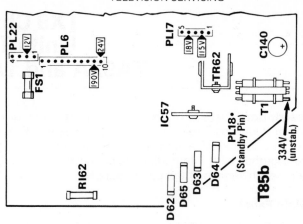

(T85b) ADJUSTMENT POINTS
(PC1044 PANEL)—TX9 CHASSIS (*PART*)

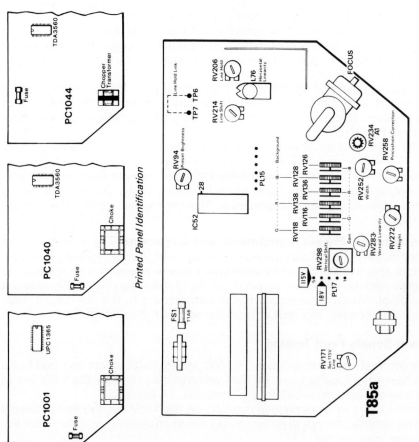

Printed Panel Identification

(T85a) ADJUSTMENT POINTS (MAIN CHASSIS)—TX9 CHASSIS (*PART*)

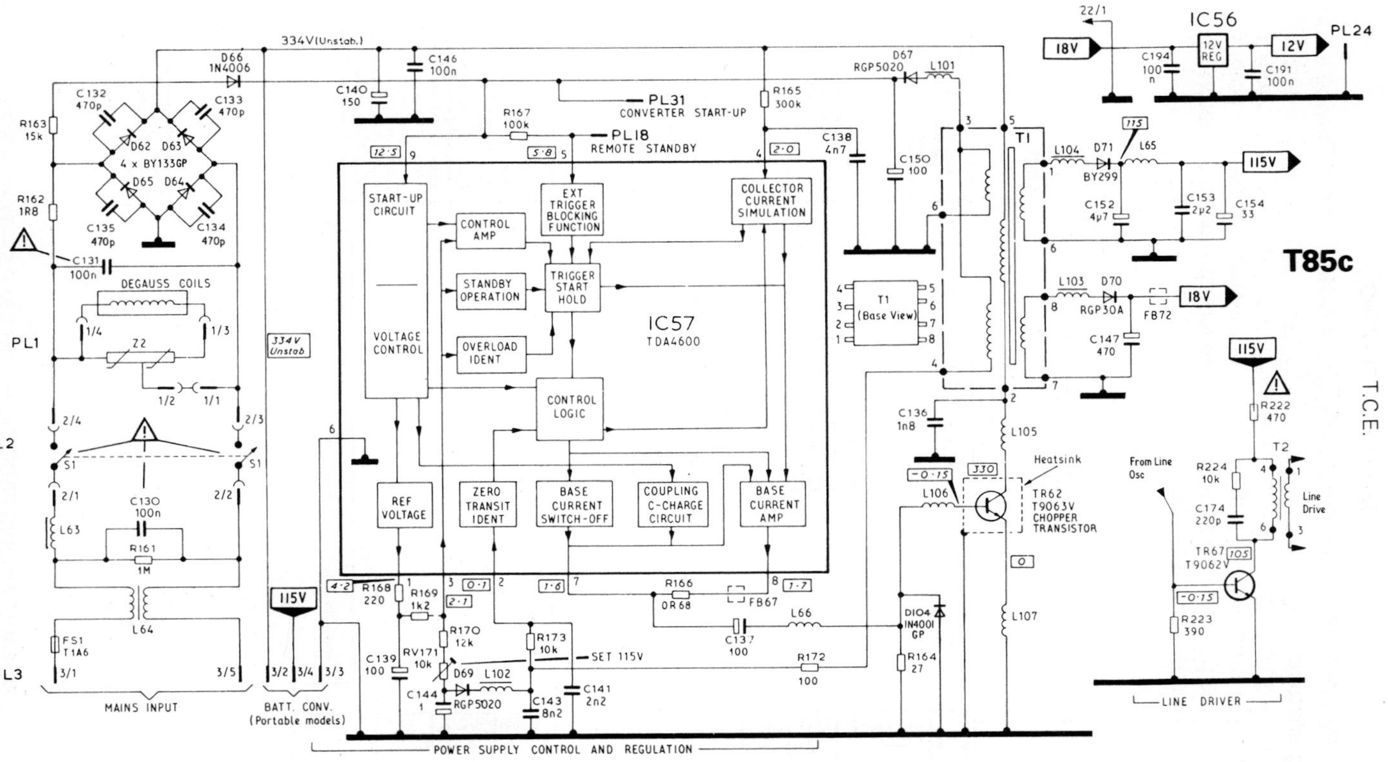

T.C.E.

T85c

(T85c) CIRCUIT DIAGRAM (S.M.P.S.)—TX9 CHASSIS

POWER SUPPLY CONTROL AND REGULATION

LINE DRIVER

The 18V rail is also protected against gross overloading by having a current limiting characteristic.

The accompanying chart (Fig. T86) suggests a step-by-step check procedure starting from a 'set dead' condition but includes a starting point from a 'Sound but no picture' condition and also describes certain symptoms related to particular component faults.

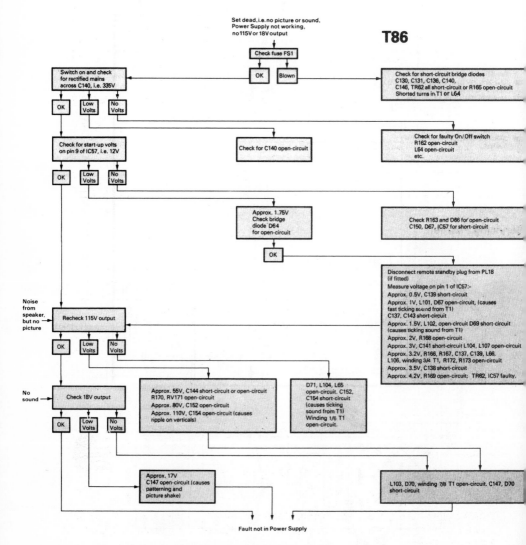

(T86) FAULT-FINDING PROCEDURE FOR S.M.P.S.—TX9 CHASSIS

Note: The chassis is live irrespective of mains polarity. An insulating transformer rated at 500VA should be used, when servicing. If not, direct contact with an earthed soldering iron or test equipment must be avoided.

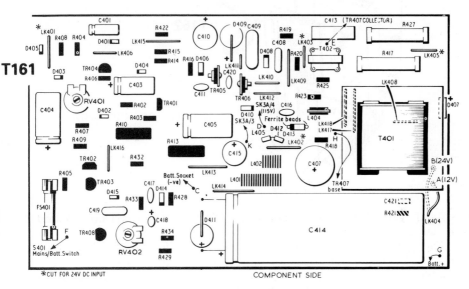

*CUT FOR 24V DC INPUT   COMPONENT SIDE

(T161) COMPONENT LAYOUT—TA92, TA92A

Battery Converter TA92, TA92A

The TA92 battery converter was designed to enable 14-in. versions of the TX9 chassis to be operated from a 12 or 24 volt battery. The circuit incorporates a low voltage shut-down circuit which operates when the supply voltage falls below a pre-determined level. Since the publication of the original service information, modifications have been introduced including a shorter connecting cable with revised dressing within the cabinet. The later version, which is described here, is designated TA92A.

Modifications (TA92 to TA92A)

Modifications: R435 and C420 are deleted (were connected between *b* and *e* of TR407); Capacitor 1·2nF added between *b* and *c* of TR405 (reference C420); R421 and C421 added in series with link L404 (fixed end); C416 increased from 470pF to 2·7nF; D412 uprated from IN4001GP to IN4008GP; R402 from 10k Ω to 4·7k Ω; Ferrite beads added on D413 leads.

Pre-set Adjustments (See Fig. T162)—Receiver connected

Set Low Battery Voltage Trip (RV401): RV401 is factory set and normally should not require readjustment even after altering the Converter Unit for 24V operation.

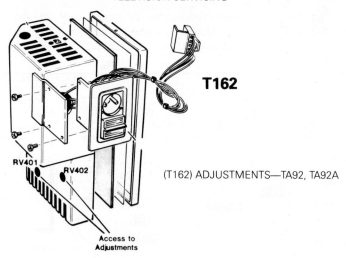

T162

(T162) ADJUSTMENTS—TA92, TA92A

RV401

RV402

Access to Adjustments

A 30V, 10A D.C. Power Supply and 20k Ω/V meter will be required if readjustment becomes necessary.

Adjust the voltage, from the 30V Power Supply feeding the Converter, to read 10·9V under full load conditions (TX9 receiver with Contrast and Brightness at maximum).

Now adjust RV401 to the point where the low voltage trip comes into operation.

Disconnect the Converter Unit from the 30V Power Supply. Re-adjust the Power Supply voltage to 12V, reconnect to the Converter and check for satisfactory receiver operation.

Set Output Volts (RV402): Before attempting this adjustment ensure that the Converter Unit is operating from a D.C. power source of exactly 12V, with the TX9 receiver set at black-level.

Connect a 20k Ω/V meter (to measure 115V D.C.) between R197 and chassis; R197 is located behind the choke on the main TX9 receiver panel. Adjust RV402 for a meter reading of 115V.

Battery Converter TA126 (For use with portable models incorporating power supply panel type PC1044). See Figs. T164—6.

Notes: This unit is fully automatic and requires no pre-set adjustments.

Circuit Diagram (Fig. T163)

Figures in rectangles are D.C. voltages measured with an Avometer Model 8 with respect to the negative lead of the main electrolytic, C414. The Converter was connected to an 11V D.C. supply and was loaded with a Dummy Load to simulate black level and full load conditions.

Except for the figures in brackets which apply to full load only, all other readings are the same for both conditions.

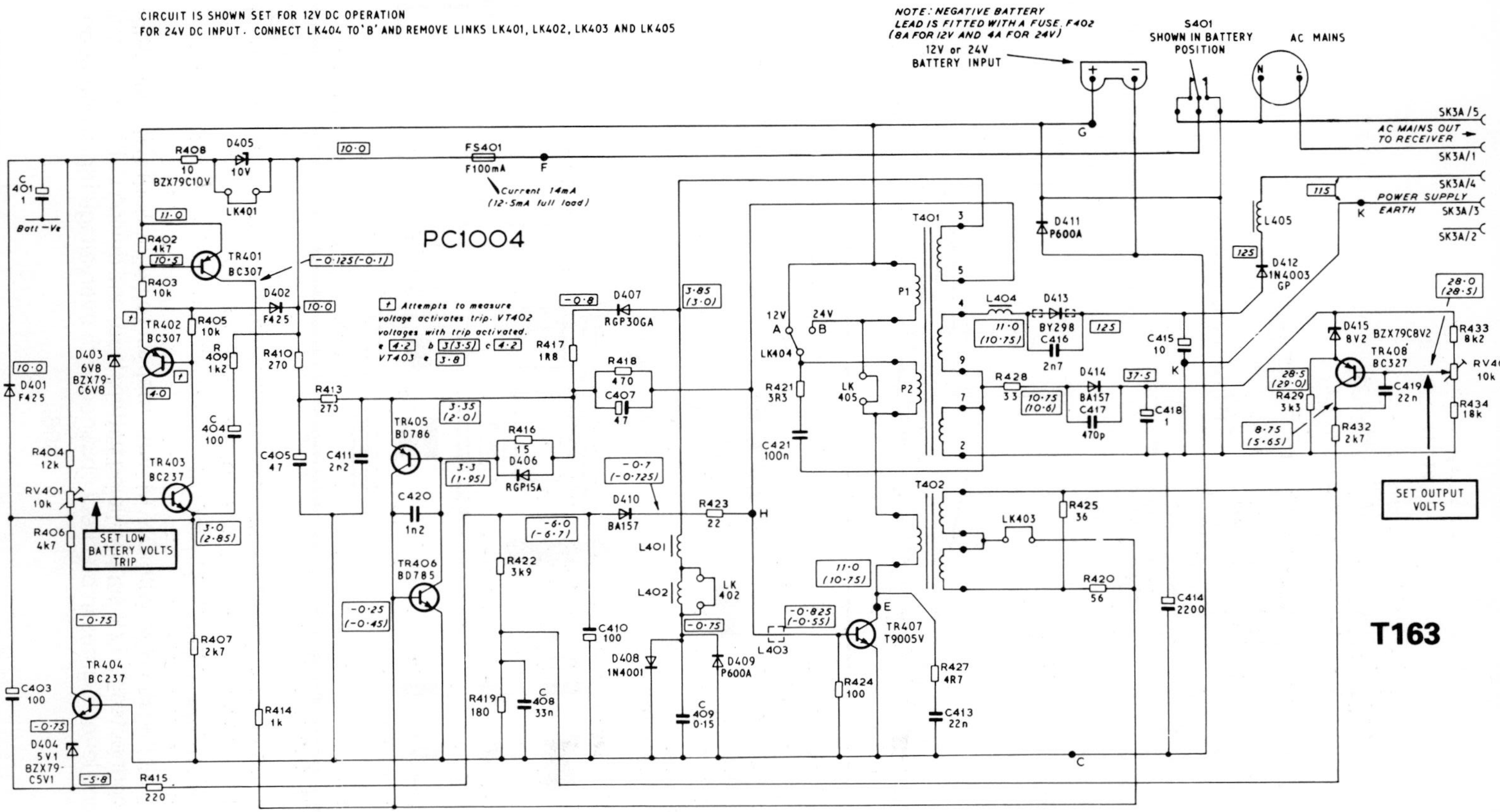

CIRCUIT IS SHOWN SET FOR 12V DC OPERATION
FOR 24V DC INPUT, CONNECT LK404 TO 'B' AND REMOVE LINKS LK401, LK402, LK403 AND LK405

NOTE: NEGATIVE BATTERY
LEAD IS FITTED WITH A FUSE F402
(8A FOR 12V AND 4A FOR 24V)

T163

(T163) CIRCUIT DIAGRAM—TA92A

PC1004

AC MAINS OUT TO RECEIVER
SK3A/5
SK3A/1
POWER SUPPLY
SK3A/4
EARTH
SK3A/3
SK3A/2

SET OUTPUT VOLTS

AC MAINS

S401 SHOWN IN BATTERY POSITION

12V or 24V BATTERY INPUT

SET LOW BATTERY VOLTS TRIP

Current 14mA
(12·5mA full load)

Batt −Ve

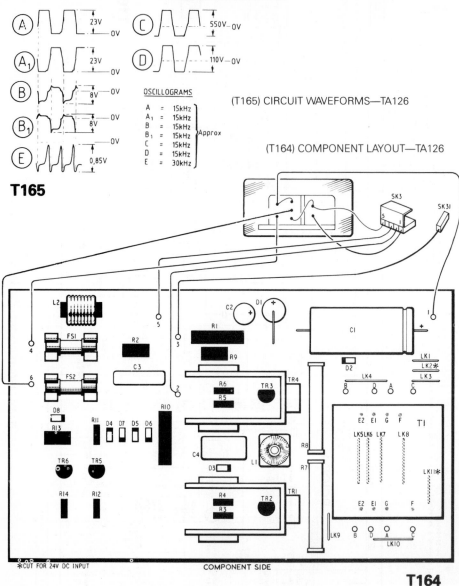

OSCILLOGRAMS

A = 15kHz
A₁ = 15kHz
B = 15kHz
B₁ = 15kHz
C = 15kHz
D = 15kHz
E = 30kHz

} Approx

(T165) CIRCUIT WAVEFORMS—TA126

(T164) COMPONENT LAYOUT—TA126

T165

T164

Circuit Diagram Notes (Fig. T166)

D.C. voltages taken with D.V.M. with respect to negative of C1. Converter was connected to 12V supply, using battery lead, and driving C.T.V. at normal viewing level. Figures in brackets obtained with overvoltage trip in operation.

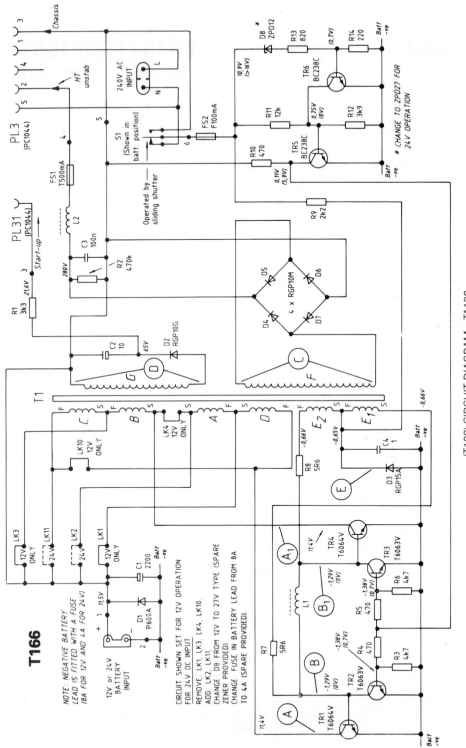

(T166) CIRCUIT DIAGRAM—TA126

T.C.E.

TX10 (1550/1551 Series Chassis)

The original information for the TX10 chassis is given in the 1981–82 volume of *Radio and Television Servicing*.

The latest development for TX10 models is the introduction of the 1550/1551 Series Chassis to improve sound performance in terms of an increase in output power from 3W R.M.S. to 10W R.M.S. (5W+5W), better tone control performance, and giving stereo sound capability. The increased audio output necessitates changing to a 5W loudspeaker or speakers.

The full benefit will only be heard in the new models which are being introduced with two loudspeakers fitted in cabinets with better acoustic design and Bass, Treble, and Balance controls. These new models will incorporate a stereo audio processing board and, when used in conjunction with a stereo video cassette recorder or video disc player, they will give true stereo sound available now from pre-recorded video cassettes and discs via audio input sockets. By means of electronic processing they will also give the present monoaural T.V. or video programmes simulated stereo (Super-sound).

Receivers fitted with the 1550/1551 Series Chassis can be easily identified by the large, double-cranked heat sink for IC562 in the centre of the signal board (PC1551). This is in addition to the similar, but smaller heat sink for IC621 which was fitted in earlier chassis.

On the signal board (PC1551) two new integrated circuits are introduced: TDA1236 (IC561, Sound Channel IC) with full-range Bass and Treble controls and TA7227P (IC562, Sound Output IC) providing two sound output channels. It should be noted that the S.A.W.F. I.F. daughter board PC1531, which incorporates automatic A.G.C. crossover adjustment, dispensing with the need for a pre-set control, is an alternative for PC991.

The new power board (PC1550) is very similar to PC1500, the minor modifications include the addition of R725 (1k Ω) in the base of the chopper driver transistor (TR721) to improve performance and fuse FS702 uprated to 1·6A to allow for increased sound power. To allow use with American as well as European V.H.D. video discs, a facility for automatically switching the field from 50Hz to 60Hz is incorporated.

Not all receivers will have Treble and Bass controls fitted as shown on the main circuit diagram.

Since the main circuit diagram was printed, pins Q and Z on PC1550 have been replaced by a plug PL90 for use in stereo sound receivers. Also R788 has been changed to 220k Ω.

Apart from the sound stages the circuit description remains basically as given previously.

368

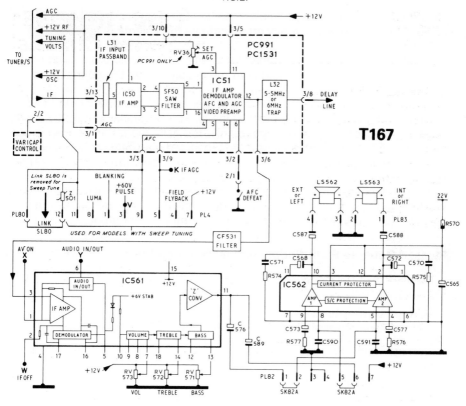

(T167) CIRCUIT DIAGRAM (SOUND STAGES)—1550 SERIES CHASSIS

Sound Stages (See Fig. T167)

The composite output of the I.F. module from PL3/6 is fed via a 6MHz ceramic filter, CF531, which removes the video content to the limiter amplifier on IC561 pin 3. This amplifier drives a symmetrical coincidence detector tuned by the C561-L561 network. R562 sets the gain and frequency to give the optimum bandwidth for the T.V. signal.

Pin 6 of IC561 provides an output signal independent of the Volume control for use in stereo receivers and with appropriate audio accessories. The connection X on pin 1 allows pin 6 to be converted into an input in a playback mode situation.

The usual D.C. Volume control is employed with the D.C. bias on pin 7 controlling the magnitude of the output signal on pin 11. IC561 also incorporates D.C. tone circuits providing separate full-range, cut and lift, Bass and Treble controls when the D.C. bias is altered on pins 13 and 14 respectively. The components on pins 12 and 18 control the crossover frequency on the Tone controls. The Volume control incorporates loudness

369

circuitry, controlled by C580 on pin 8, so that the bass and treble ends of the audio signal are not so severely attenuated as the volume is reduced.

The output on pin 11 is coupled via electrolytic capacitors C576 and C589 to PL82 which provides the connection to the stereo audio processing board PC1072 on stereo models. As drawn the circuit shows the connections for monoaural receivers.

IC562 comprises two identical output stages each capable of 5W R.M.S. output into 8 Ω loads. Each output stage is a conventional differential input amplifier and, referring to one stage only, the gain and L.F. response is set by R576-C577. The loudspeaker is A.C. coupled via C588. Zobel networks R575-C570, R574-C571 ensure total stability with both internal and external loudspeakers.

IC562 also contains safety circuits to protect against loudspeaker shorts which could cause overheating. H.T. decoupling is provided by the R570-C565 network to minimise picture bounce at maximum output.

Revised Guide to Fault-finding under 'Trip' Conditions (Fig. T168)

Procedure: Connect an AVOmeter 8 (or similar) to receiver 150V supply tag adjacent to R733, and observe the voltage reading at switch-on.

Voltage attempts to rise: Fault is on T705 secondary (depending on fault, voltage reading may vary between 10V and 40V).

Voltage does not rise: Fault is on T705 primary, or D731 could be short-circuit.

T705 Primary:
Check: D702, D703 and D704 for short- or open circuit.
L702 for open circuit.
C712, C711 for short- or open circuit.
Primary printed circuit for cracked copper.
Note: under certain fault conditions the chopper power supply may trip very quickly, and make fault finding difficult. In these circumstances if pin 8 of IC801 is connected to earth, this will prevent the duty cycle of the chopper increasing and hence limit the peak chopper collector current to a safe value.

In this mode, unless there is a direct fault on the *primary side* of T705, the power supply should function without tripping with supplies at a low level.

T705 Secondary: To locate faulty area, disconnect the following plugs in the order given until tripping stops, then follow procedure given in Fig. T169.

370

No Sound or Picture

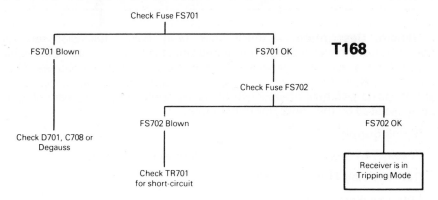

T168

Check Fuse FS701

FS701 Blown — FS701 OK

Check Fuse FS702

FS702 Blown — FS702 OK

Check D701, C708 or Degauss

Check TR701 for short-circuit

Receiver is in Tripping Mode

(T168) REVISED FAULT-FINDING PROCEDURE (POWER SUPPLY)—1550 SERIES CHASSIS

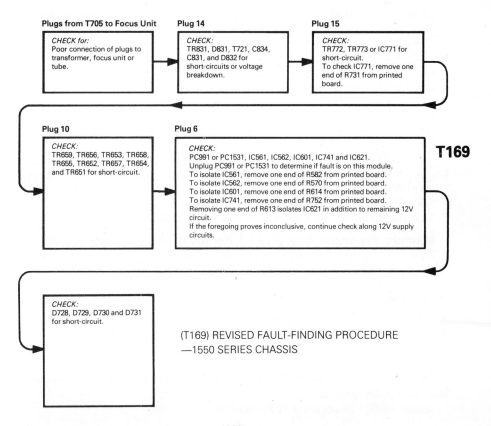

Plugs from T705 to Focus Unit

CHECK for:
Poor connection of plugs to transformer, focus unit or tube.

Plug 14

CHECK:
TR831, D831, T721, C834, C831, and D832 for short-circuits or voltage breakdown.

Plug 15

CHECK:
TR772, TR773 or IC771 for short-circuit.
To check IC771, remove one end of R731 from printed board.

Plug 10

CHECK:
TR659, TR656, TR653, TR658, TR655, TR652, TR657, TR654, and TR651 for short-circuit.

Plug 6

CHECK:
PC991 or PC1531, IC561, IC562, IC601, IC741 and IC621.
Unplug PC991 or PC1531 to determine if fault is on this module.
To isolate IC561, remove one end of R582 from printed board.
To isolate IC562, remove one end of R570 from printed board.
To isolate IC601, remove one end of R614 from printed board.
To isolate IC741, remove one end of R752 from printed board.
Removing one end of R613 isolates IC621 in addition to remaining 12V circuit.
If the foregoing proves inconclusive, continue check along 12V supply circuits.

T169

CHECK:
D728, D729, D730 and D731 for short-circuit.

(T169) REVISED FAULT-FINDING PROCEDURE
—1550 SERIES CHASSIS

General Description: A touch-tuner 14-in. colour television receiver operating from mains supplies. Integrated circuits are used in low-level signal circuits and a socket is provided for the connection of earphones. Care should be taken when servicing the power supply section as components are at mains potential irrespective of mains connection, the main chassis being isolated by the convertor transformer.

Mains Supply: 220 volts, 50Hz.

Fuses: T4A (mains); T630mA (D.C.); T400mA (line output).

Cathode Ray Tube: 370HFB22.

Loudspeaker: 16 ohms impedance.

Access for Service

Chassis Removal: Remove Main S.W. and Sound Volume knobs (without Colour, Bright, and Picture knob).

Disconnect the leads from U.H.F./V.H.F. external antenna, and pull it out from Antenna Terminal Board.

Remove the four screws which secure the rear cover of the main cabinet (two on back, two on the bottom).

Remove solder connection of hook-up wire connecting Picture Tube and Neck Board and then remove Neck Board from Picture Tube.

Remove two screws that secure the Volume assembly mounting bracket to cabinet front.

Remove Deflection Yoke Socket, Degaussing Coil Socket and Speaker Socket, Memory Pre-setter Board Socket from Main Board, then remove second Anode lead.

Take the chassis out from the cabinet front.

To install chassis, repeat the above procedure in reverse order.

Memory Pre-setter Board (S.F.) Removal: Remove four screws that secure the Memory Pre-setter Board (S.F.) to cabinet front.

Gently withdraw the assembly from the cabinet front.

Adjustments

H.T.: The +127V adj. control (VR601) is adjusted at the factory. However, if re-adjustment should be required, proceed as follows.

Operate receiver for at least 15 minutes at 220V A.C. line.

Maximise Brightness and Picture control.

Connect Positive lead of V.T.V.M. to +B Test Point of main board, negative lead to chassis ground.

Adjust VR601 to obtain +127V reading.

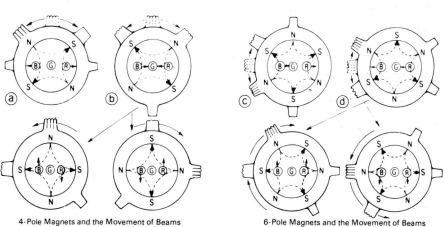

T94

4-Pole Magnets and the Movement of Beams   6-Pole Magnets and the Movement of Beams

(T94) PURITY AND STATIC CONVERGENCE ADJUSTMENTS—MODEL CPL-144

Colour Purity and Vertical Centre Adjustment (Fig. T94): For best results, it is recommended that the purity adjustment be made in the final receiver location. If the receiver can be moved, perform this adjustment with it facing east or west. The receiver must have been operating 15 minutes prior to this procedure and the faceplate of the C.R.T. must be at room temperature.

The receiver is equipped with an automatic degaussing circuit. However, if the C.R.T. shadow mask has become excessively magnetised, it may be necessary to degauss it with a manual coil.

Purity Magnets are used for Colour Purity and V. Centring. Adjustment procedure is as follows.

Set Brightness control (VR603) to maximum.

Turn Green Cut-off control (VR404) on the Neck Board fully anti-clockwise.

Turn Red and Blue Cut-off control (VR403, VR405) fully clockwise.

Pull the Deflection Yoke backward so that a vertical Magenta band appears.

Move the two Purity Magnets and bring the Magenta band to the mechanical centre of the screen. The vertical centre position should be set V.R.S.: −2 mm as shown.

373

Push the Deflection Yoke forward gradually and fix it at the place where the Magenta screen becomes uniform throughout.

Turn Cut-off control, and Drive control and confirm that each colour is uniform.

If the colour is not uniform, re-adjust it by moving Purity Magnets slightly.

Move the two Purity Magnets at the same time (do not change the angle of the pair), and adjust the vert. centre to centre of screen.

Obtain the three colours and confirm whether white uniformity is balanced.

Insert the temporary wedge as shown and adjust the angle of Deflection Yoke.

Static Convergence Adjustment: A recently developed Deflection Yoke and Electron Guns construction has been used on this equipment in combination with In-Line Guns and Black Stripe Screen to make a barrel-type magnetic-field distribution for vertical deflection and a pin-cushion-type magnetic field for horizontal deflection with which a self-converging system can be obtained. This type is different from conventional unitymagnetic field distribution type deflection yoke. 4-Pole Magnets and 6-Pole Magnets are employed for static convergence instead of a Convergence Yoke.

A crosshatch signal should be connected to the antenna terminal of the receiver.

Set Picture and Brightness knobs to obtain a visible screen.

Adjust each cut-off V.R. for minimum trace and horizontal lines are turned to white colour.

When the pole opens to the left and right 45° symmetrically, the magnetic field maximises. Red and Blue beams move to the left and right oppositely. Variation of the angle between the tabs adjusts the convergence of Red and Blue vertical lines.

When the both 4-Pole Convergence Magnet Tabs are rotated as a pair, the convergence of the Red and Blue horizontal lines is adjusted.

A pair of 6-Pole Convergence Magnets are also provided and adjusted to converge the Magenta (Red+Blue) to Green beams.

When the Pole opens to the left and right 30° symmetrically, the magnetic field is maximised. Red and Blue beams both move to the left and right.

Variation of the opening angle adjusts the convergence of magenta to green vertical lines. When both 6-Pole Convergence Magnet Tabs are rotated as a pair the convergence of magenta to green horizontal lines is adjusted.

Adjustment of Dynamic Convergence (See Fig. T95): Feed a crosshatch signal to the antenna terminals. Set Picture and Bright knobs to obtain a visible screen. Insert the temporary wedge and fix Deflection Yoke so as to obtain the best circumference convergence.

Note: The temporary wedges may need to be moved during adjustments.

Insert three rubber wedges to the position as shown to obtain the best circumference convergence.

Note: Tilting the angle of the yoke up and down adjusts the crossover of both vertical and horizontal Red and Blue.

374

Tilting the angle of the yoke sideways adjusts the parallel convergence of both horizontal and vertical lines at the edges of the screen. (See Figs. T94 and T95).

Use three rubber wedges (thick and thin rubber wedges are used for a purpose).

The angle of each rubber wedge is shown.

After three rubber wedges have been inserted, pull out the temporary wedge.

Fix the rubber wedges with chloroprene rubber adhesive.

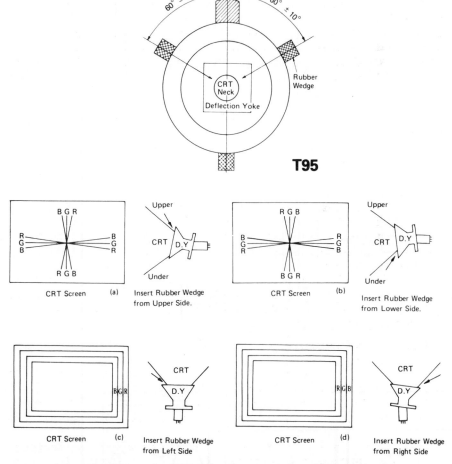

T95

(T95) DYNAMIC CONVERGENCE ADJUSTMENTS—MODEL CPL-144

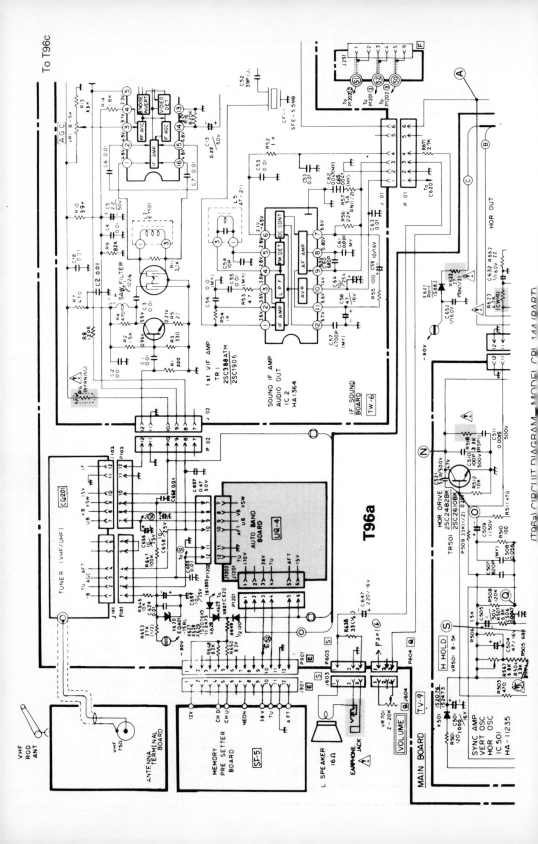

(T96) CIRCUIT DIAGRAM, MODEL CBL-144 (PART)

(T96b) CIRCUIT DIAGRAM—MODEL CPL-144 (PART

(T96c) CIRCUIT DIAGRAM—MODEL CPL-144 (PART)

378

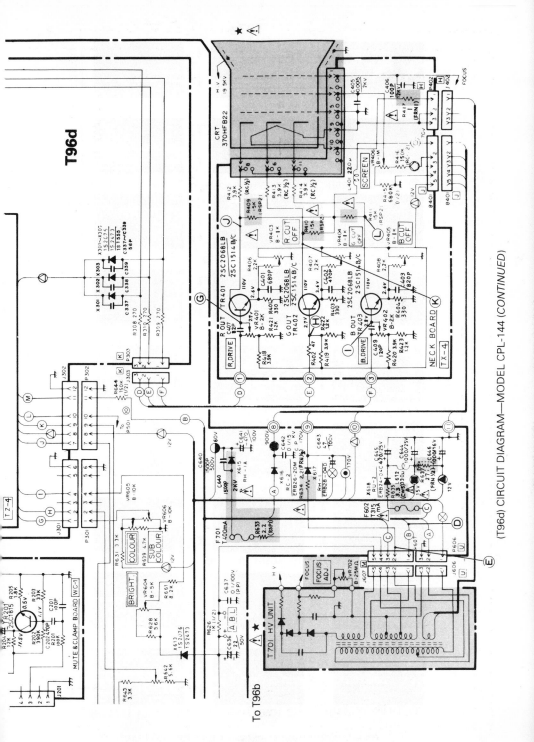

(T96d) CIRCUIT DIAGRAM—MODEL CPL-144 (CONTINUED)

379

T97a

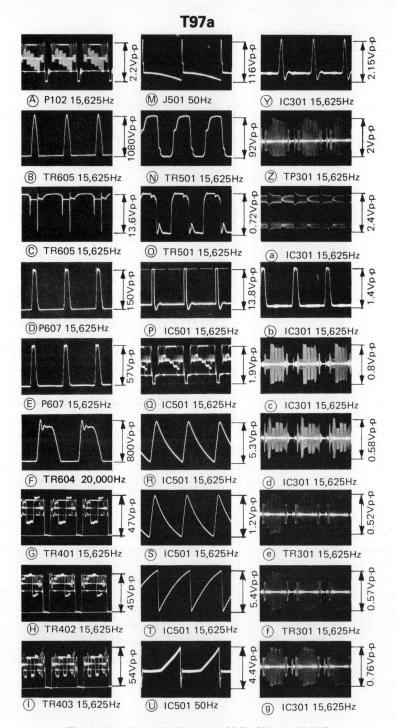

A P102 15,625Hz   2.2Vp-p

M J501 50Hz   116Vp-p

Y IC301 15,625Hz   2.15Vp-p

B TR605 15,625Hz   1080Vp-p

N TR501 15,625Hz   92Vp-p

Z TP301 15,625Hz   2Vp-p

C TR605 15,625Hz   13.6Vp-p

O TR501 15,625Hz   0.72Vp-p

a IC301 15,625Hz   2.4Vp-p

D P607 15,625Hz   150Vp-p

P IC501 15,625Hz   13.8Vp-p

b IC301 15,625Hz   1.4Vp-p

E P607 15,625Hz   57Vp-p

Q IC501 15,625Hz   1.9Vp-p

c IC301 15,625Hz   0.8Vp-p

F **TR604 20,000Hz**   800Vp-p

R IC501 15,625Hz   5.3Vp-p

d IC301 15,625Hz   0.58Vp-p

G TR401 15,625Hz   47Vp-p

S IC501 15,625Hz   1.2Vp-p

e TR301 15,625Hz   0.52Vp-p

H TR402 15,625Hz   45Vp-p

T IC501 15,625Hz   5.4Vp-p

f TR301 15,625Hz   0.57Vp-p

I TR403 15,625Hz   54Vp-p

U IC501 50Hz   4.4Vp-p

g IC301 15,625Hz   0.76Vp-p

(T97a) CIRCUIT WAVEFORMS—MODEL CPL-144 (*PART*)

T97b

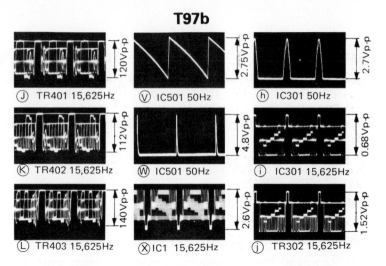

(T97b) CIRCUIT WAVEFORMS—MODEL CPL-144 (*CONTINUED*)

Grey Scale Tracking: The purpose of this procedure is to optimise the picture tube to obtain good black and white picture at all brightness levels while at the same time achieving maximum usable brightness. Normal A.G.C. setting and purity adjustments must precede this procedure.

With antenna connected to the receiver, tune in picture on a strong channel. Move the Colour control to minimum position and re-adjust the Fine Tuning control, so that the receiver will not produce a colour picture while the following adjustments are being performed.

Rotate the Red, Green, and Blue cut-off controls fully anti-clockwise.

Rotate the B. Drive and R. Drive controls to mid-range.

Short-circuit Tip on Video circuit board and on Vert/Horiz. circuit board.

Set the Brightness (VR603) and Picture (VR602) controls at maximum. Then, adjust the Sub Brightness control (VR305) to obtain the base (TP47G) voltage, $2{\cdot}1V{\pm}0{\cdot}05V$ of TR402 (G. OUT).

Slowly turn Screen V.R. (VR406) on neck circuit board counter-clockwise until colour (Colours) appears faintly on the screen.

Adjust each Cut-off V.R. so that colour becomes feint and horizontal lines are turned to white colour.

Return service tip to original position.

Set the Brightness and Picture controls at maximum and adjust R.B. Drive controls (VR401, 402) to produce a hi-lite white screen.

Set the Brightness and Picture controls at minimum.

Move the Brightness Picture until a dim raster is obtained.

If necessary, touch-up adjustment of the three Cut-off controls to obtain best white uniformity on the C.R.T. screen.

Set the Brightness and Picture controls at maximum. If necessary, adjust the R. Drive and B. Drive controls to produce a uniform black and white picture.

TELETON   Model T-122

General Description: A portable 12-in. monochrome television receiver operating from mains or battery supplies. The chassis is isolated and a socket is provided for the connection of an earphone.

Mains Supply: 240 volts, 50Hz.

Battery: 12 volts, 13 watts.

Cathode Ray Tube: 310GNB4.

Access for Service

Cabinet Removal: Remove antenna connection leads and six screws. Remove rear cover by pulling it out.

Cabinet Removal: Remove all knobs from the front of the cabinet.
Remove solder connections to speaker (white lead) and neck P.C.B. ground (black lead).
Remove C.R.T. socket, deflection yoke, and anode cap.
Remove the four screws holding the tuner bracket and one screw fixing the cord stopper.
Note: The Main P.C.B. of this set is retained by the cabinet and rear cover, when installing the rear cover, ensure that the Main P.C.B. is in rails at the side of both cabinet and rear cover.

Adjustments

H.T.: Set VR601 to obtain 11·5±0·1 volts between pins 3 and 4 of the cathode ray tube.

Line Hold: L401 through centre base of the cabinet.

Focus: Cut or make link on base of cathode ray tube.

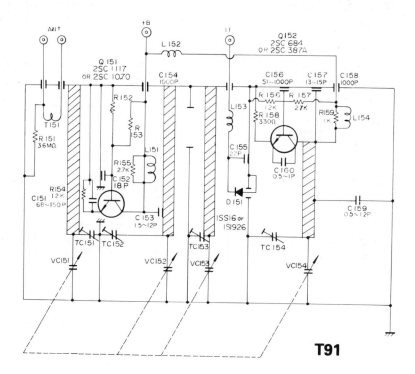

T91

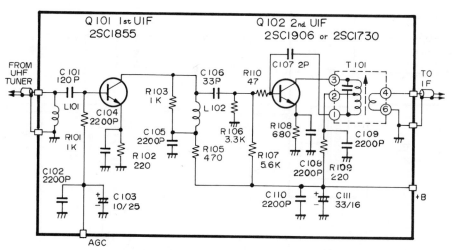

(T91) CIRCUIT DIAGRAM (TUNER AND I.F.)—Model T-122

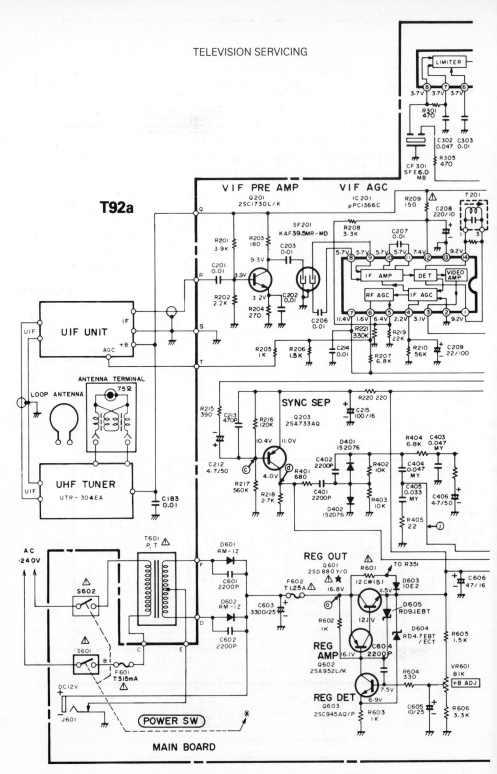

(T92a) CIRCUIT DIAGRAM—MODEL T-122 (*PART*)

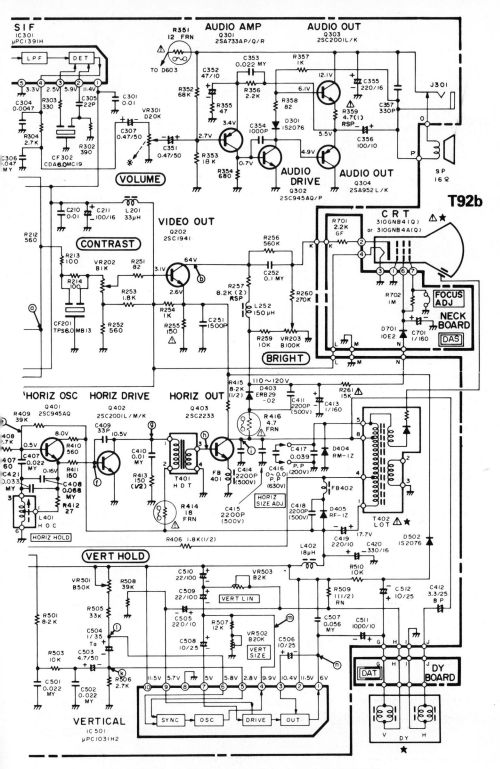

(T92b) CIRCUIT DIAGRAM—MODEL T-122 (*CONTINUED*)

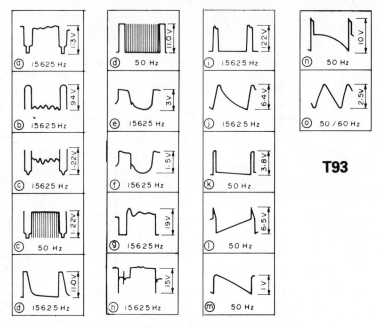

(T93) CIRCUIT WAVEFORMS—MODEL T-122

ZANUSSI   Model 20ZT505

General Description: A 20-in. mains-operated colour television receiver with precision-in-line cathode ray tube. The main chassis is isolated but care should be taken when checking the switch-mode power supplies which are live irrespective of mains connection. An electronic tuning system provides a choice of sixteen channels including V.C.R. with infra-red remote control which is described in the section on Model 22ZT905 in this volume. The circuitry associated with the main chassis is given here.

Mains Supply: 240 volts± 10%, 50Hz.   **Cathode Ray Tube:** A51-420X.

E.H.T.: 24·5KV. ˌ **Loudspeaker:** 12 ohms impedance.

Access for Service

Insert a screwdriver into the slots at the top corners of the rear cover and lever towards the centre. Stretch out the tube yoke wiring strap and disengage control panel wiring from the clamps. Remove the chassis from its seating and position vertically in the appropriate slots in the chassis rails.

T98

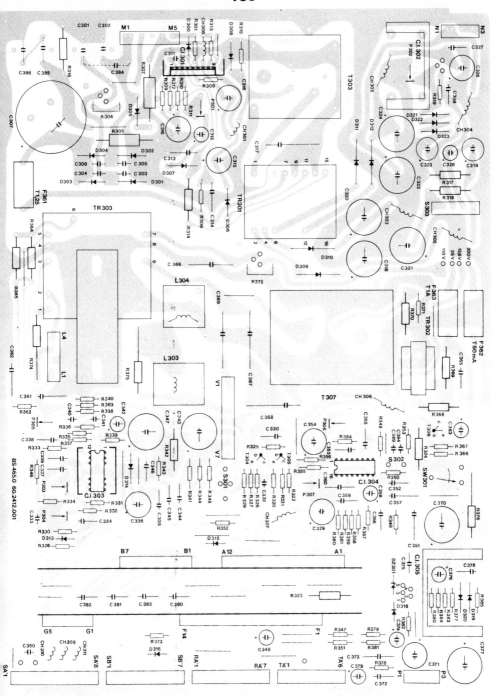

(T98) COMPONENT LAYOUT (MAIN PANEL)—MODEL 20ZT505

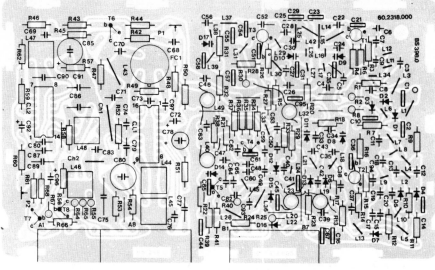

T99

(T99) COMPONENT LAYOUT (R.F./I.F. BOARD)—MODEL 20ZT505

T100

(T100) COMPONENT LAYOUT (DECODER BOARD)—MODEL 20ZT505

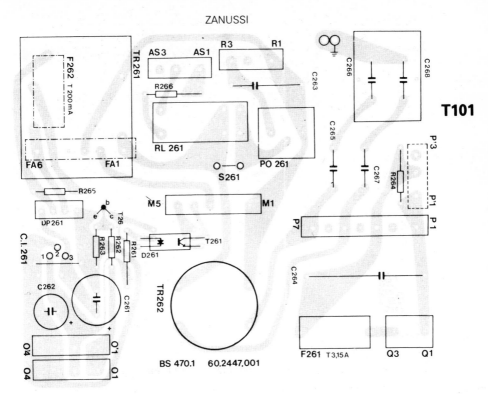

(T101) COMPONENT LAYOUT (MAINS PANEL)—MODEL 20ZT505

Adjustments

Check that the P.C.B. is correctly supplied: +12·6V±250mV at contact F12, +200V D.C.±10V at contact G5; +2·2V D.C.±100mV at contact F1; 12Vpp at contact F14; 160Vpp at contact G2; 11Vpp at contact F10.

Brightness and Contrast controls at maximum (adjustment on remote control when provided).

Colour saturation to minimum.

Grey scale signal input.

Adjust black-level potentiometers P105—P107—P109 to max. (clockwise).

Adjust gain potentiometers, P104—P106—P108 to max. (anti-clockwise).

Oscilloscope with attenuated probe on contact F2.

Check that the signal present at this point has a total height of 2·65Vpp.

Procedure

Oscilloscope probe on Red output (H4).

Adjust P105 to set the peak reference level to 175V D.C.

Oscilloscope probe on Green output (H2).

Adjust P107 to set the peak reference level to 175V D.C.

389

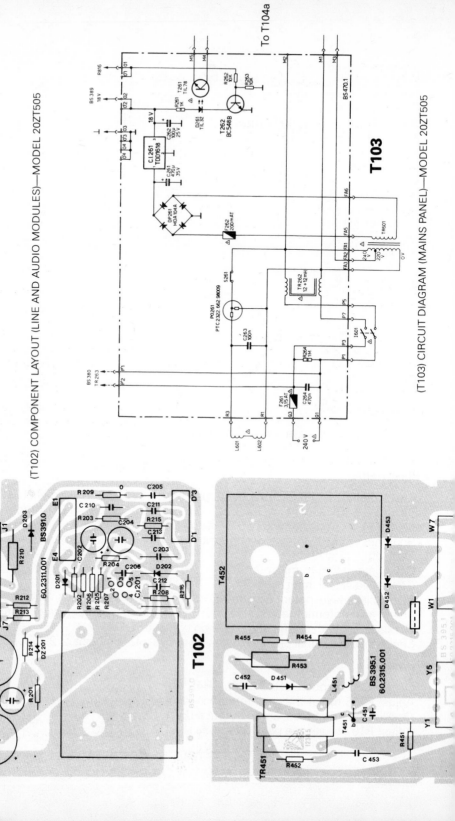

(T102) COMPONENT LAYOUT (LINE AND AUDIO MODULES)—MODEL 20ZT505

(T103) CIRCUIT DIAGRAM (MAINS PANEL)—MODEL 20ZT505

Oscilloscope probe on Blue output (H1).

Adjust P109 to position the peak reference level to 175V D.C.

Colour bar signal in antenna. Saturation to max.

Oscilloscope probe on TP102. Apply a 1V D.C. voltage at TP103.

Adjust L101 for max. amplitude of signal displayed.

Suppress voltage applied at TP103.

Plug-in jumper S103. Disconnect SW102.

Adjust C123 for minimum bar horizontal drift.

Disconnect S103. Reconnect SW102.

Test signal (delay) in antenna.

Oscilloscope probe on Red output (H4).

Adjust P103 and, if necessary, L104-L105 so as to perfectly align the bottom portion of the signal displayed.

Test signal (phase) in antenna.

Adjust L102 for best superimposition of the two wave forms displayed.

Grey Scale Alignment: Grey scale signal input. Saturation to minimum. Disconnect SW301 (501).

Slowly adjust P551 to obtain a horizontal line slightly visible.

Keeping as reference the colour of the prevailing electron gun of this line, accurately adjust Black-level potentiometers of the other two electron guns (P105 for red, P107 for green, P109 for blue) until the horizontal line is white.

Adjust P551 up to the point at which the white line disappears (cut-off).

Reconnect SW301 (501).

If a predominance of a colour is noticeable on the screen, eliminate it by re-adjusting the Black level potentiometer of the relevant electron gun involved.

Line Hold: Bridge S302 and adjust P307 for horizontal drift. Remove S302 link.

Power Supply: P301 adjusts the 204V, 119V, 25V, and 17V rails; P303 adjusts the 12·6V rail.

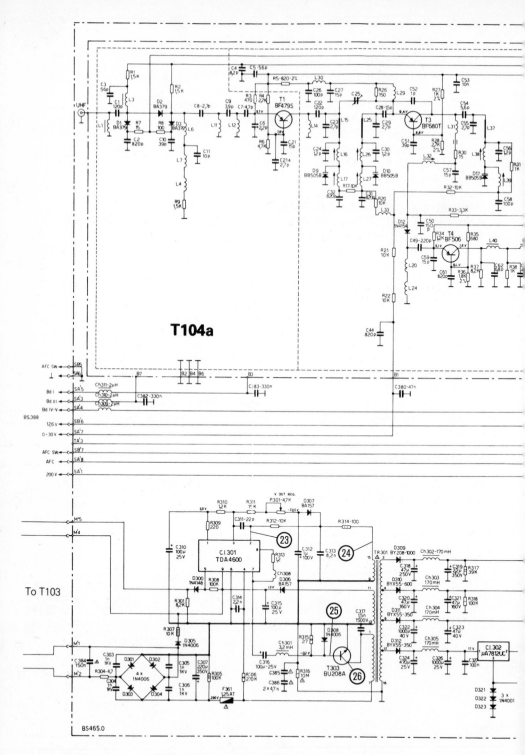

(T104a) CIRCUIT DIAGRAM—MODEL 20ZT505 (*PART*)

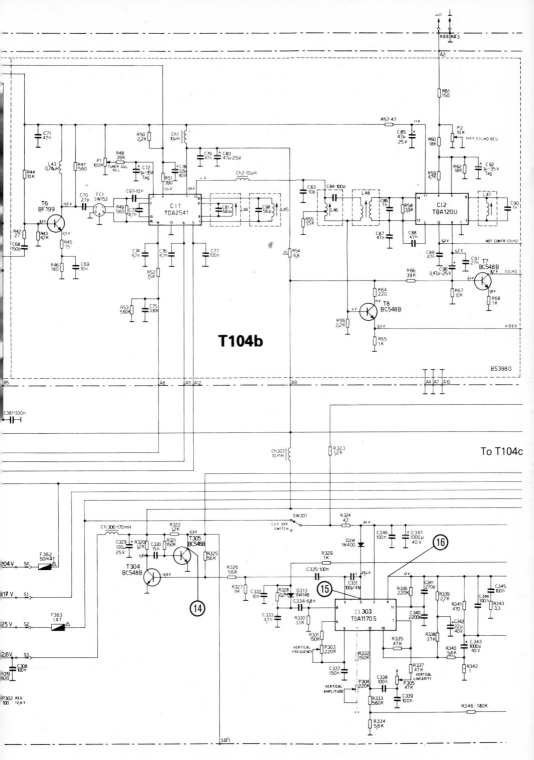

(T104b) CIRCUIT DIAGRAM—MODEL 20ZT505 (*PART*)

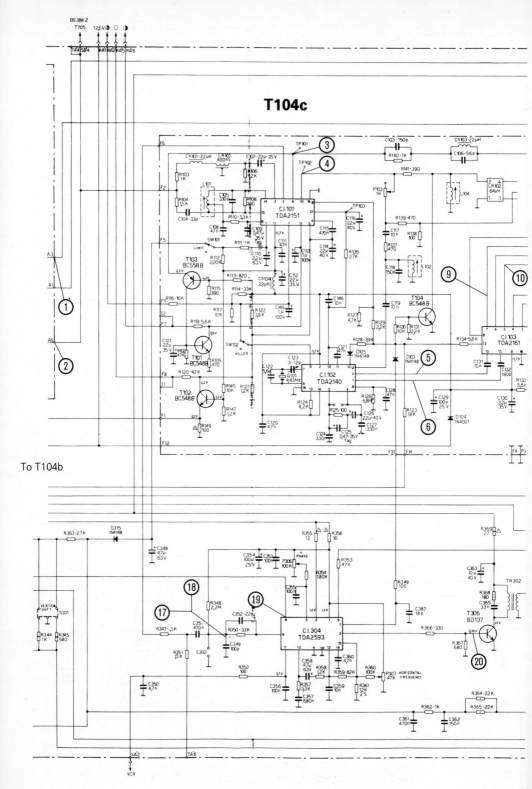

(T104c) CIRCUIT DIAGRAM—MODEL 20ZT505 (*PART*)

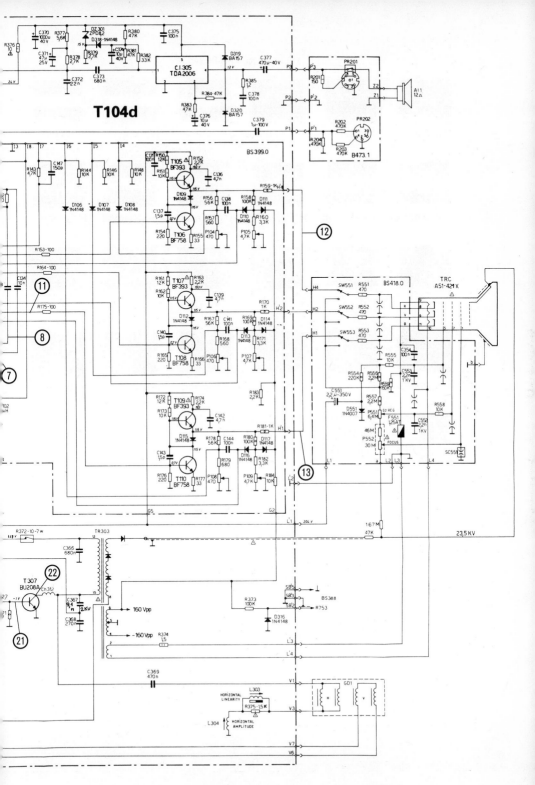

T104d

(T104d) CIRCUIT DIAGRAM—MODEL 20ZT505 (*CONTINUED*)

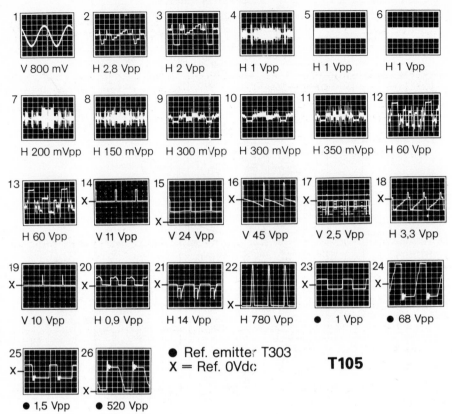

1 V 800 mV
2 H 2,8 Vpp
3 H 2 Vpp
4 H 1 Vpp
5 H 1 Vpp
6 H 1 Vpp

7 H 200 mVpp
8 H 150 mVpp
9 H 300 mVpp
10 H 300 mVpp
11 H 350 mVpp
12 H 60 Vpp

13 H 60 Vpp
14 V 11 Vpp
15 V 24 Vpp
16 V 45 Vpp
17 V 2,5 Vpp
18 H 3,3 Vpp

19 V 10 Vpp
20 H 0,9 Vpp
21 H 14 Vpp
22 H 780 Vpp
23 ● 1 Vpp
24 ● 68 Vpp

25 ● 1,5 Vpp
26 ● 520 Vpp

● Ref. emitter T303
X = Ref. 0Vdc

T105

(T105) CIRCUIT WAVEFORMS—MODEL 20ZT505

396

ZANUSSI   Model 22ZT905

General Description: A 22-in. mains-operated colour television chassis with infra-red remote control. The chassis is isolated but care should be taken when servicing the switch-mode power supply which is 'live' irrespective of mains connection. The R.F./I.F. and Decoder boards are similar to those used in Model 20ZT505 described previously.

Mains Supply: 240 volts± 10%, 50Hz.

Cathode Ray Tube: A56-701X.

E.H.T.: 24·5KV.

Access for Service

Insert a screwdriver in the slots at the outer edges of the rear cover and press down the retaining lugs.

Adjustments

Power Supplies: Adjust P301 for +150V at terminal N5.
Adjust P303 for +12·6V at terminal N1.
Check for +27V at terminals N3 and N4 and for +200V at N6.

Line Oscillator: Bridge S401 and adjust P401 for horizontal drift. Remove link from S401.

Video Adjustments: These follow the procedures outlined for Model 20ZT505, described previously in this volume.

T106

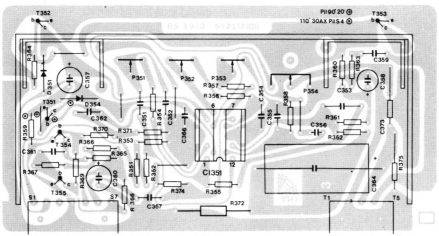

(T106) COMPONENT LAYOUT (FIELD PANEL)—MODEL 22ZT905

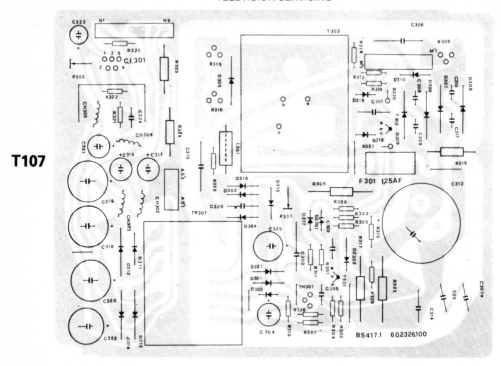

T107

(T107) COMPONENT LAYOUT (SUPPLY MODULE)—MODEL 22ZT905

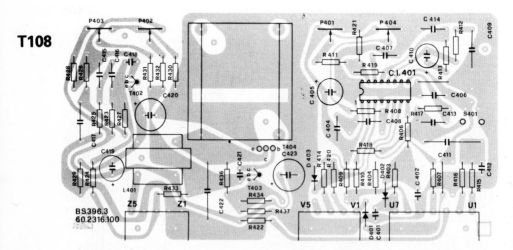

T108

(T108) COMPONENT LAYOUT (SYNC. AND PINCUSHION MODULE)—MODEL 22ZT905

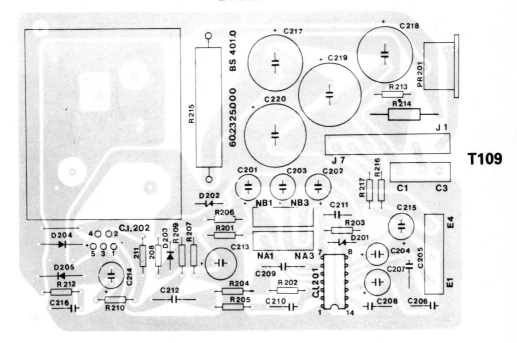

(T109) COMPONENT LAYOUT (AUDIO MODULE)—MODEL 22ZT905

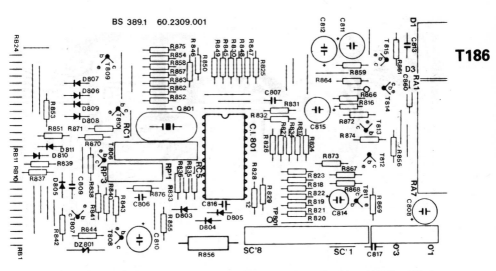

(T186) COMPONENT LAYOUT (REMOTE CONTROL RECEIVER BOARD)—MODEL 22ZT905

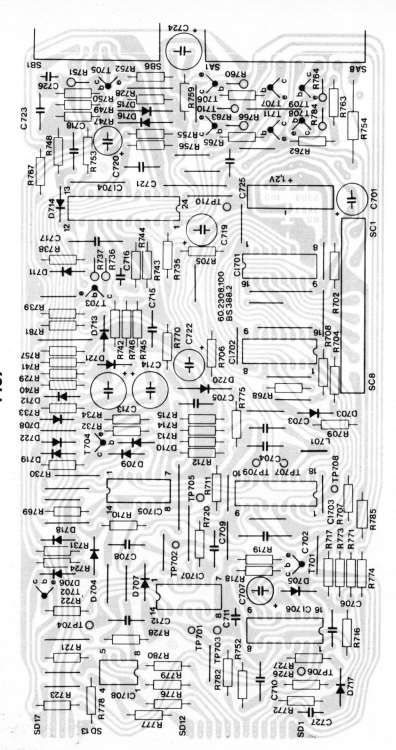

T187

(T187) COMPONENT LAYOUT (TUNING BOARD)—MODEL 22T905

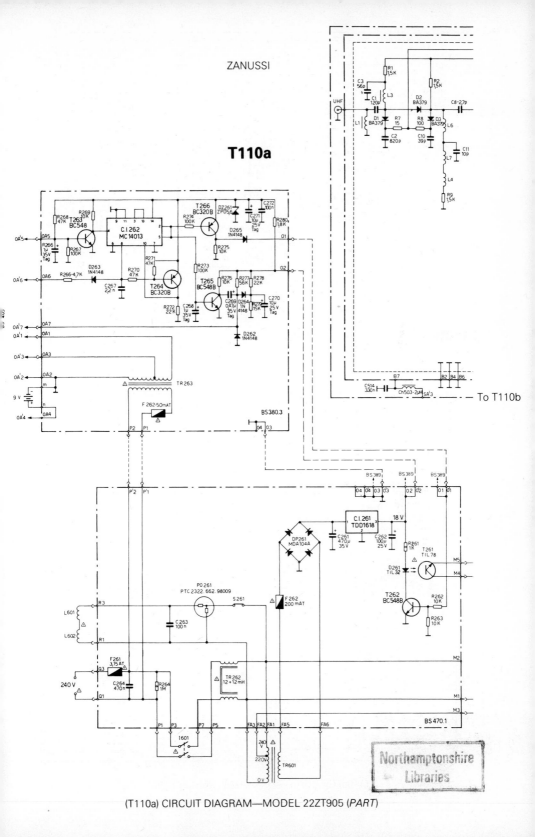

ZANUSSI

T110a

(T110a) CIRCUIT DIAGRAM—MODEL 22ZT905 (*PART*)

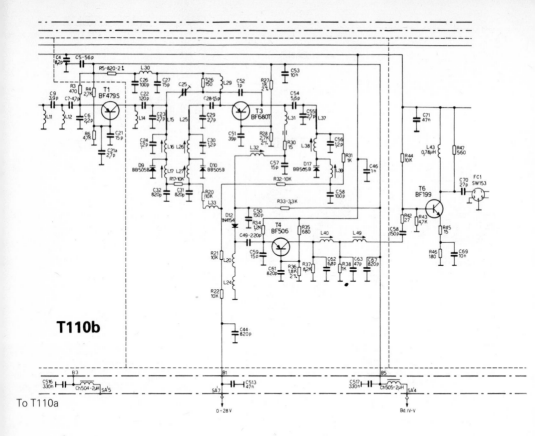

T110b

To T110a

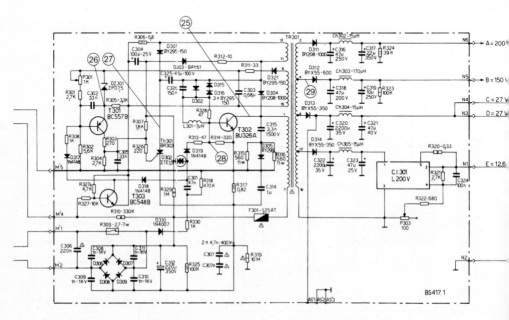

(T110b) CIRCUIT DIAGRAM—MODEL 22ZT905 (*PART*)

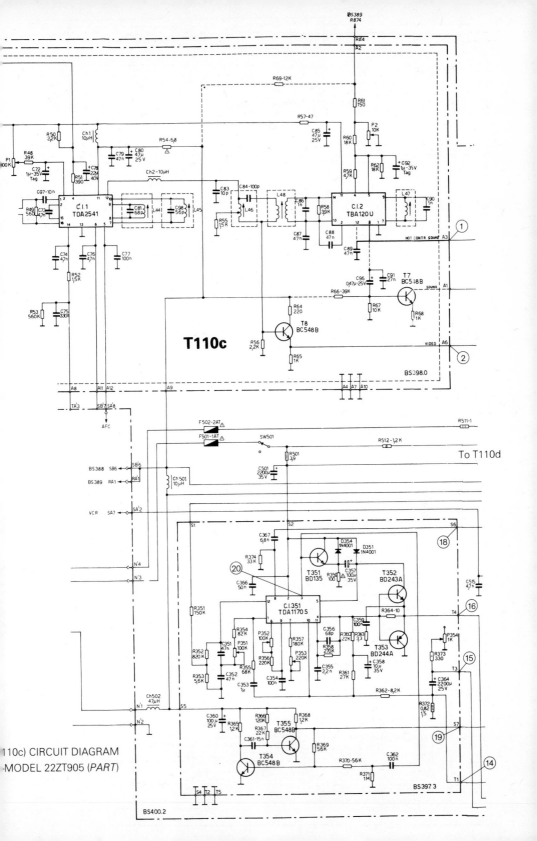

T110c

110c) CIRCUIT DIAGRAM
-MODEL 22ZT905 (*PART*)

To T110d

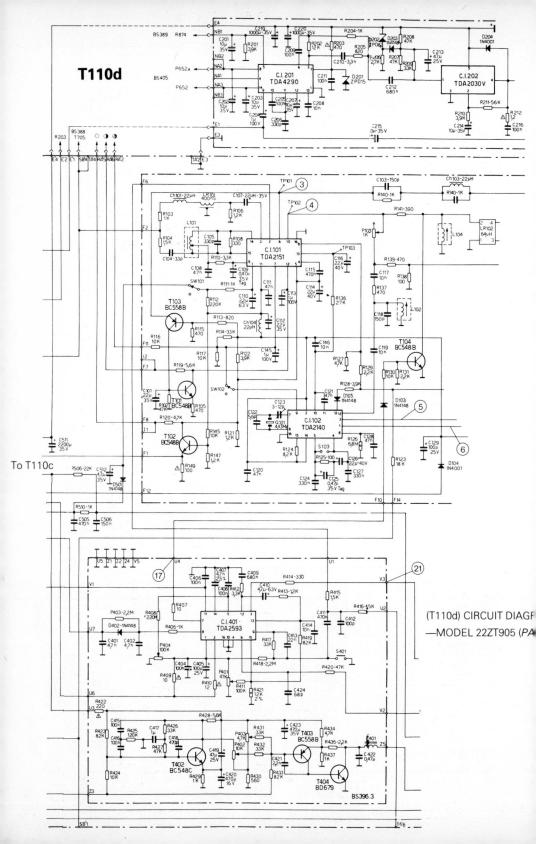

T110d

To T110c

(T110d) CIRCUIT DIAGR
—MODEL 22ZT905 (PA

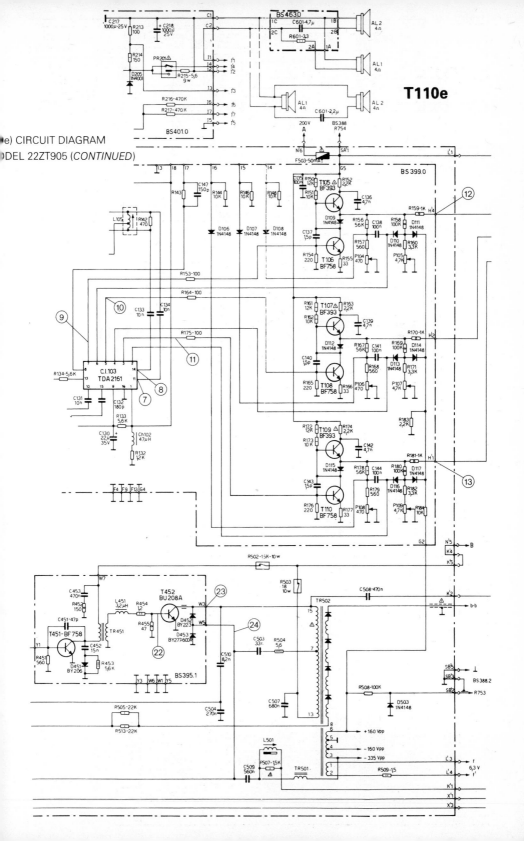

T110e

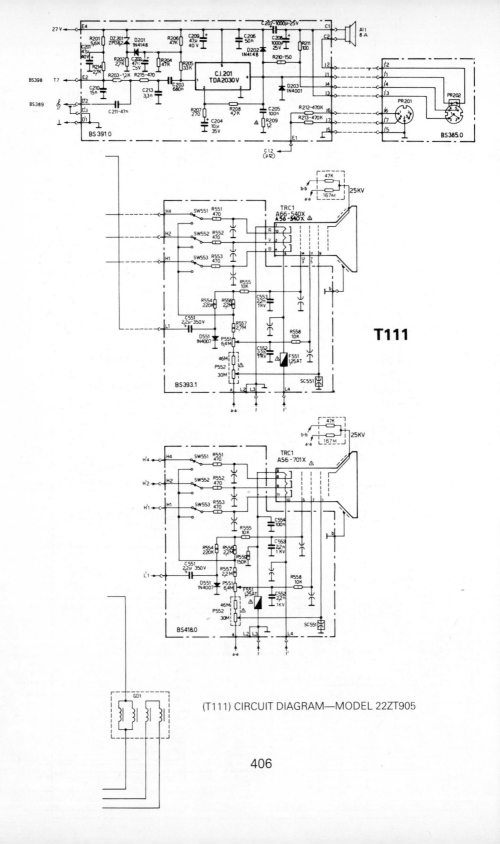

T111

(T111) CIRCUIT DIAGRAM—MODEL 22ZT905

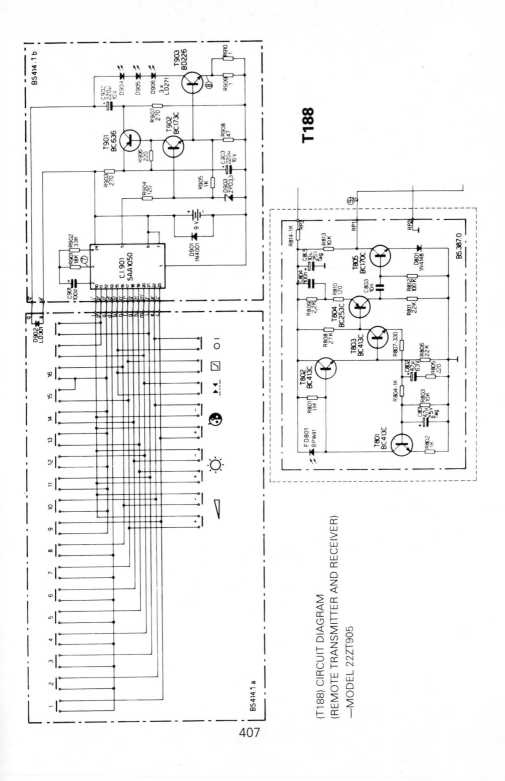

(T188) CIRCUIT DIAGRAM
(REMOTE TRANSMITTER AND RECEIVER)
—MODEL 22ZT905

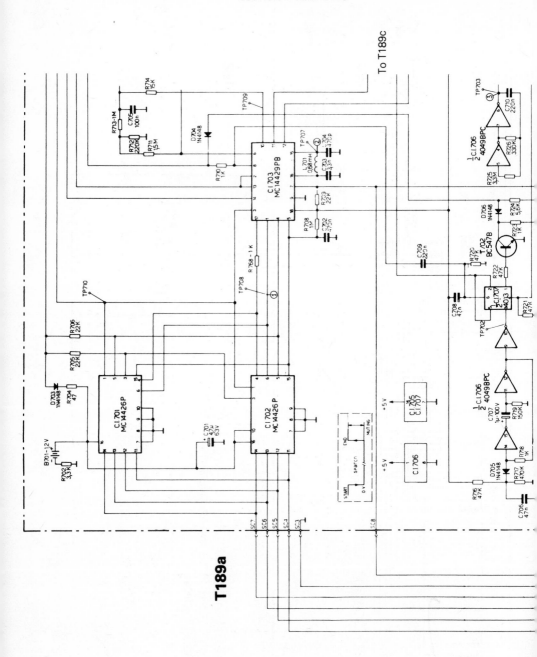

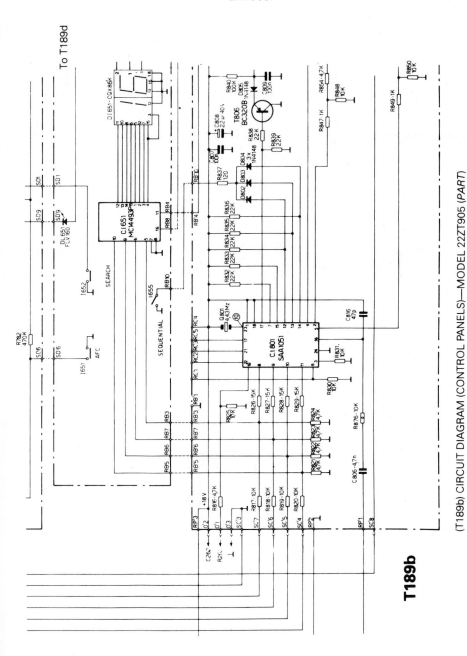

(T189b) CIRCUIT DIAGRAM (CONTROL PANELS)—MODEL 22ZT905 (*PART*)

T189b

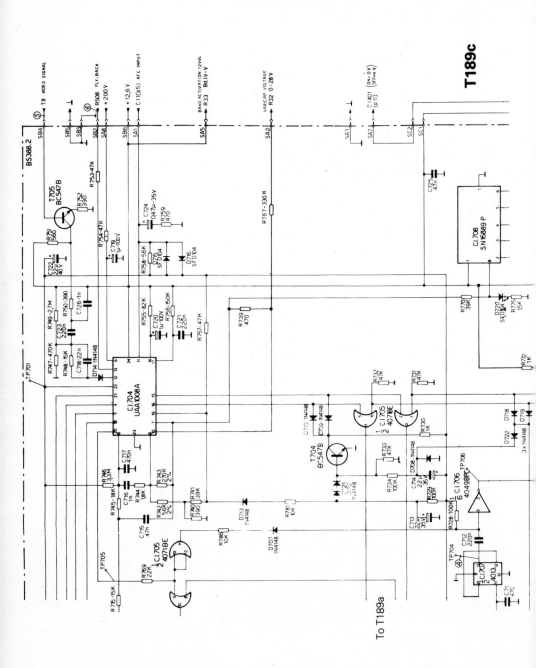

410

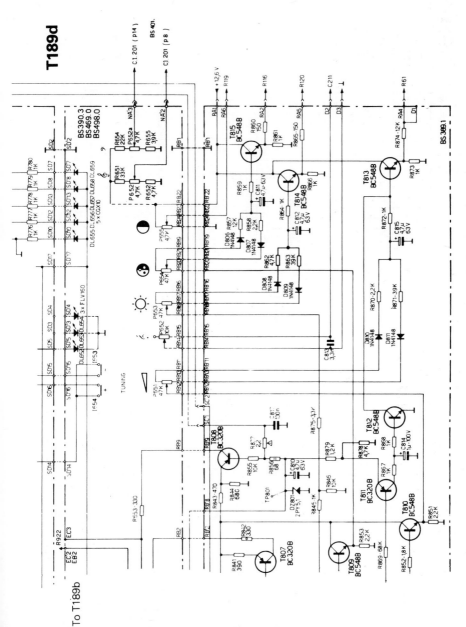

(T189d) CIRCUIT DIAGRAM (CONTROL PANELS)—MODEL 22ZT905 (CONTINUED)

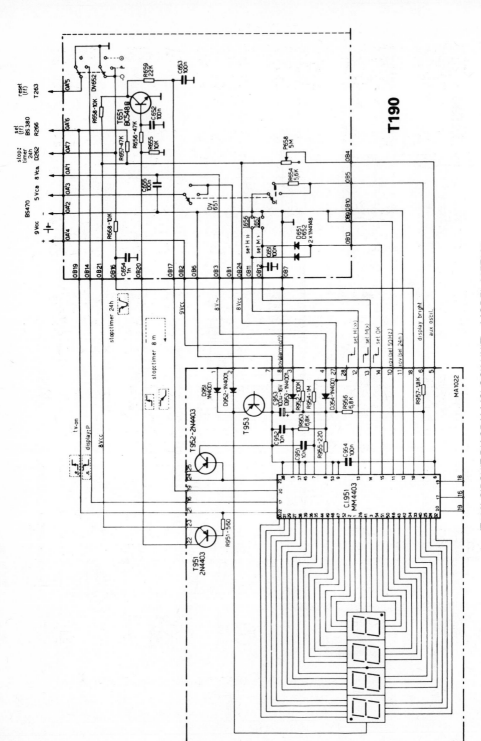

(T190) CIRCUIT DIAGRAM (CLOCK UNIT)—MODEL 22ZT905

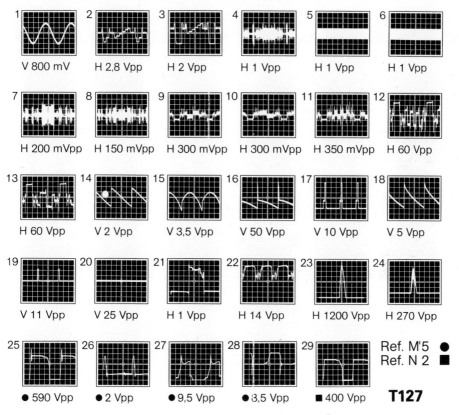

1 V 800 mV	2 H 2,8 Vpp	3 H 2 Vpp	4 H 1 Vpp	5 H 1 Vpp	6 H 1 Vpp
7 H 200 mVpp	8 H 150 mVpp	9 H 300 mVpp	10 H 300 mVpp	11 H 350 mVpp	12 H 60 Vpp
13 H 60 Vpp	14 V 2 Vpp	15 V 3,5 Vpp	16 V 50 Vpp	17 V 10 Vpp	18 V 5 Vpp
19 V 11 Vpp	20 V 25 Vpp	21 H 1 Vpp	22 H 14 Vpp	23 H 1200 Vpp	24 H 270 Vpp

Ref. M'5 ●
Ref. N 2 ■

25	26	27	28	29
● 590 Vpp	● 2 Vpp	● 9,5 Vpp	● 3,5 Vpp	■ 400 Vpp

T127

(T112) CIRCUIT WAVEFORMS—MODEL 22ZT905

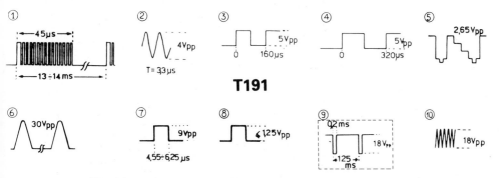

T191

(T191) CIRCUIT WAVEFORMS (CONTROL PANEL)—MODEL 22ZT905

413

RADIO SERVICING
(including Tape Recorders, Record Players, etc.)

ACKNOWLEDGEMENTS

Amstrad Consumer Electronics
Bush Radio Ltd.
Ever Ready Ltd.
Fidelity Radio Ltd.
Intersound Ltd.
J.V.C. (U.K.) Ltd.
Luxor (U.K.) Ltd.
Philips Service
Roberts Radio Ltd.
Sharp (Electronics) U.K. Ltd.
Sony (U.K.) Ltd.
Technical and Optical Equipment (London) Ltd.
Thorn-E.M.I.-Ferguson
Toshiba (U.K.) Ltd.

AMSTRAD   Model 6011

General Description: A multi-band A.M./F.M. portable radio receiver operating from mains or battery supplies. A socket is provided for the connection of an earphone.

Mains Supply: 240 volts, 50Hz.

Batteries: 6 volts (4×HP2).

Wavebands: L.W. 150–300kHz; M.W. 520–1650kHz; F.M./Air 88–135MHz; Marine 4–6MHz; S.W. 7–23MHz; C.B. 27.56–28.1MHz.

Transistor Voltages:

	E	B	C
TR1	0·9	1·0	4·3
TR2	1·55	1·1	4·35
TR3	3·8	4·1	4·35
TR4	3·7	0·9	4·35
TR5	0·3	0·85	3·3
TR6	0·3	0·9	3·15
TR7	0·45	1·1	3·45
TR8	0	0·6	3·4 (C.B.)
TR9	0	0·6	1·35 (C.B.)
TR10	0·7	1·35	4·5
TR11	4·8	0·7	4·2
TR12	0	0·6	0·8
TR13	1·5	0·8	1·5
TR14	2·7	1·1	4·9
TR15	0	0·6	2·65
TR16	3·0	2·25	0·65
TR17	0	0·65	2·95
TR18	3·6	4·2	0
TR19	3·6	3·0	0

All measurements volts D.C.
All voltages taken with AVO 8 Mk. V volume control at minimum. No signal conditions unless otherwise shown.

416

AMSTRAD

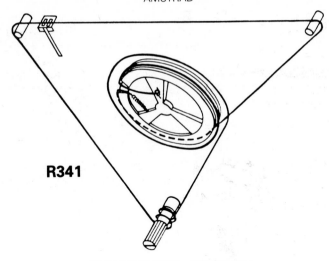

R341

(R341) DRIVE CORD—MODEL 6011

Component List

Circuit Reference	Value
Resistors	
R85	0·5 ohm
R44	4·7 ohm
R80	15 ohm
R49	33 ohm
R38, 39	56 ohm
R3, 6, 66, 76	68 ohm
R7, 41	100 ohm
R47, 82, 83	150 ohm
R22, 55, 84	220 ohm
R10, 17, 28	330 ohm
R11, 18, 65	470 ohm
R29, 43, 62	680 ohm
R13, 20, 25, 30, 33, 36, 51, 56, 59	1K
R1, 67, 70, 71, 81	1K5
R48, 58, 60, 79	2K2
R45	2K7
R4, 42	3K3
R15, 63, 69, 72, 73, 75	4K7
R14, 26, 31, 32, 34, 52, 64	5K6
R12, 19	6K8
R23, 24, 35, 46, 57, 58	10K
R16, 27	15K
R9, 53	22K
R77, 78	33K
R21	47K
R8	100K
R37	220K
R50	330K
R2, 5	470K
R40, 68, 74	680K
R61	1M

Circuit Reference	Value
Ceramic Capacitors	
C6, 25, 55	3pf
C11, 35, 37	5pf
C34	7pf
C9	10pf
C8	15pf
C1	18pf
C3	20pf
C2	30pf
C38	35pf
C36	40pf
C53	47pf
C33, 56	50pf
C39, 50, 52	80pf
C49	150pf
C57	250pf
C51, 81	300pf
C54	380pf
C7, 86	500pf
C27, 29, 31, 60, 72, 98	·001uf
C66, 68, 69	·003uf
C4, 10, 46, 47, 48	·005uf
C21, 22, 45, 65, 91, 92	·01uf
C5, 13–19, 23, 24, 26, 28, 40, 41, 43, 58, 59, 61, 70, 73	·02uf
C44, 62, 71, 82	·04uf

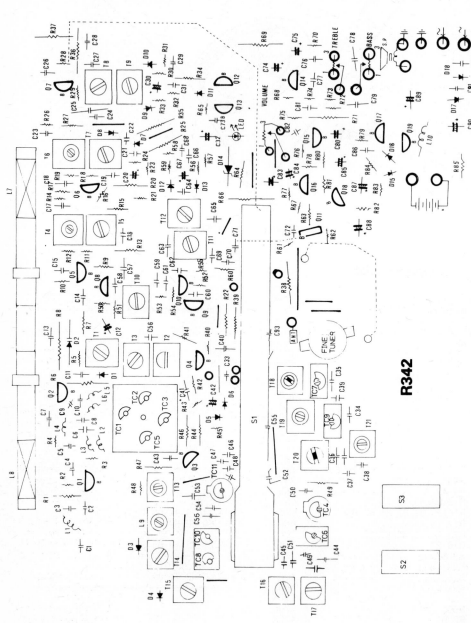

R342

(R342) COMPONENT LAYOUT—MODEL 6011

Mylar Capacitors

0·022uf	C76, 77
0·2uf	C78, 79

Polyester Capacitor

250pf	C63
650pf	C64

Electrolytic Capacitor

0·5uf/16V	C74
1uf/16V	C84
5uf/16V	C75, 80
10uf/16V	C20, 30, 42, 67
100uf/16V	C12, 83, 85, 87
220uf/16V	C32
470uf/16V	C90
1000uf/16V	C88, 89

Transistors

Q1, 2, 3, 4	1417H, BF595
Q5, 10	460C, BF595
Q11, 14, 15, 17	458C, BC237
Q12	9013H, BC237
Q13, 16	9015C, BC212
Q18	5609C, BD371
Q19	5610C, BD370

Diodes

D8–13, 17	1N60
D5, 6, 15, 16	CDG24 1S44
D3, 4, 7	CDG00 1N60
D2	varicap-diode IS2193B
D14	zener 4·5V D.C.
D17, 18	IN4001
L.E.D.	TIL220

Trimmer

TC6, TC7, TC9	8pf
TC8, TC10	2×8pf
TC4, TC11	20pf

Alignment Procedure (See Fig. R342)

A.M. Alignment: Equipment required: R.F. Signal generator. Output coupling coil. V.T.V.M.

ep Function	Signal in	Signal out	Method	Remarks
A.M. I.F. alignment	465KHz gen. via coupling coil to ferrite rod	Monitor signal on V.T.V.M. across C82 & ground	Adjust T2, T3, T5, T7, to get max. output at 465kHz	1. Ensure set switched to M.W. 2. Reduce output as necessary to avoid A.G.C. action 3. Ensure volume control at minimum and tone controls at mid position 4. Tune to bottom end of scale
M.W. osc. alignment—low frequency 525KHz	525KHz gen. via coupling coil to ferrite rod	Monitor signal on V.T.V.M. across C82 & ground	Tune pointer to 525kHz. Adjust T16 for max. output	1. Ensure set switched to M.W. 2. Reduce output as necessary to avoid A.G.C. action 3. Ensure volume control at minimum and tone controls at mid position
M.W. osc. alignment—high frequency 1650KHz	1650KHz gen. via coupling coil to ferrite rod	Monitor signal on V.T.V.M. across C82 & ground	Tune pointer to 1650kHz. Adjust TC6 for max. output	1. Ensure set switched to M.W. 2. Reduce output as necessary to avoid A.G.C. action 3. Ensure volume control at minimum and tone controls at mid position

Step	Function	Signal in	Signal out	Method	Remarks
4	Repeat steps 2 and 3 until no further improvement				
5	M.W. aerial alignment—low frequency 600KHz	600KHz gen. via coupling coil to ferrite rod	Monitor signal on V.T.V.M. across C82 & ground	Tune pointer to 600KHz. Adjust aerial coil L8 for max. output	1. Ensure set switched to M.W. 2. Reduce output as necessary to avoid A.G.C. action 3. Ensure volume contro at minimum and tone controls at mid position
6	M.W. aerial alignment—high frequency 1400KHz	1400KHz gen. via coupling coil to ferrite rod	Monitor signal on V.T.V.M. across C82 & ground	Tune pointer to 1400kHz. Adjust TC5 for max. output	1. Ensure set switched to M.W. 2. Reduce output as necessary to avoid A.G.C. action 3. Ensure volume at minimum and tone controls at mid position
7	Repeat steps 5 and 6 until no further improvement				M.W. alignment now complete. Seal L8 to ferrite rod with wax
8	L.W. osc. alignment—low frequency 145KHz	145KHz gen. via coupling coil to ferrite rod	Monitor signal on V.T.V.M. across C82 & ground	Tune pointer to 145kHz. Adjust T17 for max. on meter	1. Ensure set switched to L.W. 2. Reduce output as necessary to avoid A.G.C. action 3. Ensure volume controls at minimum position and tone controls at mid position
9	L.W. osc. alignment—high frequency 310KHz	310KHz gen. via coupling coil to ferrite rod	Monitor signal on V.T.V.M. across C82 & ground	Tune pointer to 310kHz. Adjust TC4 for max. on meter	1. Ensure set switched to L.W. 2. Reduce output as necessary to avoid A.G.C. action 3. Ensure volume controls at minimum an tone controls at mid position
10	Repeat steps 8 and 9 until no further improvement				
11	L.W. aerial alignment—low frequency 180KHz	180KHz gen. via coupling coil to ferrite rod	Monitor signal on V.T.V.M. across C82 & ground	Tune pointer to 180KHz. Adjust L7 for max. on meter	1. Ensure set switched to L.W. 2. Reduce output as necessary to avoid A.G.C. action 3. Ensure volume controls at minimum an tone controls at mid position
12	L.W. aerial alignment—high frequency 280KHz	280KHz gen. via coupling coil to ferrite rod	Monitor signal on V.T.V.M. across C82 & ground	Tune pointer to 280KHz. Adjust TC3 for max. on meter	1. Ensure set switched to L.W. 2. Reduce output as necessary to avoid A.G.C. action 3. Ensure volume controls at minimum an tone controls at mid position
13	Repeat steps 11 and 12 until no further improvement				L.W. alignment now complete—Seal L7 to ferrite rod with wax

C.B. Alignment: Equipment required: F.M. Signal generator. V.T.V.M.

tep	Function	Signal in	Signal out	Method	Remarks
	C.B. Osc. alignment—low frequency 27·56MHz	Disconnect C.B. antenna wire at point S1C on P.C.B. and inject 27·56MHz direct into P.C.B.	Monitor signal on V.T.V.M. across C82 ground	Adjust T13 for max. output at 27·56MHz	1. Ensure set switched to C.B. 2. Reduce output as necessary to avoid A.G.C. action 3. Ensure volume control at minimum and tone control to mid position
	C.B. Osc. alignment—high frequency 28·1MHz	Disconnect C.B. antenna wire at point S1C on P.C.B. and inject 27·8MHz direct into P.C.B.	Monitor signal on V.T.V.M. across C82 and ground	Adjust TC11 for max. on meter	1. Ensure set switched to C.B. 2. Reduce output as necessary to avoid A.G.C. action 3. Ensure volume control at minimum and tone control to mid position

Repeat steps 1 and 2 until no further improvement

	C.B. aerial alignment 27·69MHz	Disconnect C.B. antenna wire at point C93 on P.C.B. and inject 27·69MHz direct into P.C.B.	Monitor signal on V.T.V.M. across C82 and ground	Adjust T20 for max. on meter	1. Ensure set switched to C.B. 2. Reduce output control to avoid A.G.C. action 3. Ensure volume control at minimum and tone control to mid position
	C.B. aerial alignment 27·89MHz	Disconnect C.B. antenna wire at point C93 on P.C.B. and inject 27·89MHz direct into P.C.B.	Monitor signal on V.T.V.M. across C82 and ground	Ajust T21 for max. on meter	1. Ensure set switched to C.B. 2. Reduce output as necessary to avoid A.G.C. action 3. Ensure volume control at minimum and tone control to mid position

Repeat steps 3 and 4 until no further improvement

F.M./Air Alignment: Equipment required: R.F. Signal generator 38–108MHz). 10·7MHz Sweep generator. Oscilloscope. V.T.V.M.

tep	Function	Signal in	Signal out	Method	Remarks
	F.M. I.F. alignment 10·7MHz	Inject 10·7MHz sweep sig to base Q2	Connect osc. to junction R36/C28	Adjust T1, 4, 6, 8, 9 for symmetrical S curve	1. Ensure set switched to F.M. 2. Reduce output as necessary to avoid A.G.C. action 3. Ensure volume control at minimum and tone controls at mid position
	F.M. osc. alignment—low frequency 87·5MHz	Inject 87·5MHz sig to base Q2	Monitor V.T.V.M. across C82 and ground	Tune pointer to 87·5MHz. Adjust L5 and L6 for max. output	1. Ensure set switched to F.M. 2. Reduce output as necessary to avoid A.G.C. action 3. Ensure volume control at minimum and tone controls at mid position

Step	Function	Signal in	Signal out	Method	Remarks
3	F.M. osc. alignment—high frequency 136MHz	Inject 136MHz sig to base Q2	Monitor V.T.V.M. across C82 and ground	Tune pointer to 136MHz. Adjust TC2 for max. output	1. Ensure set switched to F.M. 2. Reduce output as necessary to avoid A.G.C. action 3. Ensure volume control at minimum and tone controls at mid position
4	Repeat steps 2 and 3 until no further improvement				
5	F.M. low frequency— 90MHz	Inject 90MHz sig to base Q2	Monitor V.T.V.M. across C82 and ground	Tune pointer to 90MHz. Adjust L2 and L3 for max. output	1. Ensure set switched to F.M. 2. Reduce output as necessary to avoid A.G.C. action 3. Ensure volume control at minimum and tone controls at mid position
6	F.M. high frequency— 130MHz	Inject 130MHz sig to base Q2	Monitor V.T.V.M. across C82 and ground	Tune pointer to 130MHz. Adjust TC1 for max. output	1. Ensure set switched to F.M. 2. Reduce output as necessary to avoid A.G.C. action 3. Ensure volume control at minimum and tone controls at mid position
7	Repeat steps 5 and 6 until no further improvement				F.M. alignment now complete

S.W./M.B. Alignment: Equipment required: R.F. Signal generator. Output coupling coil. V.T.V.M.

Step	Function	Signal in	Signal out	Method	Remarks
1	S.W. osc. alignment—low frequency 6·9MHz	Disconnect S.W. ant. wire at point S1C on P.C.B. and inject 6·9MHz direct into P.C.B.	Monitor signal on V.T.V.M. across C82 and ground	Adjust T14 for max. output at 6·9MHz	1. Ensure set switched to S.W. 2. Reduce output as necessary to avoid A.G.C. action 3. Ensure volume control at minimum and tone controls at mid position 4. Tune pointer to low frequency end
2	S.W. osc. alignment—high frequency 23·5MHz	Disconnect S.W. ant. wire at point S1C on P.C.B. and inject 23·5MHz direct into P.C.B.	Monitor signal on V.T.V.M. across C82 and ground	Adjust TC10 for max. on meter	1. Ensure set switched to S.W. 2. Reduce output as necessary to avoid A.G.C. action 3. Ensure volume control at minimum and tone controls at mid position 4. Tune pointer to high frequency end
3	Repeat steps 1 and 2 until no further improvement				

ɘp	Function	Signal in	Signal out	Method	Remarks
	S.W.1 aerial alignment—low frequency 8MHz	Disconnect S.W. ant. wire at point S1C on P.C.B. and inject 8MHz direct into P.C.B.	Monitor signal on V.T.V.M. across C82 and ground	Tune pointer to 8MHz. Adjust ant coil T19 for max. on meter	1. Ensure set switched to S.W.1 2. Reduce output as necessary to avoid A.G.C. action 3. Ensure volume control at minimum and tone controls at mid position
	S.W.1 aerial alignment—high frequency 22MHz	Disconnect S.W. ant. wire at point S1C on P.C.B. and inject 22MHz direct into P.C.B.	Monitor signal on V.T.V.M. across C82 and ground	Tune pointer to 22MHz. Adjust TC9 for max. on meter	1. Ensure set switched to S.W.1 2. Reduce output as necessary to avoid A.G.C. action 3. Ensure volume control at minimum and tone controls at mid position
	Repeat steps 4 and 5 until no further improvement				S.W. aerial alignment now complete
	M.B. osc. alignment—low frequency 3·9MHz	Disconnect M.B. ant. wire at point S1C on P.C.B. and inject 3·9MHz direct into P.C.B.	Monitor signal on V.T.V.M. across C82 and ground	Adjust T15 for max. on meter	1. Ensure set switched to M.B. 2. Reduce output as necessary to avoid A.G.C. action 3. Ensure volume control at minimum and tone controls at mid position 4. Tune pointer to low frequency end
	M.B. osc. alignment—high frequency 6·5MHz	Disconnect M.B. ant. wire at point S1C on P.C.B. and inject 6·5MHz direct into P.C.B.	Monitor signal on V.T.V.M. across C82 and ground	Adjust TC8 for max. on meter	1. Ensure set switched to M.B. 2. Reduce output as necessary to avoid A.G.C. action 3. Ensure volume control at minimum and tone controls at mid position
)	Repeat steps 1 and 2 until no further improvement				
1	M.B. aerial alignment—low frequency 4MHz	Disconnect M.B. ant. wire at point S1C on P.C.B. and inject 4MHz direct into P.C.B.	Monitor signal on V.T.V.M. across C82 and ground	Tune pointer to 4MHz. Adjust T18 for max. on meter	1. Ensure set switched to M.B. 2. Reduce output as necessary to avoid A.G.C. action 3. Ensure volume control at minimum and tone controls at mid position
2	M.B. aerial alignment—high frequency 6MHz	Disconnect M.B. ant. wire at point S1C on P.C.B. and inject 6MHz direct into P.C.B.	Monitor signal on V.T.V.M. across C82 and ground	Tune pointer to 6MHz. Adjust TC7 for max. on meter	1. Ensure set switched to M.B. 2. Reduce output as necessary to avoid A.G.C. action 3. Ensure volume control at minimum and tone controls at mid position
3	Repeat steps 4 and 5 until no further improvement				M.B. alignment now complete

R343

(R343) CIRCUIT DIAGRAM—MODEL 6011

AMSTRAD   Model 8060

General Description: An A.M./F.M. stereo radio incorporating a stereo cassette-recorder and record-player operating from mains or battery supplies. Integrated circuits are used in the I.F. decoder, power amplifier, and motor control stages. Sockets are provided for the connection of separate loud-speaker units, headphones, microphones, and power supply. Direct recording is possible by built-in electret condenser microphones.

Mains Supply: 240 volts, 50Hz.

Batteries: 12 volts.

R208a

To R208c

R208b

(R208b) CIRCUIT DIAGRAM—MODEL 8060 (*PART*)

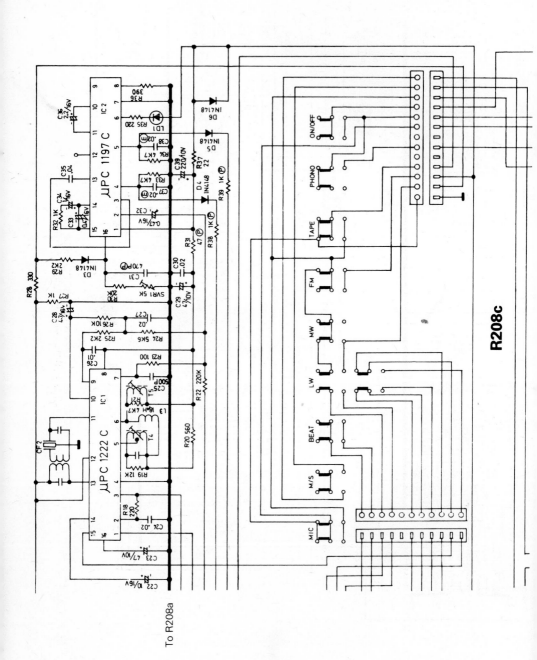

R208c

To R208a

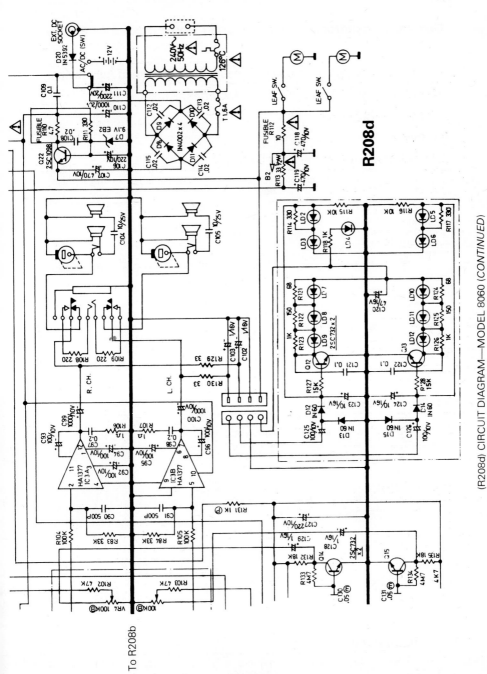

R208d

(R208d) CIRCUIT DIAGRAM—MODEL 8060 (CONTINUED)

To R208b

BUSH   Model 6090

General Description: A three-waveband stereo radio-cassette recorder with built-in microphones and operating from mains or battery supplies. Sockets are provided for auxiliary inputs, headphones, and power supplies, with signal processing by integrated circuits in decoder and A.F. stages.

Mains Supply: 240 volts, 50Hz. **Batteries:** 6 volts.

Quiescent Current: 80mA.

Alignment

L.W. Alignment (Set the Band Switch in position L.W.):

Step	Signal source connection	Signal frequency	Indicator connection	Dial setting of unit	Adjustment
1	R.F. generator to a test loop or a short wire close to the ferrite core	145KHz	V.T.V.M. to output terminal	145KHz tuning closed	
2		265KHz		265KHz tuning open	TC5 for maximum
	Repeat steps 1 and 2				
3	R.F. generator to a test loop or a short wire close to the ferrite core	180KHz	V.T.V.M. to output terminal	180KHz	L107 for maximum
4		240KHz		240KHz	
	Repeat steps 3 and 4				

F.M. Stereo Alignment:

Step	Circuits	Cassette Mode	Indicator connection	Function S.W.	Adjustment	Remarks
1	P.L.L. V.C.O. Adj.		Frequency counter at TP-7		VR-10KB	19KHz ±200
2	98MHz Ant. input 60dB F.M. S.S.G. Pilot 19KHz 8% L−0% R−45%	F.M. Ant. Terminal through dummy Ant.	L. CH V.T.V.M. and scope. Min. output −20dB	98MHz		VR-2 VR-1 if necessary for channel balance
3	98MHz Ant. input 60dB F.M. S.S.G. Pilot 19MHz 8% R−0% L−45%		R. CH V.T.V.M. and scope. Min. output −20dB			
4			Stereo indicator			Stereo indica bright at und 30dB R.F. Se

Tape Alignment Instruction:

Step	Signal source connection	Signal feed	Indicator connection	Dial setting of unit	Adjustment	Remarks
1	A.C. Bias oscillator	Play and Rec.	Frequency counter at TP-8	Tape position	L203	71KHz ±1KH

F.M. I.F. Alignment Procedures (Band Switch in F.M. position):

ep	Signal Source Connection	Signal frequency	Indicator connection	Dial setting of the unit	Adjustment
	Sweep generator to TP2 and ground	10·7MHz Marker at 10·6, 10·7 and 10·8MHz	Oscilloscope to point TP3	Quiet point on band	T101, T102, T103 for max. and symmetric response
			Oscilloscope to point TP5		T104, T105 for straight and response
	Repeat steps 1 and 2				

F.M. R.F. Alignment (Band Switch in F.M. position):

ep	Signal source connection	Signal frequency	Indicator connection	Dial setting of the unit	Adjustment
	R.F. generator to F.M. Ant. terminals TP1 through matching network	87MHz	V.T.V.M. to output terminal	87MHz with tuning gang, closed	L104 for maximum
		109MHz		109MHz with tuning gang, open	TC2 for maximum
		90MHz		90MHz	L102 for maximum
		106MHz		106MHz	TC1 for maximum
	Repeat steps 1 to 4				
	Check overall response curve and repeat above steps as necessary to obtain maximum sensitivity				

M.W. I.F. Alignment (Band Switch in Positon M.W.):

ep	Signal source connection	Signal frequency	Indicator connection	Dial setting of the unit	Adjustment
	Sweep generator to TP6	Marker 465KHz	Oscilloscope to TP4	Quiet point on band	T106, T107, T108 maximum and symmetrical
	Repeat steps 1 and 2				

M.W. R.F. Alignment (Set the Band Switch in Position M.W.):

tep	Signal source connection	Signal frequency	Indicator connection	Dial setting of the unit	Adjustment
	R.F. generator to a test loop or a short wire close to ferrite core	515KHz	V.T.V.M. to output terminal	515KHz tuning gang, closed	L106 for maximum
		1630KHz		1630KHz tuning gang, open	TC4 for maximum
	Repeat steps 1 and 2				
	R.F. generator to a test loop	600KHz	V.T.V.M. to output terminal	600KHz	L105 for maximum
	or a short wire close to	1400KHz		1400KHz	TC3 for maximum

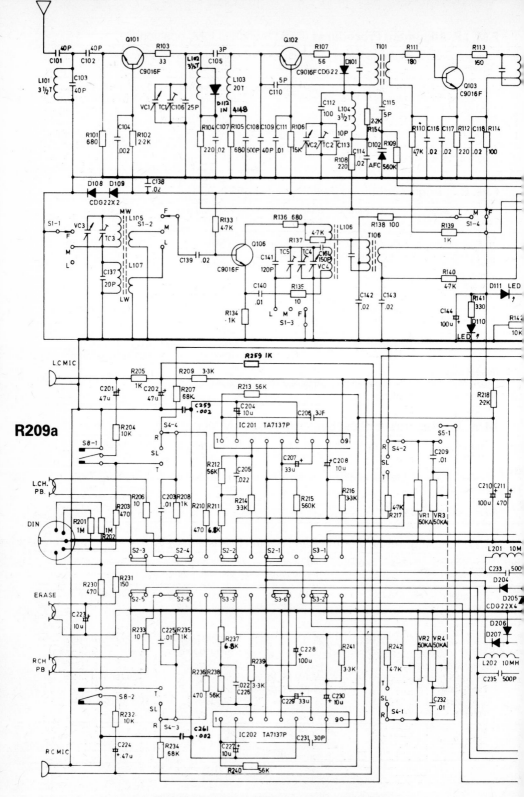

(R209a) CIRCUIT DIAGRAM—MODEL 6090 (*PART*)

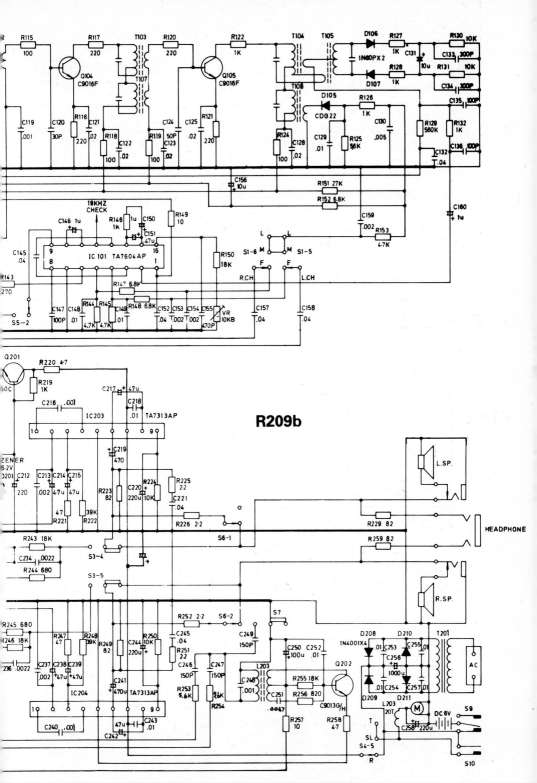

(R209b) CIRCUIT DIAGRAM—MODEL 6090 (*CONTINUED*)

BUSH   Model 9450

General Description: A portable A.M./F.M. stereo radio receiver with twin stereo cassette tape-recorders for tape transfer purposes. The instrument operates from mains or battery supplies, microphones are 'built in' and sockets are provided for the connection of stereo pick-up, microphones, and headphones.

Mains Supply: 240 volts, 50Hz.   **Batteries:** 9 volts (6×HP2).

Wavelengths: L.W.150–300kHz; M.W.540–1600kHz; F.M.87·5–108MHz.

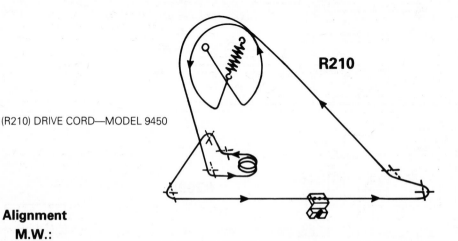

R210

(R210) DRIVE CORD—MODEL 9450

Alignment
M.W.:

Step	Signal source connected to	Set signal generator to	Set radio dial to	Signal O/P connector	Adjust	Adjust for
1	Set function switch to M.W. position					
2		I.F. frequency	gang open	A.C. voltmeter across speaker terminal	T28, T24, T27	Max.
3	Repeat step 2 to obtain max. and ideal waveform					
4	R.F. generator with standard radiating loop antenna	525KHz	gang closed	A.C. voltmeter across speaker terminal	Osc. coil L8	Max.
5	I.F. frequency accordingly	1625KHz	gang open		Osc. trimmer	Max.
6	Repeat steps 4 and 5 to obtain correct frequency coverage					
7		600KHz	600KHz rock gang	A.C. voltmeter across speaker terminal	ferrite bar antenna coil	Max.
8		1400KHz	1400KHz rock gang		M.W. antenna trimmer	Max.
9	Repeat steps 7 and 8 to obtain best tracking					

L.W.:

Step	Signal source connected to	Set signal generator to	Set radio dial to	Signal O/P connector	Adjust	Adjust for
1	Set function switch to L.W. position					
2						
3						
4	R.F. generator with standard radiating loop	145KHz	gang closed	A.C. volmeter across speaker terminal	Osc. coil (L4)	Max.
5	I.F. frequency accordingly	310KHz	gang open		Osc. trimmer (C222= 20P)	
6	Repeat steps 4 and 5 to obtain corrected frequency coverage					
7		165KHz	165KHz rock gang	A.C. voltmeter across speaker terminal	ferrite bar antenna coil (L1)	Max.
8		280KHz	280KHz rock gang		antenna trimmer (C221= 20P)	
9	Repeat steps 7 and 8 to obtain best tracking					

F.M.:

Step	Signal source connected to	Set signal generator to	Set radio dial to	Signal O/P connector	Adjust	Adjust for
1	Set function switch to F.M. position					
2	Emitter of Q2 and ground F.M. sweep generator thru. two 0·4 uf capacitors	I.F. frequency 10·7MHz	gang open	oscilloscope	T21 T22 T23 T25 T26	Max. amplitude balance curve with 10·7MHz marker
3	Repeat step 2 to obtain max. amplitude balance curve					
4		87MHz	gang closed	A.C. voltmeter across speaker terminal	Osc. coil L24 F.M.	Max.
5	Marker generator, F.M. antenna and ground	109MHz	gang open		Osc. trimmer	
6	Repeat steps 4 and 5 to obtain correct frequency coverage					
7		90MHz	90MHz rock gang	A.C. voltmeter across speaker terminal	antenna coil L22 F.M.	Max.
8		106MHz	106MHz rock gang		antenna trimmer	
9	Repeat steps 7 and 8 to obtain best tracking					

Continued on p. 440

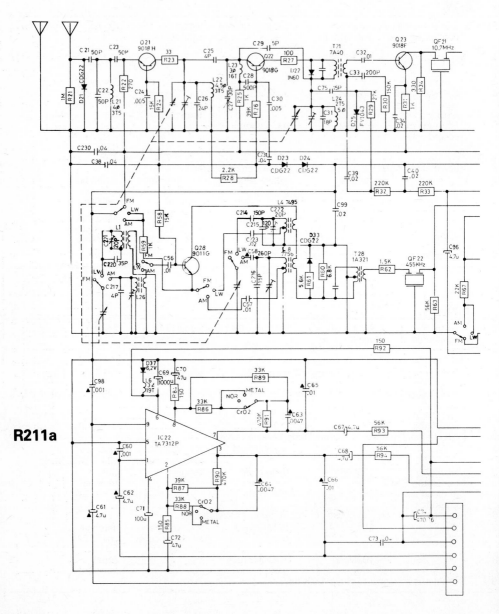

(R211a) CIRCUIT DIAGRAM (RADIO SECTION)—MODEL 9450 (*PART*)

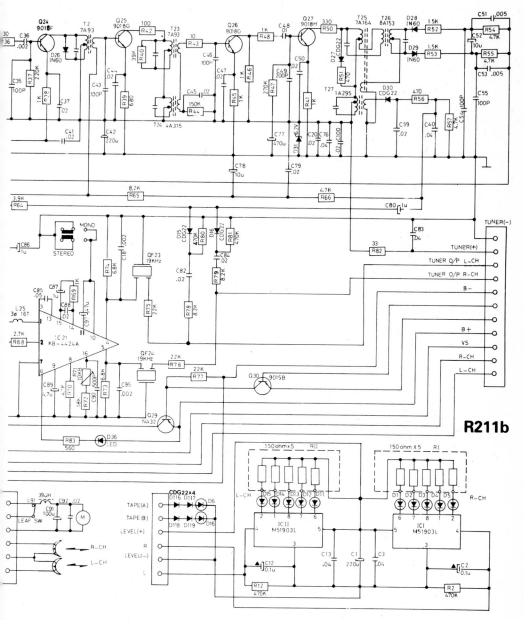

R211b

(R211b) CIRCUIT DIAGRAM (RADIO SECTION)—MODEL 9450 (*CONTINUED*)

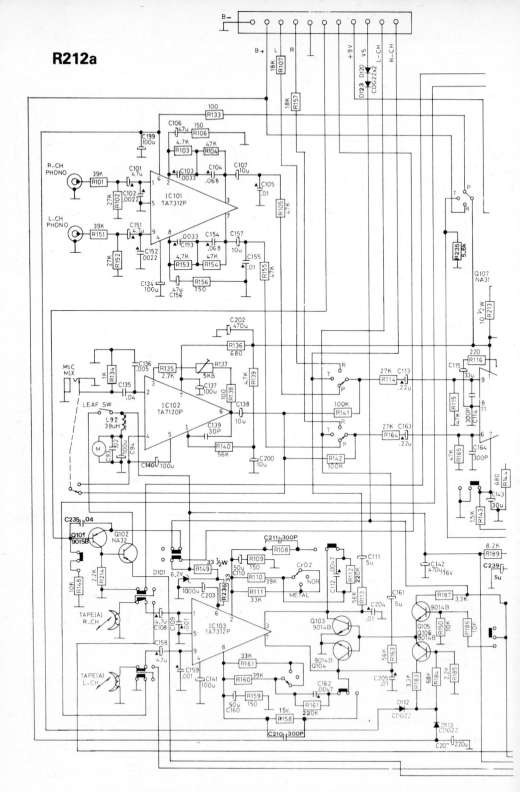

R212a

(R212a) CIRCUIT DIAGRAM (TAPE SECTION)—MODEL 9450 (*PART*)

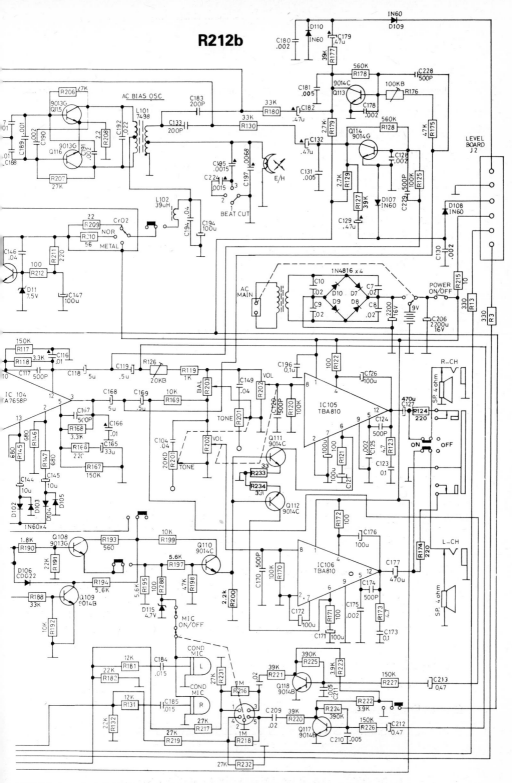

R212b

(R212b) CIRCUIT DIAGRAM (TAPE SECTION)—MODEL 9450 (*CONTINUED*)

F.M.-M.P.X.:

Step	Signal source connected to	Set signal generator to	Set radio dial to	Signal O/P connector	Adjust	Adjust for
1	Set function switch to F.M. stereo position					
2	Stereo signal generator connected to F.M. antenna and ground terminal	98MHz	98MHz rock gang	frequency counter across I.C. pin 12 and G.N.D.	10KB trimmer on I.C. pin 16	19KHz and stereo lamp ON

Head Adjustment: Connect a standard dummy load (4 ohm resistor) to the external speaker jack, and connect a high-sensitivity A.C. voltmeter or an oscilloscope across the dummy load, playback a standard test tape (6·3KHz) with the Volume control at a suitable level, and adjust the Azimuth by rotating the adjusting screw for maximum output. After completing this alignment, lock the screw with a screw lock or enamel.

EVER READY · Model 'Skytime'

General Description: A three-waveband A.M./F.M. radio receiver covering Long, Medium, and V.H.F. wavebands and incorporating a digital clock with liquid crystal display. The instrument operates from battery supplies with a second battery to operate and illuminate the display. An integrated circuit is used for audio amplification and a socket is provided for the connection of an earphone.

Batteries: 9 volts (PP9) plus 1·5 volts (clock).

Modifications: C40 deleted; C47 (20pf); C62 (0·01uF); R47 (1OK).

Circuit Diagram Notes

S.W.1 F.M. position; S.W.2 On; S.W.3 Radio; S.W.4 Run.

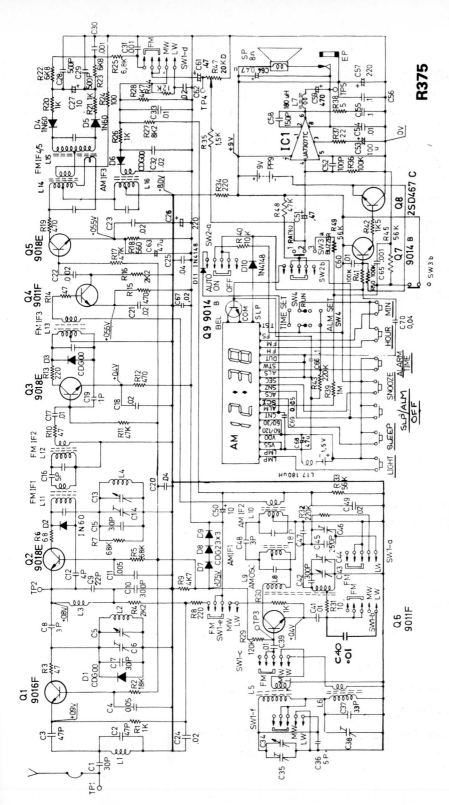

(R375) CIRCUIT DIAGRAM—MODEL 'SKYLINE'

R375

FERGUSON   Model 3R05

General Description: A three-waveband A.M./F.M. portable radio receiver for mains or battery operation. A socket is provided for the connection of an earphone.

Mains Supply: 240 volts, 50Hz.

Batteries: 6 volts (4×HP7).

Wavebands: L.W. 160–260kHz; M.W. 540–1600kHz; F.M. 88–108MHz.

Loudspeaker: 8 ohms impedance.

Access for Service

Remove 4 screws to release back cover. The printed circuit board may be withdrawn after removing 5 fixing screws and the screw from the bottom of the telescopic aerial.

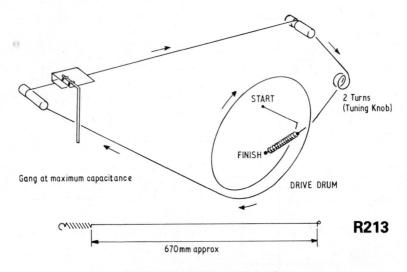

START

2 Turns
(Tuning Knob)

FINISH

Gang at maximum capacitance

DRIVE DRUM

R213

670mm approx

(R213) DRIVE CORD—MODEL 3R05

Alignment (See Fig. R214)

Tuning indication can be obtained by connecting a 20,000 Ω/V meter, set to the 10V A.C. range, across the loudspeaker terminals.

Throughout alignment the signal input to the receiver should be adjusted, as necessary, so that the meter reading does not exceed 0·63V (50mW), with maximum volume.

442

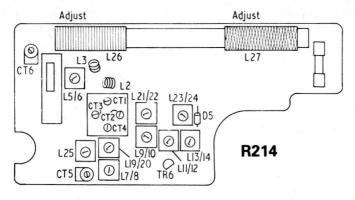

(R214) ALIGNMENT ADJUSTMENTS—MODEL 3R05

Signals should be injected via a loop loosely coupled to the ferrite rod aerial. 400Hz (30% modulated).

A.M. I.F. Circuits: Switch to M.W., set cursor to the low frequency end and the volume to minimum. Position the A.M. aerial coil close to the generator aerial and inject a 468kHz signal. Connect the oscilloscope to the anode of D5 and adjust L19/20, L21/22 and L23/24 for maximum output. Repeat procedure until no further improvement is obtained.

A.M. R.F. Circuits:

Range	Inject	Cursor position	Adjust for max.
M.W.	520kHz	Extreme left	L25
	1650kHz	Extreme right	CT3
	600kHz	600kHz	L26*
	1400kHz	1400kHz	CT4
L.W.	270kHz	Extreme right	CT5
	160kHz	160kHz	L27*
	240kHz	240kHz	CT6

* Adjust by sliding coil former along ferrite rod.

F.M. I.F. Circuits: Alignment can be achieved using a wobbulator having V.H.F./F.M. facilities and an oscilloscope. The wobbulator should be terminated with a 75 Ω resistor across the output terminals and a 10pF capacitor connected in series with the 'live' lead to the injection point.

Switch receiver to F.M., set the cursor to the low frequency end and the volume to minimum. Connect the oscilloscope to the collector of TR6 and inject a 10·7MHz signal between TP3/4. Adjust L5/6, L7/8 and L9/10 for maximum signal. Move the oscilloscope to TP5/6 and adjust L11/12 and L13/14 for overall symmetry of 'S' curve. Repeat procedure until no further improvement is obtained.

F.M. R.F. Circuits: Connect the oscilloscope across the speaker and with volume maximum inject V.H.F./F.M. signals (25 kHz deviation) between TP1/2.

Range	Inject	Cursor position	Adjust for max.
F.M.	87MHz	Extreme left	L3†
	109MHz	Extreme right	CT1
	90MHz	90MHz	L2†
	106MHz	106MHz	CT2

† Adjust by slightly opening or closing coil turns.

Repeat procedure until no further improvement is obtained.

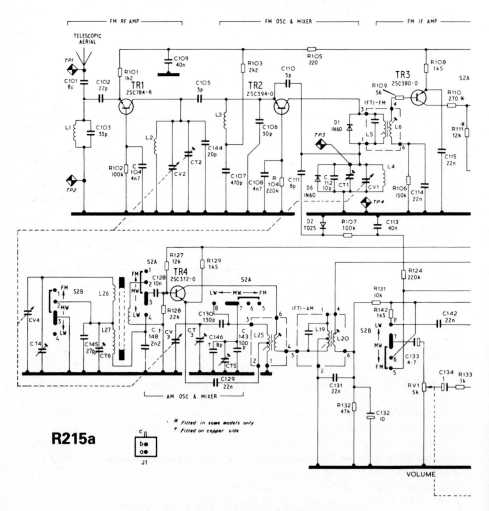

(R215a) CIRCUIT DIAGRAM—MODEL 3R05 (*PART*)

444

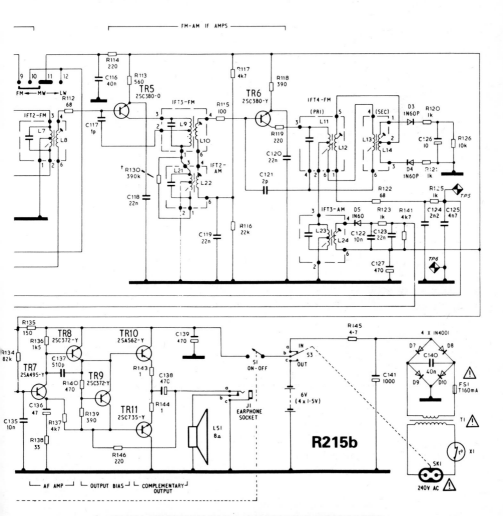

(R215b) CIRCUIT DIAGRAM—MODEL 3R05 (*CONTINUED*)

FERGUSON   Model 3R06

General Description: A three-waveband A.M./F.M. portable radio receiver operating from mains or battery supplies. A socket is provided for the connection of an earphone.

Mains Supply: 240 volts, 50Hz.

Battery: 6 volts.

Wavebands: L.W. 145–270kHz; M.W. 510–1640kHz; F.M. 87·5–104·5MHz.

Loudspeaker: 8 ohms impedance.

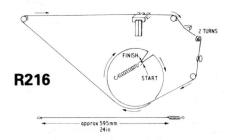

(R216) DRIVE CORD—MODEL 3R06

Access for Service

Remove 4 back cover screws and two printed circuit board screws.

Alignment

A.M. Alignment:

Range	Connections	Inject	Cursor position	Adjust for max.
I.F.	Inject signals via loop loosely coupled to ferrite rod aerial	468kHz	Extreme right	L106 L107
M.W.	Output meter in place of loudspeaker. Volume control at max. Output not to exceed 50mW	510kHz 1640kHz 600kHz 1400kHz	Extreme left Extreme right 600kHz 1400kHz	L110 C133 L108* C128
L.W.		270kHz 145kHz	270kHz Extreme left	L109* C137

* Adjust by sliding coils along ferrite rod.

Repeat adjustments until no further improvement results.

F.M. Alignment:

Range	Connections	Inject	Cursor position	Adjust for max.
I.F.	Sweep generator to TP2 with 100nF capacitor in series with 'live' lead. Oscilloscope to TP4	10·7MHz	Extreme right	L105 L111 L112*
R.F.	Signal generator to TP1	87·5MHz 104·5MHz 90MHz 102MHz	Extreme left Extreme right 90MHz 102MHz	L104 C112 L102† C104

* Adjust by symmetrical 'S' curve.
† Adjust by slightly opening or closing coil turns.

Repeat adjustments until no further improvement results.

Important Note: Earth rails are of different polarities, i.e. bottom of L101 is not the same as RV101. Care must therefore be taken when connecting test equipment.

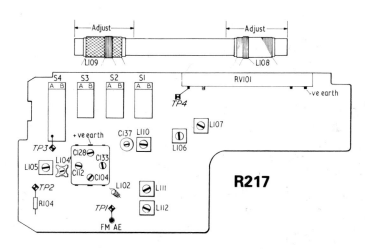

(R217) ALIGNMENT ADJUSTMENTS—MODEL 3R06

R218a

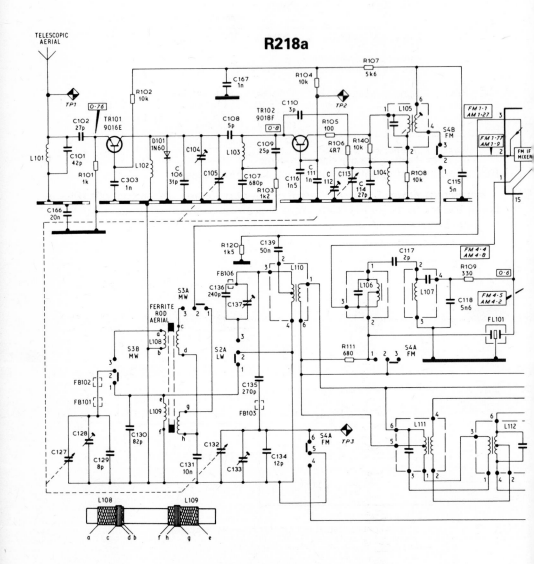

(R218a) CIRCUIT DIAGRAM—MODEL 3R06 (*PART*)

R218b

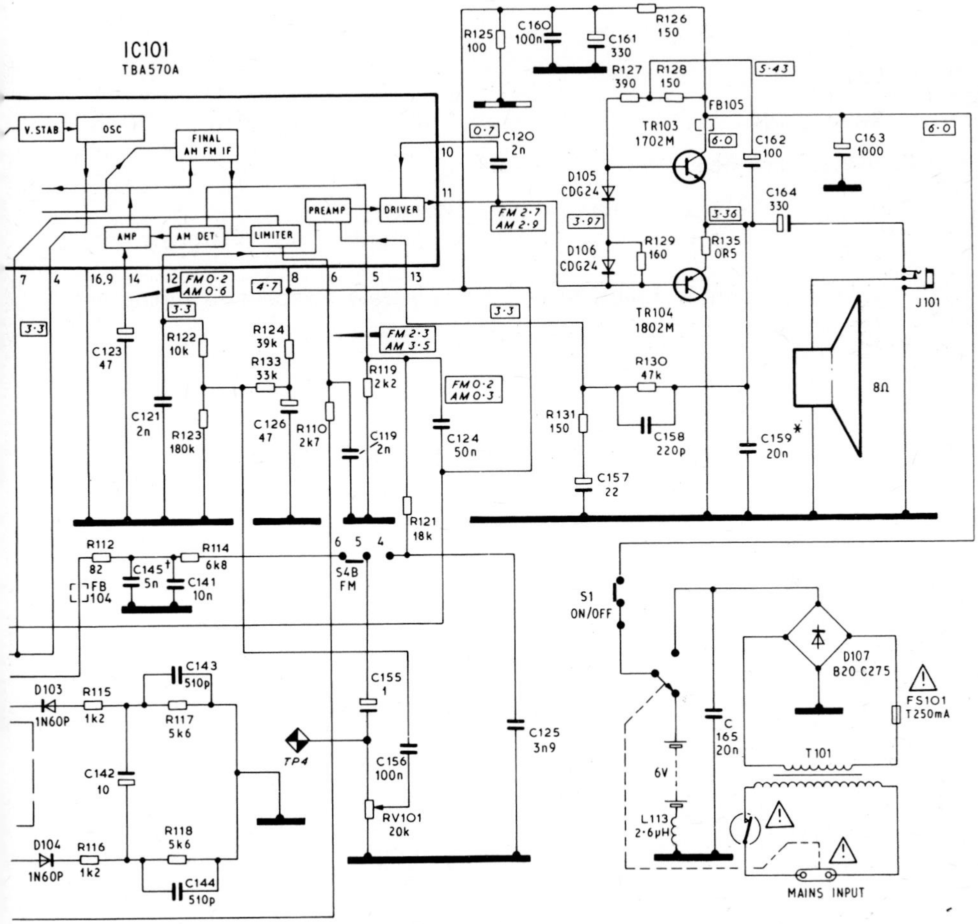

(R218b) CIRCUIT DIAGRAM—MODEL 3R06 (*CONTINUED*)

FERGUSON   Model 3R07

General Description: A three-waveband A.M./F.M. portable radio receiver operating from mains or battery supplies. A socket is provided for the connection of an earphone.

Mains Supplies: 240 volts, 50Hz.

Battery: 9 volts.

Wavebands: L.W. 145–170kHz; M.W. 510–1640kHz; F.M. 87·5–104·5MHz.

Access for Service

Remove battery cover and take out batteries.

Take out five screws (one in battery compartment) to expose printed circuit board.

To remove printed board take out one screw adjacent to tuning shaft assembly.

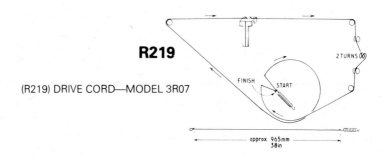

R219

(R219) DRIVE CORD—MODEL 3R07

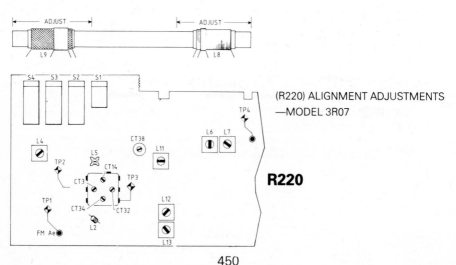

(R220) ALIGNMENT ADJUSTMENTS —MODEL 3R07

R220

Alignment

A.M. Alignment:

Range	Connections	Inject	Cursor position	Adjust for max.
I.F.	Inject signals via loop loosely coupled to ferrite rod aerial	468kHz	Extreme right	L6 L7
M.W.	Volume control at max. Output meter connected in place of loudspeaker	510kHz 1640kHz 600kHz 1400kHz	Extreme left Extreme right 600kHz 1400kHz	L11 CT34 L8* CT32
L.W.	Adjust output from signal generator to maintain 50mW	145kHz 170kHz	Extreme left 170kHz	CT38 L9*

* Adjust by sliding coils along ferrite rod aerial.

Repeat adjustments until no further improvement results.

F.M. Alignment:

Range	Connections	Inject	Cursor Position	Adjust for max.
I.F.	Connect sweep generator via 10pF capacitor to TP2, oscilloscope to TP4 via 100nF capacitor	10·7MHz	Extreme right	L4 L12 L13*
R.F.	Signal generator to TP1. Adjust output to maintain 50mW	87·5Mhz 104·5MHz 90MHz 102MHz	Extreme left Extreme right 90MHz 102MHz	L5 CT14 L2† CT3

* Adjust for symmetrical 'S' curve.
† Adjust by slightly opening or closing coil turns.

Repeat adjustments until no further improvement results.

Important Note: Earth rails are of different polarities, i.e. bottom of CV2 is not the same as RV2.
Care must be taken when connecting test equipment.

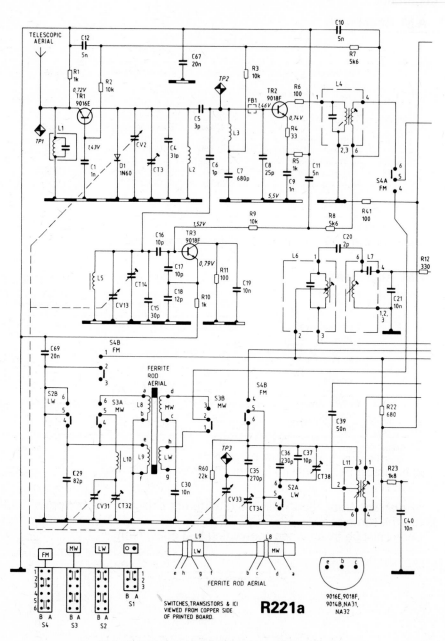

(R221a) CIRCUIT DIAGRAM—MODEL 3R07 (*PART*)

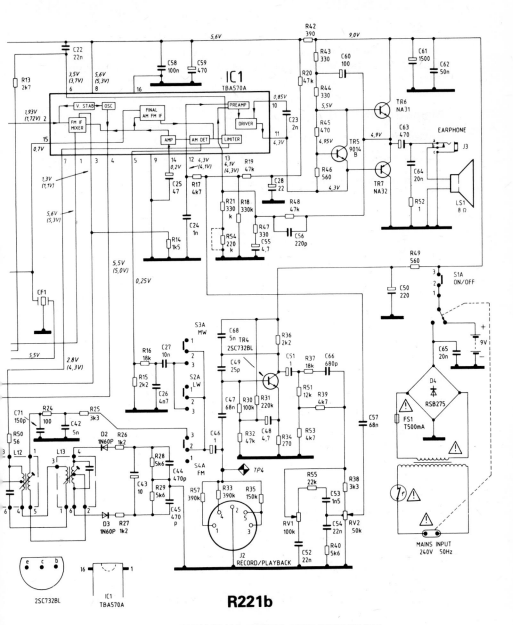

R221b

(R221b) CIRCUIT DIAGRAM—MODEL 3R07 (*CONTINUED*)

FERGUSON   Model 3R11

General Description: A three-waveband A.M./F.M. clock-radio receiver operating from mains or stand-by battery supplies. An integrated circuit stereo decoder is incorporated and integrated circuits are fitted to the audio stages. A socket is provided for the connection of stereo headphones.

Mains Supply: 240 volts, 50Hz.

Batteries (Clock stand-by)**:** 2×PP3.

Wavebands: L.W. 150–260kHz; M.W. 530–1600kHz; F.M. 88–108MHz.

Loudspeakers: 16 ohms impedance.

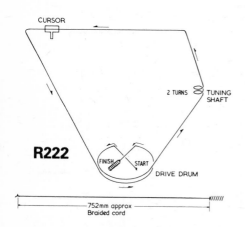

(R222) DRIVE CORD—MODEL 3R11

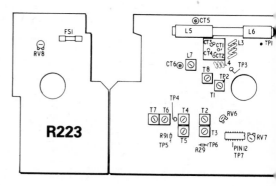

(R223) ALIGNMENT ADJUSTMENTS—MODEL 3R11

Alignment

Throughout the alignment the Volume control should be kept in the minimum position.

A.M. I.F. Circuits: Alignment can be achieved by using a wobbulator which has an A.M. facility, centred at 470kHz and with a sweep of ±3kHz. With the receiver switched to A.M. and the cursor set to the high frequency end (min. capacitance). Inject the 470kHz signal via a loop loosely coupled to the A.M. aerial coil. Connect the wobbulator input lead to TP6 and adjust T8 (Yellow), T3 (White), T5 (Black) respectively for maximum output.
Repeat until no further improvement is obtained.

A.M. R.F. Circuits: Signals are injected via a loop loosely coupled to the ferrite rod aerial.

454

Range	Inject	Cursor position	Adjust for max.
M.W.	515kHz	Extreme left	L7 (Red)
	1640kHz	Extreme right	CT4
	600kHz	600kHz	L5*
	1400kHz	1400kHz	CT3
L.W.	275kHz	Extreme right	CT6‡
	175kHz	175kHz	L6*
	245kHz	245kHz	CT5‡

* Adjust by sliding coil former along ferrite rod.
‡ Shown as C37 and C32 on board.

Repeat each stage until no further improvement is obtained.

F.M. I.F. Circuits: Switch receiver to F.M. and position the cursor at the high frequency end (min. capacitance). Short-circuit the primary coil T6 (connect pin 3 to earth).

Using a wobbulator which has a centre frequency of 10·7MHz and with a sweep of ±200kHz, connect the output lead via a 0·01μF capacitor to TP3 and the input lead to TP4. Adjust T1 (Orange), T2 (Green), T4 (Green) for maximum amplitude.

Reconnect the wobbulator input lead to TP5, remove the short-circuit on the primary coil of T6. Adjust T6 (Pink), T7 (Blue) for symmetrical 'S' curve.

Repeat each stage until no further improvement is obtained.

F.M. R.F. Circuits: Switch the receiver to F.M. terminate the F.M. signal generator with a 75 Ω resistor across the output and a 10pF capacitor in series with the 'live' lead to the injection point TP1 and earth TP2.

Range	Inject	Cursor position	Adjust for max.
F.M.	87·5MHz	Extreme left	L4†
	109MHz	Extreme right	CT2
	90MHz	90MHz	L3†
	106MHz	106MHz	CT1

† Adjust by slightly opening or closing coil turns.

Repeat each stage until no further improvement is obtained.

F.M. Stereo: For alignment of the 19·05kHz pilot tone connect a frequency counter to TP7 and with no input signal adjust RV7 for a reading of 19·02kHz–19·1kHz.

For F.M. stereo separation position an F.M. stereo generator close to the F.M. aerial terminal and set both the generator and receiver accurately to 98kHz with 1kHz modulation. Check the stereo separation of each channel and if less than 30dB re-adjust RV6 as necessary.

50Hz Stand-by Adjustment
Note: This adjustment is for battery supply only.

Connect a high impedance, low capacitance oscilloscope probe across R74. Synchronise scope timebase to mains frequency and adjust RV8 until image on scope is stationary.

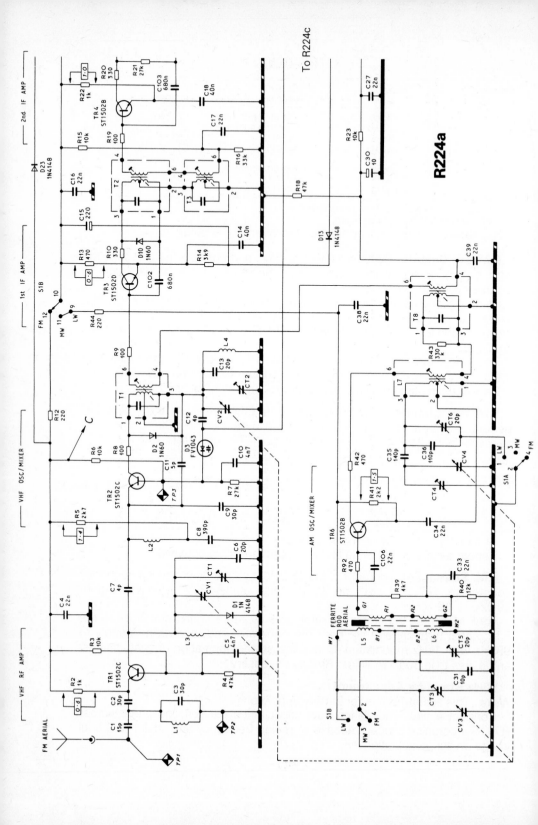

R224a

To R224c

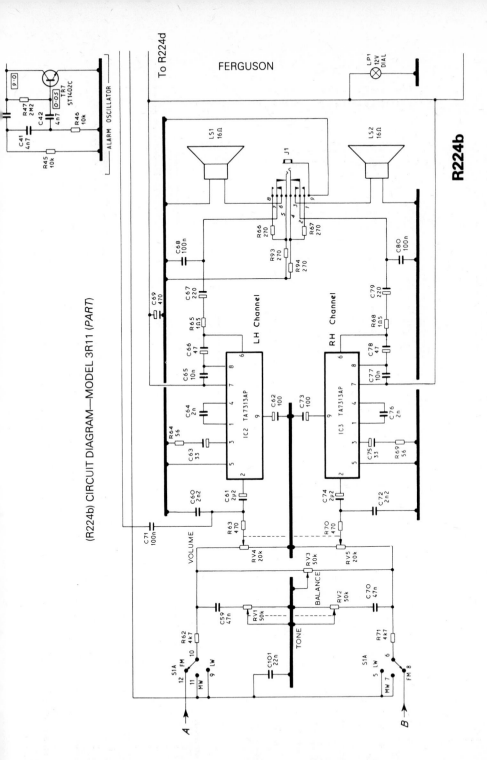

(R224b) CIRCUIT DIAGRAM—MODEL 3R11 (PART)

FERGUSON

To R224d

R224b

457

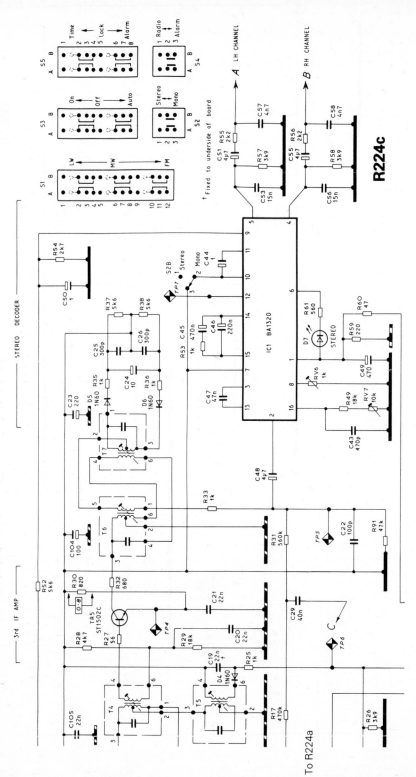

(R224c) CIRCUIT DIAGRAM—MODEL 3R11 (PART)

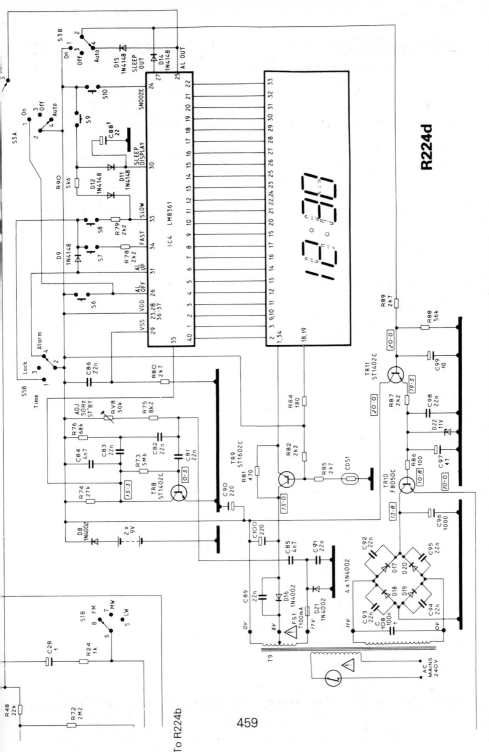

R224d

(R224d) CIRCUIT DIAGRAM—MODEL 3R11 (CONTINUED)

To R224b

459

FERGUSON   Model 3T12

General Description: A three-waveband A.M./F.M. radio receiver incorporating a public address amplifier using a built-in microphone. Sockets are provided for the connection of auxiliary inputs and earphones.

Mains Supply: 240 volts, 50Hz.

Batteries: 12 volts (8×HP11).

Wavebands: L.W. 150–250kHz; M.W. 530–1600kHz; F.M. 88–108MHz.

Bias: Two frequencies switchable approximately 46kHz, −2·4kHz (loudness button).

Erase: D.C.

Input/Output Sockets:
Five-pin DIN: Input: 1mV into 4·7k Ω for radio or audio systems sockets (pins 1–4 and 2).
Output: 800mV into 2·2k Ω for external amplifier or radio (pins 3–5 and 2).
Double Jack Socket (PA): Signal input 0·1mV into 1k Ω for dynamic microphone (3·5 mm jack socket).
Start/stop switch for remote control (2·5 mm jack socket).
3·5 mm jack socket for earphone or extension loudspeaker, impedance 4 Ω minimum (mutes internal loudspeaker).

Loudspeaker: 4 ohms impedance.

Access for Service (See Fig. R225).

Alignment (See Fig. R227).

F.M. R.F. Alignment: Signals are injected via a 10nF capacitor into the F.M. aerial socket.

Range	Inject	Cursor position	Adjust for max.
F.M.	86·5MHz	Minimum frequency	L104*
	109·5MHz	Maximum frequency	CT2
	90MHz	90MHz	L102*
	106MHz	106Mhz	CT1

* Adjust by slightly opening or closing coil turns.
Repeat adjustments until no further improvement results.

Cassette Adjustments

Head Azimuth: Connect an output meter, set to 4 ohms, in place of the loudspeaker and with Volume and Tone controls at maximum play back a

460

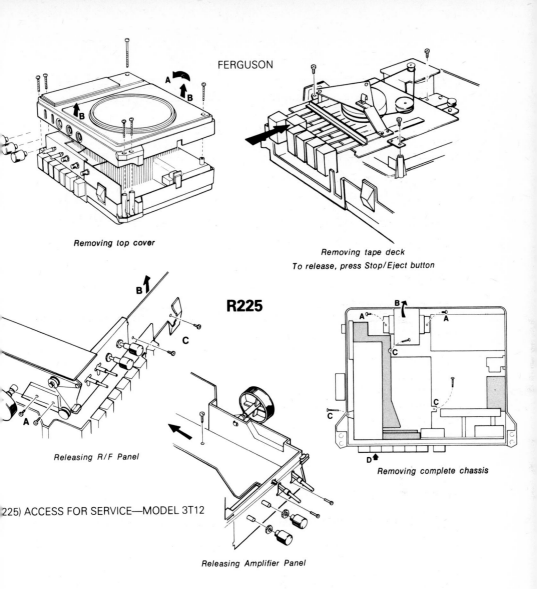

FERGUSON

Removing top cover

Removing tape deck
To release, press Stop/Eject button

R225

Releasing R/F Panel

A

Removing complete chassis

225) ACCESS FOR SERVICE—MODEL 3T12

Releasing Amplifier Panel

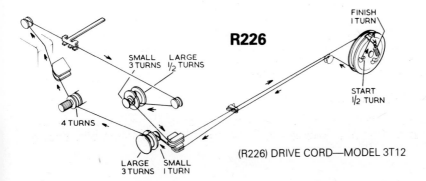

R226

SMALL
3 TURNS

LARGE
1/2 TURNS

4 TURNS

LARGE
3 TURNS

SMALL
1 TURN

FINISH
1 TURN

START
1/2 TURN

(R226) DRIVE CORD—MODEL 3T12

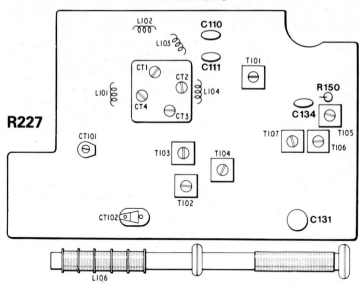

(R227) ALIGNMENT ADJUSTMENTS—MODEL 3T12

standard azimuth tape. Adjust the azimuth screw for maximum reading on output meter.

If an azimuth tape is not available play a musical tape and adjust azimuth screw for maximum treble. After adjustment the azimuth screw should be sealed with locking paint.

Note: A small hole in the bottom edge of the cassette envelope permits access to the azimuth screw with the recorder switched to the 'Play' mode.

Oscillator Coil Adjustment: Connect a frequency counter across R204 and ensure Loudness switch is released. Switch to 'Record' and adjust T201 for an indication of 45kHz on frequency counter.

After adjustment press Loudness switch and note that the frequency changes to approximately 42·5kHz.

Note: This adjustment must be made with S205 (MIC-PA switch) in the MIC position.

Bias Adjustment: Switch to 'Record' mode and connect an A.C. millivolt-meter set to a convenient range across R204. Set S205 to MIC position. Adjust RV203 for a 4mV reading, this is equivalent to 400μFA of oscillator current.

Production Modification: A 1 Ω, 1W resistor R107 has been added in series with the 12V D.C. supply to prevent surge current, under certain operating conditions, when the reservoir capacitor C231 discharges directly into C227 via switch S202.

A modification kit can be obtained from T.C.E. Service Division comprising replacement switch, 1 Ω, 1W resistor and insulating piece (00X6-378).

A.M. I.F. Alignment: Connect an output meter, set to 4 ohms impedance, in place of the loudspeaker. Switch to M.W. and turn Volume control ·to maximum, Tone control to maximum and Loudness switch released. Set tuning capacitor to minimum capacitance and inject 468kHz (30% modulated) signals via a loop loosely coupled to the ferrite rod aerial, then adjust T104 and T107 for maximum output. Throughout alignment the signal to the receiver should be adjusted, as necessary, so that the meter reading does not exceed 500mW.

Repeat adjustments until no further improvement results.

A.M. R.F. Alignment: With conditions as for I.F. alignment inject signals (30% modulated) via loop loosely coupled to the ferrite rod aerial.

Range	Inject	Cursor position	Adjust for max.
M.W.	525kHz	Minimum frequency	T103
	1650kHz	Maximum frequency	CT4
	600kHz	600kHz	L105*
	1400kHz	1400kHz	CT3
	142kHz	Minimum frequency	T102
L.W.	265kHz	Maximum frequency	CT102
	150kHz	150kHz	L106*
	250kHz	250kHz	CT101

* Adjust by sliding coils along ferrite rod.

Repeat adjustments until no further improvement results.

F.M. I.F. Alignment: Alignment is best obtained using a wobbulator with F.M. facility and a display unit. The wobbulator should be terminated with a 75 ohm resistor across the output, and a capacitor of 10nF in series with the 'live' lead to the junction point.

Switch to F.M. and with Volume and Tone controls at maximum position; Loudness switch released and tuning gang at minimum capacitance. Connect oscilloscope across C134 then inject 10·7MHz (25kHz deviation) signals to the junction of C110, C111. Adjust T101 and T105 for maximum output.

Switch to wobbulator operation and adjust T106 for a symmetrical 'S' curve.

R228a

(R228a) CIRCUIT DIAGRAM, MODEL 8716/8717

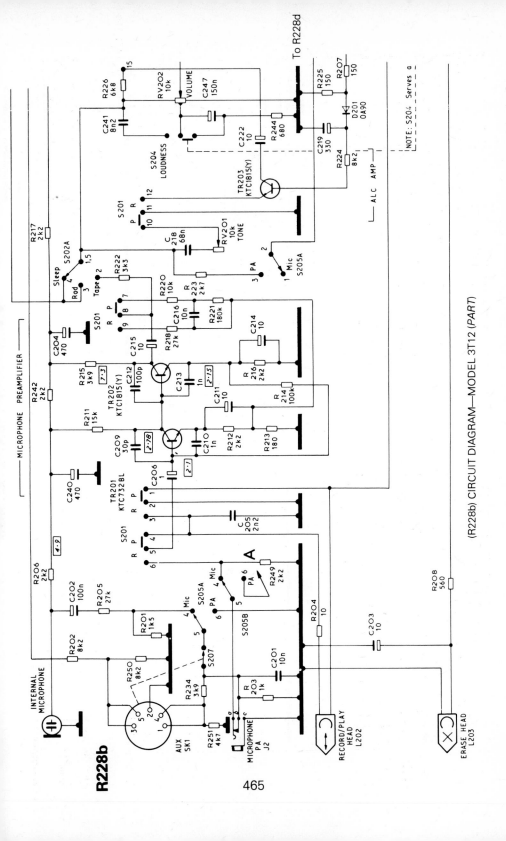

(R228b) CIRCUIT DIAGRAM—MODEL 3T12 (PART)

465

R228c

LW — MW — FM
S101

PLAY → RECORD
S201

— PA PREAMPLIFIER —

R241 2k2
C253 10
TR205 KTC1815(Y)
C251 1n5
RV2O4 10k
PA VOLUME

C235 10
R239 2k7
R240 33k
[1·38]
R238 1k
C250 400p
[7·5]
C234 1n
D203 1S2473
C236 10
TR204 KTC732BL
[1·0] [0·09]
R237 820k
R235 22k
C233 1n
R236 82
C232 220
C249 1n
R209 2k2
C239 10
[0·67]

A

MOTOR
S206
R243 1
C237 1000
C238 220
TR206 KTC1173
[8·0] [6·7]
FB
ZD202 RD-9V1EB
R248 560
[14·0]

R144 10
R145 270
ZD107 RD 6V8
C136 220
C137 22n

AMP

C117 22n

C133 1n
C132 1n
R139 10k
R138 10k
R136 1k2
C131 4μ7
R137 1k2
D104 20A90
D105 20A90
T106
6 3 1 4
2

C135 100n
R143 8k2

R142 560
C130 22n
D106 0A90
R135 220
T107
T105
TR105 KTC380(0)
[5·68]
C144 22n
R146 560
R130 2k7
C127 10
R129 1k

C128 100
C145 100

C134 22n
R150 150
R133 68k
R134 15k
R119 100k
R113 10k
C141 22n

C142 1

Hi — Lo — Off
S203

Tape ← Radio → Sleep
S202

To R228a

(R228c) CIRCUIT DIAGRAM — MODEL 2T12 (PART)

R228d

J1
REMOTE

12V DC

DC
AC
S208

S209
S

R

T

8

7

S202B

9

6,10

S

C231
2200

C504
47n

C502
200n

D504
D502
4 x 10D1
D503
D501

C503
47n

C501
47n

T501

240V 50Hz
SK2

FS1
T1A

14·0

2

1
Mic

S205B

3
PA

16
P

17

18
R
S201

EARPHONE
J3

R233
22

R
231
27

C252
10n

MONITOR
High

4Ω

2

3
Low

4
Off

R232
4Ω7

S202A

T
R

8

6,10

S

9

AUDIO AMP

14·5

C226
40n

S201
13
R
P
14

C227
470

C223
1

0·02

C208
2n2

IC201
TBA810S

1

3

4

5

12

6

7

8

9,10

14·0

C
220
6n8

0·74

C229
100

C228
1n

7·6

C230
470

R229
2Ω2

C217
20n

C225
100

6·65

C224
100

1·36

R227
22

R228
33k

C252
15n

R230
1Ω8

S203B
1,5

S201
R P
21 20 19

T201

R245
180

TR207
KTC2120(Y)
BIAS OSCILLATOR

6·9

1·0

R246
33k

C246
33n

0·3

R247
10

C245
100

C
248
22n

BIAS ADJ.

C
244
6n8

RV203
100k

S204
BEAT SWITCH

C
243
1n

R210
10k

C207
2n2

R219
10k

C242
6n8

dual function

467

To R228b

(R228d) CIRCUIT DIAGRAM—MODEL 3T12 (CONTINUED)

FERGUSON   Model 3T16

General Description: A three-waveband A.M./F.M. portable radio cassette-recorder operating from mains or battery supplies. A microphone is 'built-in' and sockets are provided for the connection of auxiliary inputs and headphones.

Mains Supply: 240 volts, 50Hz.   **Batteries:** 9 volts (6×HP7).

Wavebands: L.W. 150–250kHz; M.W. 540–1600kHz; F.M. 88–108MHz.

Bias Frequency: 35 and 38kHz.   **Loudspeaker:** 8 ohms impedance.

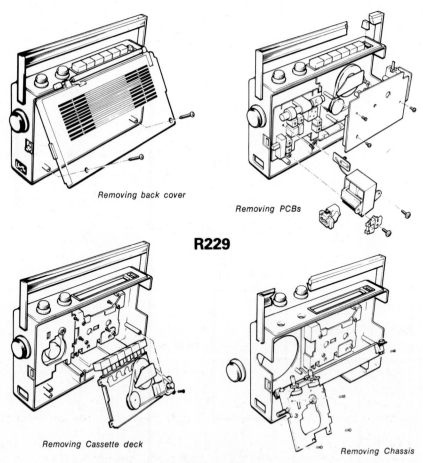

Removing back cover

Removing PCBs

R229

Removing Cassette deck

Removing Chassis

(R229) ACCESS FOR SERVICE—MODEL 3T16

FERGUSON

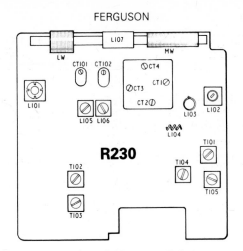

Location of Alignment Points

(R230) ALIGNMENT ADJUSTMENTS—MODEL 3T16

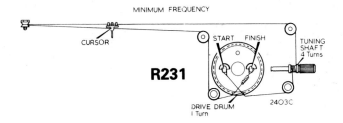

(R231) DRIVE CORD—MODEL 3T16

Alignment (See Fig. R230).

A.M. I.F. Alignment: Set up the receiver as follows: Volume control to maximum, Tone control to central position, cursor on tuning scale to minimum frequency (extreme left position). Inject 470kHz via a loop of wire loosely coiled around the ferrite rod aerial. Ensure that throughout alignment, the signal strength does not exceed 0·63V (50mW).

Connect a 20,000 Ω/voltmeter, set to the 10V A.C. range, across the loudspeaker terminals. Adjust T104 and T105 to give maximum meter reading.

A.M. R.F. Alignment: Inject signals via a loop of wire coiled loosely around the ferrite rod aerial. Volume control at maximum, Tone control set to central position and a 20,000 Ω/voltmeter connected across the loudspeaker terminals.

Range	Inject	Cursor position	Adjust for max.
M.W.	525kHz	Minimum frequency	L106
	1650kHz	Maximum frequency	TC4
	600kHz	600kHz	L107*
	1400kHz	1400kHz	TC3
L.W.	142kHz	Minimum frequency	L105
	255kHz	Maximum frequency	TC101
	160kHz	160kHz	L107*
	250kHz	250kHz	TC102

* Adjust by sliding coils along ferrite rod.

Repeat adjustments until no further improvement results.

F.M. I.F. Alignment: Connect the input of a wobbulator with an F.M. facility and display unit across C107 (0·001μF). Disconnect the input to C152 and connect the display unit to this lead. Set the receiver Volume control at maximum, Tone control to central position and cursor to maximum frequency position (right hand end). Detune T103 by unscrewing the core to just above the can.

Inject 10·7MHz ±1MHz, adjust T101 and T102 for maximum output and symmetrical waveform, with the 10·7MHz marker at the peak. Repeat the adjustments of T101 and T102 until no further improvement results, then reconnect the lead of C152.

Connect the display unit to R120 and adjust T103 to obtain an 'S' curve, symmetrical about the 10·7MHz marker.

F.M. R.F. Alignment: Inject signals via the telescopic aerial terminal. Set Volume control to maximum, Tone control to central position and connect the 20,000 Ω/voltmeter across the loudspeaker terminals.

Range	Inject	Cursor position	Adjust for max.
F.M.	66·5MHz	Minimum frequency	L104
	109·5MHz	Maximum frequency	TC2
	90MHz	90MHz	L102
	106MHz	106MHz	TC1

Repeat adjustments until no further improvement results.

Cassette Adjustments

Tape-Head Demagnetization: The tape-head will gradually become magnetised with use or if a magnetic tool is accidentally brought close to it; this will show up by an increase in tape noise associated with a decrease of

470

the high frequencies. The tape-head can be demagnetized by using a tape-head demagnetizer in accordance with the manufacturer's instructions.

Head Azimuth Alignment: Connect a high impedance voltmeter to the earphone jack. Set Volume control to maximum and play an azimuth adjustment tape. Adjust the azimuth screw to give a maximum reading on the voltmeter. If an azimuth adjustment tape is not available then play a music cassette and adjust the azimuth screw for maximum Treble. After adjustment the screw should be sealed with locking paint. The azimuth adjustment screw is the right-hand mounting screw viewed from the front of the set.

Cleaning and Lubrication: Do not use cleaning fluids such as petrol or carbon tetrachloride, which might cause damage to plastic surfaces or rubber drives. Use a soft cloth dampened with methylated spirits to clean drive surfaces. Head faces can be cleaned in the same manner or by using a special head cleaning cassette. All moving parts are lubricated during manufacture and further lubrication during service should not be necessary.

When any of the moving parts are replaced, a smear of graphite grease should be applied to the bearing surfaces. Take care that lubricant does not contaminate any of the drive surfaces.

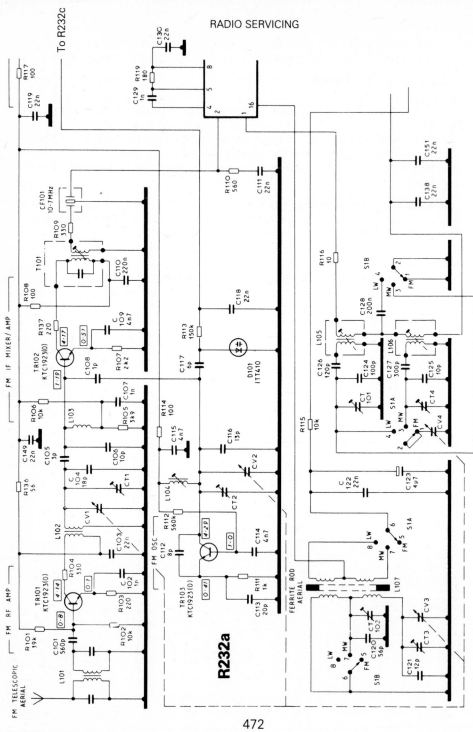

(R232a) CIRCUIT DIAGRAM—MODEL 3T16 (PART)

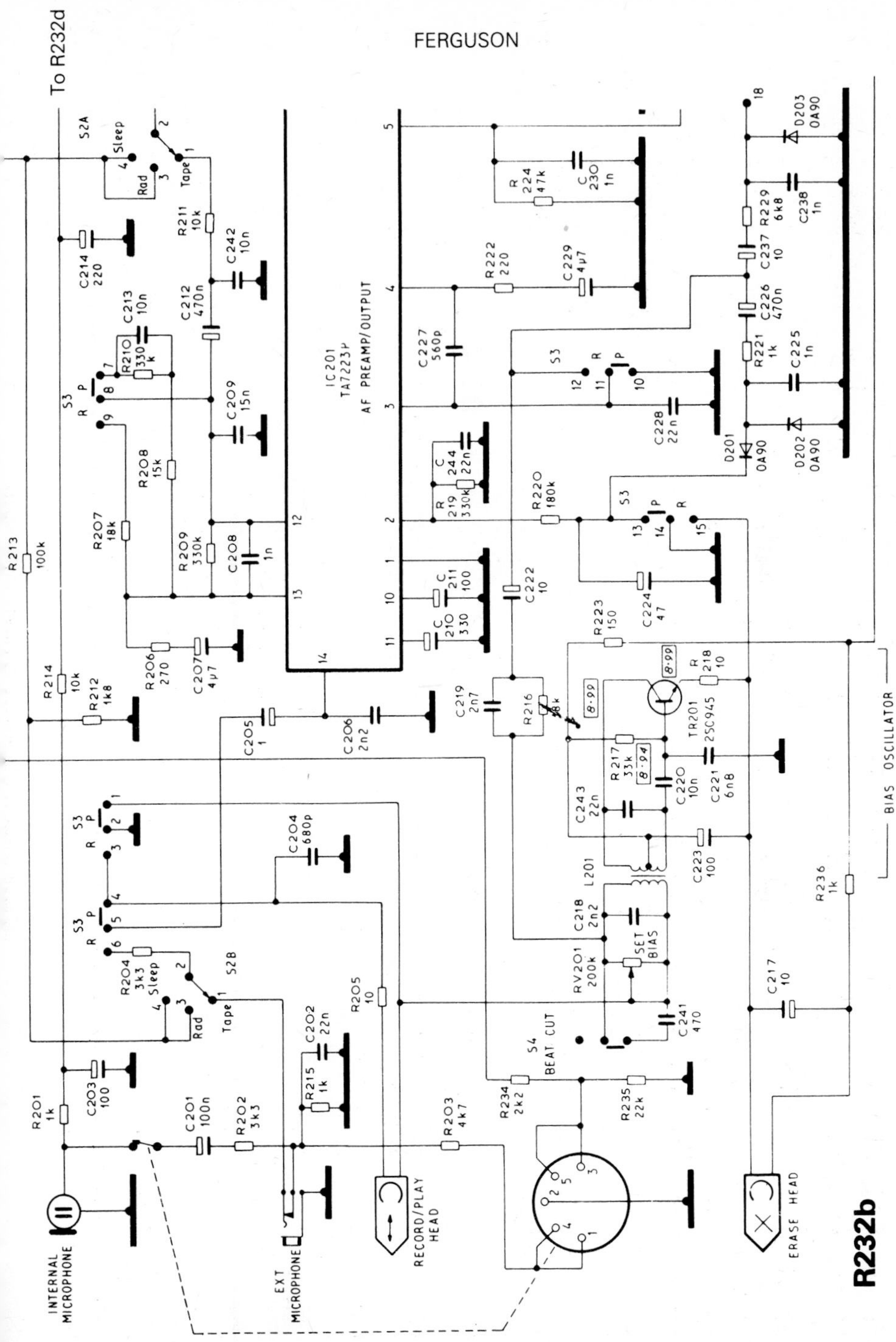

FERGUSON

(R232b) CIRCUIT DIAGRAM—MODEL 3T16 (PART)

R232b

473

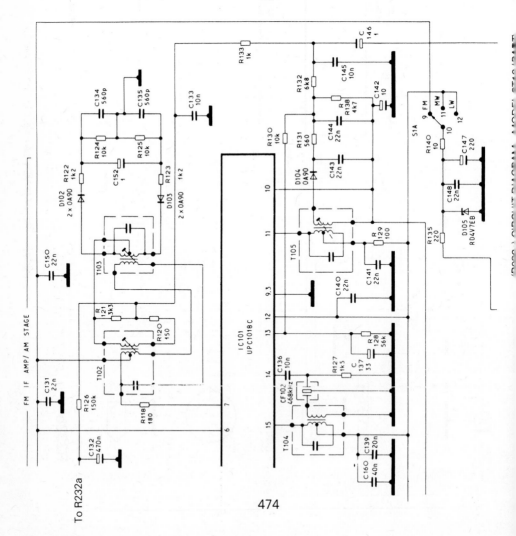

R232c

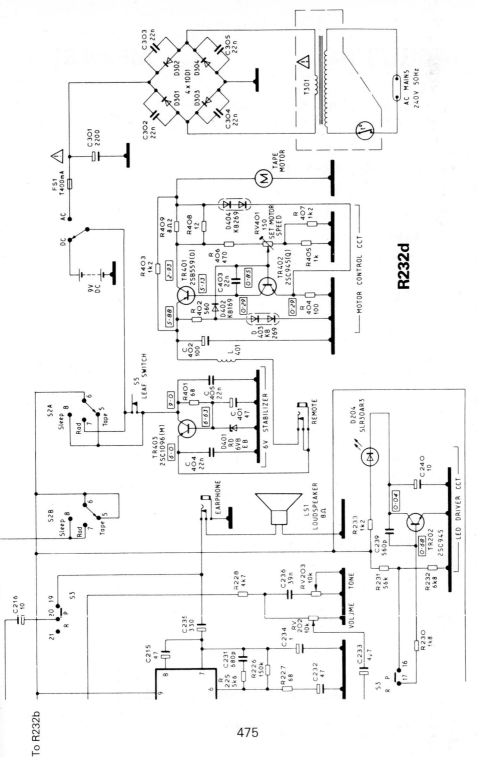

R232d

MOTOR CONTROL CCT

6V STABILIZER

LED DRIVER CCT

To R232b

(R232d) CIRCUIT DIAGRAM—MODEL 3T16 (CONTINUED)

FERGUSON   Model 3T17

General Description: A three-waveband A.M./F.M. stereo radio and cassette-recorder operating from mains or battery supplies. Integrated circuits are widely used and sockets are provided for the connection of auxiliary inputs and stereo headphones.

Mains Supply: 240 volts, 50Hz.

Batteries: 9 volts (6×HP11).

Wavebands: L.W. 150–250kHz; M.W. 540–1600kHz; F.M. 88–108MHz.

Loudspeakers: 4 ohms impedance.

Access for Service (See Fig. R233).

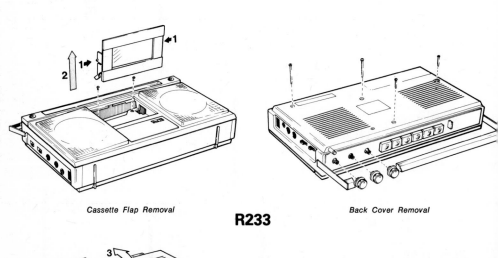

Cassette Flap Removal

R233

Back Cover Removal

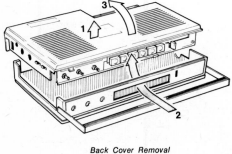

Back Cover Removal

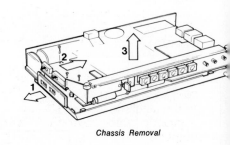

Chassis Removal

(R233) ACCESS FOR SERVICE—MODEL 3T17

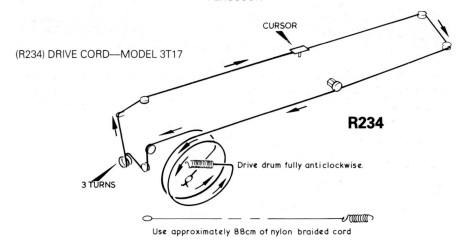

(R234) DRIVE CORD—MODEL 3T17

CURSOR

R234

Drive drum fully anticlockwise.

3 TURNS

Use approximately 88cm of nylon braided cord

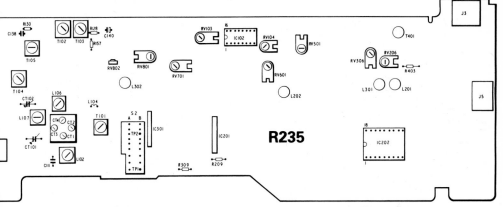

R235

(R235) ALIGNMENT ADJUSTMENTS—MODEL 3T17

Alignment

Equipment required: Wobbulator with A.M./F.M. facilities; A.M./F.M. R.F. generator; A.C. Voltmeter; Oscilloscope.

Note: Access to some alignment points may require removal of cassette-deck.

A.M. I.F.: Connect the output from the wobbulator to the ferrite rod aerial coils via a loosely coupled coil. The input to the wobbulator being connected to the junction of R130/C138.

Inject a signal centred on 468kHz and adjust T104, T105 for maximum output.

Repeat until no further improvement results.

Note: Gang must be at maximum capacitance.

A.M. R.F.: Signals (30% modulated) are injected via a loop loosely coupled to L105 (ferrite aerial rod), the voltmeter being connected to the output from the Headphone socket (J3). Turn Volume and Tone to maximum.

Note: Whilst adjusting the R.F. stages, keep the input level as low as possible. Consistent with a reasonable background noise level at about 50mW audio output.

Range	Inject	Cursor position	Adjust for max.
M.W.	520kHz	Minimum frequency	L107
	1650kHz	Maximum frequency	CT4
	600kHz	600kHz	L105A*
	1400kHz	1400kHz	CT3
L.W.	142kHz	Minimum frequency	L106
	265kHz	Maximum frequency	CT102
	160kHz	160kHz	L105B*
	250kHz	250kHz	CT101

* Adjust by sliding along ferrite rod.

Repeat adjustments until no further improvement results.

F.M. I.F.: Alignment can be best achieved by using a wobbulator centred at 10·7MHz (sweep of 1MHz). Unsolder the wire link between C140, RV802 and select F.M. Connect the wobbulator output lead, correctly terminated with a 75 ohm resistor, via a 1nF capacitor to C111. The input lead to the wobbulator being connected to the positive side of C140. Gang capacitance to minimum.

Adjust T101, T102 for maximum amplitude 'V' shape response until no further improvement is obtained.

Reconnect the lead to RV802, and also reconnect the wobbulator input lead to R157/R128. Adjust T103 for a symmetrical 'S' curve.

F.M. R.F. (A.F.C. does not need defeating): R.F. signals (22·5kHz deviation) are injected into the telescopic aerial tag, the output from the Headphone socket being connected to an A.C. voltmeter. Set Volume and Tone to maximum, and R.F. levels as low as possible. Select F.M.

Injected frequency	Cursor position	Adjust for max.
86·5MHz	Minimum frequency	L104*
109·5MHz	Maximum frequency	CT2
90MHz	90MHz	L102
106MHz	106MHz	CT1

* Adjust by opening or closing turns of coil.

Repeat adjustments until no further improvement results.

478

Stereo Decoder: Connect a suitable high impedance frequency counter between pin 16 IC102 and a convenient earth point. With the receiver tuned to a monophonic V.H.F. transmission, adjust RV103 for a reading of 19kHz. Re-tune receiver to a stereo generator set for a right-hand (1kHz modulation) only, and adjust RV104 for minimum left-hand loudspeaker output.

Head Azimuth: For azimuth adjustment, play back a standard azimuth tape. With output meters connected in place of the speakers adjust the azimuth screw (nearest the erase head) for maximum meter reading, whilst keeping the output at a low level.
Finally check for satisfactory reproduction from both channels.

Motor Speed Adjustment: The motor speed control is set to give correct frequency playback of a tape pre-recorded with a known signal.
The 50Hz speed test cassette and 50Hz null detector can be used for this purpose.
Unsolder one loudspeaker lead and connect to the 50Hz null detector. Playback the 50Hz speed test cassette and adjust RV501 for a null in the meter reading.
If a 50Hz null detector is not available it is possible to carry out the adjustment by injecting the 50Hz tape signal into the X amplifier of an oscilloscope and comparing it with a 50Hz mains frequency signal injected into the Y amplifier.
A stationary image on the oscilloscope is sufficient to establish that the speed of the motor is correct.

Bias Frequency Adjustment: With the bias oscillator shift switch (S4) in position 1 (bias shift off), connect a suitable frequency counter via a $0.047\mu F$ capacitor, to the base of TR401. Adjust T401 for a reading of 55kHz.

A.L.C. Adjustment: Select Record/Play and tape bias normal. Inject a signal of 1kHz of $15\mu V$ into the microphone socket (J2). Disconnect one end of R403 (bias oscillator supply rail). With an electronic voltmeter monitor the voltages developed across R209 and R309 and if necessary adjust RV601 for equal readings. Reconnect R403.

Bias Current Adjustment: Connect an electronic voltmeter across R209. Select tape Record/Play, tape bias to normal, then adjust RV206 for a meter reading of 3·5mV. Transfer the voltmeter to R309 and similarly adjust RV306 for 3·5mV.
This voltage corresponds to approximately $350\mu A$ bias current.

Bias Trap Adjustment: With an oscilloscope connected across R209, select tape Record/Play (bias normal) and adjust L201 for maximum. Similarly adjust L301, with the oscilloscope connected across R309. With the oscilloscope connected between TP1 (S2B radio/tape switch, pin 1), under the same conditions as above, and chassis adjust L202 for minimum display. Similarly adjust L302 for minimum, with oscilloscope connected to TP2 (S2B pin 7). Continued on p. 484

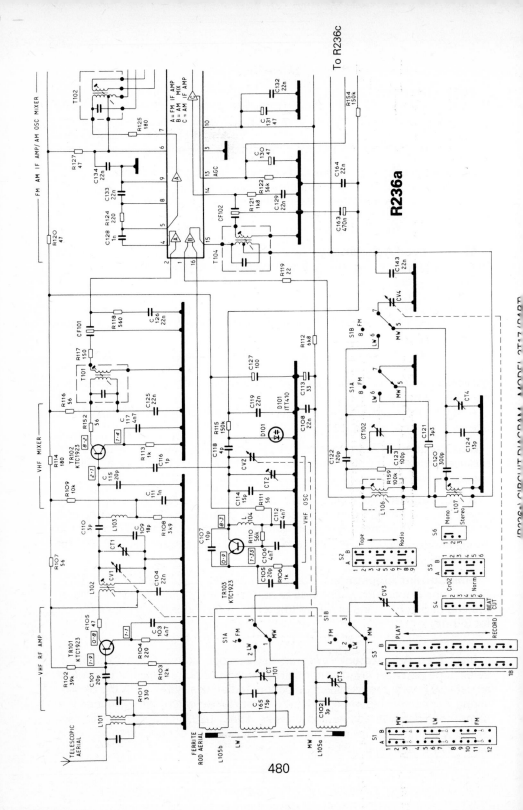

R236a

To R236c

480

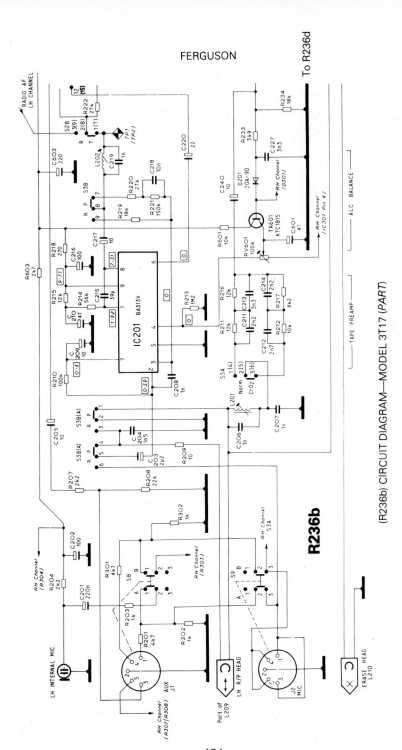

(R236b) CIRCUIT DIAGRAM—MODEL 3T17 (PART)

R236b

(R236c) CIRCUIT DIAGRAM—MODEL 3T17 (PART)

482

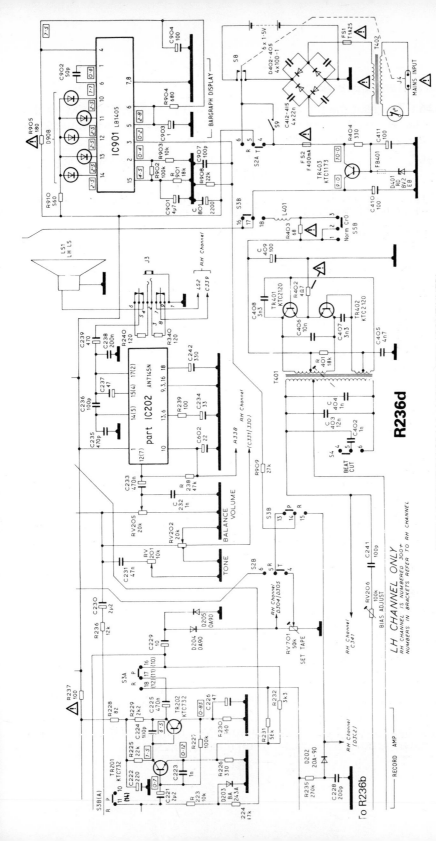

R236d

(R236d) CIRCUIT DIAGRAM—MODEL 3T17 (CONTINUED)

Level Meter Adjustments

A.M.: To adjust the pre-set level, inject an R.F. signal of 5mV/M via the ferrite aerial rod (30% modulated at 1kHz). Select Radio and M.W. functions and adjust RV801 until the 0dB L.E.D. is glowing.

V.H.F.: Inject a signal into the F.M. aerial tag (22·5kHz deviation) at a level of 1mV. Select V.H.F. and adjust RV802 until the 0dB L.E.D. glows.

Tape: Select the Tape function and playback a DOLBY Test tape and adjust RV701 until the 0dB L.E.D. is just glowing.

I.C. 101 Voltages

Pin	F.M.	A.M.	Pin	F.M.	A.M.
1	0·3	9·2	9	0	0·7
2	0·7	0	10	0·03	9·2
3	0	0	11	0	0·66
4	6·2	0	12	0	0·64
5	6·7	0	13	0	9·2
6	7·8	0	14	0	9·0
7	7·4	0	15	0·03	0·7
8	6·7	0	16	0	0

FERGUSON Model 3T18

General Description: A three-waveband A.M./F.M. stereo radio cassette-recorder with Dolby noise reduction and operating from mains or battery supplies. A digital clock is incorporated and sockets are provided for auxiliary inputs and stereo headphones.

Mains Supply: 240 volts, 50Hz.

Batteries: 12 volts (8×HP2).

Wavebands: 150–270kHz; M.W. 520–1620kHz; F.M. 87·5–108MHz.

Bias: 85kHz pprox.

Access for Service

Remove 6 screws from the back cover and pull-off aerial lead from P.C.B. To remove front cover, take out two screws either side of cassette compartment and lift-off cover and envelope. Remove control knobs and prise top edge away from clips. Pull off attached socket.

FERGUSON

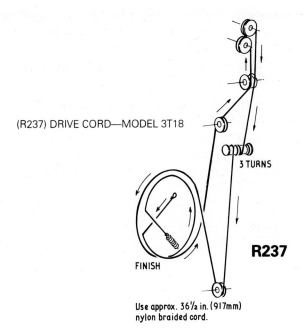

(R237) DRIVE CORD—MODEL 3T18

3 TURNS

R237

FINISH

Use approx. 36½ in. (917mm)
nylon braided cord.

I.C. 801 Voltages:

Pin	M.W. (1000kHz) Display		M.W. (200kHz) Display		F.M. (MHz) Display	
	Off	*On*	*Off*	*On*	*Off*	*On*
7	2·3	1·8	3·8	1·8	3·8	1·7
2	2·3	1·8	3·8	1·8	3·8	1·7
3	1·8	—	2·4	—	1·7	—
4	1·8	1·8	2·0	1·8	2·2	1·7
5	1·8	1·8	2·0	1·8	1·7	1·7
6	1·7	1·1	3·0	1·1	1·7	1·1
7	2·5	0	1·9	0	1·7	7·2
8	1·2	0	1·9	0	1·7	1·7
9	1·7	0	1·9	0	1·7	0
10	1·7	—	1·7	—	1·4	—
11	1·7	7·9	1·7	7·9	1·4	7·8
12	2·2	0·6	2·7	0·6	1·4	0·6
13	1·7	1·7	1·7	1·7	1·4	1·7
14	1·7	—	1·7	—	1·4	—
15	2·2	0	2·7	0	1·4	0
16	1·7	7·9	1·7	7·9	1·4	7·8
17	1·7	7·9	1·7	7·9	1·4	7·8
18	1·7	—	1·7	—	1·7	—
19	1·7	0	1·7	0	1·7	0
20	2·2	4·1	2·2	4·1	2·2	4·1

Alignment (See Fig. R239)

A.M. I.F. Alignment: Switch to M.W.; Loudness control (S106) OFF; Volume control to minimum and Bass and Treble controls to mid-position.

Inject wobbulator signals, centred on 460kHz, via a loop loosely coupled to the ferrite rod aerial.

Connect oscilloscope to TP3 (R41) and set tuning gang to minimum capacity.

Adjust T5 and T6 for maximum peak then switch on 460kHz marker and readjust T5 and T6 to bring the marker to the peak of the curve.

A.M. R.F. Alignment: Signals, 30% modulated, are injected from an A.M. signal generator via loop loosely coupled to the ferrite rod aerial. Set Bass and Treble controls to mid position and Volume control to maximum. Switch to Mono and connect output meters, set to 4 ohms minimum, via suitable plugs to the extension loudspeaker sockets. Inject and tune to frequencies as detailed below, keeping signal generator output low enough to maintain audio output below 50mW.

Range	Frequencies	Adjust for max.
M.W.	515kHz	T4
	1650kHz	CT4
	600kHz	L8*
	1400kHz	CT3
L.W.	145kHz	T3
	280kHz	CT6
	160kHz	L7*
	250kHz	CT5

* Adjust by sliding coils along ferrite rod.

Repeat adjustments on both ranges until no further improvement results.

F.M. I.F. Alignment: Signals are injected from a V.H.F. wobbulator, terminated with a 75 ohm resistor, between pins 2 and 3 of PL1. Set wobbulator to sweep 200kHz either side of 10·7MHz with a marker at 10·7MHz. Set tuning gang to minimum capacitance; Volume control at minimum, Loudness control switched off and Bass and Treble controls at mid-position.

Switch receiver to V.H.F. and connect an oscilloscope to TP4 (R52) then adjust T1 for a symmetrical response with the 10·7MHz marker at the peak of the curve. Adjust T2 for a symmetrical 'S' curve with the 10·7MHz marker at the centre of the straight section of the curve. Repeat adjustments until no further improvement results.

F.M. R.F. Alignment: Signals, modulated 1kHz (22·5kHz deviation), are injected from an F.M. signal generator, terminated with a 75 ohm resistor, and a 10nF capacitor in series with 'live' lead between pins 2 and 3 of PL1. Connect, with suitable plugs, output meters set to 4 ohms impedance to the extension loudspeaker sockets.

Set Volume control at maximum with Bass and Treble controls at mid-

486

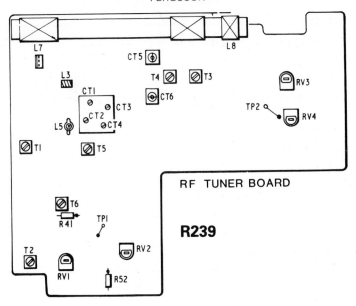

RF TUNER BOARD

R239

(R239) ALIGNMENT ADJUSTMENTS—MODEL 3T18

position and Loudness control switched off. Inject and tune to frequencies as detailed below keeping signal generator output as low as possible so that audio output does not exceed 50mW.

Range	Frequencies	Adjust for max.
F.M.	87MHz	L5
	109MHz	CT2
	90MHz	L3*
	106MHz	CT1

* Adjust by slightly opening or closing coil turns.

Repeat adjustments until no further improvement results.

Adjustments (See Fig. R238)

Tuning Meter Adjustment: Set Function switch (S101), Meter switch (S102) and Mode switch (S103) to Radio, V.H.F. and Tuning positions respectively. Inject a 1mV 98MHz unmod. signal into pin 2 of PL1. Tune receiver to 98MHz and set RV3 until the green centre L.E.D. is fully illuminated.

Head Azimuth: Connect an electronic voltmeter across TP101 and a convenient chassis point.

Set Tape selector (S104) to Normal and Dolby switch (S105) to Off. Play back a standard azimuth tape and adjust the azimuth screw for maximum reading on meter.

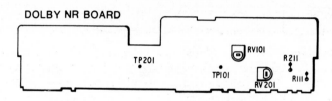

DOLBY NR BOARD

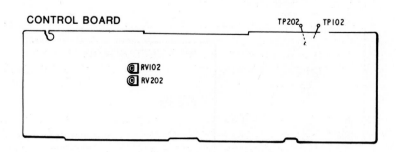

CONTROL BOARD

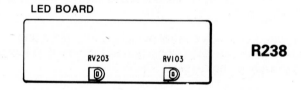

LED BOARD

R238

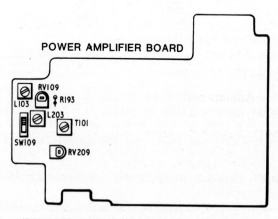

POWER AMPLIFIER BOARD

(R238) ELECTRICAL ADJUSTMENTS—MODEL 3T18

Transfer meter 'live' lead to TP201 and check output. A slight difference may be noticed due to 'Playback Level' not yet having been adjusted.

Playback Level: With test conditions as above, play back a Dolby standard test tape (400Hz, 200nWb/m) and adjust RV101 for a reading of 580mV at TP101. Transfer meter lead to TP201 and adjust RV201 for 580mV. Operate Dolby switch (S105) and check for a change in levels of less than $\pm$1dB. Recheck 'Head Azimuth' to ensure left-hand and right-hand outputs are equal, if not repeat 'Head Azimuth' procedure.

Bias Frequency: Disconnect any input to the recorder, turn Record Level controls (R104; R204), on front panel, to minimum, remove electronic voltmeter and set Tape selector switch (S104) to Normal. Connect a frequency counter to TP103 (R193), switch recorder to Record and Dolby switch (S105) to Off, then set Oscillator switch (S109) to Off. Adjust T101 for a reading of 85kHz.

Bias Current: Disconnect frequency counter and reconnect electronic meter across R111. Set Function switch (S101) to Phono and Tape switch (S104) to CrO_2, switch recorder to Record and adjust RV109 for a reading of 5·4mV. Transfer electronic voltmeter across R211 adjusting RV209 for 5·4mV.

Bias Trap: Return Function switch (S101) to Tape and switch recorder to Record. Connect an oscilloscope to TP102 and adjust L103 for minimum display.

Transfer oscilloscope to TP202 and adjust L203 for minimum display.

***Note*:** These test points are two short black leads soldered to the Control Board to facilitate connections.

Recording Current: Connect the electronic voltmeter to TP101 with Function switch (S101) and Tape switch (S104) switched to Tape and Metal respectively. Set recorder to Record and inject a 400Hz signal to the microphone socket (J101).

Adjust Record Level control (RV104), on front panel, to centre of its track. Reset 400Hz signal so that 580mV appears at TP101.

Make a recording at this level using Metal tape then playback and check the replay output at TP102. If 580mV is not shown at TP102 adjust RV102 for this value.

Make similar recordings on Normal and Chrome tapes and note that the same level is obtained on replay.

Signal Strength Meter Adjustment: Set Function switch (S101) to Phono and Meter switch (S102) to Level positions. Connect the electronic voltmeter to TP101. Inject a 400Hz signal to pin 3 of Phono socket (J104). Adjust RV104 to produce 580mV at TP101. Finally adjust RV103 until the +3 L.E.D. is fully illuminated.

Repeat operations for right-hand channel with meter connected to TP201. Injecting signal to pin 5 and adjusting RV204, RV203 to illuminate the right-hand channel L.E.D.

(R240a) CIRCUIT DIAGRAM - MODEL 3T18 (PART)

R240a

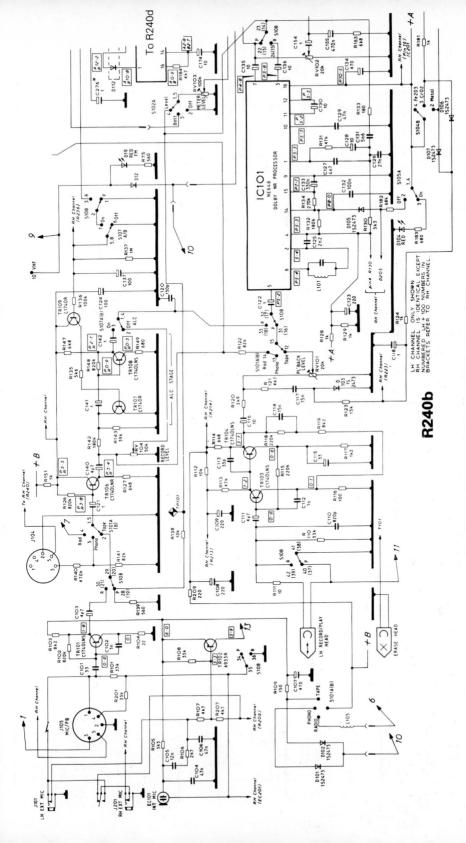

(R240b) CIRCUIT DIAGRAM—MODEL 3T18 (PART)

R240b

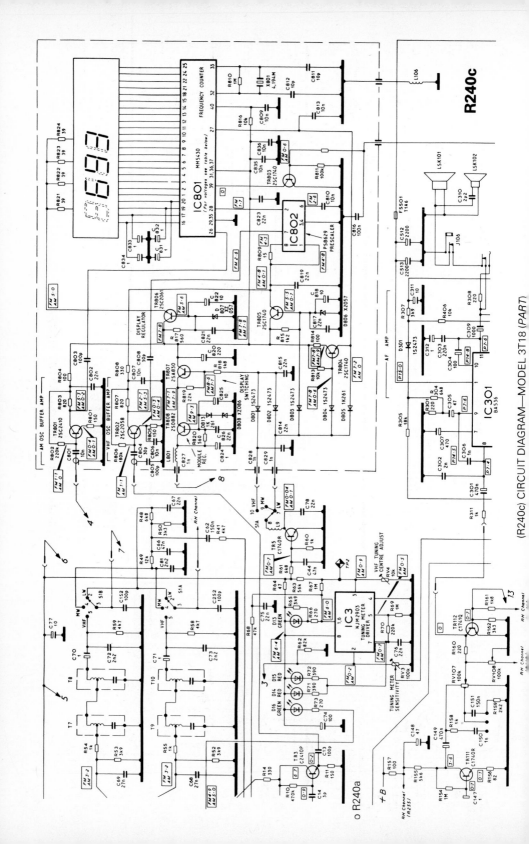

(R240c) CIRCUIT DIAGRAM—MODEL 3T18 (PART)

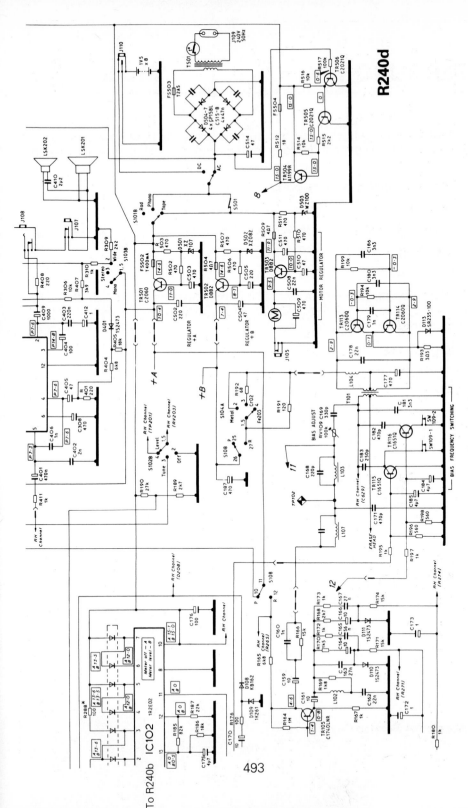

R240d

(R240d) CIRCUIT DIAGRAM—MODEL 3T18 (CONTINUED)

493

FERGUSON   Model 3T19

General Description: A three-waveband A.M./F.M. radio cassette-recorder operating from mains or battery supplies. A microphone is built into the case and sockets are provided for the connection of auxiliary inputs and earphone.

Mains Supply: 240 volts, 50Hz.

Batteries: 6 volts (4×HP2).

Wavebands: L.W. 150–350kHz; M.W. 550–1500kHz; F.M. 88–104MHz.

Bias Frequency: 30kHz.

Loudspeaker: 8 ohms impedance.

Important Note: Earth rails are of different polarities, i.e. the anode of D8 is not the same as TP6. Care must therefore be taken when connecting test equipment.

IC1 Voltages:

Pin No.	F.M.	A.M.	Rec.
1	0·7		
2	1·4		
3	0		
4	1·3		
5	1·4		
6	0		
7	4·8	5·0	
8	4·4	4·9	
9	4·0		4·9
10	0·3		4·9
11	1·0		4·9
12	1·0		4·9
13	4·3		4·9
14	0·7		
15	0·4		
16	0		

IC2 Voltages:

Pin No.	F.M.	Rec.
1	1·7	1·8
2	0·5	
3	·001	
4	0	
5	0	
6	0	
7	0·4	
8	2·5	2·9
9	4·2	4·7

Access for Service (See Fig. R241)

FERGUSON

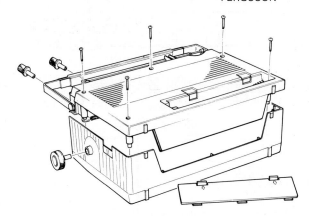

Removing rear cover

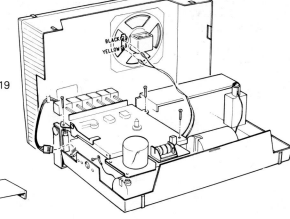

Releasing rear cover and tape deck

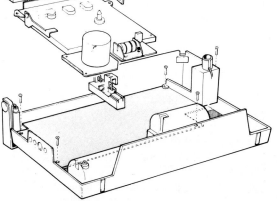

R241

Removing printed board assembly

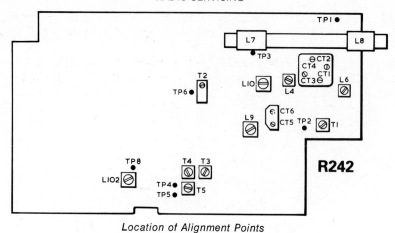

Location of Alignment Points

(R242) ALIGNMENT ADJUSTMENTS—MODEL 3T19

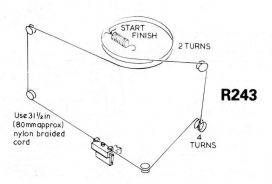

(R243) DRIVE CORD—MODEL 3T19

Alignment (See R242)

A.M. I.F. Circuits: Alignment can be achieved using a wobbulator terminated with a 75 Ω resistor across the output and a 0·01μF capacitor in series with the 'live' lead to the injection point.

Switch receiver to A.M., set Volume control to minimum, Tone control to maximum and tune gang to minimum capacitance (maximum frequency).

Inject a signal centred at 470kHz between TP3 and TP6, connect the wobbulator display unit across TP4 and TP6. Adjust T2 and T3 for maximum response.

Repeat until no further improvement is obtained.

496

A.M. R.F. Circuits: Throughout A.M./F.M. R.F. alignment connect a 20,000 Ω/voltmeter across the loudspeaker terminals and adjust the input signal so as not to exceed 50mW (0·63V).

Signals are injected via a loop loosely coupled to the ferrite rod aerial. Volume and Tone controls should be set to maximum (30% modulated).

Range	Inject	Cursor position	Adjust for max.
M.W.	500kHz	Extreme left	L9
	1650kHz	Extreme right	CT5
	620kHz	620kHz	L7*
	1400kHz	1400kHz	CT2
L.W.	145kHz	Extreme left	L10
	360kHz	Extreme right	CT6
	150kHz	150kHz	L8*
	350kHz	350kHz	CT1

* Adjust by sliding coil former along ferrite rod.

Repeat until no further improvement is obtained.

F.M. I.F. Circuits: With the wobbulator correctly terminated, switch receiver to F.M., set Volume control to minimum, Tone control to maximum and tune gang to minimum capacitance (maximum frequency).

Inject a signal centred at 10·7MHz between TP2 and TP6. Connect the display unit between TP5 and TP6. Detune T5 by unscrewing the core for an 'M' curve. Adjust T1 and T4 for maximum output. Re-adjust T5 and T4 to give an 'S' curve which is symmetrical about the 10·7MHz marker.

F.M. R.F. Circuits: With the R.F. signal generator correctly terminated, inject F.M. signals (22·5kHz deviation) between TP1 and TP6. Volume and Tone controls should be set to maximum.

Range	Inject	Cursor position	Adjust for max.
F.M.	87·5MHz	Extreme left	L6
	105MHz	Extreme right	CT3
	90MHz	90MHz	L4
	105MHz	104MHz	CT4

Repeat until no further improvement is obtained.

Cassette Adjustments

Bias Frequency Adjustment: Connect an oscilloscope between TP8 and D8 anode (earth).

Adjust L102 for minimum on display.

Head Azimuth Adjustment: For azimuth adjustment, play back a standard azimuth tape. Connect an output meter in place of the speaker. Adjust the azimuth screw (nearest the erase head) for maximum reading, whilst keeping the output at a low level.

Finally check for satisfactory reproduction.

Tape Head Demagnetization: The tape-head will gradually become magnetized with use, or if a magnetic tool is accidentally brought close to it; this will show up by an increase in tape noise associated with a decrease of the high frequencies. The tape-head can easily be demagnetized by using a tape head demagnetizer in accordance with the manufacturer's instructions.

Cleaning and Lubrication: Do not use cleaning fluids such as petrol or carbon tetrachloride, which might damage plastic surfaces or rubber drives. A soft cloth dampened with methylated spirit should be used to clean drive surfaces. Head faces can be cleaned either by using a soft cloth dampened in methylated spirit or with a special head cleaning cassette. All moving parts are lubricated during manufacture and further lubrication during service should rarely be required. If it becomes necessary to replace any of the moving parts, only the slightest amount of graphite grease should be applied to the bearing surfaces, ensuring that it does not find its way on to the drive surfaces.

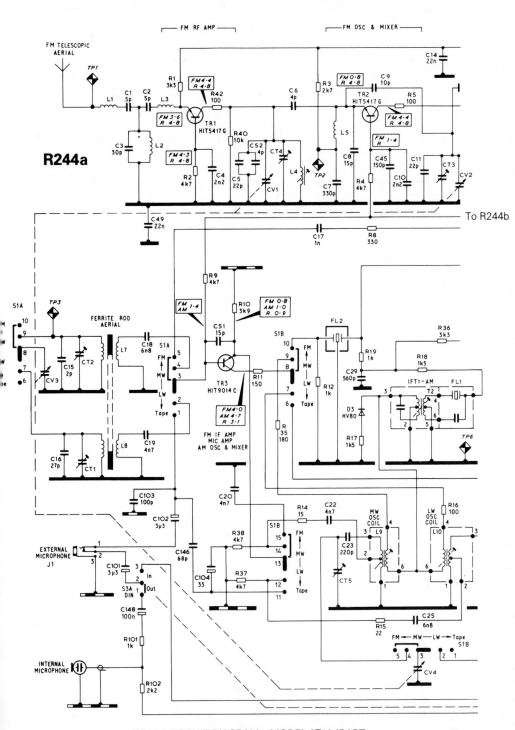

(R244a) CIRCUIT DIAGRAM—MODEL 3T19 (*PART*)

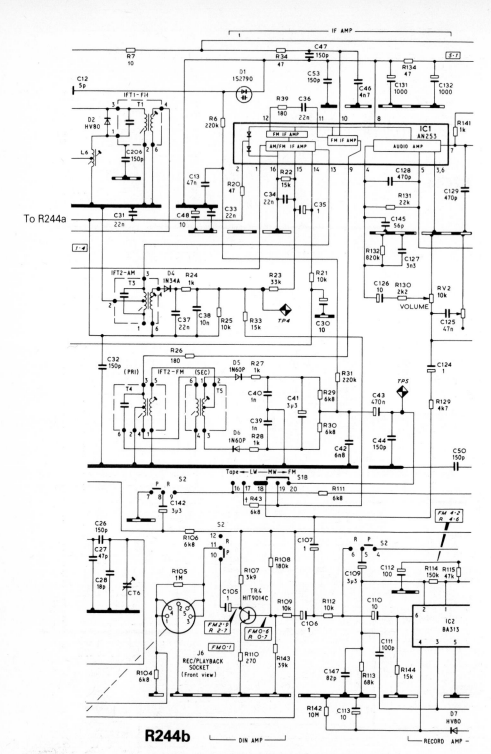

R244b

└─ DIN AMP ─┘   └─ RECORD AMP ─┘

(R244b) CIRCUIT DIAGRAM—MODEL 3T19 (*PART*)

500

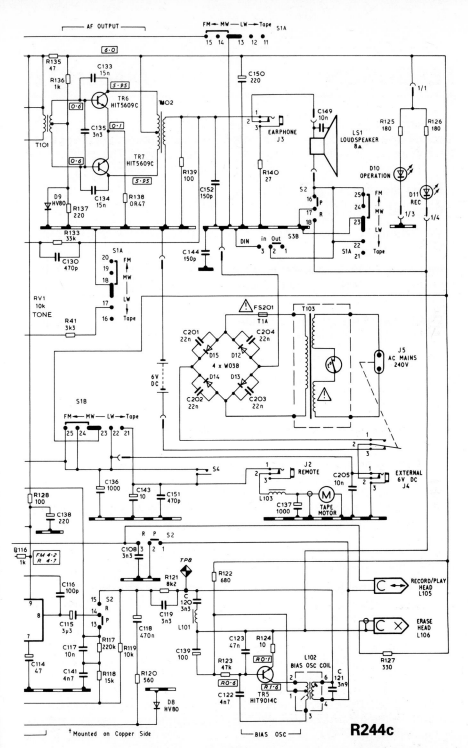

(R244c) CIRCUIT DIAGRAM—MODEL 3T19 (*CONTINUED*)

FERGUSON   Model 3T20

General Description: A four-waveband A.M./F.M. stereo radio-cassette-recorder operating from mains or battery supplies. Sockets are provided for the connection of auxiliary inputs and stereo headphones. The circuit shown here illustrates the left-hand channel, the right-hand channel is similar.

Mains Supply: 240 volts, 50Hz.

Batteries: 9 volts (6×HP2).

Wavebands: L.W. 150–300kHz; M.W. 520–1620kHz; S.W. 5·9–16MHz; F.M. 87·5–108MHz.

Input and Output Sockets: Five-pin Din socket; Input 0·3mV into 3·3k Ω, for external microphones; Output up to 2V into 2·2k Ω (adjustable via Volume control).

Headphone Socket: Output for stereo headphones (6·3 mm R.C.A. jack). Impedance 8–600 ohms (mutes internal loudspeakers).

Loudspeakers: 8 ohms impedance.

Access for Service (See Fig. R245)

Alignment

IC301 Bias Alignment: Connect a D.C. voltmeter between pins 4 and 6 of IC301 and adjust R314 to give a reading of 0·5V.

A.M. I.F. Alignment: Connect an A.M. signal generator, with the sweep centred at 470kHz modulated, to the ferrite rod aerial, using a standard radiation loop. Connect an oscilloscope between TP1 and earth.

Select M.W. on the function switch, set cursor to maximum frequency end of scale, Tone control to maximum and Volume control to minimum.

Adjust T303 for maximum reading on the oscilloscope, consequently adjust T304 and T305 to give maximum readings. Repeat the adjustment until no further improvement is obtained.

F.M. I.F. Alignment: Connect an F.M. signal generator, with the sweep centred at 10·7MHz modulated, between TP1 and TP2. Connect an oscilloscope between TP6 and earth.

Select F.M. on the function switch, set cursor to maximum frequency, Tone control to maximum and Volume control to minimum.

Adjust T301 to give maximum waveform with the 10·7MHz marker at the peak and symmetrical about that marker. Consequently adjust T302, T306, T307 for maximum symmetrical waveform.

Adjust T308 to give a symmetrical 'S' curve about the 10·7MHz marker and centred on the datum line. Repeat the adjustments as necessary to obtain 'S' curve linearity.

502

FERGUSON

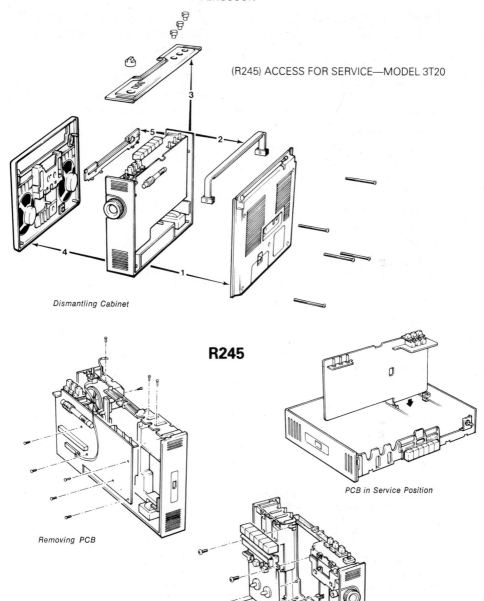

(R245) ACCESS FOR SERVICE—MODEL 3T20

Dismantling Cabinet

R245

Removing PCB

PCB in Service Position

Removing Tape Deck

503

A.M. R.F. Alignment: Connect an A.M. signal generator to the ferrite rod aerial via a standard radiation loop. Obtain tuning indication by connecting an electronic voltmeter across the loudspeaker terminals. Set Volume and Tone controls to maximum and adjust as follows.

Range	Inject	Cursor position	Adjust for max.
L.W.	145kHz	Min. frequency	L209
	310kHz	Max. frequency	C226
	160kHz	160kHz	L211*
	300kHz	300kHz	C209
M.W.	510kHz	Min. frequency	L206
	1650kHz	Max. frequency	C220
	600kHz	600kHz	L203*
	1400kHz	1400kHz	C207
S.W.	5·7MHz	Min. frequency	L204
	16·5MHz	Max. frequency	C215
	7MHz	7MHz	L202
	15MHz	15MHz	C205

* Adjust by sliding coils along ferrite rod.

Repeat adjustments until no further improvement results.

F.M. R.F. Alignment: Connect an F.M. signal generator, terminated with a 75 ohm resistor, across TP1 and TP2. Obtain tuning indication by connecting an electronic voltmeter across the loudspeaker terminals. Set Volume and Tone controls to maximum and adjust as follows.

Range	Inject	Cursor position	Adjust for max.
F.M.	87·35MHz	Min. frequency	L4
	109MHz	Max. frequency	C11
	90MHz	90MHz	L3*
	106MHz	106MHz	C8

* Adjust by slightly opening or closing coil turns.

Repeat adjustments until no further improvement results.

Stereo Decoder Alignment (IC302): Adjust the F.M. signal generator to give an unmodulated 96MHz, 1mV signal, and inject across TP1 and TP2. Connect a frequency counter between TP3 and earth. Adjust RV334 to give a frequency readout of 19kHz with the cursor set to the 96MHz position.

Inject a stereo signal via an encoder and F.M. signal generator to TP1 and TP2. With an input signal of 96MHz, 1mV, and the receiver correctly tuned in, set the Balance control to the 'rnid' position.

Switch signal source to Left only (or Right only) and adjust RV332 to give minimum output on the unwanted channel.

Tape Circuit Alignment

Tape Speed Adjustment: Playback a 50Hz speed test tape and connect the X amplifier of an oscilloscope to the terminals of one of the loudspeakers

504

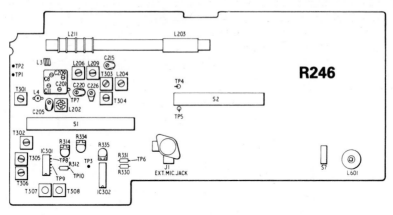

(R246) ALIGNMENT ADJUSTMENTS—MODEL 3T20

(L or R). Inject the 50Hz standard mains frequency into the Y amplifier and compare the two waveforms.

A variable resistor is fitted in the motor governor circuit and access to this is via a small aperture in the base of the motor (normally sealed with an adhesive label to protect the motor from dust). Adjust this variable resistor until a stationary image is obtained on the oscilloscope.

Recording Bias Oscillator Alignment: Connect a frequency meter between TP4 (or TP5) and earth. Set the tape-deck to the record mode and the beat cut switch (S7) in position 1. Adjust L601 to give a readout of 87kHz for bias oscillator frequency.

Azimuth Alignment: Connect an electronic voltmeter to the terminals of one loudspeaker (L or R) and playback a standard azimuth adjustment tape. Turn the azimuth adjustment screw to give maximum reading on the voltmeter.

Note: The output voltage shows three peaks whilst adjustment is being made. Be sure to find the maximum (centre) peak.

Cleaning and Lubrication: Do not use cleaning fluids such as petrol or carbon tetrachloride, which might damage plastic surfaces or rubber drives. A soft cloth dampened with methylated spirit should be used to clean drive surfaces. Head faces can be cleaned either in the same way or with a special head cleaning cassette.

All moving parts are lubricated during manufacture and further lubrication should rarely be required. If it becomes necessary to replace any of the moving parts, only the slightest amount of graphite grease should be applied to the bearing surfaces, taking care to ensure that it does not find its way on to the drive surfaces.

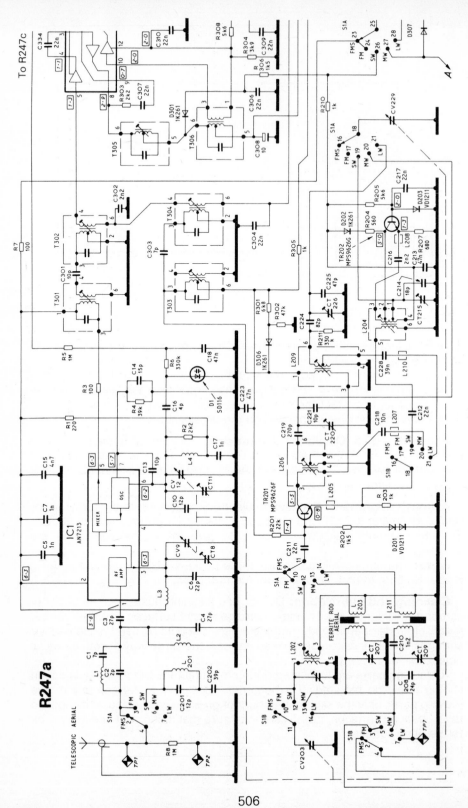

(R247a) CIRCUIT DIAGRAM—MODEL 3T20 (*PART*)

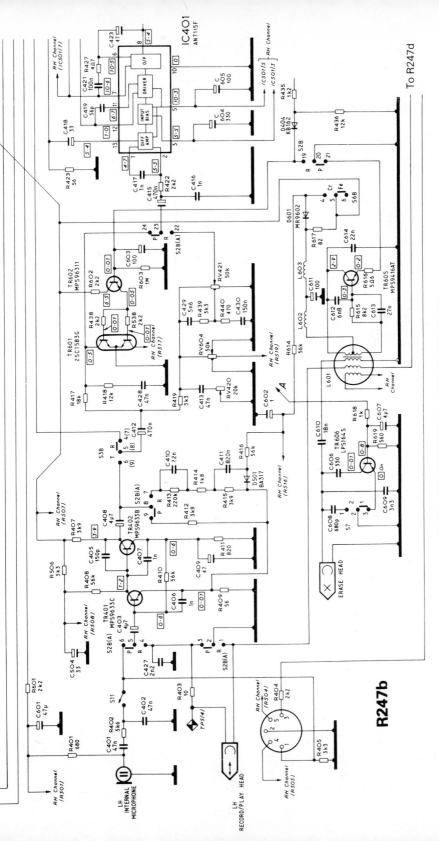

(R247b) CIRCUIT DIAGRAM—MODEL 3T20 (PART)

To R247d

R247b

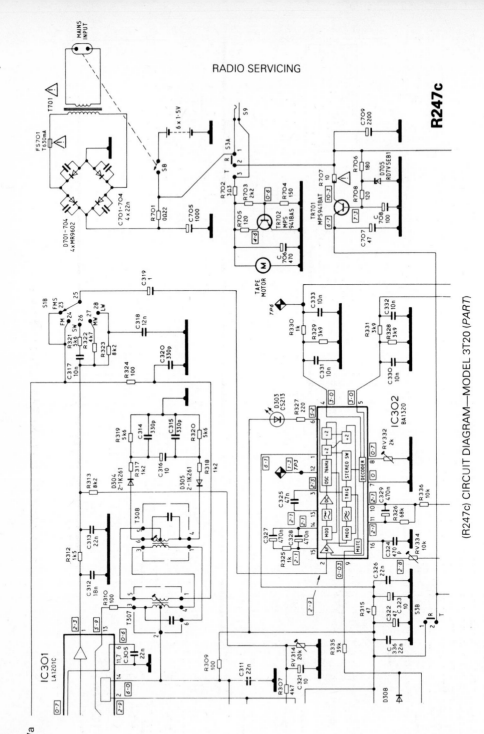

(R247c) CIRCUIT DIAGRAM—MODEL 3T20 (PART)

R247c

To R247a

R247d

FERGUSON

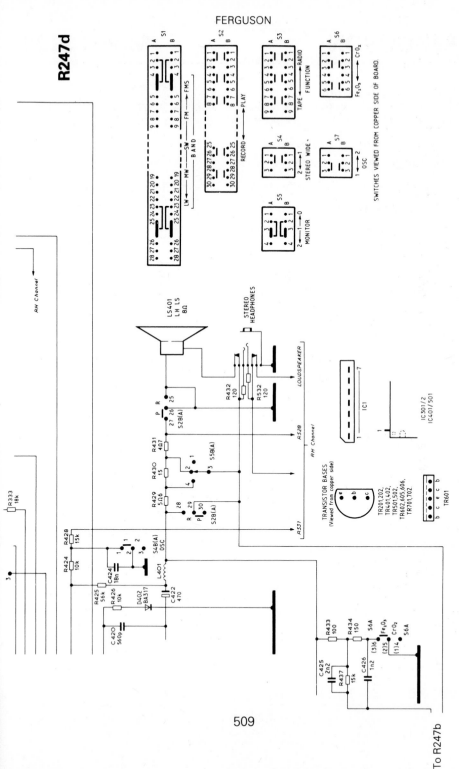

(R247d) CIRCUIT DIAGRAM—MODEL 3T20 (CONTINUED)

SWITCHES VIEWED FROM COPPER SIDE OF BOARD

509

To R247b

FERGUSON   Model 3T21

General Description: A three-wave band A.M./F.M. radio cassette-recorder operating from mains or battery supplies. A microphone is built into the case and a socket is provided for the connection of an earphone.

Mains Supply: 240 volts, 50Hz.

Batteries: 3 volts (6×HP11).

Wavebands: L.W. 150–265kHz; M.W. 520–1620kHz; F.M. 87·5–108MHz.

Bias and Erase: D.C.

Loudspeaker: 8 ohms impedance.

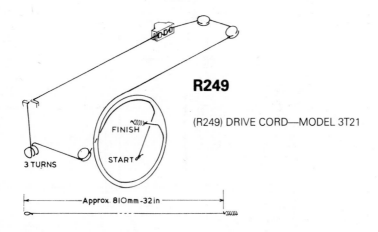

(R249) DRIVE CORD—MODEL 3T21

Access for Service (See Fig. R248)

IC1 Voltages:

Pin	F.M.	A.M.	Pin	F.M.	A.M.
1	—	—	10	—	—
2	1·9	1·1	11	0	0
3	1·9	1·1	12	1·2	1·2
4	0	0	13	0	0
5	9·0	9·0	14	4·7	4·7
6	9·0	9·0	15	9·0	9·0
7	0	1·1	16	—	—
8	0	1·0	17	—	—
9	2·0	1·5	18	—	—

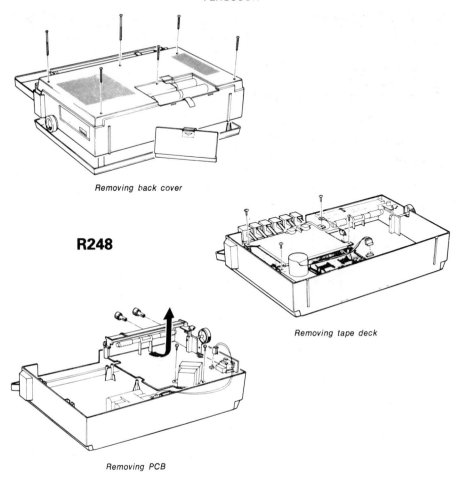

Removing back cover

R248

Removing tape deck

Removing PCB

(R248) ACCESS FOR SERVICE—MODEL 3T21

Alignment

A.M. I.F. Alignment: Switch to M.W. and connect A.M. R.F. signal generator to TP5 and nearest convenient earth point. Disconnect loudspeaker and replace with an output meter set to 8 ohms impedance. Set Tone control to 'high', Volume control to minimum and gang to maximum capacitance.

Inject 468kHz signals, modulated 30%, and adjust T2 and then T4 for maximum output.

During alignment keep the output from the signal generator as low as possible so as not to exceed 50mW audio output.

Repeat adjustments until no further improvement results.

511

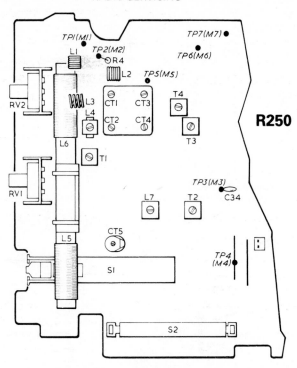

(R250) ALIGNMENT ADJUSTMENTS—MODEL 3T21

A.M. R.F. Alignment: With conditions as for I.F. alignment inject signals, modulated 30%, as shown below.

Range	Inject	Cursor position	Adjust for max.	?
M.W.	510kHz	Extreme left	L7	
	1650kHz	Extreme right	CT4	
	600kHz	600kHz	L6*	
	1400kHz	1400kHz	CT3	
L.W.	270kHz	Extreme right	CT5	
	250kHz	250kHz	L5*	

* Adjust by sliding coil along ferrite rod.

Repeat adjustments until no further improvement results.

F.M. I.F. Alignment: Switch to V.H.F./F.M. and connect an output meter set to 8 ohms in place of loudspeaker. Signals are injected from a wobbulator terminated 75 Ω, with marker set for 10·7MHz into TP1 and TP2. Connect oscilloscope to TP3 and TP4, inject 10·7MHz (22·5MHz deviation) signals and adjust T1 for maximum output with marker at the top of the waveform.

Adjust T3 for symmetrical 'S' curve with the 10·7MHz marker at the centre of the straight portion of the curve.

Repeat adjustments until no further improvement results.

F.M. R.F. Alignment: Signals are injected from a V.H.F./F.M. signal generator, terminated 75 Ω, to TP1 and TP2. Output from the signal generator is adjusted during alignment so as not to exceed 50mW audio output.

Range	Inject	Cursor position	Adjust for max.
F.M.	87MHz	Extreme left	L4*
	109MHz	Extreme right	CT2
	90MHz	90MHz	L2*
	106MHz	106MHz	CT1

* Adjust by slightly opening or closing coil turns.

Repeat adjustments until no further improvement results.

Electrical Adjustments

Azimuth Adjustment: This is the only adjustment required on replacing a Record/Play head. Play back a standard azimuth tape and, with an output meter connected across the loudspeaker, adjust the azimuth screw which can be reached through a small hole in the cabinet front, adjacent to the cassette door, for maximum meter reading keeping the Volume control as low as possible.

The azimuth screw is the left-hand mounting screw viewed from the rear of the head.

Tape Motor Speed Adjustment: The speed control is set to give correct frequency playback of a tape pre-recorded with a known signal. Play a 50Hz speed test tape and inject the output from the loudspeaker into the X amplifier of an oscilloscope.

Inject the 50Hz standard mains frequency into the Y amplifier and compare the two waveforms. Adjust the variable resistor in the motor governor circuit until a stationary image is obtained on the oscilloscope. The variable resistor can be adjusted by inserting a small insulated screwdriver through a hole in the top of the motor. The hole is sealed with a plastic cover to protect the motor from dust; ensure that the hole is re-sealed after adjustment.

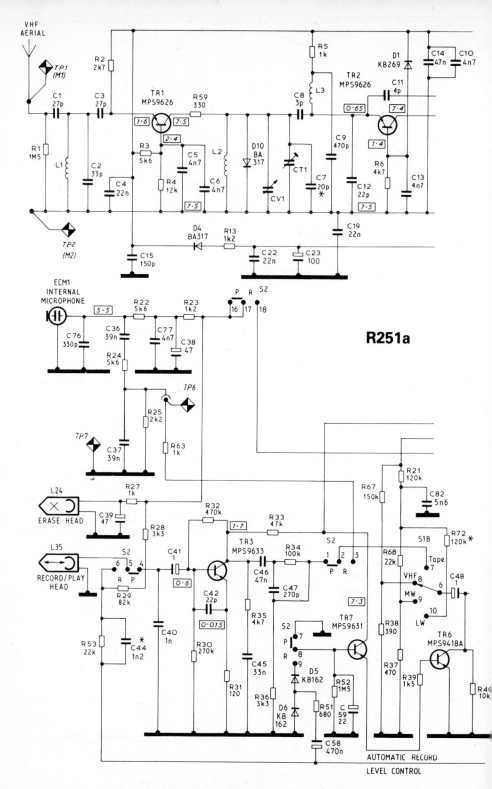

R251a

(R251a) CIRCUIT DIAGRAM—MODEL 3T21 (*PART*)

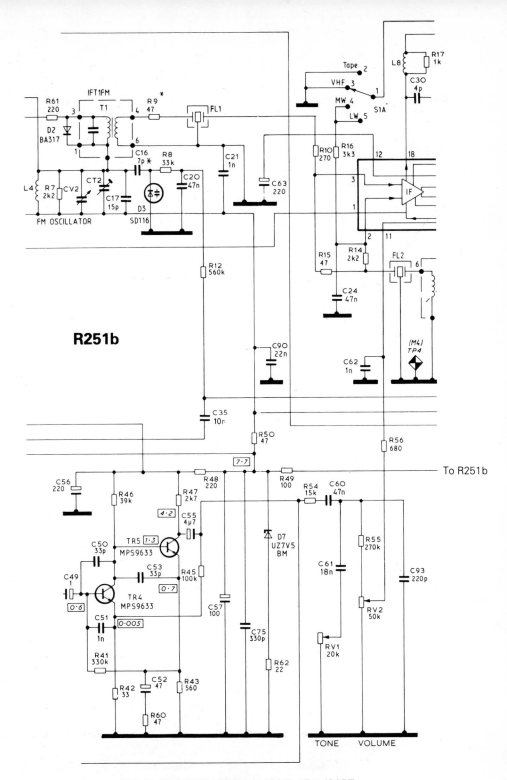

R251b

(R251b) CIRCUIT DIAGRAM—MODEL 3T21 (*PART*)

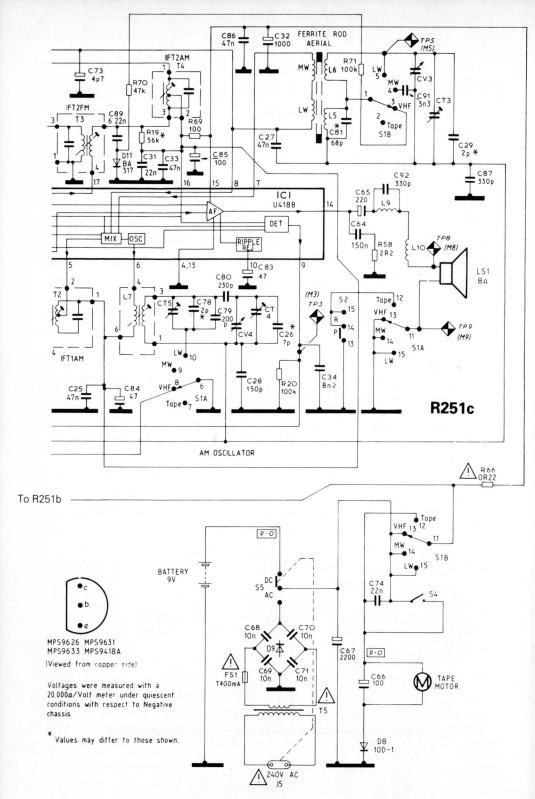

(R251c) CIRCUIT DIAGRAM—MODEL 3T21 (*CONTINUED*)

FERGUSON

Music Centre 100D, Models 3967, 3967B, 3967C, 3967D

General Description: A series of high fidelity music-centres comprising A.M./F.M. radio receiver, record-player, and stereo cassette-recorder with Dolby noise reduction system. Some variations occur in various sub-units used in these models and are detailed as applicable. The receiver section of 3967C and 3967D is similar to that used in Model 3967B. Model 3967 has an alternative receiver. All instruments operate on mains supplies.

Fundamental differences between 3967C and 3967B

1. The A.C. power supply to the cassette and rectifying components on the P.C.B. have been removed (includes FS3).
The D.C. supply now being taken from the RX91'C' chassis.
2. The motor and A.F. signal earth returns have been separated on the cassette P.C.B.
3. The bias oscillator shift switch is dual function with the A.F.C. switch.
4. Primary mains fuse and mains interference by-pass capacitors removed. *Some models may still have by-pass capacitors fitted.*
5. Cabinet presentation changes.
The cassette deck becomes: MG4799DM.   Radio/Amp. becomes: RX91'C'.
Alignment of the Radio and Cassette electronics are similar to 3967B (MG4699DM, RX91). Except for;
Radio Alignment: In the alignment procedure, F.M. R.F., the 108MHz cursor position and frequency of signal generator should be changed to align at 104MHz.

Cassette Electrical Adjustment: Note, that when referring to the bias oscillator switch, this has now become dual function with the A.F.C. switch.

3967D Differences

The main difference of the 3967D (as compared to 3967C) is that the Record Player has been changed to a B.S.R. unit. This unit is basically a P207 with a P259 Turntable.

RX91C

Electrically the RX91C chassis is similar to the RX91. A suffix being added to indicate that the power to the cassette is derived from the chassis. There is also an addition of a capacitor across C15 (C91 100μF. mounted on the copper side of the board at component location G1).

Mains Supply: 240 volts, 50Hz.

Wavebands: (3967) L.W. 148–267kHz; M.W. 530–1625kHz; F.M. 87·5–104MHz.
(3967B) L.W. 148–267kHz; M.W. 530–1625kHz; F.M. 87·5–108MHz.

Pick-up Cartridge: CZ-680-2.

Stylus: 6-2070-2.

Access for Service (See Fig. R344)

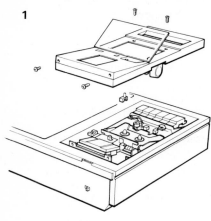

Removing cassette covers

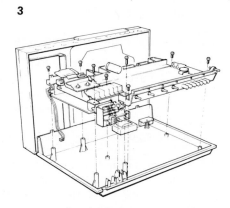

Removing chassis from cabinet
3967 is similar

R344

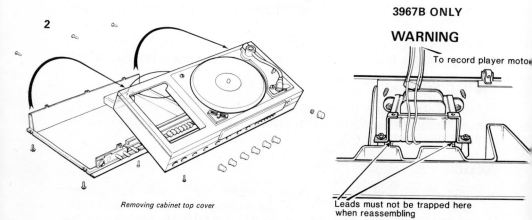

Removing cabinet top cover

3967B ONLY

WARNING

To record player motor

Leads must not be trapped here
when reassembling

(R344) ACCESS FOR SERVICE—MODELS 3967, 3967B

FERGUSON

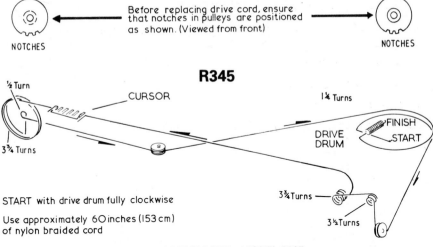

Before replacing drive cord, ensure that notches in pulleys are positioned as shown. (Viewed from front)

NOTCHES

NOTCHES

R345

½ Turn

CURSOR

1¼ Turns

3¾ Turns

DRIVE DRUM

FINISH

START

START with drive drum fully clockwise

Use approximately 60 inches (153 cm) of nylon braided cord

3¾ Turns

3½ Turns

(R345) DRIVE CORD—MODEL 3967

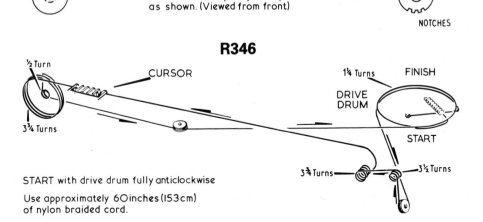

NOTCHES

Before replacing drive cord, ensure that notches in pulleys are positioned as shown. (Viewed from front)

NOTCHES

R346

½ Turn

CURSOR

1¼ Turns

FINISH

3¾ Turns

DRIVE DRUM

START

START with drive drum fully anticlockwise

Use approximately 60 inches (153 cm) of nylon braided cord.

3¾ Turns

3½ Turns

(R346) DRIVE CORD—MODEL 3967B

Alignment (See Figs. R347, 348, 349)

Cassette Adjustments (See Fig. R347)

These will not normally require attention except when a pre-set component or a fixed component affecting an adjustment is replaced.

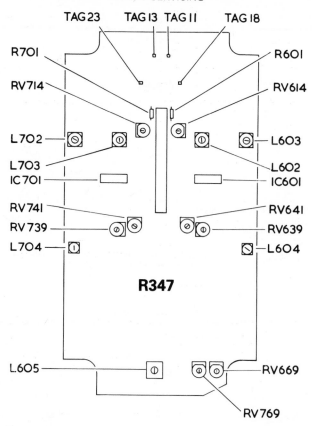

(R347) ALIGNMENT ADJUSTMENTS—MODELS 3967, 3967B

Motor Speed Adjustment: The motor speed should not normally require any adjustment unless the motor itself is replaced. The Speed control is set to give correct frequency playback off a tape pre-recorded with a known signal. The 50Hz speed test cassette and 50Hz null detector can be used for this purpose.

The playback signal from a test cassette is taken from the loudspeaker socket and is compared with the Null Detector, i.e. 50Hz mains frequency via the Null Detector is compared with the test cassette.

If a 50Hz Null Detector is not available it is possible to carry out the adjustment by injecting the 50Hz tape signal into the X amplifier of an oscilloscope and comparing it with 50Hz mains frequency which is injected into the Y amplifier.

A stationary image on the oscilloscope is sufficient to establish that the speed of the motor is correct.

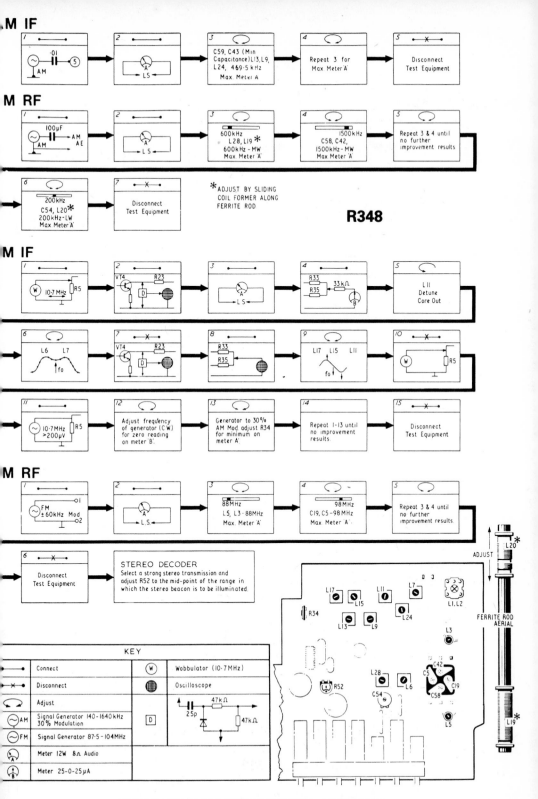

(R348) RADIO ALIGNMENT PROCEDURE—MODEL 3967

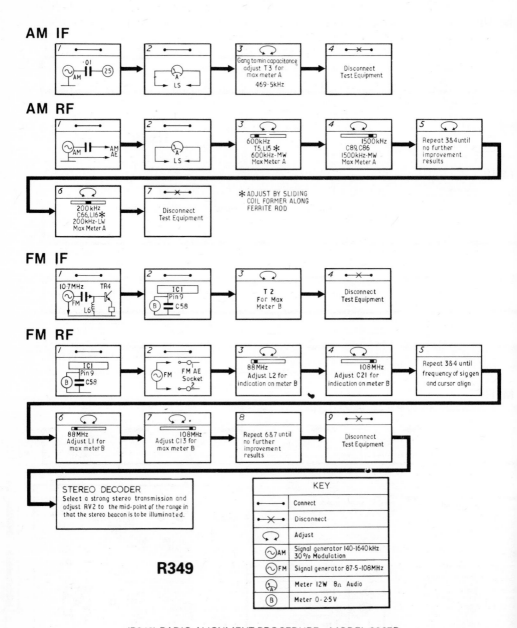

(R349) RADIO ALIGNMENT PROCEDURE—MODEL 3967B

The Motor Speed control is located at the base of the motor. It can be adjusted by using an insulated screw-driver blade and turning the slotted screw to increase or decrease motor speed as desirable.

Check that the dust seals are closed after any adjustment so avoiding dust entering the motor.

Multiplex, Filter Coils (L602, L702, L603 and L703): Select Record and switch Dolby N.R. off; inject a 19KHz signal into tag 11 of radio input. Set L.H. Record Level control to maximum and connect electronic voltmeter to pin 3 IC601. Tune L603 for a null in the signal, which should not be less than −30dB. Repeat for R.H. channel, injecting signal into tag 13 and adjusting L703, L602, L702 and 38KHz low pass filters and should be set for an inductance of 36mH. This corresponds approximately to the turntable cores being flush with the top edge of the screening can.

Record Level Pre-sets (RV641, RV741): Inject 1KHz signal at 5MV into tages 11 and 13. Select record and switch Dolby N.R. on. Set L.H. Record Level control (RV623) to give 580mV reading on an electronic voltmeter connected to pin 3 IC601. Transfer voltmeter to pin 3 IC701 and similarly adjust R.H. Record Level control (RV723).

With input signals and record level settings still adjusted as above, connect the voltmeter to tag 18 and chassis and adjust RV641 for a meter reading of 550μV. Transfer voltmeter to tag 23 and adjust RV741 for a meter reading of 550μV.

Playback Level (RV614, RV714): Connect voltmeter to pin 3 IC601; insert Dolby N.R. test cassette: Select play, switch Dolby N.R. on and adjust RV614 for a reading of 580mV. Transfer meter to pin 3 IC701 and similarly adjust RV714.

Meter Calibration (RV639, RV739): Insert Dolby N.R. test cassette; Select, play, switch Dolby N.R. on, adjust RV639 and RV739 for 0dB readings on record level meters.

Record Equalization: Select record and switch Dolby N.R. off, inject a 1KHz signal at 4mV into tag 11 of radio input. Adjust Record Level control to give 0dB indication of L.H. record level meter. Change input signal to 10KHz and reduce input level to 850μV. Adjust L604 to restore meter indication to 0dB. Repeat procedure for R.H. channel, injecting signals into tag 13 and adjusting L704.

Bias Oscillator Tuning (L605): Connect an oscilloscope or frequency counter across L605. Ensure the oscillator shift switch (S5) is 'off'. Select record and adjust L605 to give 56·6KHz. Switch 'on' the oscillator shift and check that the frequency changes to 55·4KHz (±0·2KHz). Return the oscillator shift switch to the 'off' position.

Bias Current Adjustment: Connect voltmeter across R601, select record and play. Adjust RV669 for 3·5mV across R601. Transfer the voltmeter to R701 and adjust RV769 for 3·5mV. These settings are for typical signal heads and will normally give overall frequency response within the model specification using the type of tape supplied with the machine. This corresponds to approximately 350μA bias current.

The overall frequency response may be optimised for a particular head by slight adjustment from this figure. This can be checked by taking the following frequency response measurements: At a level −25dB below peak, make a frequency response recording using a Super Ferric cassette (similar to the type supplied with the machine), then switch to play and check that the frequency response falls within the falling limits relative to the level at 1kHz.

(A) 5kHz+1−2dB (B) 12·5kHz+0−6dB.

If the result is not within these limits, RV669 and RV769 must be re-adjusted. If the response is too high, increase the bias; if too low, reduce bias to within tolerance limits 2·5mV−6mV across R601/R701. If the correct frequency response cannot be achieved within these limits then it may be assumed that either the record/play head is faulty or that a fault exists which affects the normal frequency response of the record amplifier, such as misalignment of L602, L702.

Circuit Description

Autostop Circuit Operation: The tape motion is sensed by a rotating magnet (mounted on the underside of the R.H. spool carrier) which activates a reed switch. The resulting pulse is applied to a differentiating circuit R758, C738. A capacitor C737 reduces the undesired effects of contact bounce.

The diode D706 allows positive pulses to be applied to the base of control transistor TR711 and suppresses the negative pulses which could exceed the reverse V$_{BE}$ rating of the transistor. C739 integrates the pulses. R759 limits the base current to TR711.

In normal working operation the voltage across C739 is low, the collector voltage of TR711 being low and TR712 being saturated. The application of each positive pulse discharges C739 by applying the supply voltage through the circuit R756, C738, D706 (turns TR711 'on'). R758 limits the time of this discharge. When the pulse has decayed, C739 supplies the base current to maintain TR711 in conduction and charges up in the process. TR711 collector current biases the series transistor TR712 into conduction, thus maintaining the supply to the motor control circuit. C740 is included to supply sufficient current through R761 to maintain the bias on TR712, when the magnet is rotating slowly.

When the train of positive pulses stops, C739 continues to be charged from V.C.E. so that the volts across R761 reduces. i.e. there is a reduction in TR711 collector current. The series transistor TR712 is therefore supplied with a decaying base current. TR712 comes out of saturation and the supply current to the motor decreases until, finally, it stops.

Motor drive is maintained during operation of the Pause control by S6 which shorts TR711 collector to the negative supply and so maintains TR712 in saturation. At the same time C739 is discharged so that the autostop is ready to resume normal operation immediately the Pause control is released.

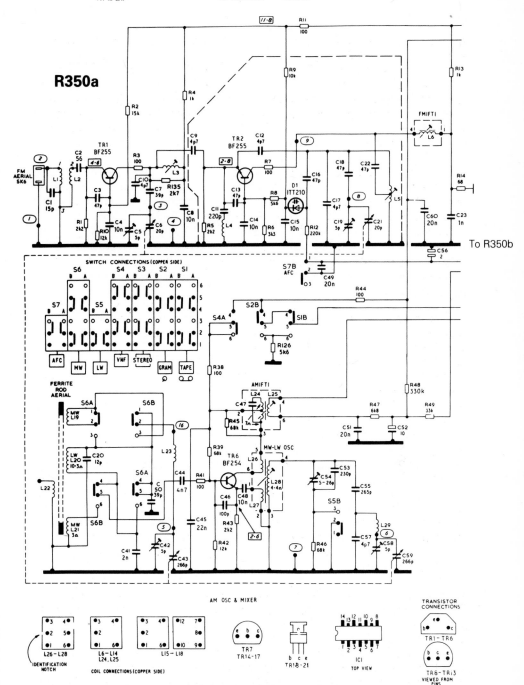

(R350a) CIRCUIT DIAGRAM (RADIO SECTION)—MODEL 3967 (*PART*)

525

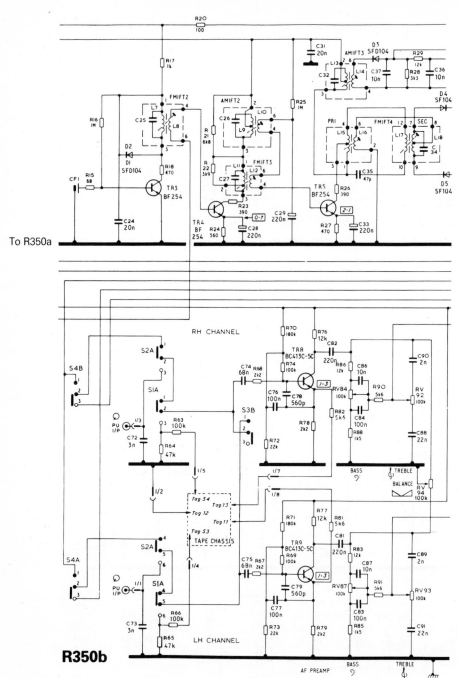

To R350a

R350b

(R350b) CIRCUIT DIAGRAM (RADIO SECTION)—MODEL 3967 (*PART*)

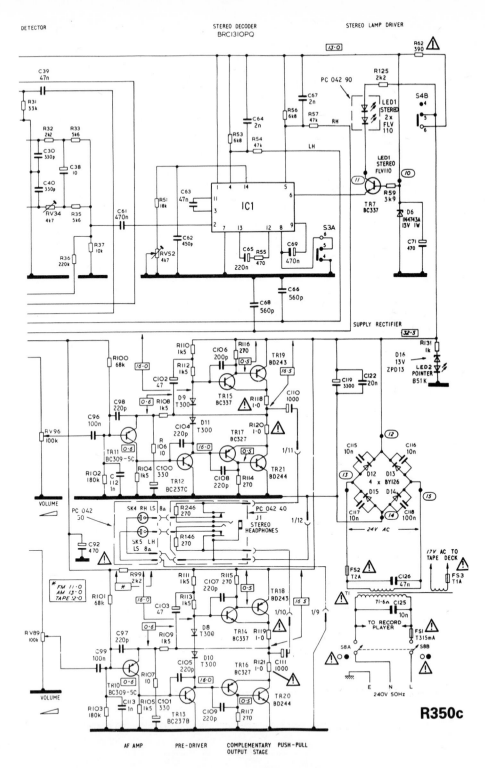

(R350c) CIRCUIT DIAGRAM (RADIO SECTION)—MODEL 3967 (*CONTINUED*)

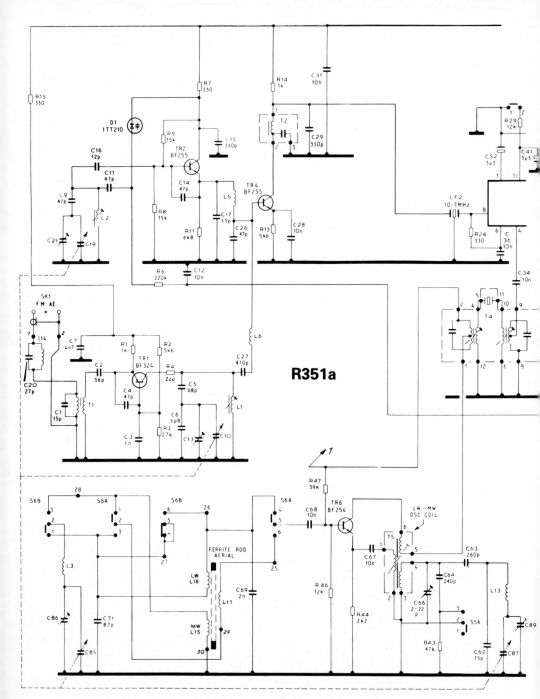

R351a

(R351a) CIRCUIT DIAGRAM (RADIO SECTION)—MODEL 3967B (*PART*)

528

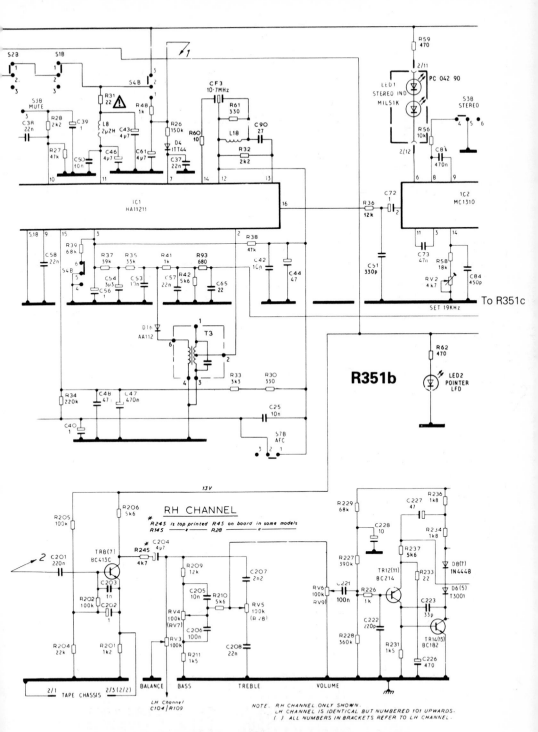

(R351b) CIRCUIT DIAGRAM (RADIO SECTION)—MODEL 3967B (*PART*)

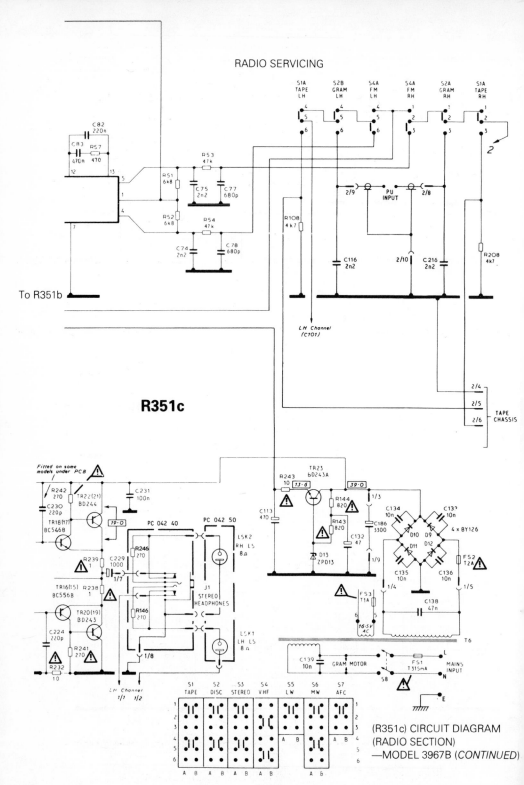

R351c

To R351b

LH Channel
(C101)

TAPE
CHASSIS

(R351c) CIRCUIT DIAGRAM
(RADIO SECTION)
—MODEL 3967B (CONTINUED)

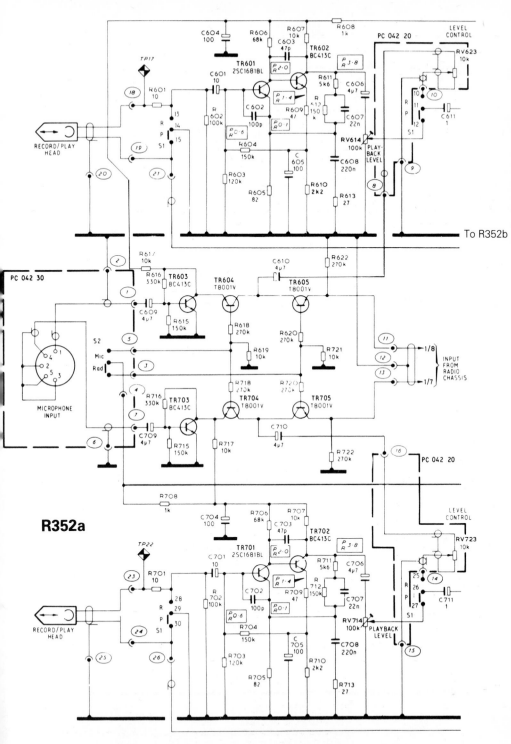

R352a

(R352a) CIRCUIT DIAGRAM (CASSETTE SECTION)—MODEL 3967 (*PART*)
(MODEL 3978B SIMILAR LESS TR10/710, D603/704, S4)

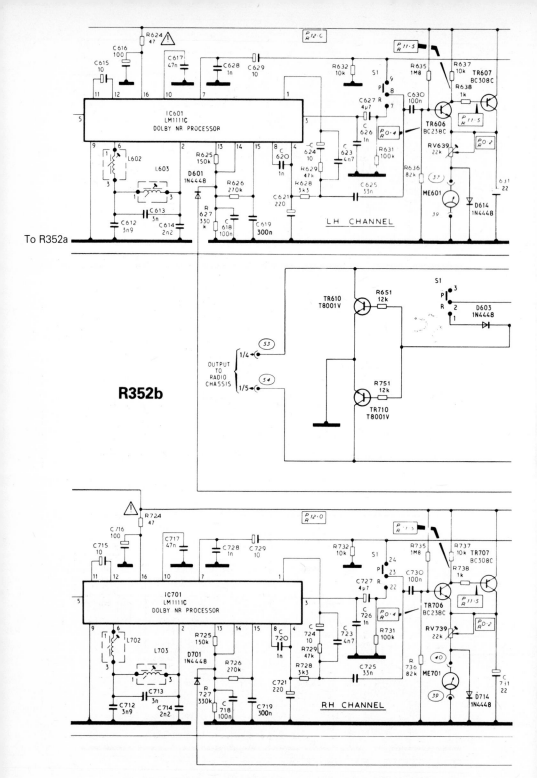

(R352b) CIRCUIT DIAGRAM (CASSETTE SECTION)—MODEL 3967 (*PART*)
(MODEL 3978B SIMILAR LESS TR610/710, D603/704, S4)

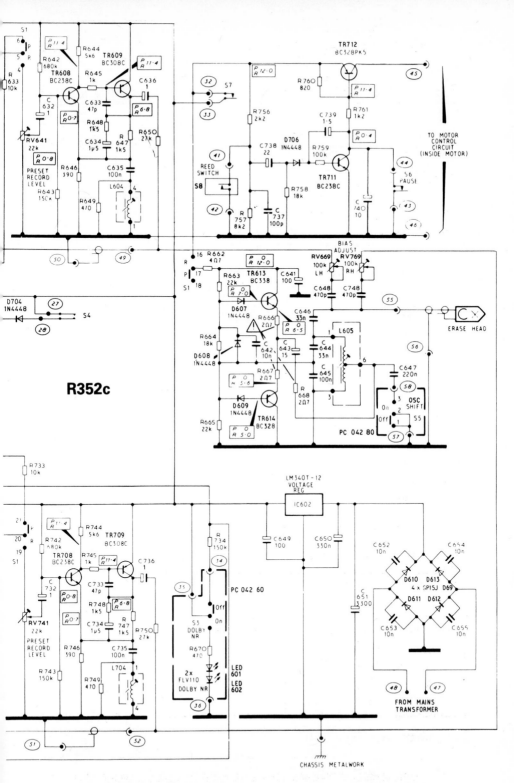

R352c

(R352c) CIRCUIT DIAGRAM (CASSETTE SECTION)—MODEL 3967 (*CONTINUED*)
(MODEL 3978B SIMILAR LESS TR610/710, D603/704, S4)

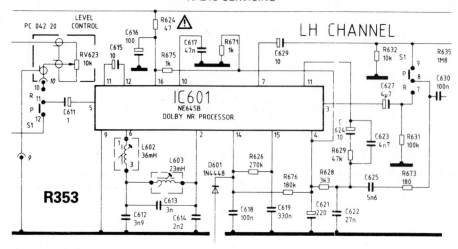

(R353) CIRCUIT DIAGRAM (ALTERNATIVE DOLBY I.C.)—MODEL 3967 ETC.

FERGUSON

Music Centre 90, Models 3968A, 3968B

General Description: This music-centre comprises a three-waveband A.M./F.M. stereo radio receiver, record-player, and cassette-recorder with separate loudspeaker enclosures. The instrument operates from mains power supplies and differences occur in the radio circuitry. Model 3968 uses the RX91 'C' radio chassis which is similar to that fitted to model 3967B and is described elsewhere in this volume. The sections of model 3968B are all included on a common chassis designated RX92.

Mains Supply: 240 volts, 50Hz.

Wavebands: L.W. 148–267kHz; M.W. 530–1625kHz; F.M. 87·5–104MHz.

Turntable: BSR P207.

Pick-up Cartridge: SC12H.

Stylus: ST20.

Loudspeakers: 8 ohms impedance.

Alignment (See Fig. R354).

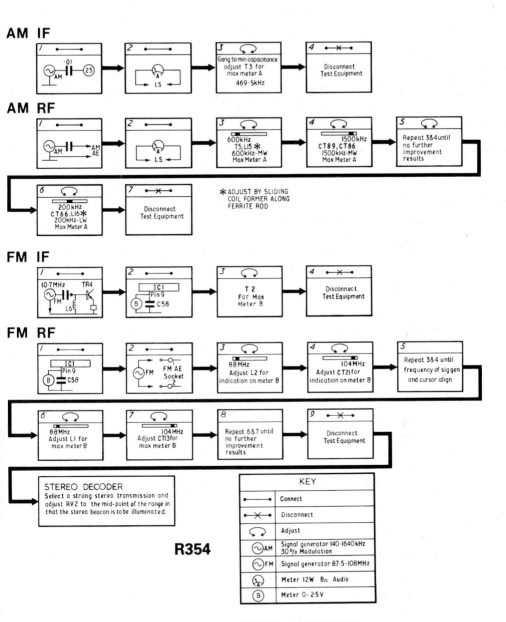

AM IF

AM RF

* ADJUST BY SLIDING
  COIL FORMER ALONG
  FERRITE ROD

FM IF

FM RF

STEREO DECODER
Select a strong stereo transmission and
adjust RV2 to the mid-point of the range in
that the stereo beacon is to be illuminated.

KEY		
•——•	Connect	
•—✕—•	Disconnect	
↻	Adjust	
∿AM	Signal generator 140-1640 kHz 30 % Modulation	
∿FM	Signal generator 87·5-108 MHz	
Ⓐ	Meter 12W 8Ω Audio	
Ⓑ	Meter 0-2·5 V	

R354

(R354) ALIGNMENT PROCEDURE—MODEL 3968

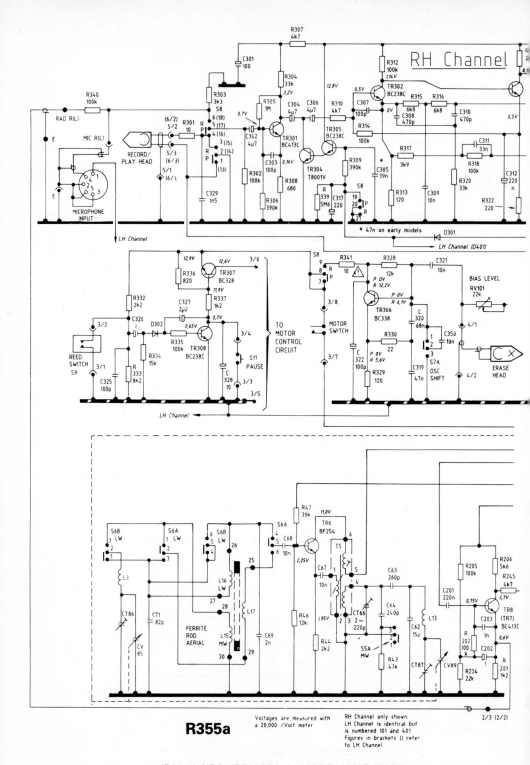

R355a

Voltages are measured with a 20,000 Ω/Volt meter

RH Channel only shown.
LH Channel is identical but is numbered 101 and 401
Figures in brackets () refer to LH Channel

(R355a) CIRCUIT DIAGRAM—MODEL 3968B (*PART*)

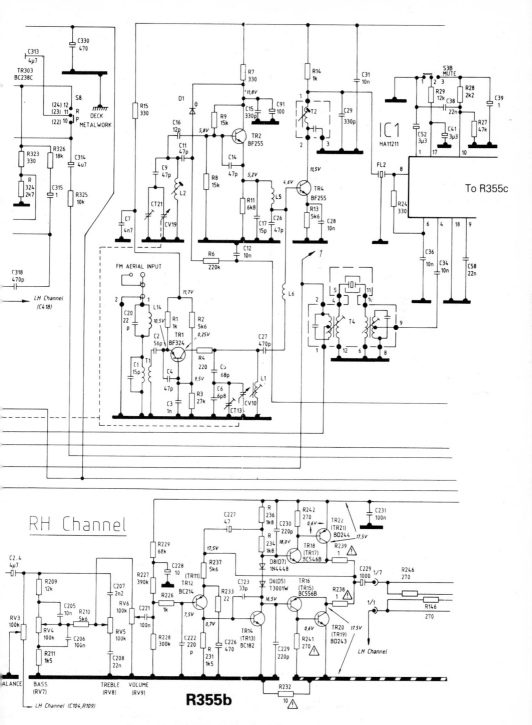

(R355b) CIRCUIT DIAGRAM—MODEL 3968B (*PART*)

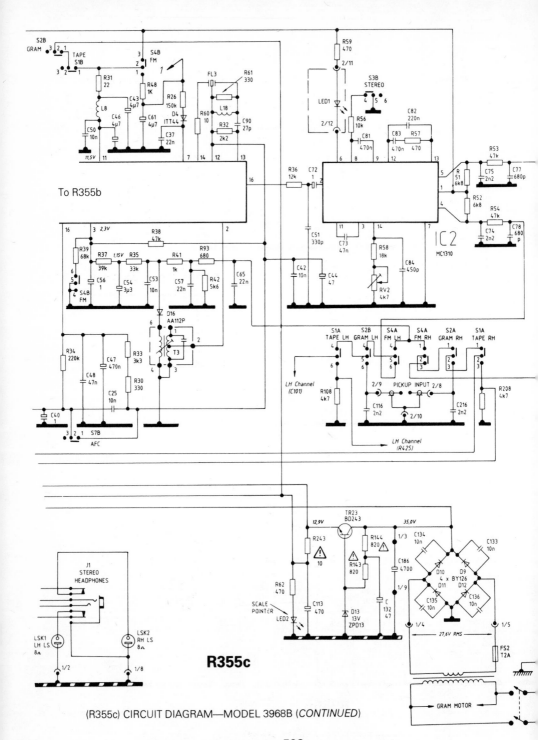

To R355b

R355c

(R355c) CIRCUIT DIAGRAM—MODEL 3968B (*CONTINUED*)

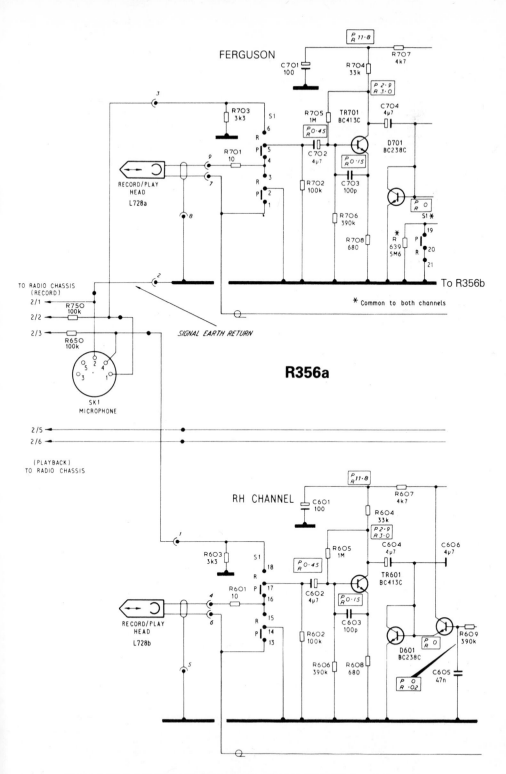

(R356a) CIRCUIT DIAGRAM (CASSETTE RECORDER)—MODEL 3968A (*PART*)

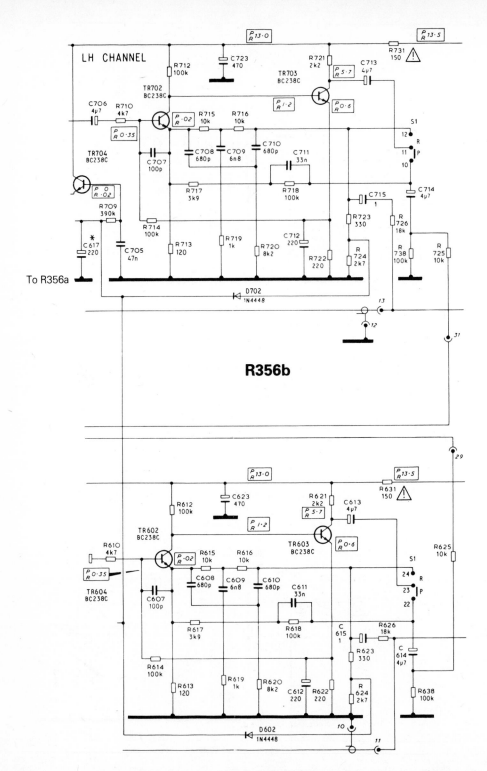

(R356b) CIRCUIT DIAGRAM (CASSETTE RECORDER)—MODEL 3968A (*PART*)

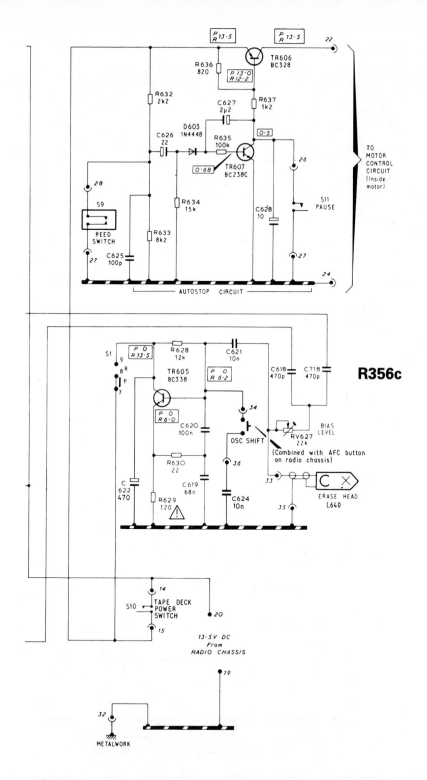

(R356c) CIRCUIT DIAGRAM (CASSETTE RECORDER)—MODEL 3968A (*CONTINUED*)

FERGUSON

Music Centre 101, Model 3977

General Description: This instrument is generally similar to the Ferguson model 3968 described previously. It is distinguished by the inclusion of an electronic clock/timer enabling programmes to be automatically recorded. The record-player unit is a Matsushita FU121-ER-6 with cartridge CZ-680 and stylus 6-2098-2.

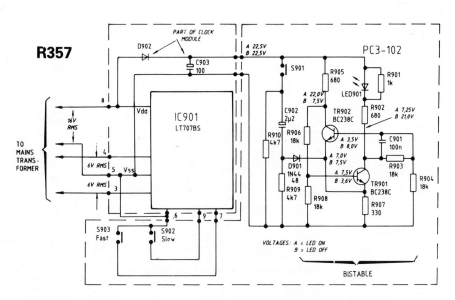

(R357) CIRCUIT DIAGRAM (CLOCK AND MESSAGE LAMP)—MODEL 3977

General Description: A mains-operated stereo music-centre comprising three-waveband A.M./F.M. radio receiver, record-player and cassette-recorder with separate loudspeaker enclosures. L.E.D. record level and power indicators are fitted and sockets are provided for the connection of auxiliary inputs and headphones.

Mains Supplies: 220–250 volts, 50Hz.

Wavebands: L.W. 150–250kHz; M.W. 525–1625kHz; F.M. 87·5–108MHz.

Pick-up Cartridge: SC12H.

Stylus: ST20.

Loudspeakers: 4 ohms impedance.

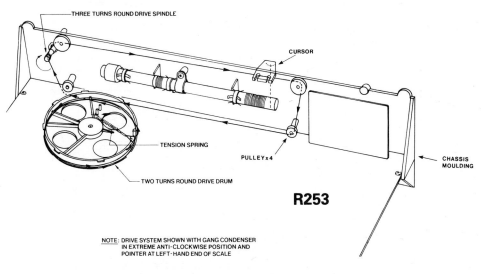

R253

NOTE: DRIVE SYSTEM SHOWN WITH GANG CONDENSER
IN EXTREME ANTI-CLOCKWISE POSITION AND
POINTER AT LEFT-HAND END OF SCALE

(R253) DRIVE CORD—1.S. SYSTEM

screws marked 'C' (diagram 1) from the rear of the unit. Gently draw out the front which will slide out together with the boards allowing access for servicing.

Note: To remove the cabinet completely, it is necessary to remove the mains cord clamp and its mounting plate from the rear of the unit. Then unplug the main connector and unsolder the green earth wire from the tag on the transformer bracket.

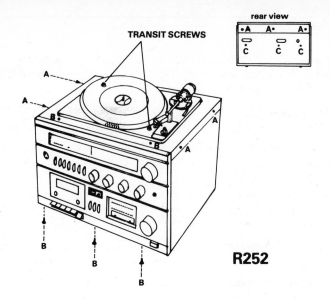

TRANSIT SCREWS

rear view

R252

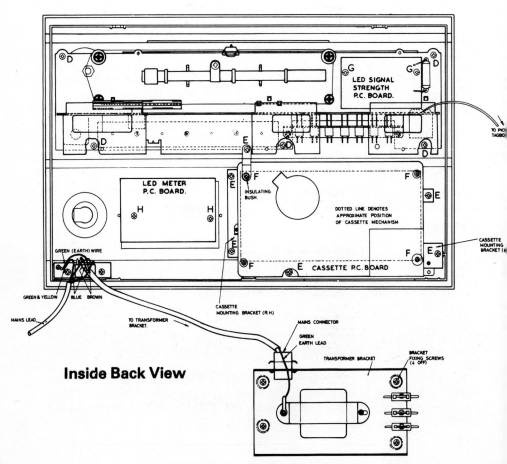

Inside Back View

LED SIGNAL
STRENGTH
P.C. BOARD.

LED METER
P.C. BOARD.

INSULATING
BUSH.

DOTTED LINE DENOTES
APPROXIMATE POSITION
OF CASSETTE MECHANISM

CASSETTE P.C. BOARD

CASSETTE
MOUNTING
BRACKET (

TO PICK
TAGBO

GREEN (EARTH) WIRE

GREEN & YELLOW   BLUE   BROWN

MAINS LEAD

TO TRANSFORMER
BRACKET.

CASSETTE
MOUNTING BRACKET (R.H.)

MAINS CONNECTOR

GREEN
EARTH LEAD

TRANSFORMER BRACKET

BRACKET
FIXING SCREWS
(4 OFF)

(R252) ACCESS FOR SERVICE—1.S. SYSTEM

Audio Tuner Chassis: First remove the headphone socket nut and the tuning, volume, balance, bass, and treble control knobs. Then remove five screws marked 'D' (diagram 2).

Tape Mechanism with Board: Remove the cassette door by unscrewing the two screws on the flap. Then remove six screws marked 'E' (diagram 2).

Access for Service (See Fig. R252)

Record-Player Mounting Plate: First clamp the record-player to its mounting plate by turning the two transit screws fully anti-clockwise, then remove the seven screws marked 'A' and two screws marked 'B' from the top as in the diagram. Then lift the plate upwards and unsolder the pick-up and the turntable motor leads from the Audio Tuner board.

Unit Cabinet: Remove the record-player mounting plate as above, then remove three screws marked 'B' from the bottom of the unit and three

Tape Board: To remove the tape board from the tape mechanism, remove four screws marked 'F' (diagram 2).

Note: The bracket fixing the tape mechanism to the Audio Tuner Board heatsink has an insulating plastic bush. This bush must be re-inserted into the bracket when assembling again. Failure to do so will result in a loud hum on tape playback.

L.E.D. Signal Strength Board: Remove two screws marked 'G' (diagram 2) and unsolder the dial lamp from the two tags on the board.

L.E.D. Meter Board: Remove two screws marked 'H' (diagram 2).

Transformer Bracket: Remove the four screws fixing the bracket to the wooden cabinet.

Adjustments of the Tuner Section

Note: Always use a high frequency insulated screwdriver or trimmer to make the adjustments. These should be done very carefully.

A.M. Alignment

I.F.: Place a coupling loop in proximity to the ferrite rod and radiate a 470KHz signal. Select M.W. band and with gang set to minimum capacitance at the H.F. end align L6 (Black) for maximum output. The 470KHz Filter (Red and Blue) should not require alignment but may be peaked if necessary.

R.F.: Check that the scale pointer is correctly positioned as indicated by the 'O' mark on the Log Scale on the dial. With receiver switched to M.W. at the L.F. end (maximum capacitance) radiate a signal of 525KHz. Align the osc. coil L5 (Red) for the signal. Tune receiver to the H.F. end (minimum capacitance) and radiate a signal of 1625KHz and align TC4 for the signal. Repeat for optimum results.

Tune receiver to 600KHz, radiate this signal and adjust M.W. aerial coil for maximum output.

Tune receiver to 1500KHz, radiate this signal and align aerial trimmer TC5 for maximum output. Repeat alignment of M.W. aerial coil and TC5 for optimum results.

Switch receiver to L.W. and tune to 150KHz, radiate this signal and adjust trimmer TC3 for the signal. Tune to 200KHz and radiate this signal and adjust the L.W. aerial coil for maximum output. Repeat above operations until no further improvement is achieved.

During the adjustments ensure that the output of the signal source is low enough to prevent the A.G.C. of the set from coming into operation.

F.M. Alignment

I.F.: Connect a wobbulator via a 2pf capacitor to the base of TR2 and earth. View the 'S' curve at pin 1 of the stereo decoder IC2. Adjust L8 (Brown) for a symmetrical 'S' curve and L4 (Brown) for maximum amplitude.

R.F.: Inject an 87·5MHz signal into the aerial socket and with the gang at maximum capacitance (L.F. end) align the osc. coil L3 for the signal and aerial coil L2 for maximum output. Tune to the H.F. end, i.e. minimum capacitance and inject a 108MHz signal. Align the osc. trimmer TC2 for the signal and adjust the aerial trimmer TC1 for maximum output. Repeat the above procedure (for R.F.) until no further improvement can be achieved.

Decoder Alignment: Inject a Multiplex R.F. signal at the aerial input and adjust PR2 for L.E.D. illumination. Reduce the R.F. signal whilst adjusting PR2 to obtain the optimum setting.

Circuit Voltages

Tape Board:

Voltages on play and record position

	TR1 2101 BC149S	TR2 2102 BC548	TR58 103 BC548	TR48 104 BC549C	TR5 2105 BC549C	TR203 BC147C
Collector	3V8	—	—	1V6	5V5	—
Base	·4V	—	—	·8V	1V5	·6V
Emitter	—	—	—	·3V	·9	—

Voltages on Record Position

	TR62 106 BC548	TR201 BC548C	TR202 BC559C
Collector	—	15V	—
Base	−3V4	7V7	6V8
Emitter	—	7V6	6V9

Left & Right Record Level Power Indicator Voltages with all L.E.D. s on Full Display:

Calibration	IR2 EO2 IC3/303		11 Way Molex to take Board from R.F. Board			6 Way Molex from Tape Deck to Record Level Power Indicator 1	
1	6V5	1	Brown	0V			
2	8V3	2	Green	0V	1	Yellow	0V
3	10V5	3	Pink	0V	2	Black	0V
4	8V3	4	Orange	0V	3	Blue	0V
5	10V5	5	Green	15V	4	Mauve	0V
6	8V4	6	Yellow	0V	5	Red	15V
7	10V5	7	Mauve	15V	6	Green	15V
8	—	8	Black	0V			
9	12V	9	Blue	·01V			
10	12V5	10	Red	T/Screen			
11	—	11	Blue	Lead – 0V			
12	·5V						
13	3V						
14	2V7						
15	—						
16	1V						

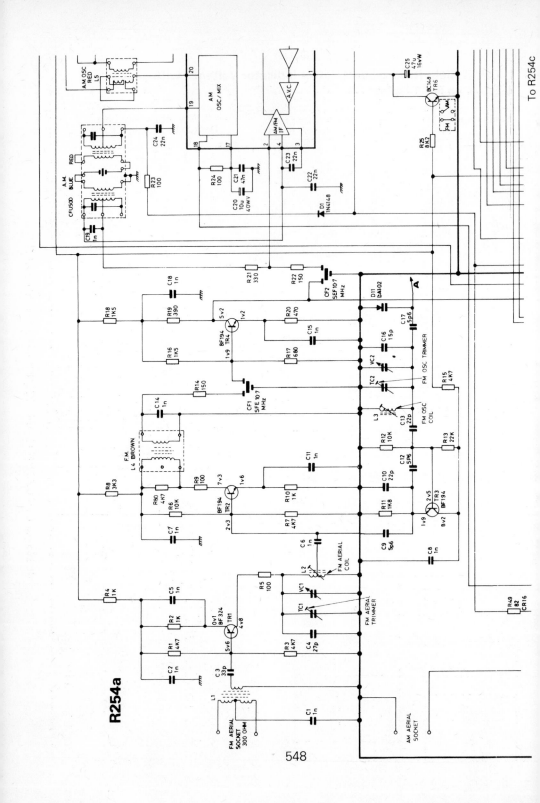

R254a

To R254c

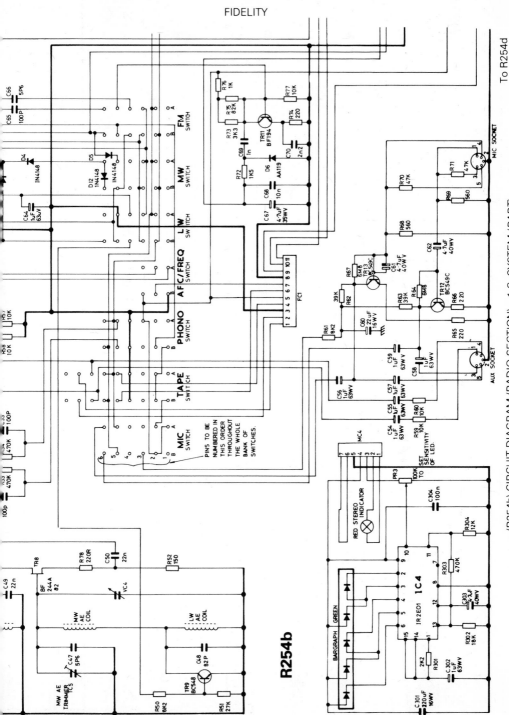

(R254b) CIRCUIT DIAGRAM (RADIO SECTION)—1.S. SYSTEM (*PART*)

R254b

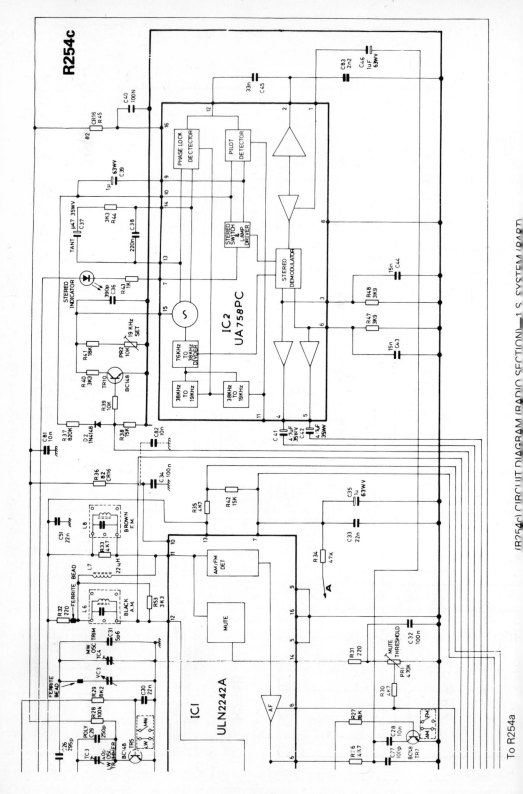

(R254c) CIRCUIT DIAGRAM (RADIO SECTION) — 1 S. SYSTEM (PART

To R254a

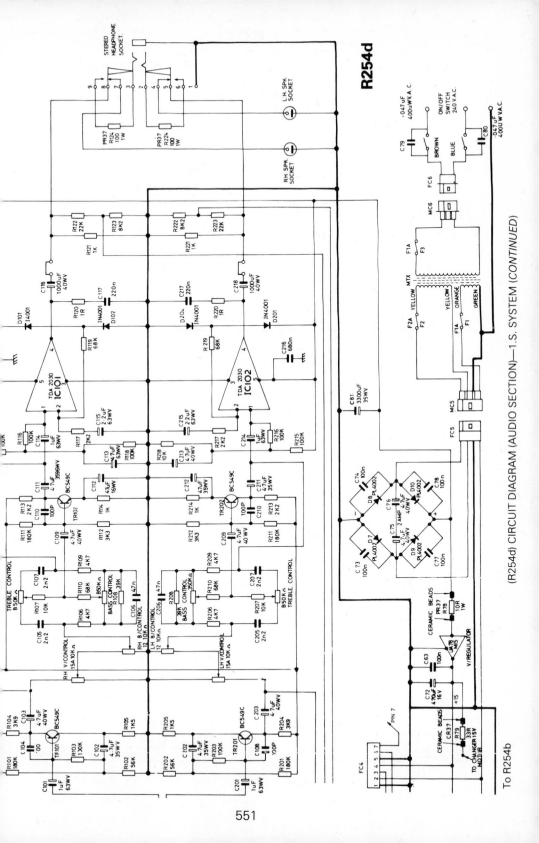

R254d

(R254d) CIRCUIT DIAGRAM (AUDIO SECTION)—1.S. SYSTEM (CONTINUED)

To R254b

R255a

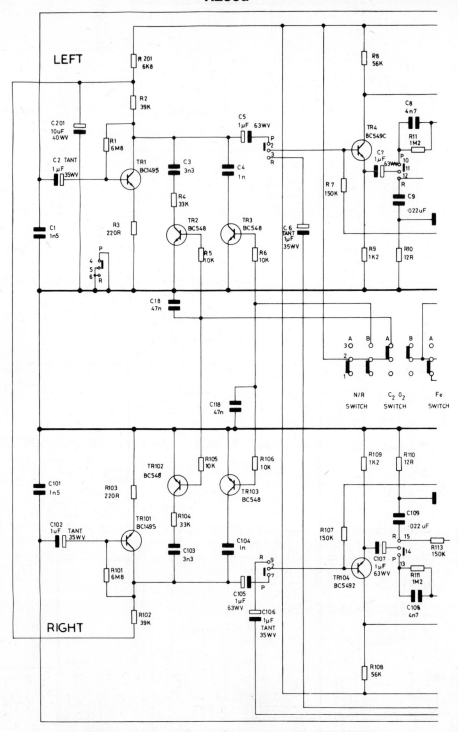

(R255a) CIRCUIT DIAGRAM (TAPE SECTION)—1.S. SYSTEM (*PART*)

R255b

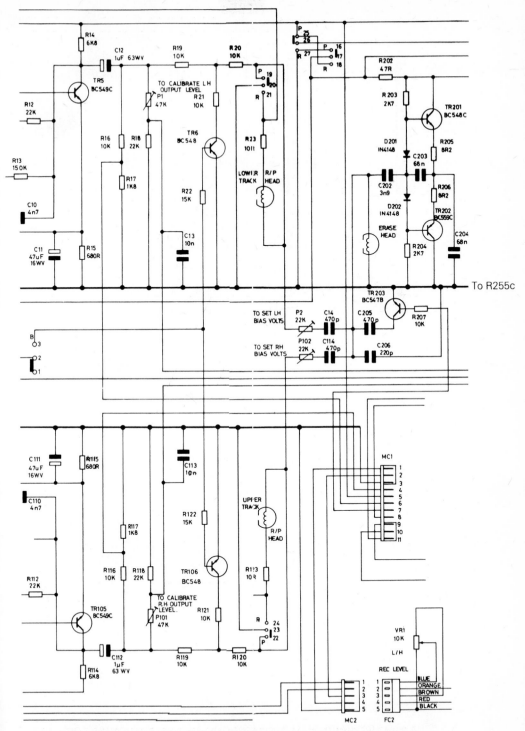

(R255b) CIRCUIT DIAGRAM (TAPE SECTION)—1.S. SYSTEM (*PART*)

R255c

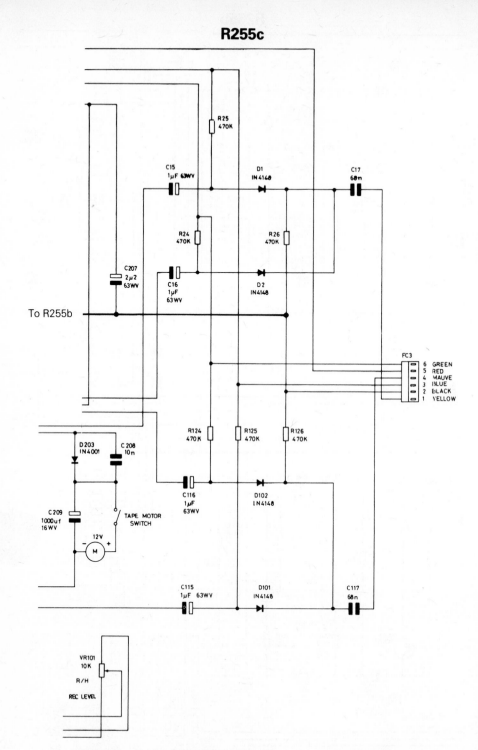

(R255c) CIRCUIT DIAGRAM (TAPE SECTION)—1.S. SYSTEM (*CONTINUED*)

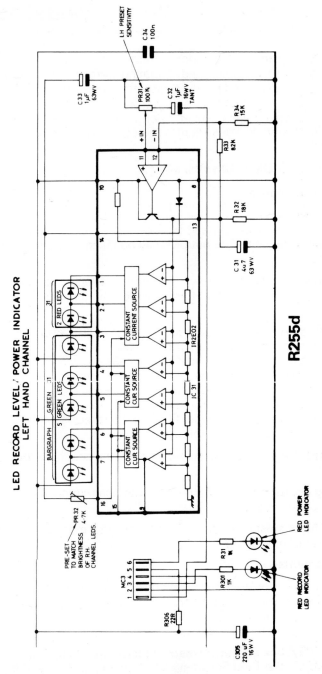

LED RECORD LEVEL / POWER INDICATOR
LEFT HAND CHANNEL

R255d

(R255d) CIRCUIT DIAGRAM (DISPLAY CIRCUIT—LEFT)—1.S. SYSTEM (R.H. CHANNEL IS SIMILAR)

H.M.V.   Model 4000C

General Description: A high-fidelity stereo cassette tape-recorder incorporating Dolby noise reduction and operating from mains supplies. Other facilities include a record indicator, lit cassette compartment, full electronic autostop, 2VU meters operating on playback and record. 3 Digital counters with memory. Damped eject cassette door. Variable output level. Independant record level controls. Pause control. Electronically speed controlled D.C. motor.

Mains Supplies: 220–240 volts, 50Hz.:

Dismantling Procedure

Bottom: Carefully turn the unit upside down and support it, avoiding damage to the front extrusion and controls. (We suggest the use of thick foam to support the unit.) The bottom may be taken off after removing the eight screws.

Top Cabinet: Remove six screws: two from left hand side, two from right hand side and two from rear top. Carefully slide the top cabinet upwards to remove it.

Front Extrusion: It is first necessary to remove the bottom and the top, then take all the knobs off except the black cassette buttons.

Then remove the two chrome nuts on the cassette window, unscrew the microphone socket nuts and the three bottom screws holding the front extrusion to the chassis bracket. Gently pull out the extrusion from the bottom and slide it upwards to remove it completely. It is now possible to service virtually any of the components without further dismantling, however the chassis may be removed as follows:

Main Board: Remove the six screws around the periphery to release the chassis.

Switching Board: To remove the switching board remove the screw and unscrew the lever switches.

Cassette Mechanism: To remove the cassette mechanism assembly, remove four bracket screws.

Tape Adjustments

Ensure that the record/playback and erase heads are clean by using either a head cleaning tape or by using a soft fluffless cloth or soft brush moistened with pure alcohol or methylated spirits.

Note: Dolby Level 100mV (0dB) at TP34 (left) and TP35 (right).

1. Record/Playback Azimuth Head Adjustment: Dolby N.R. switch off, select Fe tape, on the Tape Select switch. Monitor TP34 and TP35, with an oscilloscope. Use a 10KHz test tape and adjust the height adjustment screw for maximum output and in phase.

2. Playback Calibration: Dolby N.R. switch off, select Fe tape on the Tape Select switch. Monitor TP34 (1) with a VV meter. Use a Dolby tone test tape and adjust P1 (L) to read 100mV (±1dB). Repeat for R.H.C. monitor TP35 and adjust P101.

3. Oscillator Frequency Adjustment: Dolby N.R. switch off select CrO2 on the Tape Select switch. Monitor TP46 (L, pin 16 of Dolby I.C.) and TP47 (R) with an oscilloscope. Switch to record. Adjust L201 Osc. coil for minimum output on both test points. Connect the output of a suitable frequency counter to a coupling coil. Place the coupling coil in proximity to the oscillator coil L201, the frequency counter should read 82–85KHz.

4. Bias Trap Check: Dolby N.R. switch on. Select CrO_2 tape, set Record Level control to mid-way. Monitor TP46 (L) with a VV meter. Feed a sine wave signal (1·5KHz @ 20mV R.M.S.) through a 470K Resistor into pin 1 of the Din socket. Switch to record and read the output on the VU meter. When oscillator is switched off (by S/C the erase head terminals) the reading on the VU meter should not drop more than 0·5dB. Repeat for R.H.C. monitoring TP47 and feeding signal through 470K. Resistor to Pin 4 of Din socket.

5. M.P.X. Filter Adjustment: Dolby N.R. switch off. Select Fe tape, monitor TP34, with a VU meter, switch to record. Feed a signal (350mV r.m.s. @ 400Hz) through a 470K resistor to pin 1 of din socket. Switch audio generator to 19KHz ±10Hz and adjust L1 (L) to give a reading of −30dB min. (3mV max.). Repeat for R.H.C. monitoring TP35, signal to Pin 4 and adjust L101.

6. Treble Boost Check: Dolby N.R. switch on. Select Fe tape, monitor TP46 with a VU meter. Switch to record. Feed a signal (20mV R.M.S. @ 15KHz) through a 470K resistor to Pin 1 of Din socket, the meter should read 300mV ±25mV. Repeat for FeCr and CrO_2 for R.H.C. monitoring TP47, signal to Pin 4.

7. Record Level Meter Adjustment: Dolby N.R. switch off. Select Fe tape. Monitor TP34, switch to record, feed a signal (400Hz) through a 470K resistor to Pin 1 of Din socket and increase the output of audio generator until VV meter reads 100mV. Adjust P2 until the left VU meter reads +3dB. Repeat for right monitoring TP35 and adjusting P102.

8. Record Calibration Adjustment: Dolby N.R. switch off. Select Fe tape. Monitor TP34, switch to record. Feed a signal (400Hz) through 470K resistor to Pin 1 of Din socket to give a reading of −10dB (31·5mV) at monitor TP34. Use a Fe tape maxell UD and record signal. When playing back recorded signal should read −10dB (31.5mV) ±1dB. Adjust P3 to meet the above limits. Check for CrO_2. Repeat for right monitoring TP35 and adjusting P103.

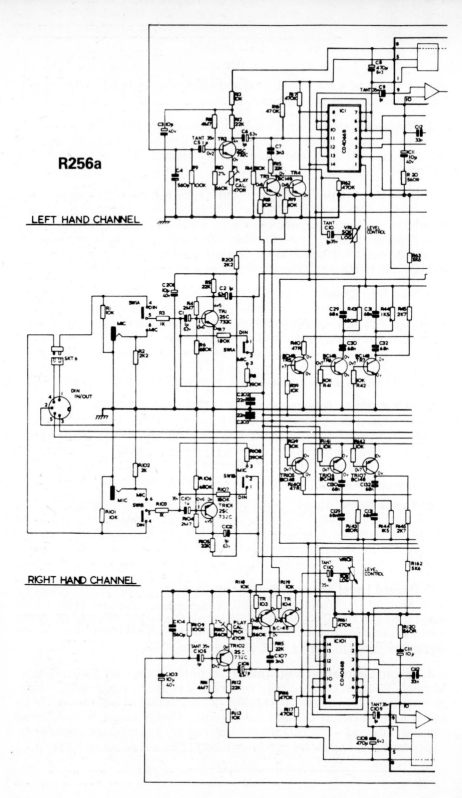

R256a

LEFT HAND CHANNEL

RIGHT HAND CHANNEL

(R256a) CIRCUIT DIAGRAM—MODEL 4000C (*PART*)

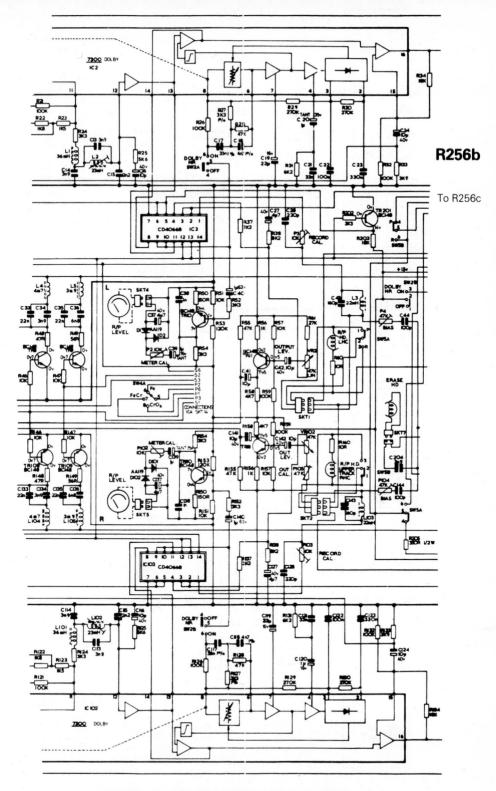

R256b

To R256c

(R256b) CIRCUIT DIAGRAM—MODEL 4000C (*PART*)

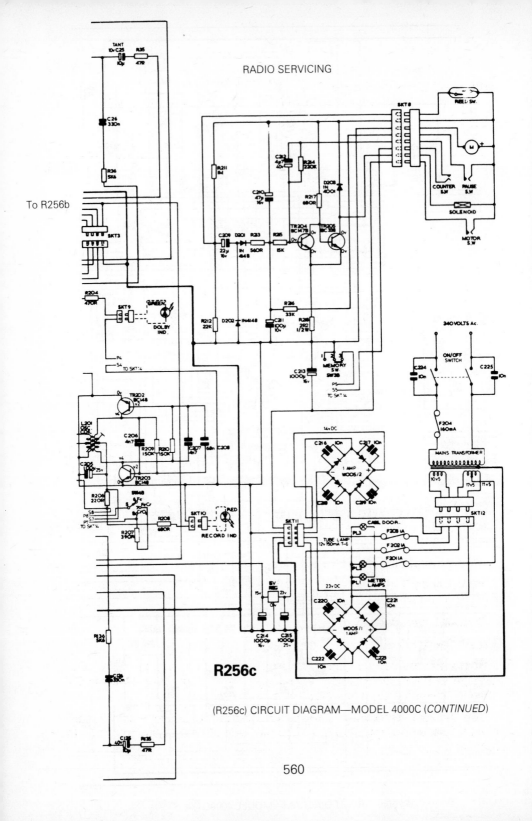

R256c

(R256c) CIRCUIT DIAGRAM—MODEL 4000C (*CONTINUED*)

To R256b

Integrated Circuit Voltages

Pin Numbers	IC1 IC101 CD4066B		IC2 IC102 UA 7300PC		IC3 IC103 CD4066B	
	Play	Record	Play	Record	Play	Record
1	—	—	—	—	—	—
2	—	—	5V8	5V8	—	—
3	5V7	6V65	6V4	6V4	—	—
4	6V7	6V7	7V	7V	—	—
5	—	14V5	14V5	14V5	14V2	—
6	—	14V5	6V4	6V4	14V2	—
7	—	—	6V7	6V7	—	—
8	3V3	3V35	6V7	6V7	7V5	6V7
9	3V5	3V35	4V9	4V9	7V5	6V5
10	3V5	3V35	6V7	6V7	7V5	6V5
11	3V5	3V5	5V8	6V7	5V7	6V5
12	14V2	—	5V75	6V55	—	14V5
13	—	14V5	5V7	6V5	—	14V5
14	14V5	14V5	7V5	6V6	14V5	14V5
15	—	—	6V9	6V1	—	—
16	—	—	7V5	6V9	—	—

All voltages positive with respect to chassis measured with an AVO-Meter.

H.M.V.   Model 4000T

General Description: A four-waveband stereo radio receiver tuner unit for use with high-fidelity power amplifiers. Six pre-set stations may be selected on the F.M. band. A.M. tuning is also by varicaps.

Mains Supplies: 220–250 volts, 50Hz.

Wavebands: L.W. 150–270kHz; M.W. 525–1625kHz; S.W. 5·9–8MHz; F.M. 87·5–108MHz.

Output Level: Adjustable 0–1·5 volts.

Access for Service

Bottom: Carefully turn the unit upside down and support it, avoiding damage to the front extrusion and controls. The bottom may be taken off after removing the eight screws.

Top Cabinet: Remove six screws: two from left hand side, two from right hand side and two from rear top. Carefully slide the top cabinet upwards to remove it.

Front Extrusion: It is first necessary to remove the bottom and the top, then take all the knobs off except the push buttons.

Then remove the three bottom screws holding the front extrusion to the chassis bracket. Gently pull out the extrusion from the bottom and slide it upwards to remove it completely. It is now possible to service virtually any of the components without further dismantling, however the main board may be removed by unscrewing eight screws from the board, unscrewing the lever switches and the nut on band selector switch.

Alignment

Stereo Decoder Alignment: Using a counter and low capacity probe align PR4 for 76KHz at test point 24.

F.M. Alignment: Inject a signal from a broadcast quality stereo encoder into the aerial socket. 98MHz 67% deviation at 1KHz. Tune the receiver for maximum deflection of the tuning meter. Increase R.F. input to 1mV. Align Q2 for the voltage on test point 26 to equal that on test point 25. This voltage should be monitored on a high impedance digital voltmeter using a 100K ohm resistor as a probe. Align Q1 for minimum distortion on a spectrum analyser or distortion meter.

Repeat operations until no improvement can be gained.

Maintaining these input conditions and tuning point align PR2 for minimum crosstalk. Select A.F.C. and align PR1 for the same distortion point. Align PR3 for minimum 19KHz at the output using the spectrum analyser. Align MPF1 and MPF2 for minimum 19KHz and 38KHz. Reduce input level for a poor signal to noise ratio and select mute. Align PR5 to just operate. Check that the pointer is aligned to the small line break to the right of 108. Tune to 108MHz, align PR6 for signal. Tune to 88MHz and align PR7 for signal.

A.M. Alignment: Using a digital voltmeter check that the tuning voltage range at test point 17 is from 0·87V to 8·37V. If incorrect reset PR10 (0·87V) and PR8 (8·37).

M.W.: With the receiver tuned to the max. L.F. end align the M.W. osc. coil, violet, for a radiated 525KHz signal. Tune to max. H.F. and align PC4 to receive a radiated 1625KHz signal. Repeat for optimum. Tune to 600KHz and align L9 for maximum signal. Tune to 1500KHz and align PC1 for maximum signal.

Repeat for optimum.

L.W.: Tune to 200KHz and align PC3 for signal, L11 for maximum.

S.W.: Inject a 6MHz signal into the A.M. aerial socket. Tune to the 6MHz calibration and align the S.W. osc. coil, orange, for signal and the S.W. aerial coil, yellow, for max. Tune to the 8mHz calibration, inject 8mHz and align PC5 for signal, PC2 for max.

Repeat for optimum.

R257

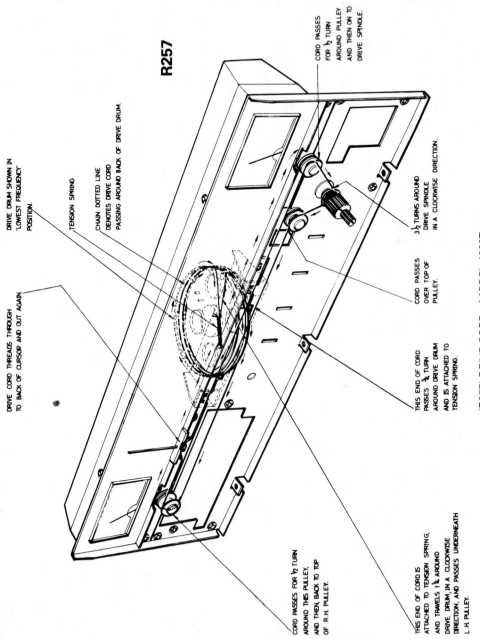

DRIVE DRUM SHOWN IN 'LOWEST FREQUENCY' POSITION.

TENSION SPRING.

CHAIN DOTTED LINE DENOTES DRIVE CORD PASSING AROUND BACK OF DRIVE DRUM.

CORD PASSES FOR ½ TURN AROUND PULLEY AND THEN ON TO DRIVE SPINDLE.

3½ TURNS AROUND DRIVE SPINDLE IN A CLOCKWISE DIRECTION.

CORD PASSES OVER TOP OF PULLEY.

DRIVE CORD THREADS THROUGH TO BACK OF CURSOR AND OUT AGAIN.

THIS END OF CORD PASSES ¾ TURN AROUND DRIVE DRUM AND IS ATTACHED TO TENSION SPRING.

CORD PASSES FOR ½ TURN AROUND THIS PULLEY, AND THEN BACK TO TOP OF R.H. PULLEY.

THIS END OF CORD IS ATTACHED TO TENSION SPRING, AND TRAVELS 1½ AROUND DRIVE DRUM, IN A CLOCKWISE DIRECTION, AND PASSES UNDERNEATH L.H. PULLEY.

(R257) DRIVE CORD—MODEL 4000T

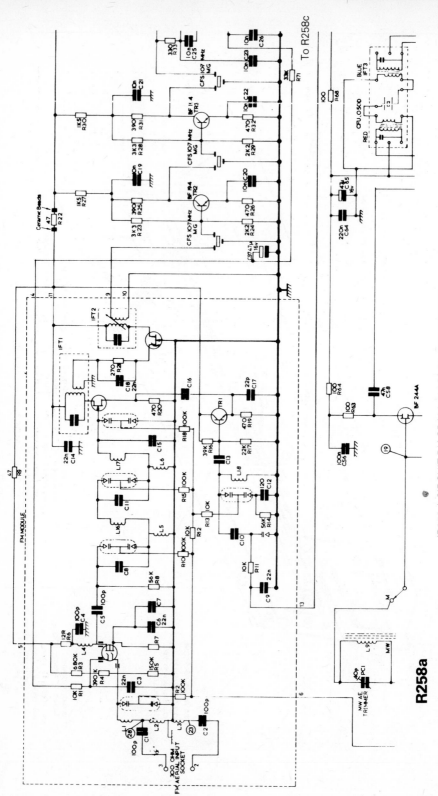

(R258a) CIRCUIT DIAGRAM—MODEL 4000T (PART)

R258a

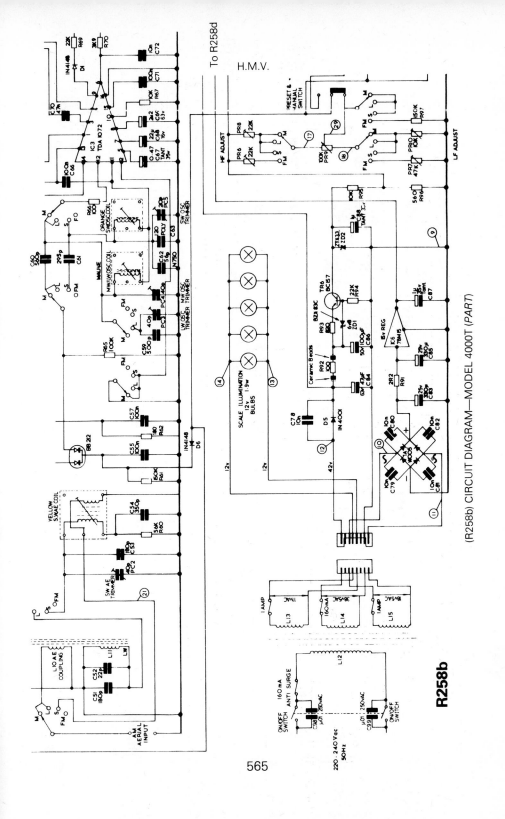

(R258b) CIRCUIT DIAGRAM—MODEL 4000T (*PART*)

R258b

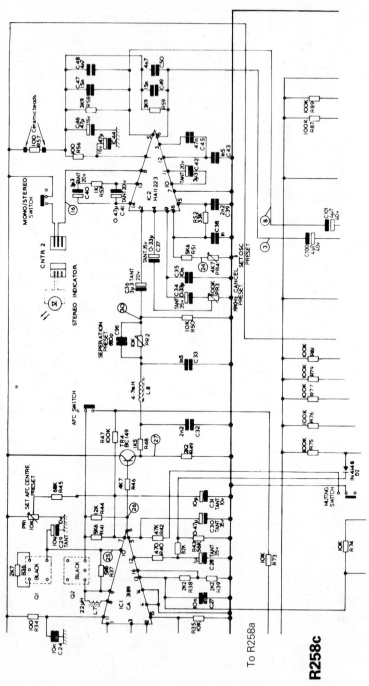

(R258c) CIRCUIT DIAGRAM—MODEL 4000T (PART)

R258c

To R258a

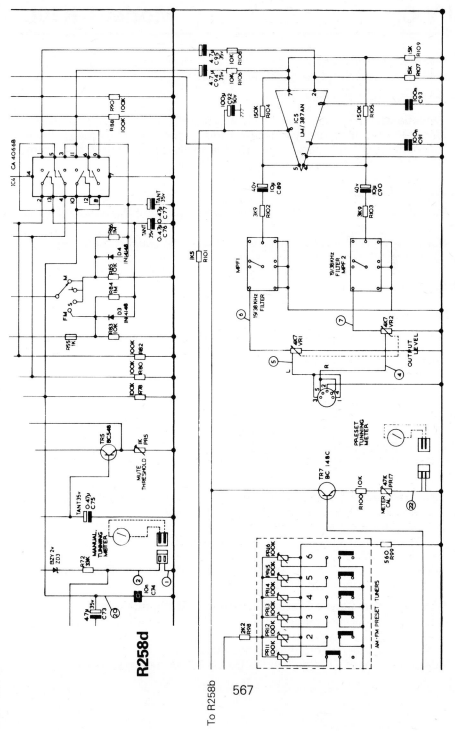

R258d

(R258d) CIRCUIT DIAGRAM—MODEL 4000T (CONTINUED)

J.V.C.   Model T-K10L

General Description: A three-waveband, mains-operated stereo radio tuner for use with high-fidelity power amplifiers.

Mains Supply: 240 volts, 50Hz.

Wavebands: L.W. 150–350kHz; M.W. 525–1605kHz; F.M. 87·5–108MHz.
Note: Printed resistors are used on this circuit board. To repair these components it is necessary to remove the faulty component with a sharp tool and fit a discrete component on the print side.

Access for Service

Remove 2 screws from each side and from the rear of the case and draw the cover to the rear.
Note: When re-installing the cover, rest it on the support moulding at an angle and pull the sides outwards as the top is pressed down.

Alignment

F.M. Section

Note: Keep in the muting pushbutton off during this procedure.

Low Frequency: Connect an R.F. generator, 1kHz modulation and 75kHz deviation, to the antenna terminals on the rear panel through a dummy antenna.
Set the R.F. Generator to 88MHz, a modulation of 1kHz and a deviation of 75kHz, to provide an input of 2μV.
Connect a V.T.V.M. and oscilloscope to Signal Cord.
Set the dial pointer to 88MHz.
Adjust the three coils L104, L102 and L101 in the tuning gang to maximise the output.

High Frequency: Set the R.F. Generator to 108MHz, a modulation of 1kHz and a deviation of 75kHz, to provide an input of 2μV.
Set the dial pointer to 108MHz.
Adjust the F.M. Trimmers TC103, TC102 and TC101 in the tuning gang to maximise the output.
Repeat these high and low frequencies adjustment alternately until maximum sensitivity is obtained.

Discriminator, Centre Meter, Distortion and Signal Gain: Connect an oscilloscope, distortion meter and A.C. V.T.V.M. to Signal Cord.
Connect a D.C. V.T.V.M. to TP-1.
Tune to frequency where there is no broadcasting.
Adjust the core of T131 so that the D.C. V.T.V.M. indicates '0' (zero).

568

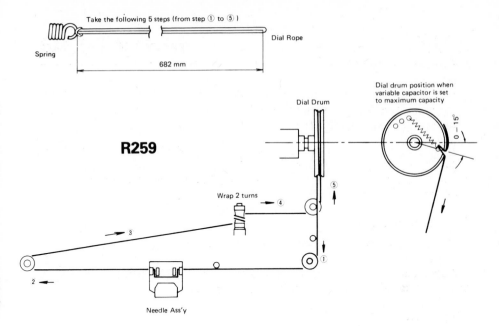

Take the following 5 steps (from step ① to ⑤)

Spring

682 mm

Dial Rope

R259

Dial Drum

Dial drum position when variable capacitor is set to maximum capacity

Wrap 2 turns

Needle Ass'y

(R259) DRIVE CORD—MODEL T-K10L

Set the generator to 98MHz.

Set the dial pointer to 98MHz.

Adjust the core of T132 so that the distortion is minimised at a value less than 0·4%.

Multiplex and Stereo Separation-Multiplex: Set the stereo signal generator as follows: 400Hz modulation frequency, 7·5kHz deviation pilot, 67·kHz main and sub carriers. Connect its output to an R.F. generator.

Connect the R.F. Generator to the antenna terminal through a dummy antenna.

Connect a V.T.V.M., an Oscilloscope and a Distortion Meter to Signal Cord.

Set the R.F. Generator to the 98MHz and an output of 1mV.

Set the dial pointer to 98MHz.

Connect the Frequency Counter to the TP-3.

Switch off the pilot signal of Stereo Modulator.

Adjust R155 so that the Frequency Counter indicates 19kHz (+0, −50Hz).

M.W./L.W. Section

A.M. Tracking and Sensitivity:

Low Frequency: Connect the R.F. Generator to the antenna terminal on the rear panel, set this to 600kHz with 30% modulation at 400Hz.

Connect an A.C. V.T.V.M. and an Oscilloscope to Signal Cord.

Set the dial pointer to the 600kHz.

Adjust Osc. Transformer L202 and L203 adjusting the M.W. coil to maximise the output signal.

High Frequency: Set the R.F. Generator to 1400kHz with 30% modulation at 400Hz.

Set the dial pointer to 1400kHz.

Adjust the trimmers TC201 and TC202 in the tuning gang so that the output signal is maximized.

Repeat these high and low frequencies adjustment procedures alternately until maximum sensitivity is obtained.

M.W. (L.W.) Tracking and Sensitivity:

Low Frequency: Connect the R.F. Generator to the antenna terminal on the rear panel, set this to 600kHz (160kHz) with 30% modulation at 400Hz.

Connect an A.C. V.T.V.M. and an Oscilloscope to Signal Cord.

Set the dial pointer to the 600kHz (160kHz).

Adjust Osc. Transformer L251 (L252) and L253 (L254) adjusting the M.W. (L.W.) coil to maximize the output signal.

High Frequency: Set the R.F. Generator to 1400kHz (350kHz) with 30% modulation at 400Hz.

Set the dial pointer to 400kHz (350kHz).

Adjust the trimmers TC251 (TC252) and TC254 (TC253) in the tuning gang so that the output signal is maximized.

Repeat these high and low frequencies adjustment procedures alternately until maximum sensitivity is obtained.

Integrated Circuit Voltages

	IC201	IC131	IC161
	LA1245	*HA1137W*	*μPC1161C*
1	5·5V	2·2V	12·1V
2	2·2V	2·0V	2·8V
3	2·8V	2·0V	5·6V
4			10·0V
5	8·5V	1·6V	10·0V
6	2·1V	5·6V	2·0V
7	8·9V	5·6V	3·3V
8	8·9V	5·5V	
9	2·8V	5·5V	12·5V
10	8·3V	5·5V	2·3V
11	0·7V	12·1V	2·3V
12		0·02V	2·3V
13	2·1V		2·3V
14	12·4V		2·3V
15	1·6V	4·9V	3·4V
16			(12·3V)
17	2·1V		
18	5·5V		
19	5·5V		
20	2·7V		

R260a

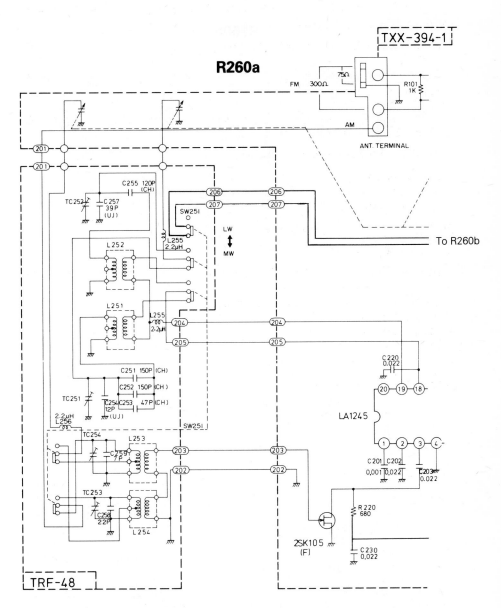

(R260a) CIRCUIT DIAGRAM—MODEL T-K10L (*PART*)

R260b

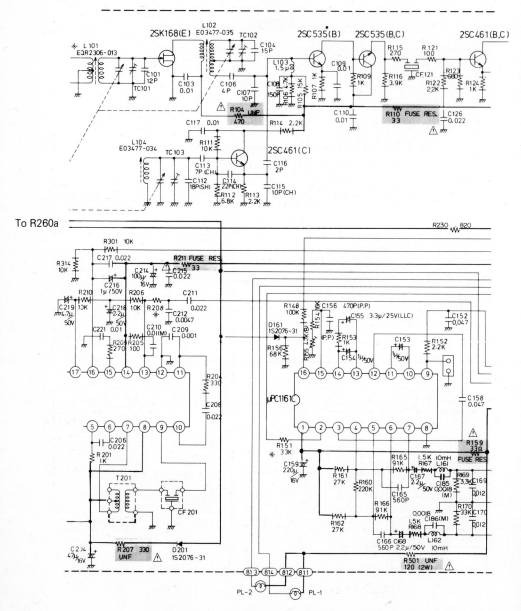

(R260b) CIRCUIT DIAGRAM—MODEL T-K10L (*PART*)

R260c

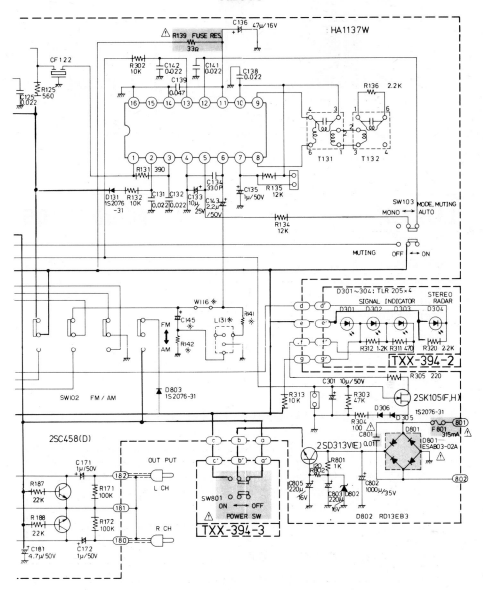

(R260c) CIRCUIT DIAGRAM—MODEL T-K10L (*PART*)

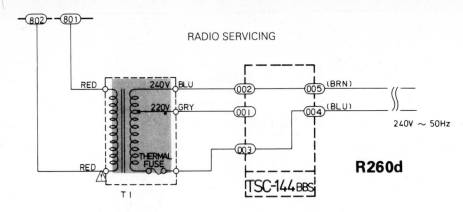

(R260d) CIRCUIT DIAGRAM (MAINS TRANSFORMER) (*CONTINUED*)

LUXOR   Model 9105

General Description: A four-waveband A.M./F.M. radio cassette-recorder operating from mains or battery supplies. A microphone is fitted to the case and sockets are provided for the connection of auxiliary inputs and earphone.

Mains Supply: 220 volts, 50Hz.

Battery: 6 volts.

Wavebands: L.W. 150–300kHz; M.W. 525–1606kHz; S.W. 6–18MHz; F.M. 88–104MHz.

Access for Service (Fig. R344)

Rear Cabinet: Open the cassette compartment by pressing the Eject button and take out the cassette tape from the unit.

Remove the battery cover and take out 4 batteries from the battery compartment.

Remove 5 screws fastening the rear cabinet and take out the rear cabinet.

Pull out the rod antenna connector and power connector.

Reassembly to be made in the reverse order.

Tuner Unit: Take out the rear cabinet and pull out the volume/tone and tuning knobs.

Remove 3 screws (A) and take out the tuner unit.

Amplifier Unit: Take out the rear cabinet.

Remove 2 screws (B) to take out the amplifier unit.

Mechanism Unit: Take out the rear cabinet and amplifier unit. Remove 3 screws (C) and take out the mechanism unit.

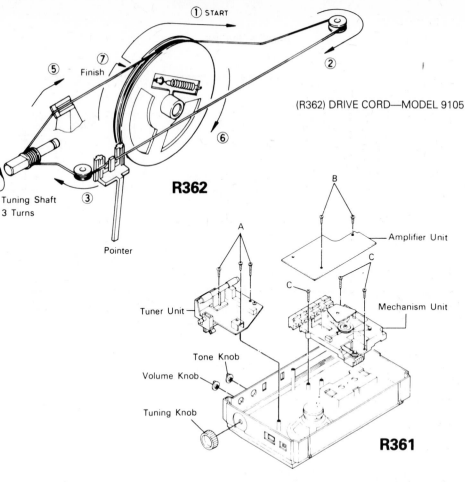

① START

⑦ Finish

⑤

(R362) DRIVE CORD—MODEL 9105

⑥

③

②

R362

Tuning Shaft
3 Turns

Pointer

B

A

Amplifier Unit

C

C

Tuner Unit

Mechanism Unit

Tone Knob

Volume Knob

Tuning Knob

R361

(R361) ACCESS FOR SERVICE—MODEL 9105

Cassette Compartment: Open the cassette compartment by pressing the Eject button.

Hold the cassette compartment and press it from both sides so that the protruded sections of the cassette compartment are to be taken out from the front cabinet.

When the cassette compartment is reset, fit the left-side protruded section into the front cabinet and then press the right-side compartment.

Dial Cord Stringing (Fig. 345): Connect the dial cord to the extension spring and hook it at the drum.

Suspend the cord in order as illustrated.

Connect the pointer.

Turn the drum counter-clockwise.

Slide the pointer to the 'O' position on the dial scale and apply a bonding material at the pointer and dial cord in order to stick the pointer.

575

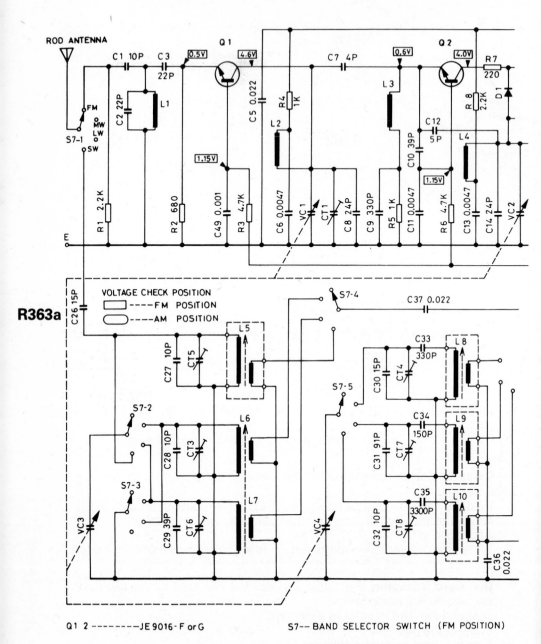

Q1 2 --------- JE 9016-F or G   S7-- BAND SELECTOR SWITCH (FM POSITION)

(R363a) CIRCUIT DIAGRAM (RADIO SECTION)—MODEL 9105 (*PART*)

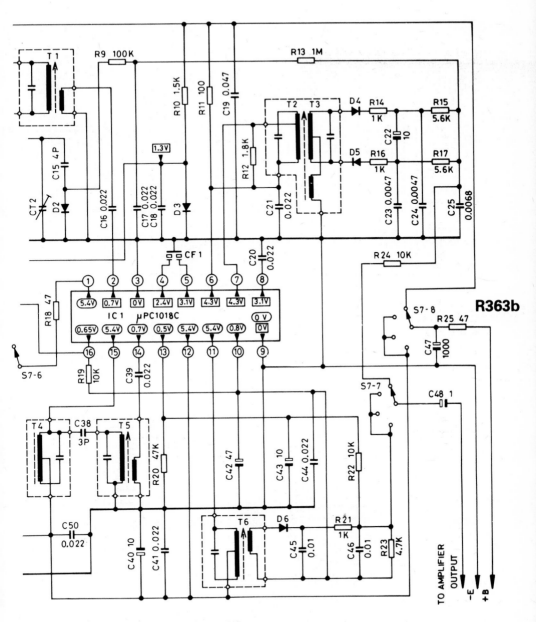

(R363b) CIRCUIT DIAGRAM (RADIO SECTION)—MODEL 9105 (*CONTINUED*)

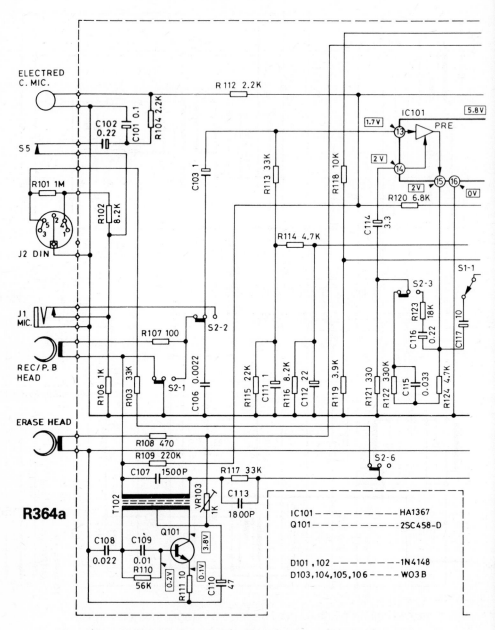

(R364a) CIRCUIT DIAGRAM (CASSETTE SECTION)—MODEL 9105 (*PART*)

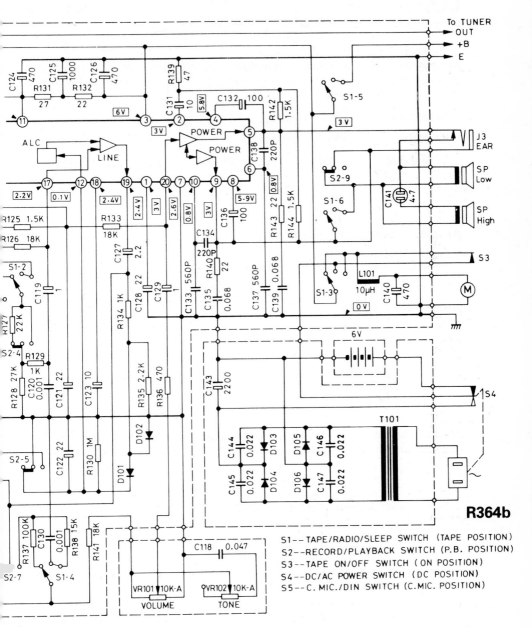

R364b

S1 -- TAPE/RADIO/SLEEP SWITCH (TAPE POSITION)
S2 -- RECORD/PLAYBACK SWITCH (P.B. POSITION)
S3 -- TAPE ON/OFF SWITCH (ON POSITION)
S4 -- DC/AC POWER SWITCH (DC POSITION)
S5 -- C. MIC./DIN SWITCH (C.MIC. POSITION)

(R364b) CIRCUIT DIAGRAM (CASSETTE SECTION)—MODEL 9105 (*CONTINUED*)

LUXOR   # Model 9111

General Description: A four-waveband A.M./F.M. stereo radio cassette-recorder operating from mains or battery supplies. Stereo speakers are fitted to a portable case and sockets are provided for the connection of auxiliary inputs and headphones.

Mains Supply: 220 volts, 50Hz.

Batteries: 9 volts (6×1·5 volts).

Wavebands: L.W. 150–300kHz; M.W. 525–1605kHz; S.W. 6–18MHz; F.M. 87–105MHz.

Access for Service (Fig. R348)

Rear Cabinet: Open the cassette compartment by pressing the Eject button and take out the cassette from the unit.
Remove the battery cover and take out 6 batteries from the battery compartment.
Remove 7 screws (A) fastening the rear cabinet and take out the rear cabinet.
Pull out the wire connector (212) and fider pin (154).
Re-assembly to be made in the reverse order.

Amplifier Unit: Take out the rear cabinet.
Remove 5 screws (B) and unsolder the lead wires to take out the amplifier unit.
When re-assembly is made, be sure to insert the protruded section of the record lever (C) into the Rec/Play slide switch (S101) of the amplifier unit.

Mechanism Unit: Take out the rear cabinet and counter belt.
Remove 3 screws (C) and take out the mechanism unit.

Tuner Unit: Take out the rear cabinet, amplifier unit, mechanism unit and pull out the volume/tone/balance knobs.
Remove 2 screws (D) and take out the tuner unit.
Remove 3 screws (E) and take out the tuner P.C. board.

Cassette Compartment: Open the cassette compartment by pressing the Eject button.
Hold the cassette compartment and press it from the both sides so that the protruded sections of the cassette compartment are to be taken out from the front cabinet.
When the cassette compartment is reset, fit the left-side protruded section into the front cabinet and then press the right-side compartment.

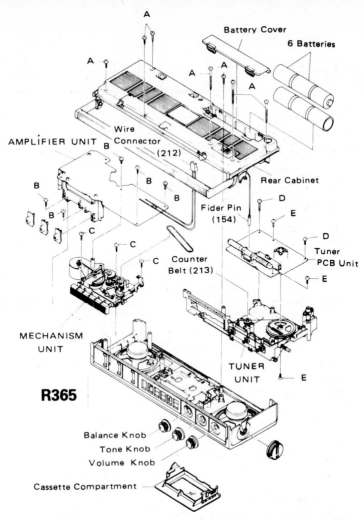

R365

A
A
A
A
A
A
Battery Cover
6 Batteries
Wire Connector (212)
AMPLIFIER UNIT
B
B
B
B
B
B
Rear Cabinet
C
C
C
Fider Pin (154)
D
D
E
E
E
Counter Belt (213)
Tuner PCB Unit
MECHANISM UNIT
TUNER UNIT
Balance Knob
Tone Knob
Volume Knob
Cassette Compartment

(R365) ACCESS FOR SERVICE—MODEL 9111

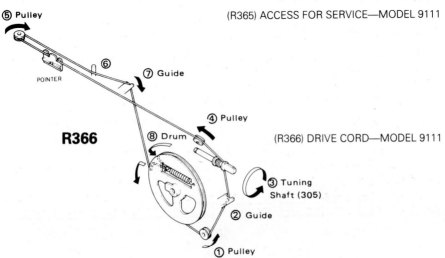

⑤ Pulley
⑥
POINTER
⑦ Guide
④ Pulley
R366
⑧ Drum
③ Tuning Shaft (305)
② Guide
① Pulley

(R366) DRIVE CORD—MODEL 9111

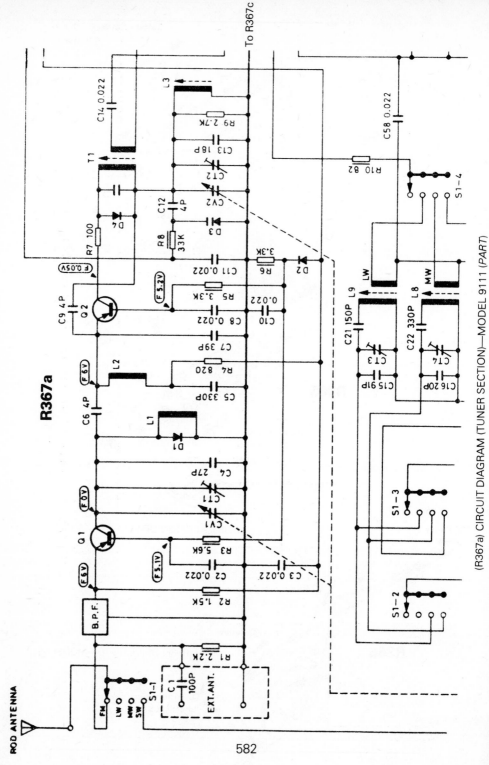

(R367a) CIRCUIT DIAGRAM (TUNER SECTION)—MODEL 9111 (PART)

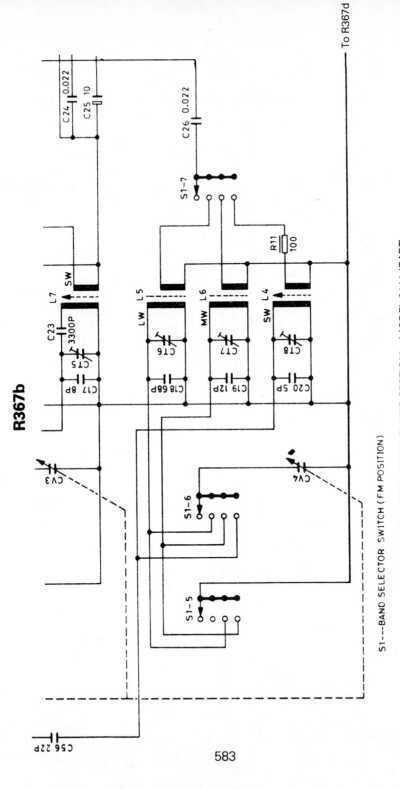

R367b

S1---BAND SELECTOR SWITCH (FM POSITION)

(R367b) CIRCUIT DIAGRAM (TUNER SECTION)—MODEL 9111 *(PART)*

583

RADIO SERVICING

(R367c) CIRCUIT DIAGRAM (TUNER SECTION)—MODEL 9111 (PART)

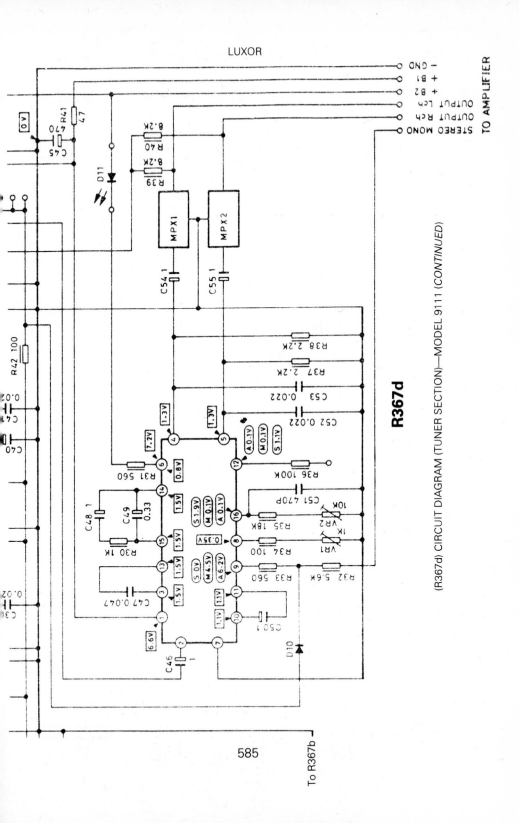

LUXOR

R367d

(R367d) CIRCUIT DIAGRAM (TUNER SECTION)—MODEL 9111 (CONTINUED)

585

To R367b

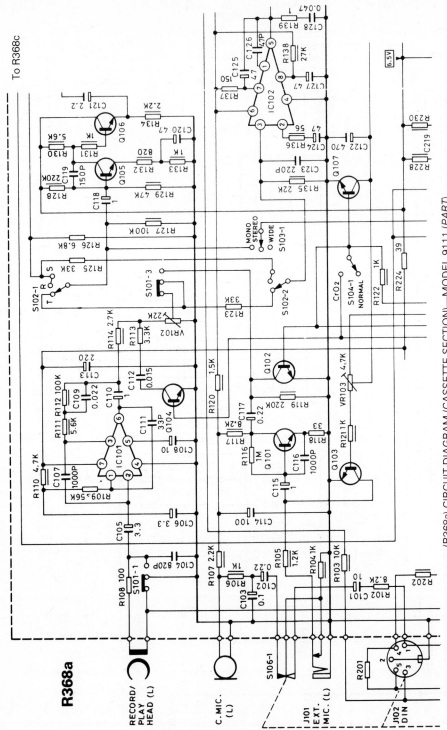

(R368a) CIRCUIT DIAGRAM (CASSETTE SECTION)—MODEL 9111 (*PART*)

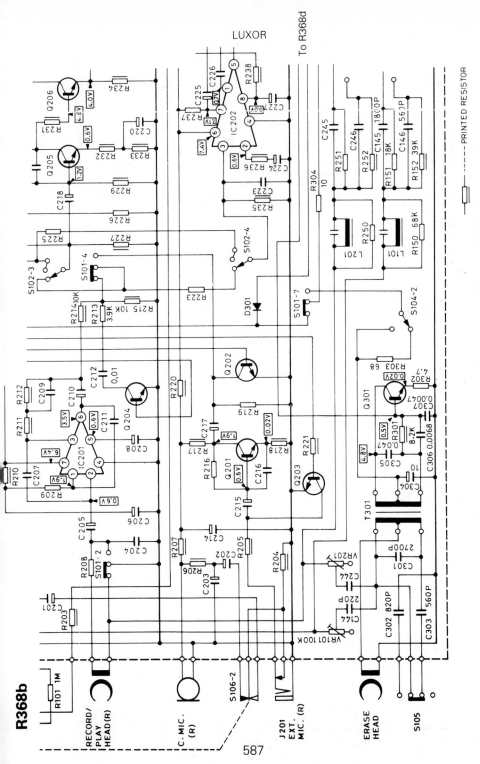

(R368b) CIRCUIT DIAGRAM (CASSETTE SECTION)—MODEL 9111 (PART)

587

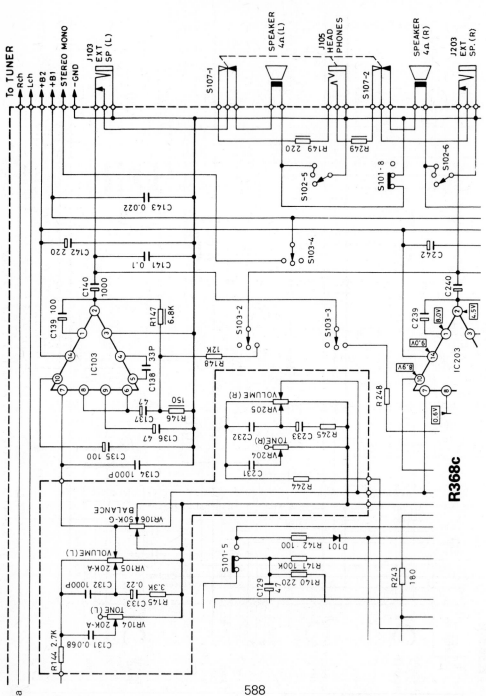

(R368c) CIRCUIT DIAGRAM (CASSETTE SECTION)—MODEL 9111 (PART)

R368c

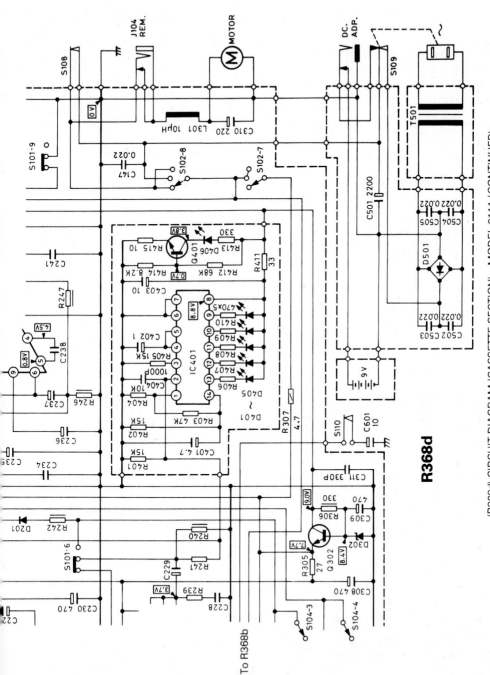

(R368d) CIRCUIT DIAGRAM (CASSETTE SECTION)—MODEL 9111 (CONTINUED)

R368d

Semiconductors:

IC1	μPC1018C	D1, 4	HV80
IC2	μPC1197C	D2	KB-265
Q1, 2	2SA1005-L or K	D3	SD115
Q3	2SC945-P	D5, 6, 7, 9, 10	1N60P
IC101, 201	TA7120P	D8	1S2076
IC102, 202	TBA820M	D11, 12	LN217RP-E
IC103, 203	LA4138	D101, 201, 301	1N4148
IC401	TA7654P	D302	UZ-8·2B
Q101, 201	2SC1335-D	D501	2WO2
Q102~107, 401 }	2SC1684-S or 2SC458-D	D401~405	LN05202P
Q202~206		D406	LN217RP-E
Q301, 302	2SD467-C		

Tape Switches:

S101. Record Play switch (Play position).
S102. Tape Radio/Sleep switch (Tape position).
S103. Mono/Stereo Wide switch (Stereo position).
S104. Normal/CrO₂ switch (Normal position).
S105. Beat Cancel switch.
S106. C.Mic/Din switch (C.Mic. position).
S107. Head-phone/Speaker switch (SP. position).
S108. Tape on/off switch (Off position).
S109. A.C./D.C. Power switch (D.C. position).
S110. Muting switch.

PHILIPS Model D1610

General Description: A pocket portable A.M./F.M. radio receiver operating from battery supplies with low level output for stereo headphones. Integrated circuits are used in I.F., decoder and A.F. stages.

Batteries: 4·5 volts (3×1·5 volts).

Wavebands: M.W. 520–1605kHz; F.M. 87·5–108MHz.

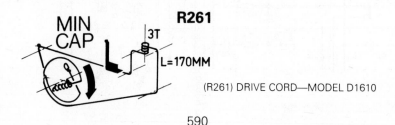

(R261) DRIVE CORD—MODEL D1610

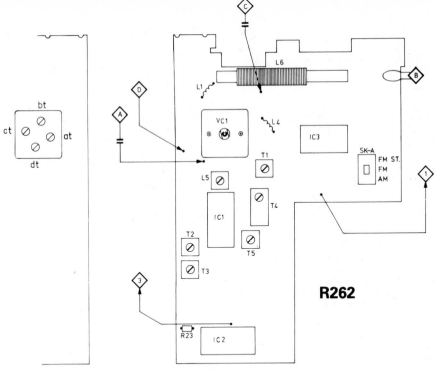

(R262) ALIGNMENT ADJUSTMENTS—MODEL D1610

	⊗→	◇	≠	⊣⊢	⌀	◇	▭	⋀
	468KHz * via 10nF	Ⓐ	min.		T5	↕ ◇₁	max.	symm.
					T4			
MW/PO (520-1605KHz)	512KHz		max.		L5	↕ ◇₁	max.	
	1635KHz		min.		VC1ct			
	550KHz	Ⓑ	⌀		L6			
	1500KHz				VC1dt			
FM 87.5-108MHz	10.7MHz via 10nF	Ⓒ	min.	T3	T1	↕ ◇₁	max.	symm.
					T2			
					T3	↕ ◇₁	▭	max.lin. + symm.
	86.5MHz +		max.		L4	↕ ◇₁	max.	
	109MHz +	Ⓓ	min.		VC1bt			
	88MHz		⌀		L1			
	106MHz				VC1at			
FM Stereo					R23	◇₃	f◇₃=19±0.2KHz	

↕ Repeat * /01/10 = 455 KHz
+ /02 = 87.4,107.5 MHz

R263

(R263) ALIGNMENT PROCEDURE—MODEL D1610

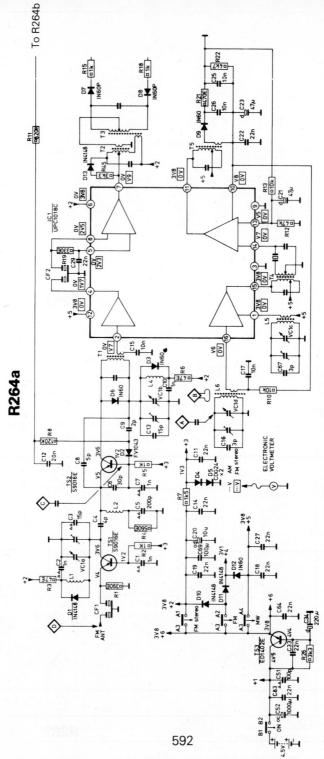

(R264a) CIRCUIT DIAGRAM—MODEL D1610 (*PART*)

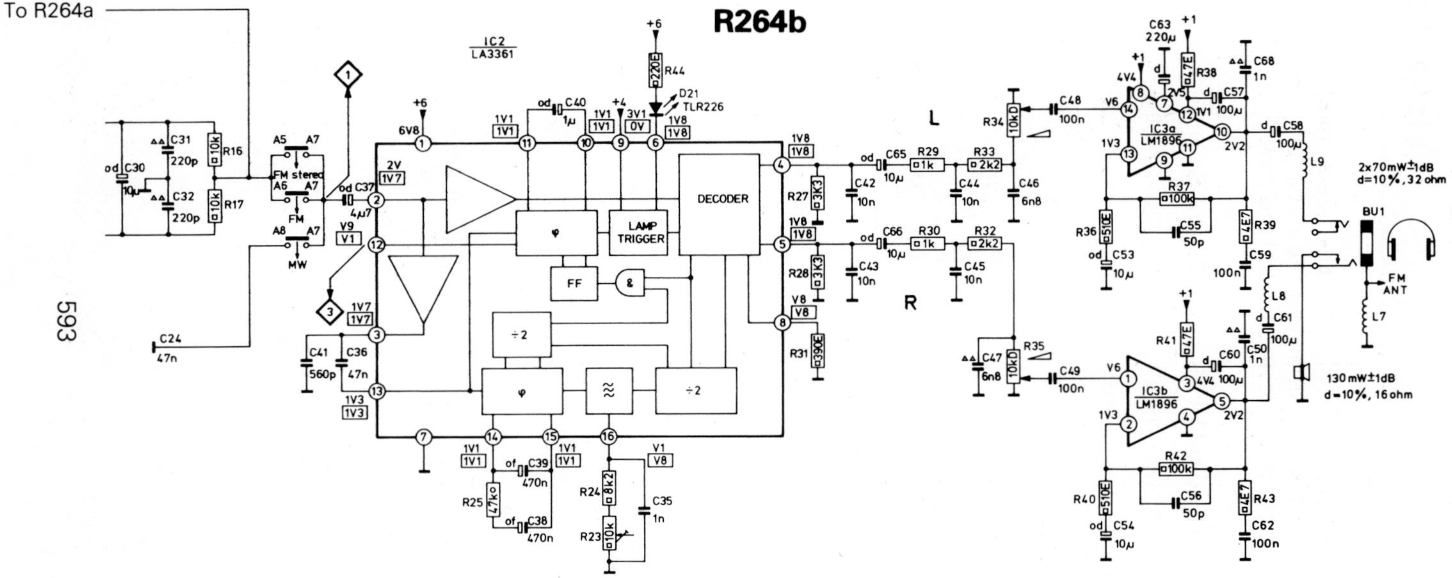

R264b

To R264a

(R264b) CIRCUIT DIAGRAM—MODEL D1610 (CONTINUED)

PHILIPS   Model D8000

General Description: A portable A.M./F.M. stereo radio cassette-recorder with electronic digital-clock operating from battery supplies. A stereo microphone is fitted to the case and sockets are provided for line inputs and headphones.

Batteries: Radio 6 volts (4×1·5 volts); Clock 1·5 volts.

Wavebands: M.W. 520–1650kHz; F.M. 87·3–109MHz.

Loudspeakers: 8 ohms impedance.

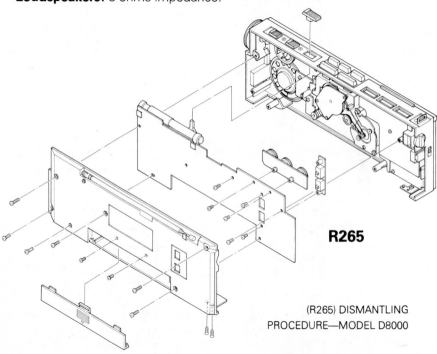

R265

(R265) DISMANTLING
PROCEDURE—MODEL D8000

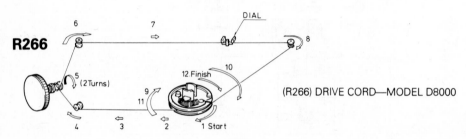

R266

(R266) DRIVE CORD—MODEL D8000

594

SK-202	(signal generator) →	◇	▭	C	(meter)	(cassette)
AM	455 kHz 468 kHz Δf 10 kHz :	Ⓐ via 33 nF	Min Cap	S109 S110	② 1	② Max
MW 525-1605 kHz	512 kHz	Ⓑ	Max Cap	S106		② Max
	1635		Min Cap	C601-4		
	600 kHz		▭	S105		
	1400 kHz		▭	C601-3		
FM	10.7 MHz Δf 300 kHz (50 Hz)	Ⓒ Via 22 nF	Min Cap	S107 S108	⑦ 1 ① 2	
FM 87.5 - 108 MHz	87.35 MHz	Ⓓ	Max Cap	S103		① Max
	90 MHz			S101		
	109 MHz		Min Cap	C601-2		
	106 MHz			C601-1		

↕ Repeat

1 Adjust curve for symmetry and max. height.

2 Adjust the "S"-curve for symmetry and max. linearity.

R267

STEREO DECODER

FM 87.5 - 108 MHz				R604	Frequency Counter ③ 19 ± 0.2 kHz	
FM 87.5 - 108 MHz	Mpx 100 MHz MOD-R 1 kHz	Ⓓ	▭	R605		④ Min

BIAS

SK201	C	(cassette)
Rec/play	R606	⑤ ≈ 6 mV
	R607	⑥ ≈ 6 mV
Rec/play	S201	⑤ max.
	S202	⑥ max.

(R267) ALIGNMENT PROCEDURE—MODEL D8000

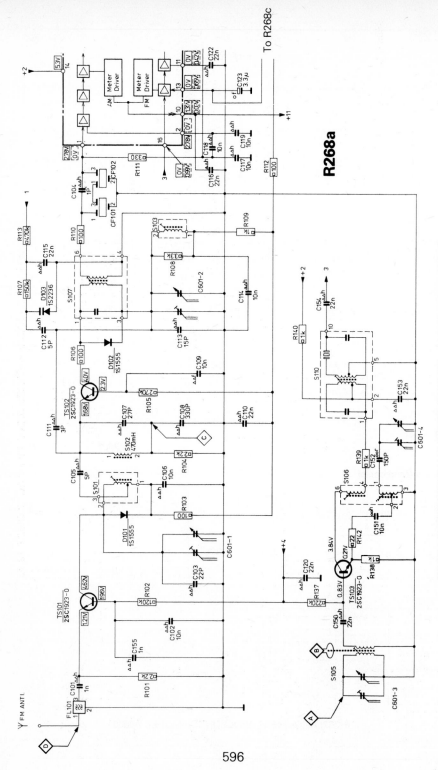

(R268a) CIRCUIT DIAGRAM (RADIO SECTION)—MODEL D8000 (*PART*)

R268a

To R268c

596

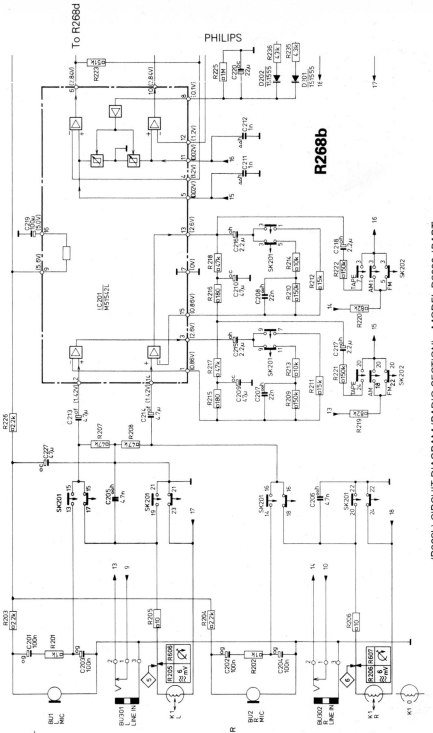

(R268b) CIRCUIT DIAGRAM (RADIO SECTION)—MODEL D8000 *(PART)*

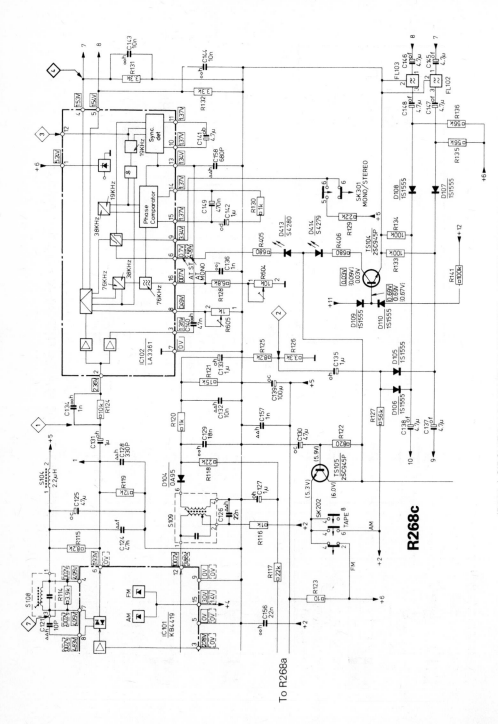

(R268c) CIRCUIT DIAGRAM (TAPE SECTION)—MODEL D8000 *(PART)*

R268c

To R268a

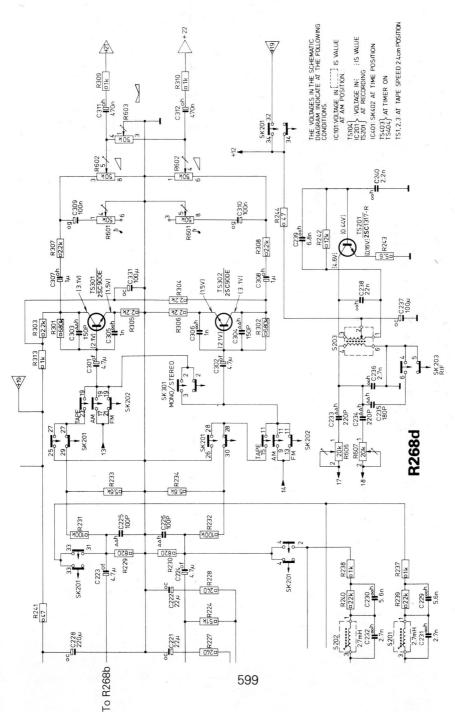

(R268d) CIRCUIT DIAGRAM (TAPE SECTION)—MODEL D8000 (CONTINUED)

R268d

THE VOLTAGES IN THE SCHEMATIC
DIAGRAM INDICATE AT THE FOLLOWING
CONDITIONS:

IC101 VOLTAGE IN [ ] IS VALUE
AT AM POSITION

TS104 } VOLTAGE IN: } IS VALUE
TS201 } AT RECORDING

IC401: SK402 AT TIME POSITION

TS403 } AT TIMER ON
TS404 }

TS1,2,3 AT TAPE SPEED 2.4cm POSITION

To R268b

(R269a) CIRCUIT DIAGRAM (CLOCK SECTION)—MODEL D8000 (PART)

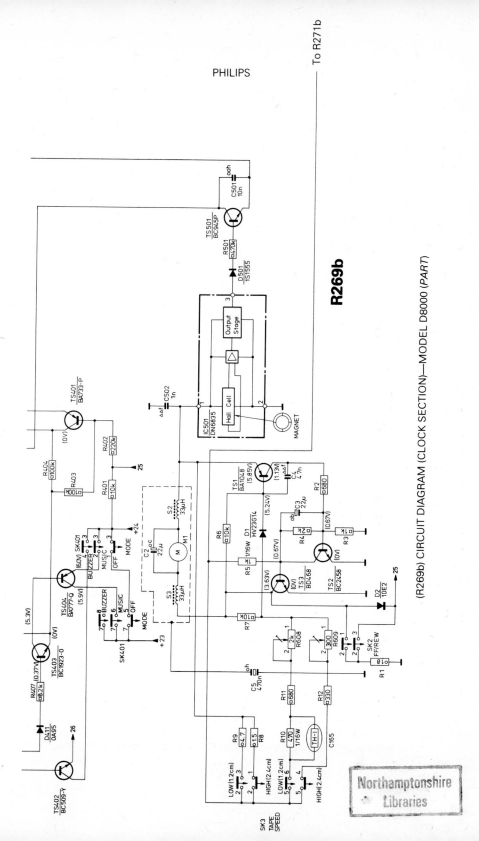

PHILIPS

R269b

(R269b) CIRCUIT DIAGRAM (CLOCK SECTION)—MODEL D8000 (*PART*)

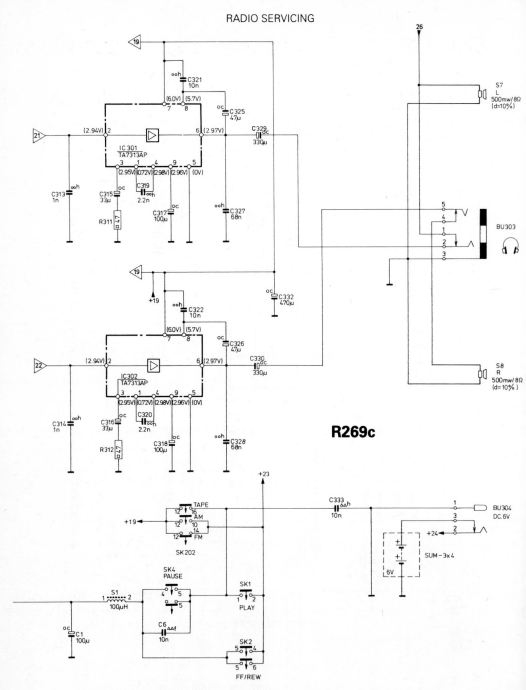

R269c

(R269c) CIRCUIT DIAGRAM (A.F. OUTPUT)—MODEL D8000 (*CONTINUED*)

PHILIPS

Models D8310, D8312

General Description: A two- or three-waveband A.M./F.M. stereo radio cassette-recorder operating from mains or battery supplies. Integrated circuits are used in the decoder and A.F. stages and some circuit variations occurring in the two models are detailed in the circuit diagram. Microphones are fitted to the case and sockets are provided for auxiliary inputs, outputs, and headphones.

Mains Supplies: 110–127, 220–240 volts, 50/60Hz.

Batteries: 9 volts (6×1·5 volts).

Wavebands: L.W. 150–255kHz (D8312 only); M.W. 520–1605kHz; F.M. 87·5–108MHz.

Loudspeakers: 4 ohms impedance.

Access for Service

Remove the 6 screws in the backplate (do not loosen the black aerial screw) and loosen the aerial wire.

After removing the knobs and 2 screws the printed panel can be raised to the lower side and can be tilted (pay attention to the wiring and the frame near the socket unit).

After removing 2 screws the tape transport can be tilted.

For removing the flap of the cassette compartment, first open the flap and then slightly press lugs in cassette flap inwards.

It is only possible to remove handgrip after the black part of the cabinet has been removed. Turn handgrip to the back. Then slightly bend both sides of handgrip outwards and the handgrip will be released.

SK-3	⊗→	◇	⊏⊐	⊣⊢	∅	[A B]	[cassette]
MW	468 kHz Δf 1 kHz	Ⓐ			S205 S207 S204		◇ Max
MW 520 ÷ 1605 kHz	1635 kHz	Ⓑ	Min Cap		C801h		◇ Max
	512 kHz		Max Cap		S203		
MW	560 kHz 1500 kHz	Ⓑ	⊏⊐		S802 C801f		◇ Max
LW* 150 ÷ 255 kHz	149 kHz	Ⓑ	Max Cap		S203 S801		◇ Max
FM	10.7 MHz Δf ± 180 kHz	Ⓒ	Max Cap	S208	S105 S206	[1] ◇	
					S208	[2]	
FM	87,35 MHz	Ⓓ	Max Cap		C801d C801b	[1]	◇ Max
87,5 ÷ 108 MHz	108 MHz		Min Cap		S104 S102	◇	

↕ Repeat

*Only for D8312

[1] Adjust for symmetry and max. height.

[2] Adjust the "S"-curve for symmetry and max. linearity.

Stereo decoder

SK...	∅		Frequency counter	
FM 87.5 - 108 MHz	R707		②	19 ± 0.2 kHz

R270

(R270) ALIGNMENT PROCEDURE—MODELS D8310, D8312

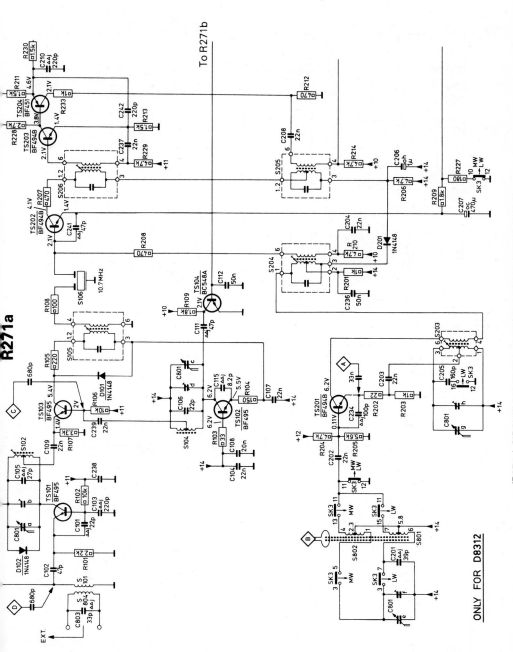

R271a

To R271b

(R271a) CIRCUIT DIAGRAM—MODELS D8310, D8312 (*PART*)

ONLY FOR **D8312**

RADIO SERVICING

R271b

ONLY FOR D8310

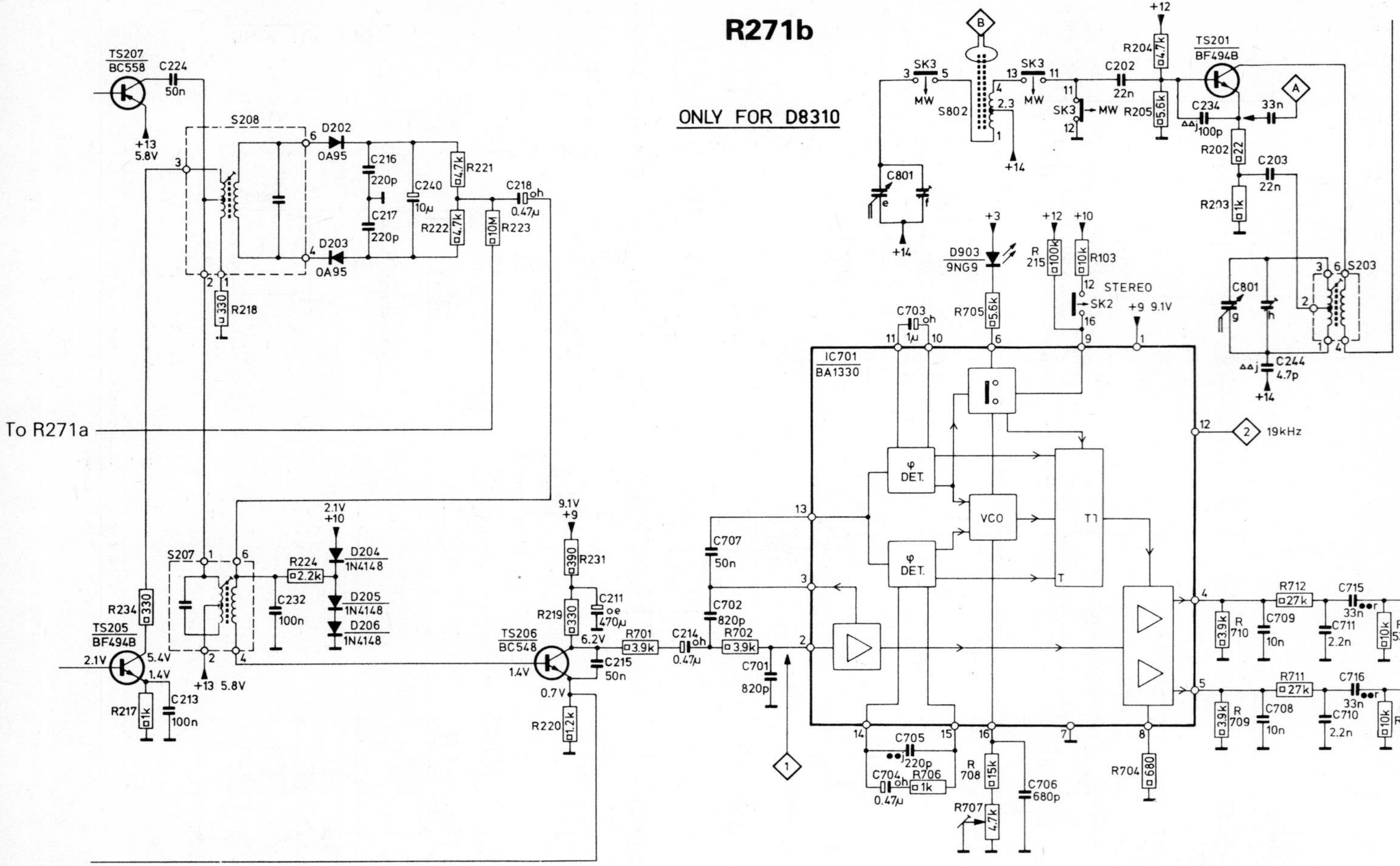

(R271b) CIRCUIT DIAGRAM—MODELS D8310, D8312 (CONTINUED)

To R271a

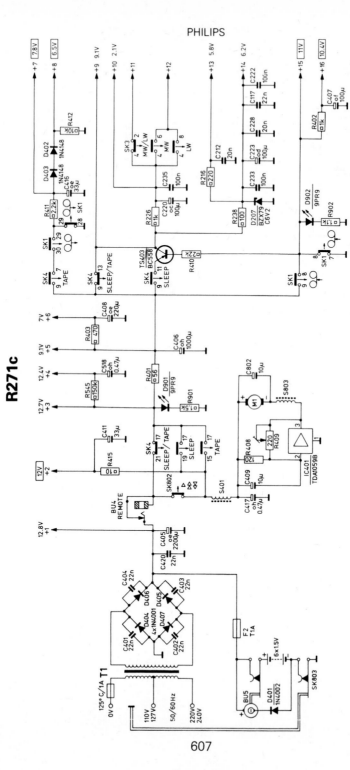

(R271c) CIRCUIT DIAGRAM (POWER SUPPLIES)—MODELS D8310, D8312

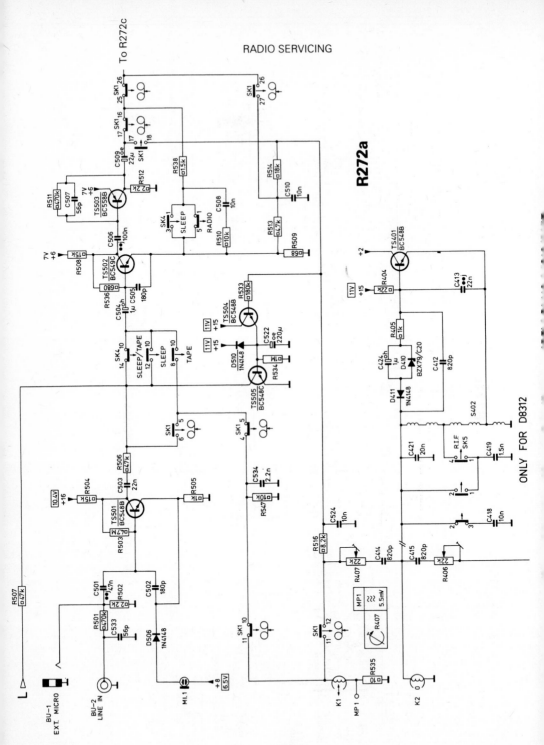

R272a

ONLY FOR D8312

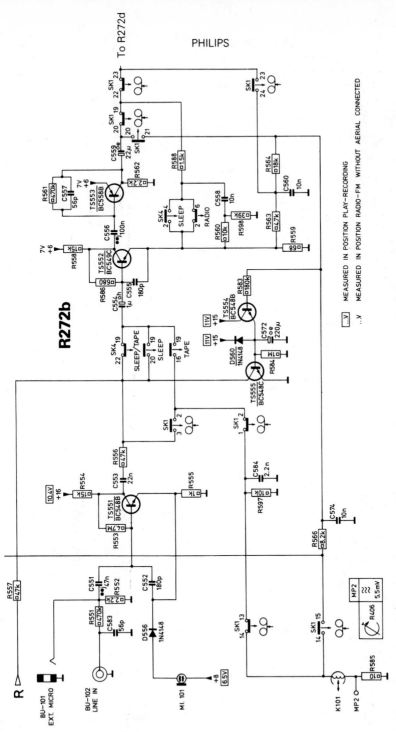

PHILIPS

(R272b) CIRCUIT DIAGRAM (A.F. SECTION)—MODELS D8310, D8312 (PART)

609

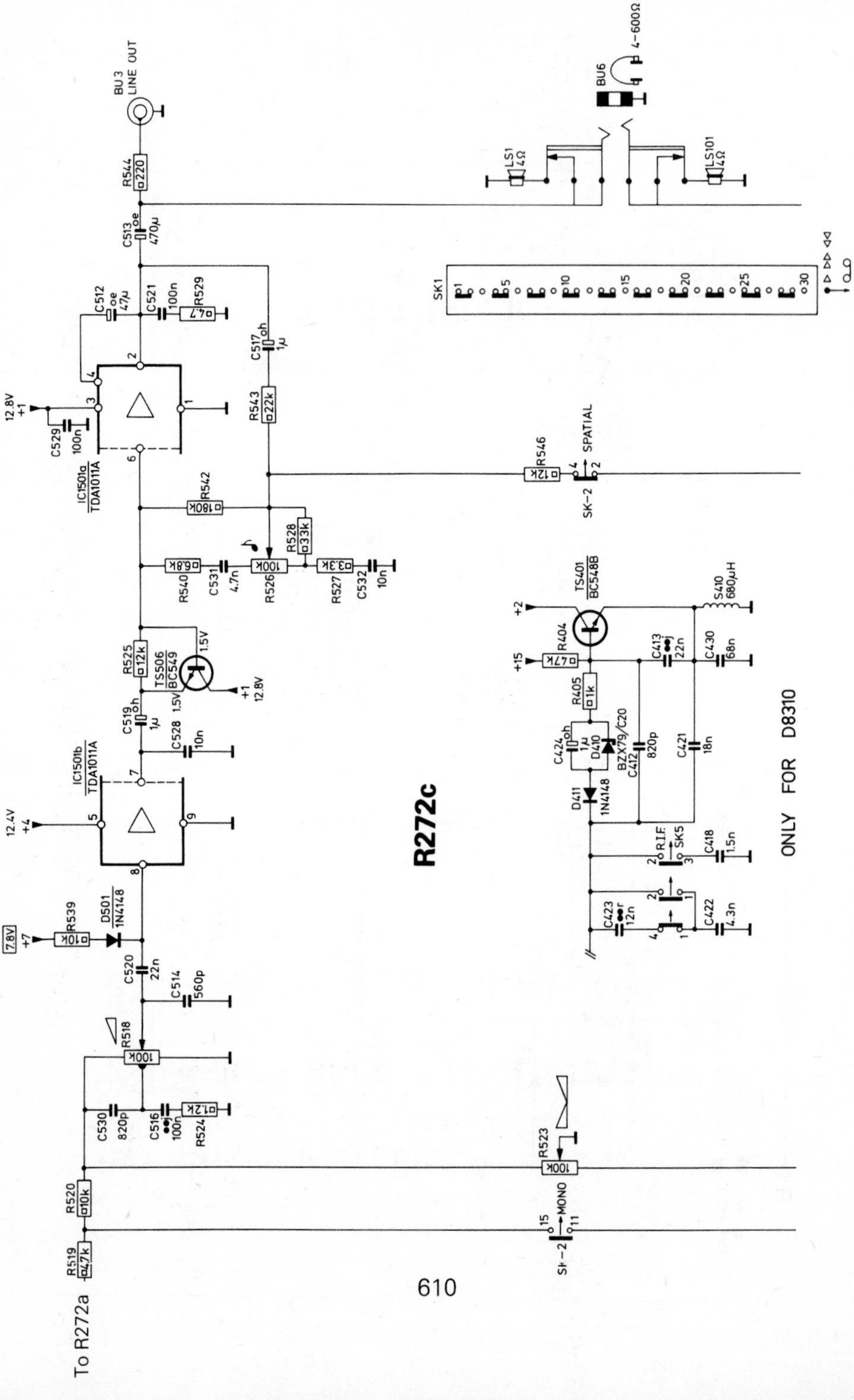

(R272c) CIRCUIT DIAGRAM (A.F. SECTION)—MODELS D8310, D8312 (PART)

R272c

ONLY FOR D8310

To R272a

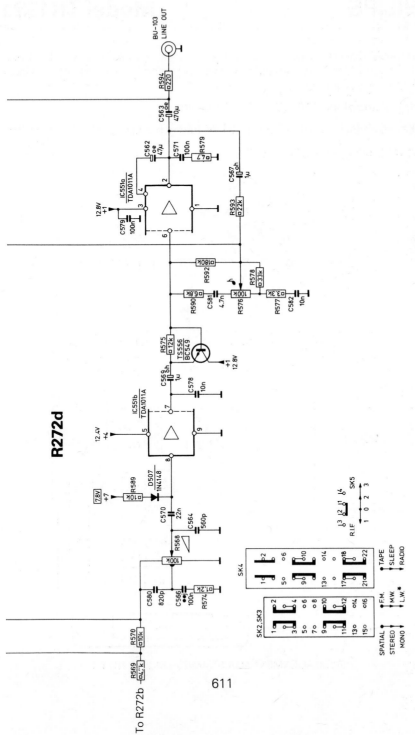

R272d

(R272d) CIRCUIT DIAGRAM (A.F. SECTION)—MODELS D8310, D8312 (CONTINUED)

PHILIPS   Model TR1321

General Description: A mains-operated clock-radio receiver for Long, Medium and V.H.F. wavebands. Alarm and snooze facilities are available from the clock module.

Mains Supplies: 110–127, 220–240 volts.

Wavebands: L.W. 150–255kHz; M.W. 520–1605kHz; F.M. 87·5–108MHz.

Loudspeaker: 8 ohms impedance.

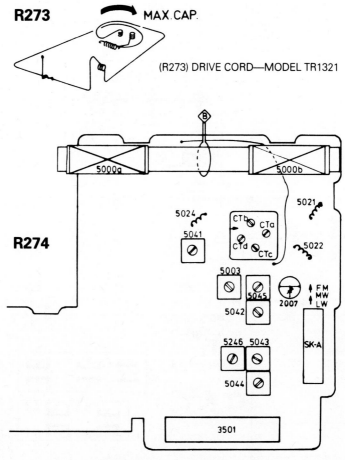

(R273) DRIVE CORD—MODEL TR1321

(R274) ALIGNMENT ADJUSTMENTS—MODEL TR1321

SK-A	⊗→	◇	≠	⌀	▭
MW/PO 520-1605KHz	468KHz via 40nF	⟨A⟩	Min Cap	5045 5046	⟨1⟩ Max + Sym
	512KHz		Max cap	5003	⟨1⟩ Max
	1635KHz	⟨B⟩	Min cap	CTd	
	600KHz		▭	5000a	
	1400KHz			CTc	
LW/GO 150-255KHz	200KHz	⟨B⟩	▭	5000b	⟨1⟩ Max
FM 87.5-108MHz	10.7MHz via 22nF	⟨C⟩	Min cap	5041 5042 5043 5044	⟨1⟩ Max + Sym
	86.5MHz		Max cap	5024	⟨1⟩ Max
	109MHz	⟨D⟩	Min cap	CTb	
	86.5MHz		▭	5022	
	109MHz			CTa	

↕ Repeat

R275

(R275) ALIGNMENT PROCEDURE—MODEL TR1321

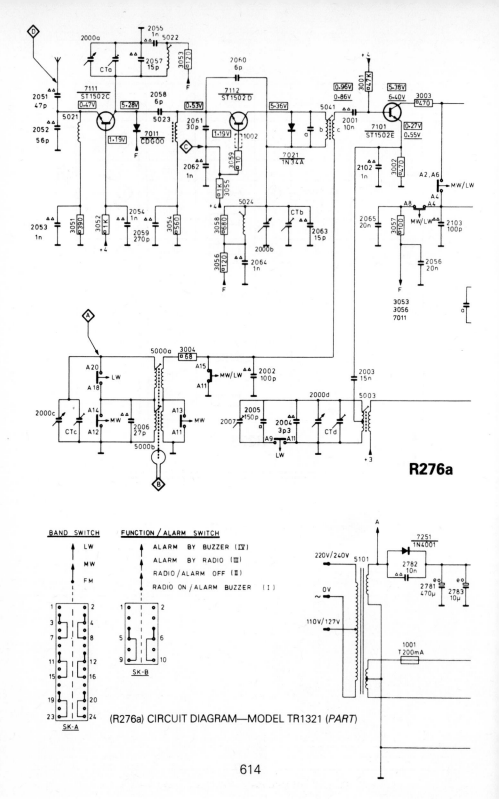

R276a

BAND SWITCH	FUNCTION / ALARM SWITCH
LW	ALARM BY BUZZER (IV)
MW	ALARM BY RADIO (III)
FM	RADIO/ALARM OFF (II)
	RADIO ON / ALARM BUZZER (I)

SK-A

SK-B

(R276a) CIRCUIT DIAGRAM—MODEL TR1321 (*PART*)

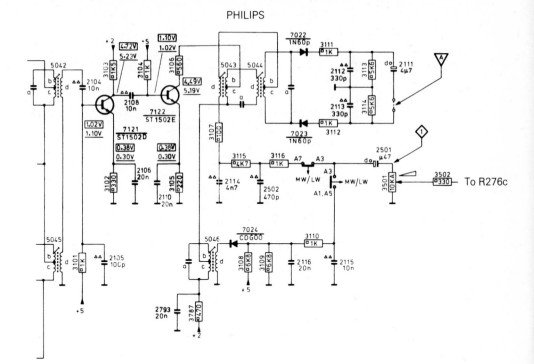

PHILIPS

R276b

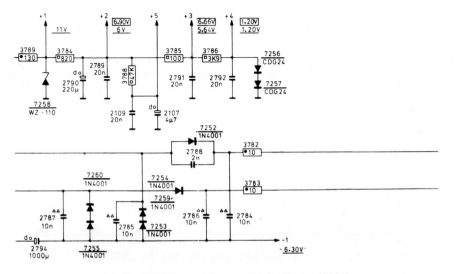

(R276b) CIRCUIT DIAGRAM—MODEL TR1321 (*PART*)

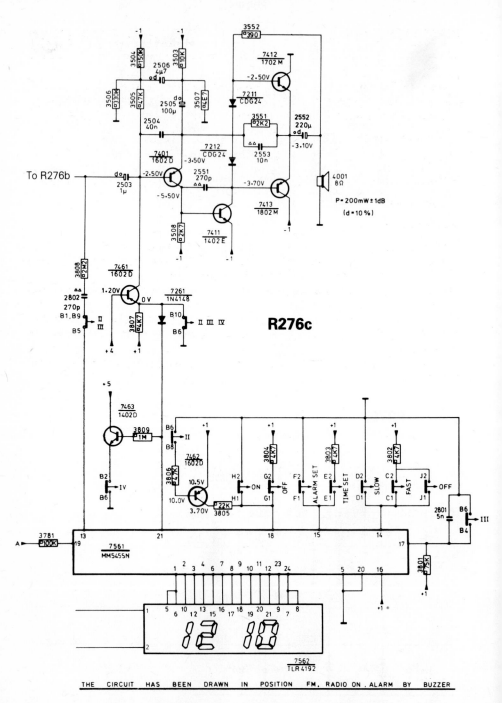

To R276b

R276c

THE CIRCUIT HAS BEEN DRAWN IN POSITION FM, RADIO ON, ALARM BY BUZZER

(R276c) CIRCUIT DIAGRAM—MODEL TR1321 (*CONTINUED*)

PHILIPS   Model TR1720

General Description: A three-waveband portable radio cassette-recorder operating from mains or battery supplies. A microphone is fitted to the case and a socket provided for the connection of an earphone.

Mains Supplies: 110–127, 220–240 volts, 50/60Hz.

Batteries: 6 volts (4×1·5 volts).

Wavebands: L.W. 150–255kHz; M.W. 520–1605kHz; F.M. 87·5–108MHz.

Loudspeaker: 4 ohms impedance.

Access for Service

Remove control knobs, battery, lid, and screws in back cover.

The back cover may now be removed by pulling upwards at the bottom side (pay attention to slide knob and F.M. aerial connection).

The P.C.B./tape-deck can be lifted out after removal of 6 screws and scale pointer.

To remove handle push clamping brackets outwards through the 'keyhole'.

To remove the cassette lid, push the lugs carefully inwards (while cassette lid is opened).

On re-assembly, be attentive to slide knob, scale pointer, and F.M. aerial connection.

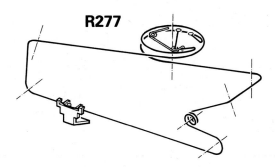

R277

(R277) DRIVE CORD—MODEL TR1720

SK-1	⊘→	◇	▭	⊣⊢	⊘	⊡	⊟
MW	468 kHz ± 1 kHz	Ⓔ via 100 nF	Min Cap		S211 S208 S206	1	① Max
MW 520-1605 kHz	1635 kHz	Ⓒ	Min Cap		C199f	1	① Max
	560 kHz		▭▭		S202		
	1500 kHz		▭▭		C199b		
LW 150-255 kHz	147 kHz	Ⓒ	Max Cap		S204 S203		① Max
FM		Ⓕ via 100 nF		S210	S209	1	① Max
	10,7 MHz ± 0,09 MHz				S207	2	
		Ⓖ via 100 nF			S205 S201	1	
					S210	2	
FM 87,5-108 MHz	108 MHz	Ⓑ via 100 nF	Min Cap		C199h C199d	1	① Max
	87,35 MHz		Max Cap		S101 S102		① Max

⇕ Repeat

1 Adjust for symmetry and max. height.

2 Adjust the "S"-curve for symmetry and max. linearity.

R278

(R278) ALIGNMENT PROCEDURE—MODEL TR1720

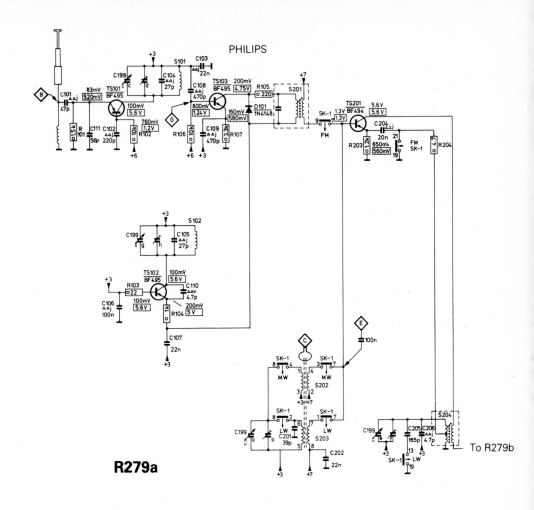

PHILIPS

R279a

To R279b

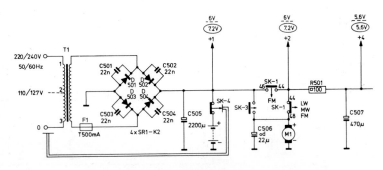

(R279a) CIRCUIT DIAGRAM—MODEL TR1720 (*PART*)

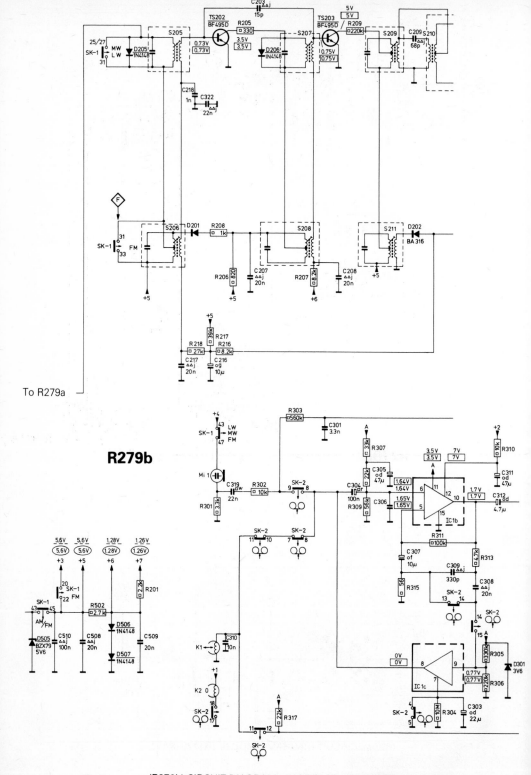

(R279b) CIRCUIT DIAGRAM—MODEL TR1720 (*PART*)

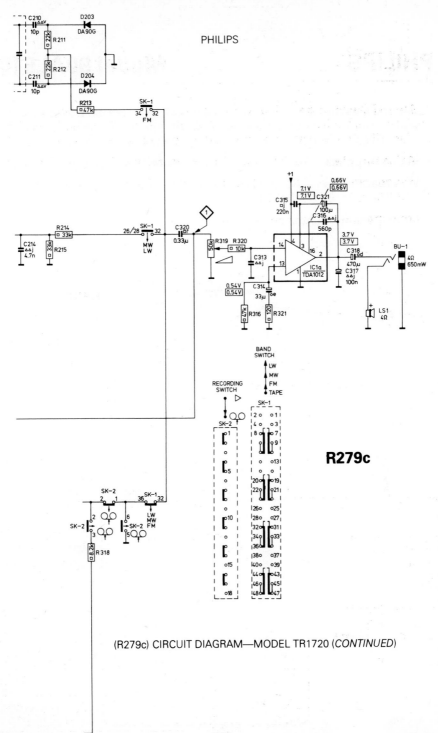

PHILIPS

R279c

(R279c) CIRCUIT DIAGRAM—MODEL TR1720 (*CONTINUED*)

PHILIPS   Model 90AL510

General Description: A four-waveband A.M./F.M. portable radio receiver operating from mains or battery supplies. An integrated circuit is used for A.F. amplification and a socket provided for the connection of earphones.

Mains Supplies: 110, 220 volts, 50Hz.   **Batteries:** 6 volts (4×1·5 volts).

Wavebands: L.W. 150–255kHz; M.W. 520–1605kHz; S.W. 5·95–15·45MHz; F.M. 87·5–104MHz.

Loudspeaker: 4 ohms impedance.

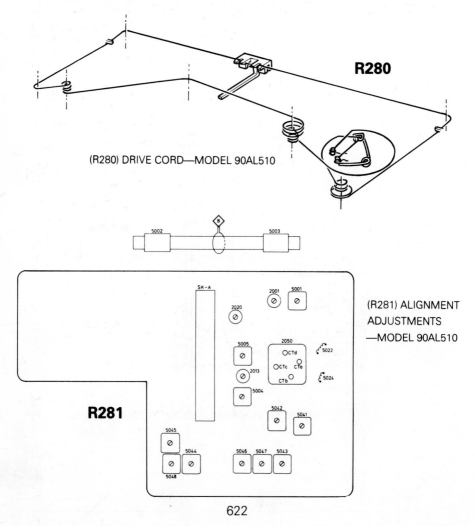

R280

(R280) DRIVE CORD—MODEL 90AL510

(R281) ALIGNMENT ADJUSTMENTS —MODEL 90AL510

R281

Wave range SK-A	Signal to	◇	Var.cap. ≠	Detune	Adjust ∅	Indication
	468 kHz via 39 nF	Ⓐ	Min.cap.	5005,5046 5047,5048	5048	↑ ⟨1⟩ Max.
					5047	
					5046	
MW (520-1605 kHz)	512 kHz	Ⓑ	Max.cap.		5005	↑ ⟨1⟩ Max.
	1635 kHz		Min.cap.		CTc	
	600 kHz		☐		5002	
	1400 kHz		Tune in		CTd	
LW (150-255 kHz)	147 kHz	Ⓑ	Max.cap.		2020	⟨1⟩ Max.
	200 kHz		Tune in ☐		5003	
SW (5.95-15.45 MHz)	5.8 MHz	Ⓒ	Max.cap.		5004	↑ ⟨1⟩ Max. ☐1
	15.9 MHz		Min.cap.		2013	
	6.5 MHz		☐		5001	
	14.5 MHz		Tune in		2001	
FM (87.5-104 MHz)	10.7 MHz via 22 nF	Ⓓ	Min.cap.	5041,5042, 5043,5044, 5045	5044	↑ ⟨2⟩ ☐2
					5043	
					5042	
					5041	
					5045	⟨1⟩ ☐3
	87.5 MHz	Ⓔ	Max.cap.		5024	↑ ⟨1⟩ Max.
	104 MHz		Min.cap.		CTb	
	88 MHz		☐		5022	
	102 MHz		Tune in		CTa	

↕ Repeat

☐1 Telescopic aerial pushed in

☐2 Open bridge ⟨A⟩. Connect an oscilloscope to ⟨2⟩ via a 100 kΩ resistor. Adjust the FM-IF curve for maximum height and symmetry.

☐3 Close bridge ⟨A⟩. Connect an oscilloscope to ⟨1⟩ via a 100 kΩ resistor. Adjust the S-curve for maximum symmetry and linearity.

R282

(R282) ALIGNMENT PROCEDURE—MODEL 90AL510

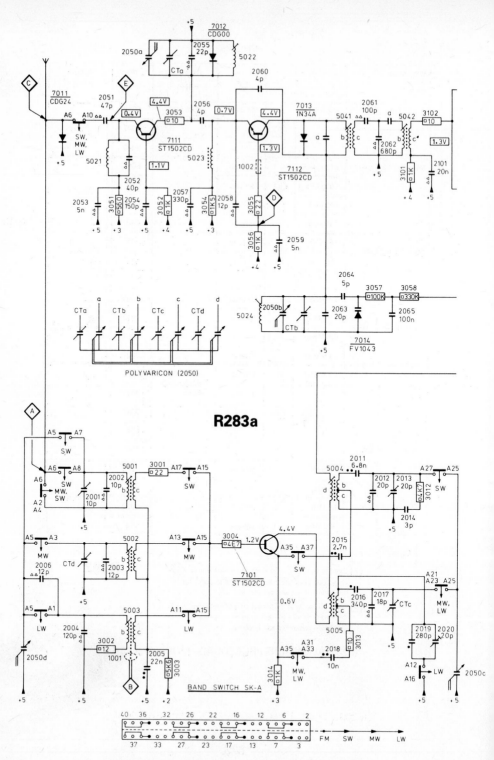

R283a

POLYVARICON (2050)

BAND SWITCH SK-A

(R283a) CIRCUIT DIAGRAM—MODEL 90AL510 (*PART*)

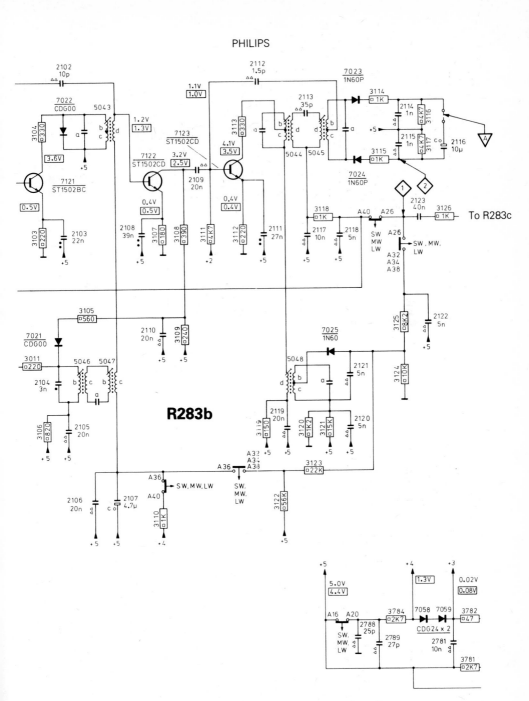

(R283b) CIRCUIT DIAGRAM—MODEL 90AL510 (*PART*)

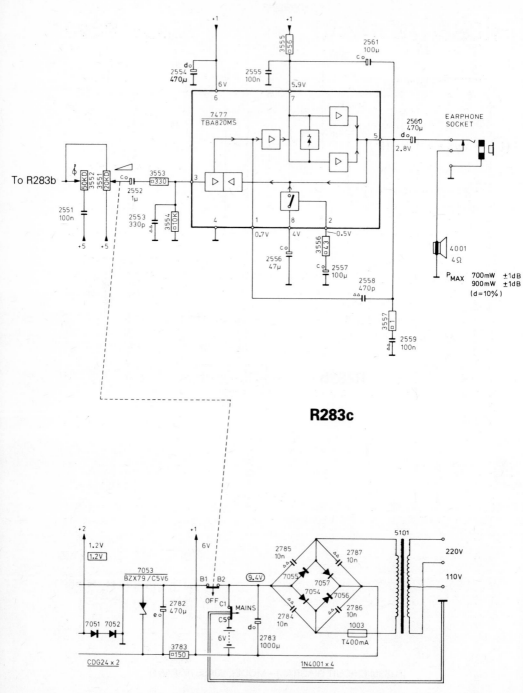

R283c

(R283c) CIRCUIT DIAGRAM—MODEL 90AL510 (*CONTINUED*)

ROBERTS RADIO   Model R800

General Description: A three-waveband A.M./F.M. portable radio receiver operating from mains or battery supplies. An integrated circuit is used for A.M. and I.F. signal processing and sockets are provided for the connection of a tape recorder and earphones.

Mains Supply: 240 volts, 50Hz.   **Battery:** 9 volts (PP9).

Wavebands: L.W. 150–260kHz; M.W. 520–1610kHz; F.M. 88–104MHz.

Loudspeaker: 8 ohms impedance.

Access for Service

Remove base, disconnect loudspeaker, remove screw securing lower and telescopic aerial.

Remove flange head screw at either end of case (above handle fixing).

Ease chassis out from top of case, to extent of leads. Disconnect power leads, B+Red and B−Black. Disconnect leads from Din socket D Yellow E Black, chassis may now be removed.

Power unit is secured by two flange head screws in case end and one nut in case back.

Alignment

With gang at max. pointer should coincide with datum marks at right-hand end of scale. All signals injected via coupling loop.

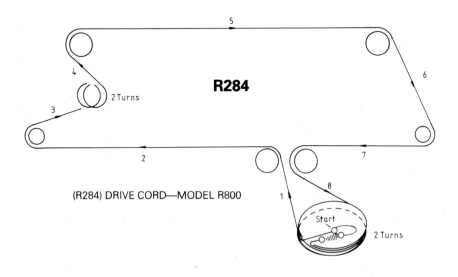

(R284) DRIVE CORD—MODEL R800

Wave-band	Sweep/Signal generator		Pointer	Indicator	Adjust	Indication
1 M.W.	468kHz	25kHz Deviation	L.F. End	Scope across C126	L106 (Yellow) L107 (White)	Max amplitude and symmetry
2 M.W.	518kHz	30% A.M.	L.F. End	Scope across C126 or output meter	L111 (Red)	Max. O/P
3 M.W.	1615kHz	30% A.M.	H.F. End	– as 2 –	C134	Max. O/P
4 M.W.	560kHz	30% A.M.	Tune to signal	– as 2 –	L108	Max. O/P
5 M.W.	1500kHz	30% A.M.	Tune to signal	– as 2 –	C132	Max. O/P

Repeat steps 2 to 5 for optimum results

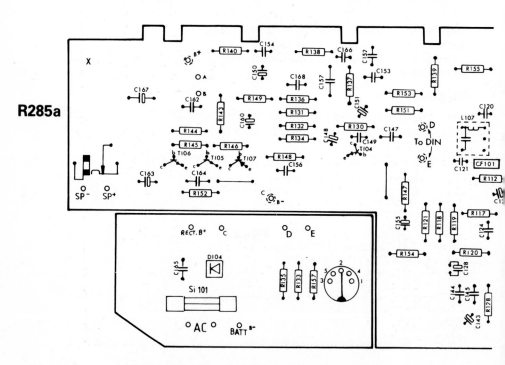

(R285a) COMPONENT LAYOUT—MODEL R800 (*PART*)

Wave-band	Sweep/Signal generator		Pointer	Indicator	Adjust	Indication
L.W.	265kHz	30% A.M.	H.F. End	– as 2 –	C138	Max. O/P
L.W.	200kHz	30% A.M.	200kHz	– as 2 –	L109	Max. O/P
For following adjustments maintain I/P below limiting						
V.H.F.	88MHz	1Mhz Deviation	Tune to signal	Scope across C142	L104, L112 and L113 Orange, Blue, Pink	Max. amplitude and symmetry of 'S' curve
V.H.F.	87·8MHz	22·5kHz Deviation	L.F. End	Scope across C142 or output meter	L105 and L102	Max. O/P
0 V.H.F.	104·5MHz	22·5kHz Deviation	H.F. End	– as 9 –	C114 and C103	Max. O/P
Repeat steps 9 and 10 for optimum results						

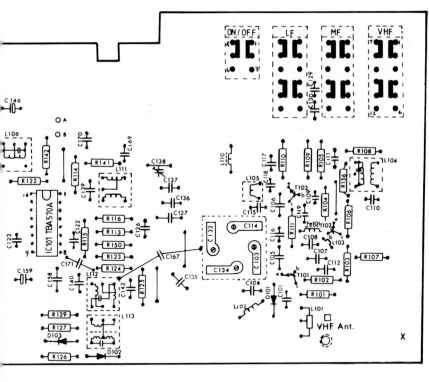

R285b

(R285b) COMPONENT LAYOUT—MODEL R800 (*CONTINUED*)

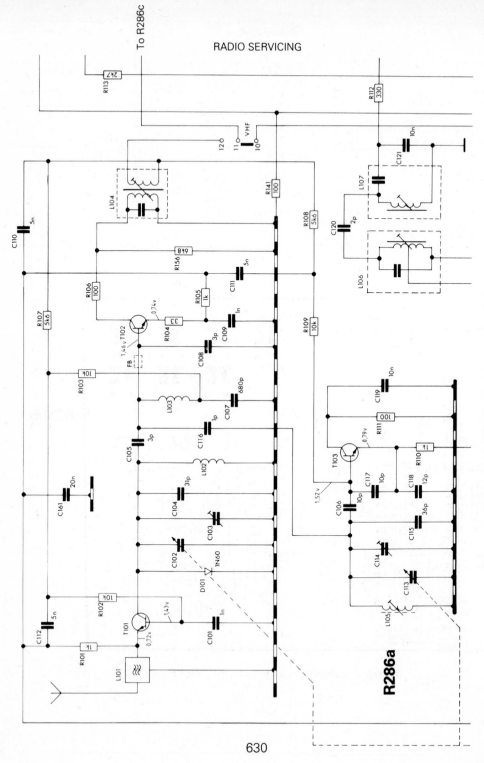

(R286a) CIRCUIT DIAGRAM—MODEL R800 (*PART*)

R286a

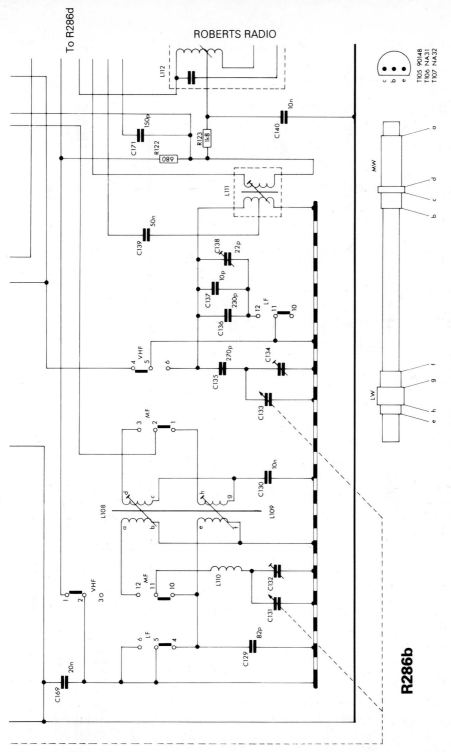

ROBERTS RADIO

(R286b) CIRCUIT DIAGRAM—MODEL R800 (PART)

R286b

R286c

(R286c) CIRCUIT DIAGRAM—MODEL R800 (*PART*)

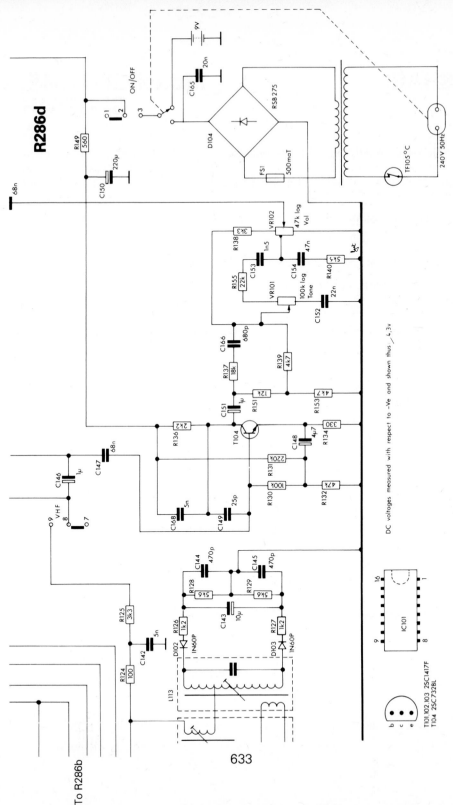

(R286d) CIRCUIT DIAGRAM—MODEL R800 (CONTINUED)

633

SHARP
Model GF-1754E, GF-1754H

General Description: A three-waveband A.M./F.M. portable radio receiver and cassette-recorder. An integrated circuit is used for A.F. signal processing and sockets are provided for the connection of auxiliary inputs and head-phones.

Mains Supplies: 110–220 volts (GF-1754H), 240 volts (GF-1754E), 50Hz.

Batteries: 6 volts (4×1·5 volts).

Wavebands: L.W. 150–285kHz; M.W. 520–1620kHz; F.M. 87·6–108MHz.

Loudspeaker: 4 ohms impedance.

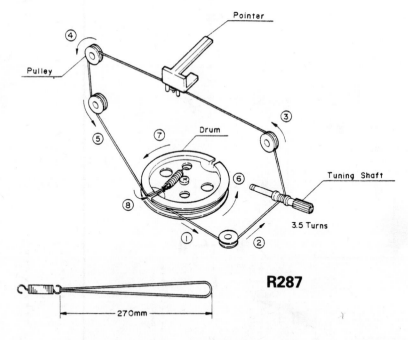

R287

(R287) DRIVE CORD—MODEL GF-1754E

Access for Service

Cabinet bottom removal: Take out the cassette from the cassette compartment.

Remove the battery compartment cover and take out batteries.

Remove the Tuning and Volume control knobs.

Remove five screws retaining the cabinet back and lightly open it.

Disconnect the antenna lead wire and from the battery terminals.

Then the cabinet back can be removed.

Adjustment of Bias Current and Check of Bias Oscillation Frequency

Connect V.T.V.M. to the resistor (R201) located on the printed wiring board.

Set the mode switch (SW3) to 'Tape'.

Place the unit in 'Record' mode.

Check the record oscillator bias frequency is 50 ±3kHz.

Adjustment the resistors (R224, R225, R226) so that the V.T.V.M. reads 60 ±5mV.

Alignment

Connect an output meter across the speaker voice coil.

Set the Volume control at maximum.

Attenuate the signals from the generator enough to swing the most sensitive range of the output meter.

Use a non-metallic alignment tool.

Repeat adjustments to insure good results.

A.M. Alignment:

Step	Band	Test Stage	Signal generator Connector to receiver	Input signal frequency	Receiver Dial setting	Remarks	Adjustment
1	M.W.	I.F.	Connect signal generator through a 10k ohm resistor to the antenna tuning capacitor. Ground lead to the receiver chassis	455kHz (400Hz, 30%, A.M. modulated)	High end of dial (minimum capacity)	Adjust for maximum output on speaker voice coil lugs	T6 T7 T8
2	M.W.		Use radiation loop. Loop of several turns of wire, or place generator lead close to receiver for adequate signal pick-up. Connect generator output to one end of this wire	Exactly 510kHz, (400Hz, 30%, A.M. modulated)	Low end of dial (maximum capacity)	Same as step 1	Adjust the M.W. oscillator coil (L7)

Step	Band	Test Stage	Signal generator Connector to receiver	Input signal frequency	Receiver Dial setting	Remarks	Adjustment
3	M.W.	Band Cover-age	Same as step 2	Exactly 1650kHz (400Hz, 30%, A.M. modulated)	High end of dial (minimum capacity)	Same as step 1	Adjust the M.W. oscillator trimmer (TC7)
4	M.W.	Track-	Same as step 2	Exactly 600kHz (400Hz, 30%, A.M. modulated)	600kHz	Same as step 1	Adjust the M.W. antenna coil (L9). (See Note A)
5	M.W.	ing	Same as step 2	Exactly 1400kHz (400Hz, 30%, A.M. modulated)	1400kHz	Same as step 1	Adjust the M.W. antenna trimmer (TC8) (See Note A)
6	M.W.		Repeat steps 2, 3, 4 and 5 until no further improvement can be made.				
7	L.W.	Band	Same as step 2	Exactly 145kHz (400Hz, 30%, A.M. modulated)	Low end of dial (maximum capacity)	Same as step 1	Adjust the L.W. oscillator coil (L8)
8	L.W.	Cover-age	Same as step 2	Exactly 295kHz (400Hz, 30%, A.M. modulated)	High end of dial (minimum capacity)	Same as step 1	Adjust the L.W. oscillator trimmer (TC9)
9	L.W.	Track-	Same as step 2	Exactly 160kHz (400Hz, 30%, A.M. modulated)	160kHz	Same as step 1	Adjust the L.W. antenna coil (L10)
10	L.W.	ing	Same as step 2	Exactly 260kHz (400Hz, 30%, A.M. modulated)	260Hz	Same as step 1	Adjust the L.W. antenna trimmer (TC10)
11	L.W.		Repeat steps 7, 8, 9 and 10 until no further improvement can be made				

Note A. Check the alignment of the receiver antenna coil by bringing a piece of ferrite (such as a coil slug) near the antenna loop stick, then a piece of brass. If ferrite increase output, loop requires more inductance. If brass increases output, loop requires less inductance. Change loop inductance by sliding the bobbin toward the centre of ferrite core to increase inductance, or away to decrease inductance.

F.M. Alignment Chart: Set the Band Selector switch at F.M. position.

Step	Test stage	Signal generator		Receiver		Adjustment
		Connection to receiver	Input signal frequency	Dial setting	Remarks	
1	I.F.	Connect signal generator through a 5PF capacitor to the emitter of Q2. Connect generator ground lead to the receiver chassis	Exactly 10·7MHz (400Hz, 30%, F.M. modulated)	Tuning gang fully closed. (maximum capacity)	Connect V.T.V.M. across the resistor (R129)	Detune T4, T5. Tune T1, T2 and T3 for maximum indication
2	Ratio detector	Same as step 1	Exactly 10·7MHz (Unmodulated)	Same as step 1	See Note B	See Note B
3	Band coverage	Connect signal generator through a dummy, including output impedance of signal generator to the F.M. antenna. Ground lead of generator connected to the receiver chassis	Exactly 87MHz (400Hz, 30%, F.M. modulated)	Tuning gang fully closed. (maximum capacity)	Adjust for maximum output speaker voice coil lugs	Adjust the F.M. oscillator coil L5 and L6
4	Band coverage	Same as step 3	Exactly 109MHz (400Hz, 30%, F.M. modulated)	Tuning gang fully open (minimum capacity)	Same as step 3	Adjust the F.M. oscillator trimmer TC5
5	Tracking	Same as step 3	Exactly 88MHz (400Hz, 30%, F.M. modulated)	88MHz	Same as step 3	Adjust the F.M. R.F. coil L2 and L3
6	Tracking	Same as step 3	Exactly 108MHz (400Hz, 30%, F.M. modulated)	108MHz	Same as step 3	Adjust the F.M. R.F. trimmer TC6
7		Repeat steps 3, 4, 5 and 6 until no further improvement can be made				

Note B.
1. Connect V.T.V.M. (0·1 volts range D.C. scale) across the resistor (R139, TP-3).
2. Adjust T5 for 0 volt on V.T.V.M.
3. Change signal generator frequency 10·7MHz+100kHz and −100kHz approximately.
4. Adjust T4 for balanced peaks. Peak separation should be approximately 200kHz.

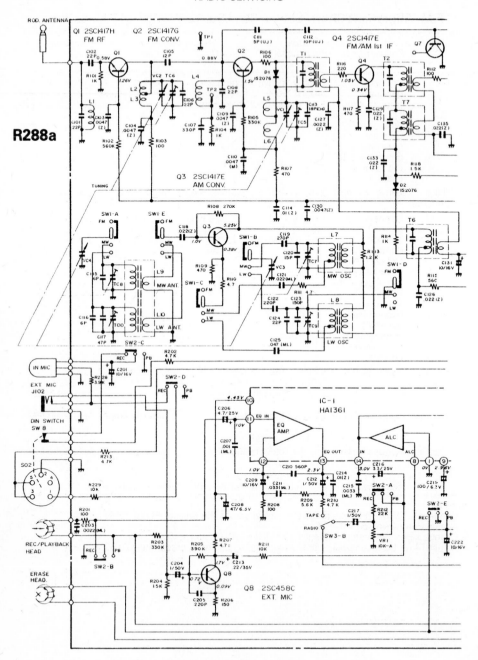

R288a

(R288a) CIRCUIT DIAGRAM—MODEL GF-1754E (*PART*)

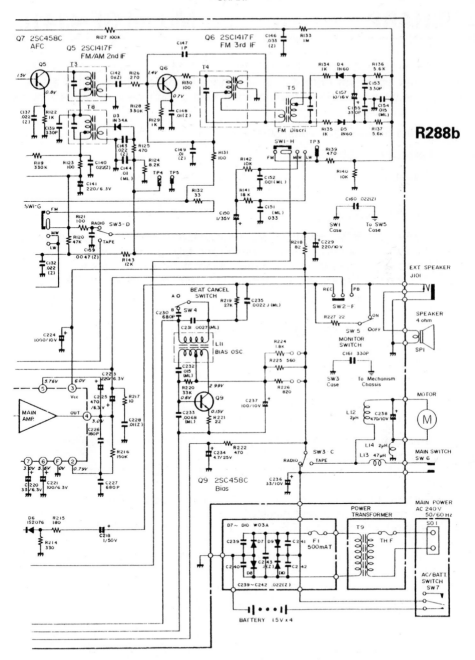

R288b

(R288b) CIRCUIT DIAGRAM—MODEL GF-1754E (*CONTINUED*)

SHARP   Model GF-4141E,

General Description: A three-waveband, A.M./F.M. stereo radio cassette-recorder operating from mains or battery supplies. Microphones are built into the case and sockets are provided for the connection of auxiliary inputs and headphones.

Mains Supply: 240 volts, 50Hz.

Batteries: 9 volts (6×HP2).

Wavebands: L.W. 150–285kHz; M.W. 520–1620kHz; F.M. 87·6–108MHz.

Access for Service

Cabinet Back Removal: Take out the cassette from the cassette holder. Remove the battery compartment lid and take out batteries.
Remove five screws retaining the cabinet back and lightly open it.
Disconnect three tips from the battery terminal, telescopic rod antenna and power supply P.W.B.
Then the cabinet back can be removed.

Cabinet Front Removal: Remove the microphone and external speaker jacks plate.
Remove six knobs.
Remove two stays and four screws retaining the cabinet front and lightly lift up the frame assembly.
Unsolder the lead wires connected to the speakers.
Then the cabinet front can be removed.

Mechanism Block Removal: Remove the screw retaining the variable tuning capacitor and turn over the set.
Remove four nuts retaining tone, volume, and balance controls and band selector switch.
Remove two screws retaining the function selector switch.
Remove seven screws retaining the P.W.B.
Unsolder the capacitor (C64) connected between mechanism chassis and P.W.B.
Pulling the P.W.B. in the arrow direction whole holding it upward.
Detach and turn over the P.W.B.
Remove four screws retaining the mechanism block. Then the mechanism block can be removed.

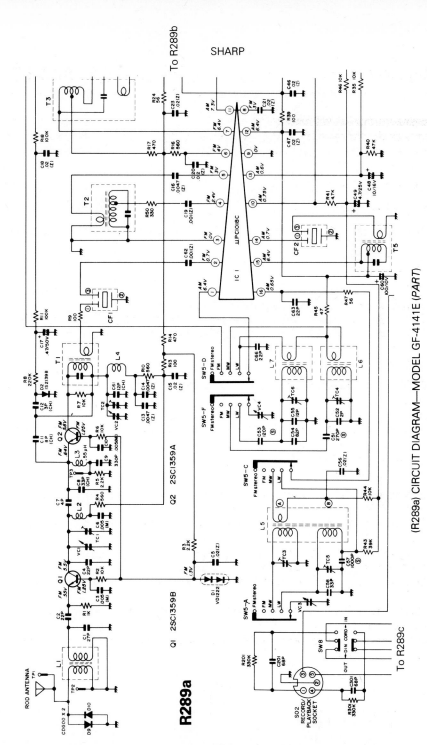

SHARP

R289a

(R289a) CIRCUIT DIAGRAM—MODEL GF-4141E (*PART*)

To R289b

To R289c

Q1 2SC1359B

Q2 2SC1359A

641

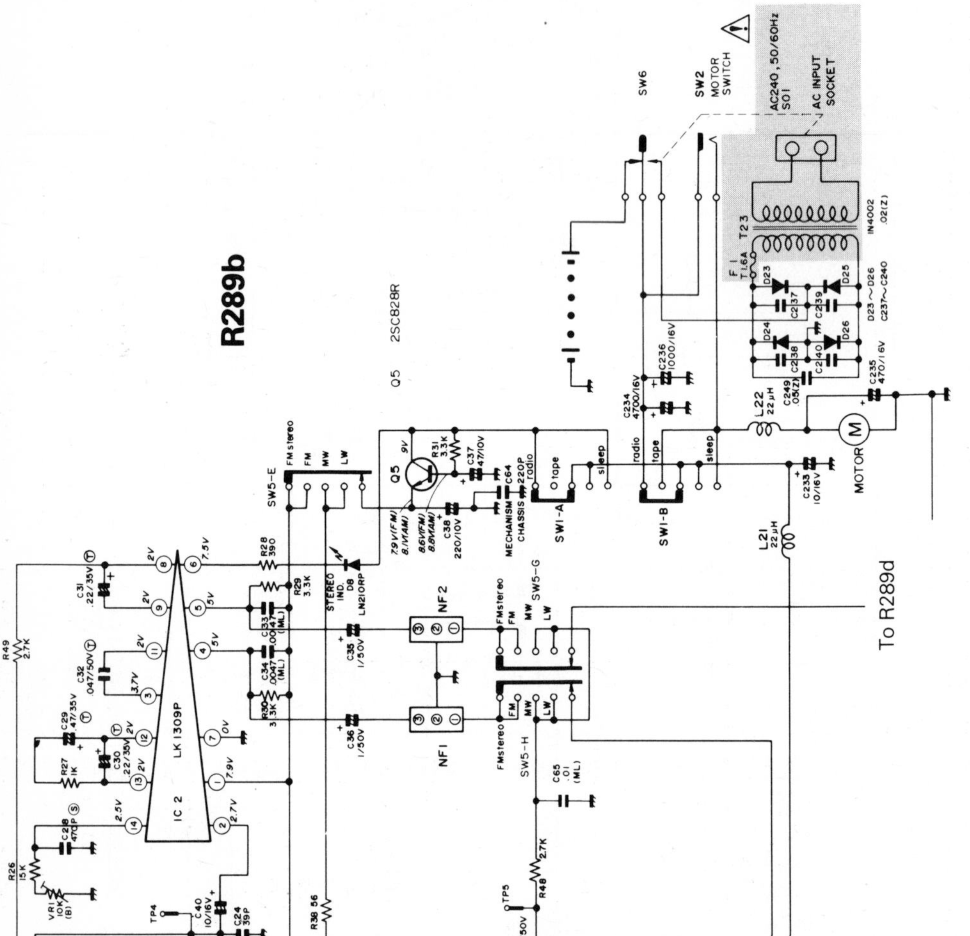

R289b

(R298b) CIRCUIT DIAGRAM—MODEL GF-4141F (PART)

To R289a

To R289d

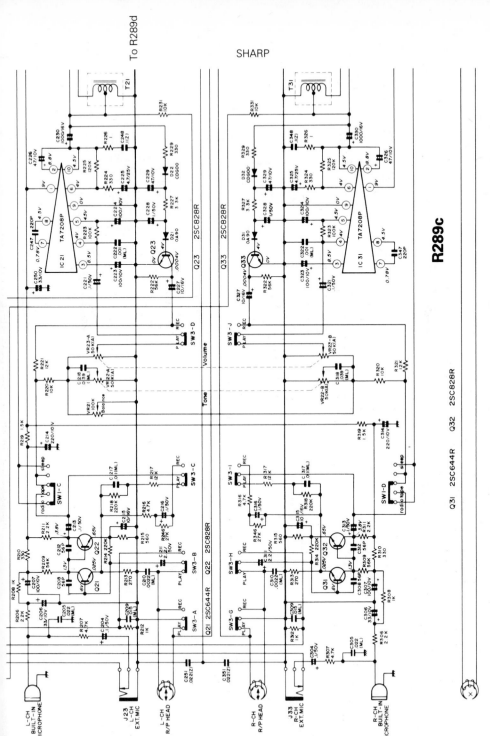

SHARP

To R289d

R289c

(R289c) CIRCUIT DIAGRAM—MODEL GF-4141E (PART)

Q31 2SC644R   Q32 2SC828R

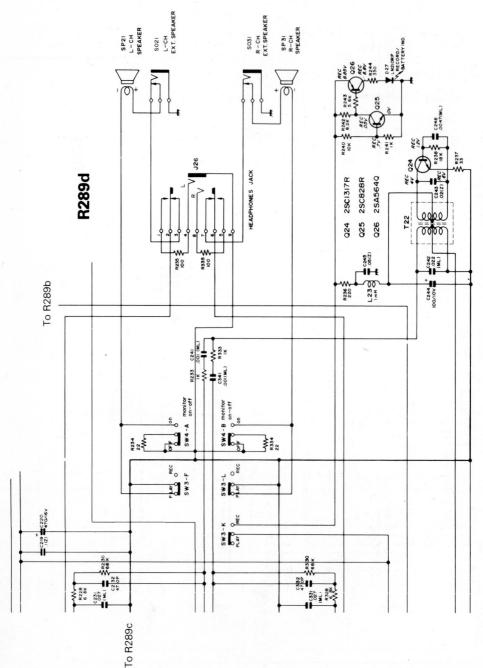

(R289d) CIRCUIT DIAGRAM—MODEL GF-4141E (CONTINUED)

SHARP   Model GF-6060E

General Description: A four-waveband A.M./F.M. stereo radio cassette tape-recorder operating from mains or battery supplies. Microphones are built into the case and sockets are provided for the connection of auxiliary inputs, extension loudspeakers and headphones.

Mains Supplies: 110–120/220–240 volts, 50/60Hz.

Batteries: 12 volts (8×HP2).

Wavebands: L.W. 150–285kHz; M.W. 520–1620kHz; S.W. 5·95–18MHz; F.M. 87·6–108MHz.

Loudspeakers: 8 ohms impedance.

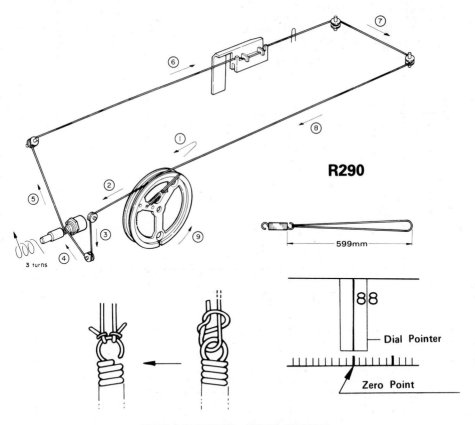

(R290) DRIVE CORD—MODEL GF-6060E

Access for Service

Front Cabinet Removal: Remove the seven screws at the rear of the unit retaining the front cabinet.

Push the stop/eject key to open the cassette holder. Detach the front cabinet, and withdraw the three tips.

Chassis Removal: Take the six knobs off.

Disconnect the three tips from the P.W.B.

Remove the seven screws retaining the amp./tuner P.W.B., then it can be removed from the back cabinet.

Mechanism Block Removal: Detach the dial scale plate.

Remove the four screws retaining the mechanism block.

Disconnect the three sockets (CNS101, CNS102, CNS103) and the one tip from the P.W.B., then the mechanism block can be removed.

Alignment Procedure

Set the Volume control to maximum.

Attenuate the signals from the generator enough to swing the most sensitive range of the output meter.

Use a non-metallic alignment tool.

Repeat adjustments to insure good results.

Set the Function Selector Switch (SW101) to 'radio' position.

A.M. I.F./R.F. Alignment:

Step	Band	Test Stage	Signal generator Input signal frequency	Receiver Dial setting	Remarks	Adjustment
1	M.W.	I.F.	Exactly 455kHz: GF-6060H Exactly 468kHz: GF-6060E	High end of dial (minimum capacity)	Adjust for best 'I.F.' curve	Adjust the M.W. I.F. transformers (T3) (T4)
2	L.W.	Band Coverage	Exactly 145kHz (400Hz, 30%, A.M. modulated)	Low end of dial (maximum capacity)	Adjust for maximum output	Adjust the L.W. oscillation coil (L7)
3	L.W.		Exactly 295kHz (400Hz, 30%, A.M. modulated)	High end of dial (minimum capacity)	Same as step 2	Adjust the L.W. oscillation trimmer (TC6)
4	L.W.	Tracking	Exactly 170kHz (400Hz, 30%, A.M. modulated)	170kHz	Same as step 2	Adjust the L.W. bar antenna coil (L5)
5	L.W.		Exactly 270kHz (400Hz, 30%, A.M. modulated)	270kHz	Same as step 2	Adjust the L.W. antenna trimmer (TC3)
6	L.W.	Repeat steps 2, 3, 4, and 5 until no further improvement can be made				
7	M.W.	Band Coverage	Exactly 510MHz (400Hz, 30%, A.M. modulated)	Low end of dial (maximum capacity)	Same as step 2	Adjust the M.W. oscillation coil (L8)
8	M.W.		Exactly 1650kHz (400Hz, 30%, A.M. modulated)	High end of dial (minimum capacity)	Same as step 2	Adjust the M.W. oscillation trimmer (TC7)

Step	Band	Test stage	Signal generator input signal frequency	Receiver Dial setting	Remarks	Adjustment
9	M.W.		Exactly 600kHz (400Hz, 30%, A.M. modulated)	600kHz	Same as step 2	Adjust the M.W. bar antenna coil (L5)
10	M.W.	Track-ing	Exactly 1400kHz (400Hz, 30%, A.M. modulated)	1400kHz	Same as step 2	Adjust the M.W. antenna trimmer (TC4)
11	M.W.		steps 6, 7, 8 and 9 until no further improvement can be made			
12	S.W.		Exactly 5·85MHz (400Hz, 30%, A.M. modulated)	Low end of dial (maximum capacity)	Same as step 2	Adjust the S.W. oscillation coil (L9)
13	S.W.	Band Cover-age	Exactly 18·5MHz (400Hz, 30%, A.M. modulated)	High end of dial (minimum capacity)	Same as step 2	Adjust the S.W. oscillation trimmer (TC8)
14	S.W.		Exactly 6·5MHz (400Hz, 30%, A.M. modulated)	6·5MHz	Same as step 2	Adjust the S.W. antenna coil (L6)
15	S.W.	Track-ing	Exactly 16MHz (400Hz, 30%, A.M. modulated)	16MHz	Same as step 2	Adjust the S.W. antenna trimmer (TC5)
16	S.W.		Repeat steps 12, 13, 14 and 15 until no further improvement can be made			

F.M. I.F./R.F. Alignment:

Step	Band	Test stage	Signal generator input signal frequency	Receiver Dial setting	Remarks	Adjustment
1	F.M.	I.F.	Exactly 10·7MHz (400Hz, 30%, F.M. modulated)	High end of dial (minimum capacity)	Adjust for best 'S' curve	Adjust the F.M. I.F. transformers (T1) (T2)
2		Band Cover-age	Exactly 87·1MHz (400Hz, 30%, F.M. modulated)	Low end of dial (maximum capacity)	Adjust for maximum output	Adjust the F.M. oscillation coils (L4)
3	F.M.		Exactly 109MHz (400Hz, 30%, F.M. modulated)	High end of dial (minimum capacity)	Same as step 2	Adjust the F.M. oscillation trimmer (TC2)
4		Track-ing	Exactly 88MHz (400Hz, 30%, F.M. modulated)	88MHz	Same as step 2	Adjust the F.M. R.F. coils (L2, L3)
5	F.M.		Exactly 108MHz (400Hz, 30%, F.M. modulated)	108MHz	Same as step 2	Adjust the F.M. R.F. trimmer (TC1)
6	F.M.		Repeat steps 2, 3, 4 and 5 until no further improvement can be made			

647

R291a

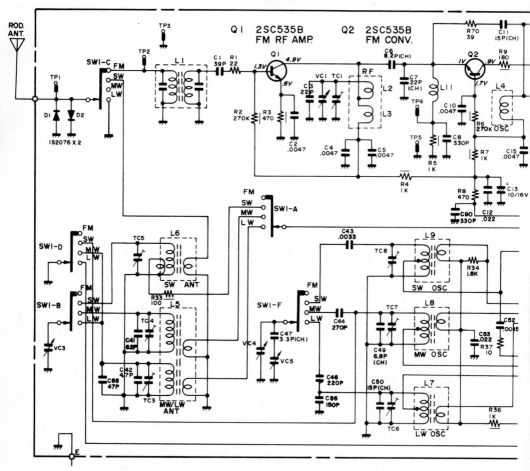

To R291d

(R291a) CIRCUIT DIAGRAM—MODEL GF-6060E (*PART*)

R291b

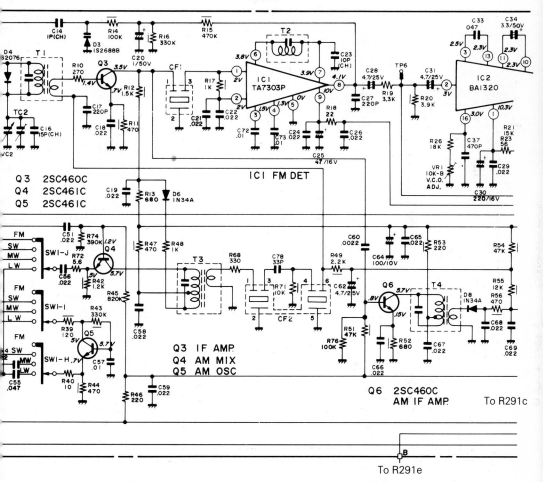

(R291b) CIRCUIT DIAGRAM—MODEL GF-6060E (*PART*)

R291c

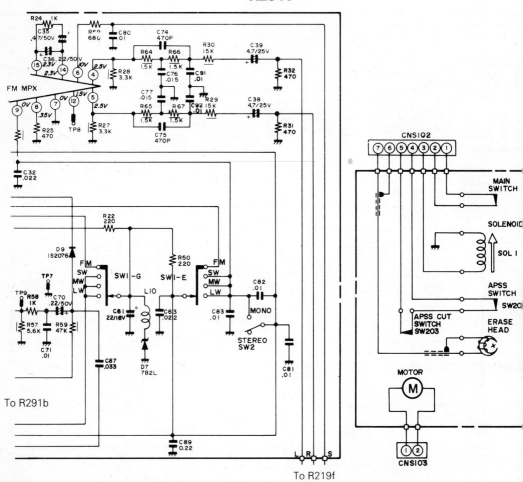

(R291c) CIRCUIT DIAGRAM—MODEL GF-6060E (*PART*)

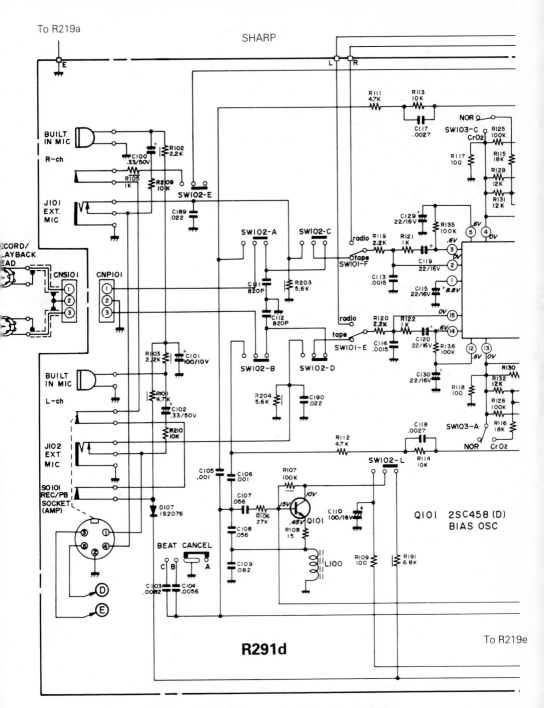

(R291d) CIRCUIT DIAGRAM—MODEL GF-6060E (*PART*)

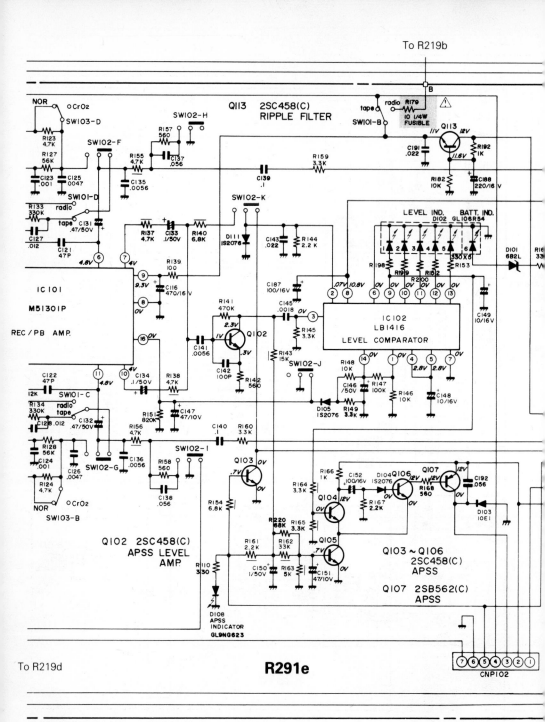

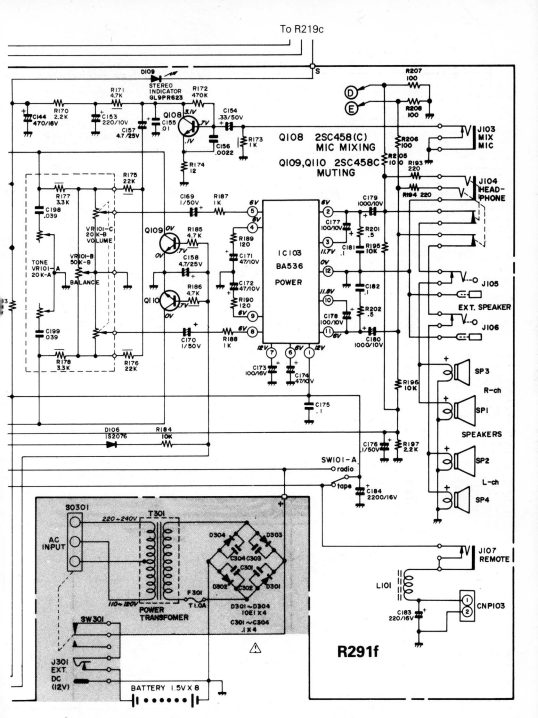

(R291f) CIRCUIT DIAGRAM—MODEL GF-6060E (*CONTINUED*)

SONY   Model CFM-23L

General Description: A four-waveband radio cassette-recorder operating from battery supplies. A microphone is fitted to the case and sockets are provided for the connection of auxiliary inputs and headphones.

Batteries: 6 volts (4×1·5 volts).

Wavebands: L.W. 150–350kHz; M.W. 530–1605kHz; S.W. 6–18MHz; F.M. 87·5–108MHz.

Loudspeakers: 4 ohms impedance.

Access for Service (See Fig. R292)

Remove 6 screws to release the rear of the cabinet.

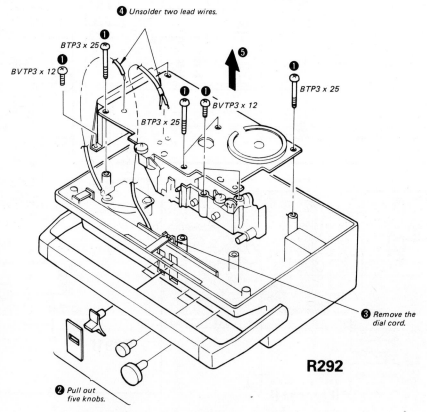

(R292) DISMANTLING PROCEDURE—MODEL CFM-23L

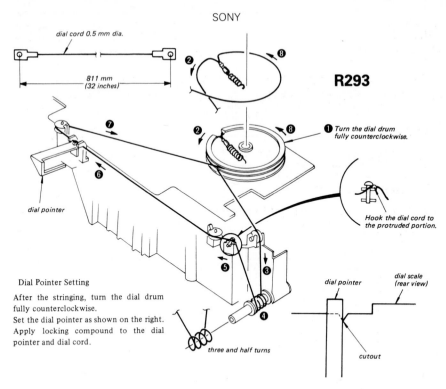

SONY

R293

dial cord 0.5 mm dia.

811 mm
(32 inches)

❶ *Turn the dial drum fully counterclockwise.*

Hook the dial cord to the protruded portion.

dial pointer

Dial Pointer Setting

After the stringing, turn the dial drum fully counterclockwise.
Set the dial pointer as shown on the right.
Apply locking compound to the dial pointer and dial cord.

three and half turns

dial pointer

dial scale (rear view)

cutout

(R293) DRIVE CORD—MODEL CFM-23L

Alignment

A.M. I.F. 468kHz (T3, T8); M.W. 1400kHz (CT5), 620kHz (L5), 1680kHz (CT8), 520kHz (T7); L.W. 160kHz (L5), 330kHz (CT4), 145kHz (T6), 365kHz (CT7); S.W. 5·8MHz (T4, T5), 18·4MHz (CT3, CT6).
F.M. I.F. 10·7MHz (T1, T2); 87·5MHz (L2, L1), 108MHz (CT8, CT1).

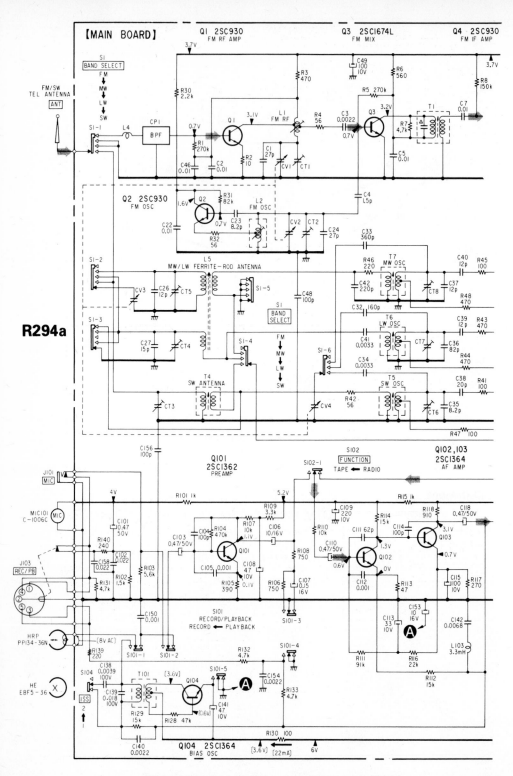

(R294a) CIRCUIT DIAGRAM—MODEL CFM-23L (*PART*)

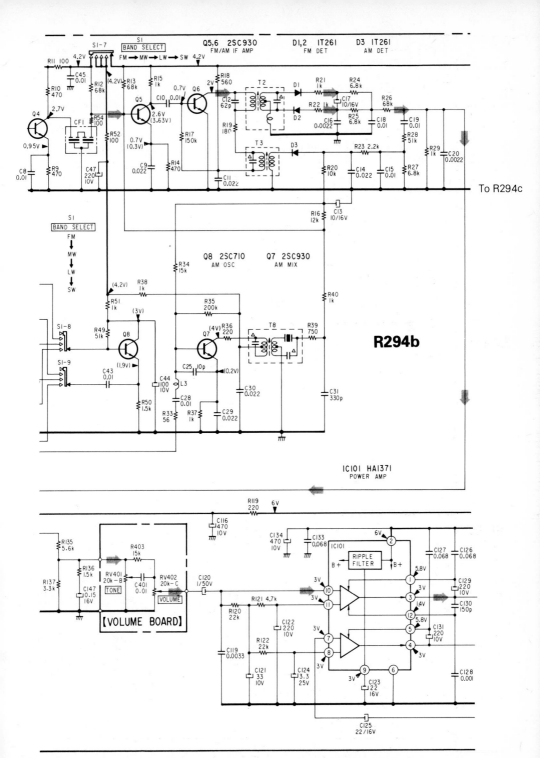

(R294b) CIRCUIT DIAGRAM—MODEL CFM-23L (*PART*)

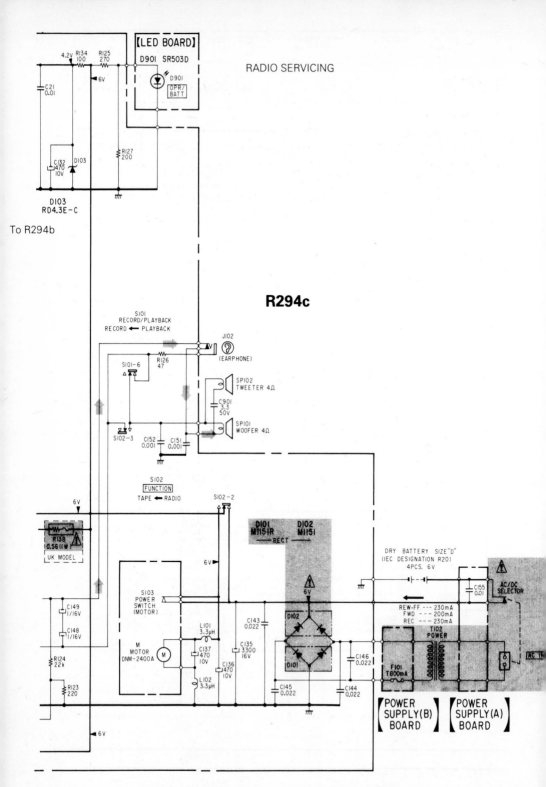

RADIO SERVICING

R294c

SONY   Model CFM-33L

General Description: A four-waveband A.M./F.M. portable radio cassette tape-recorder operating from mains or battery supplies. Sockets are provided for the connection of auxiliary inputs and earphones and an integrated circuit is used in the A.F. power output stage.

Mains Supply: 240 volts, 50Hz.   **Batteries:** 6 volts (4×1·5 volts).

Wavebands: L.W. 150–350kHz; M.W. 530–1605kHz; S.W. 6–18MHz; 87·5–108MHz.

Loudspeakers: 4 ohms impedance.

Access for Service

Removal of 5 screws from the rear of the cabinet allows the front case to be withdrawn.

Alignment

A.M. I.F. 468kHz (CFZ, T3); M.W. 520kHz (T8), 620kHz (L3), 1680kHz (CT8), 1400kHz (CT5); L.W. 145kHz (T7), 160kHz (L5), 365kHz (CT7), 330kHz (CT4); S.W. 5·8MHz (T6, T4), 18·4MHz (CT6, CT3).
F.M. I.F. 10·7MHz (T1, T2); 87·5MHz (L2, L1), 108MHz (CT2, CT1).

Bias Adjustment: Connect 100 ohm resistor in live side of Record/Replay head. Connect a frequency meter across this resistor and adjust T101 for a reading of 35kHz ±1kHz.

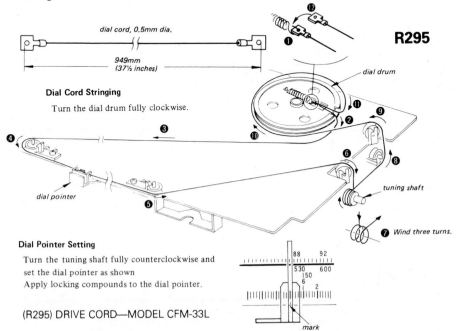

R295

dial cord, 0.5mm dia.

949mm
(37½ inches)

dial drum

Dial Cord Stringing

Turn the dial drum fully clockwise.

dial pointer

tuning shaft

Wind three turns.

Dial Pointer Setting

Turn the tuning shaft fully counterclockwise and set the dial pointer as shown
Apply locking compounds to the dial pointer.

(R295) DRIVE CORD—MODEL CFM-33L

88   92
530   600
150
6   2

mark

659

R296a

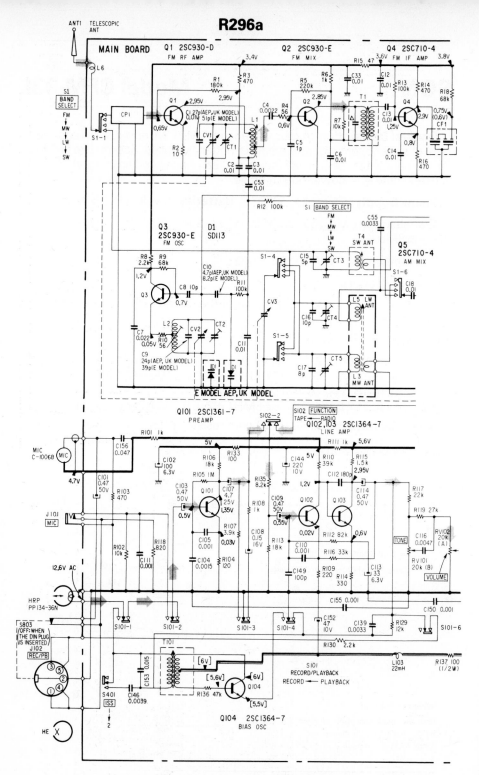

(R296a) CIRCUIT DIAGRAM—MODEL CFM-33L (*PART*)

R296b

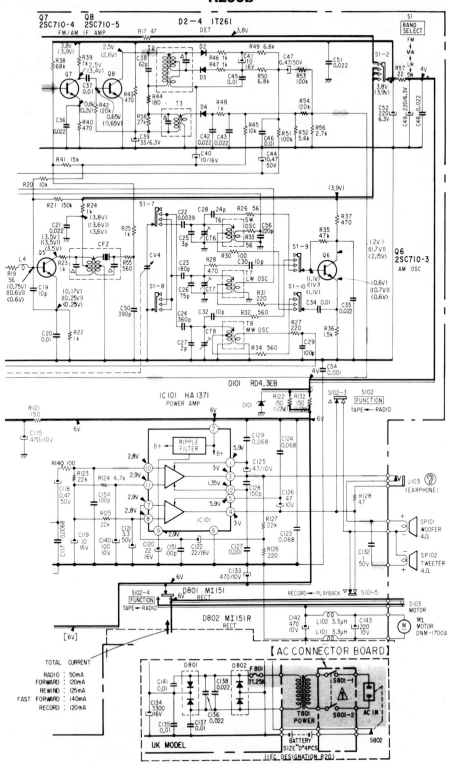

(R296b) CIRCUIT DIAGRAM—MODEL CFM-33L (*CONTINUED*)

SONY   Model ICF-C12L

General Description: A three-waveband A.M./F.M. clock-radio operating from mains supplies only and presented in a cubic case. Alarm facilities are fitted and a socket is provided for the connection of an earphone.

Mains Supply: 240 volts, 50Hz.

Wavebands: L.W. 150–255kHz; M.W. 530–1605kHz; F.M. 87·5–108MHz.

Loudspeaker: 16 ohms impedance.

Alignment

A.M. I.F. 468kHz (T5); M.W. 520kHz (T2), 620kHz (L4), 1680kHz (CT2), 1400kHz (CT1). L.W. 265kHz (CT6), 250kHz (CT5), 145kHz (T2), 160kHz (L4). F.M. I.F. 10·7MHz (T1, T3, T4); 87·1MHz (CT4, L1), 108·5MHz (CT4, CT3).

Dial Cord Stringing

Turn the tuning drum fully clockwise.
String the dial cord in the numerical order given. (❶ – ❻)
Then, install the pointer guide with pointer holder. (❼ , ❽)

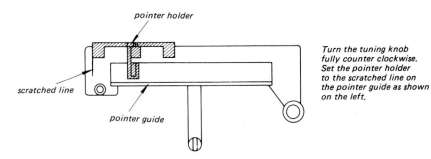

300mm (11¹³/₁₆ inches)

dial cord
0.3mm dia

pointer holder

pointer guide

❸

❺ tuning drum

❼

❷

❻

❶

❹ 3¹/₂ turns

❽

R297

Dial Pointer Setting

pointer holder

scratched line

pointer guide

Turn the tuning knob
fully counter clockwise.
Set the pointer holder
to the scratched line on
the pointer guide as shown
on the left.

(R297) DRIVE CORD—MODEL ICF-C12L

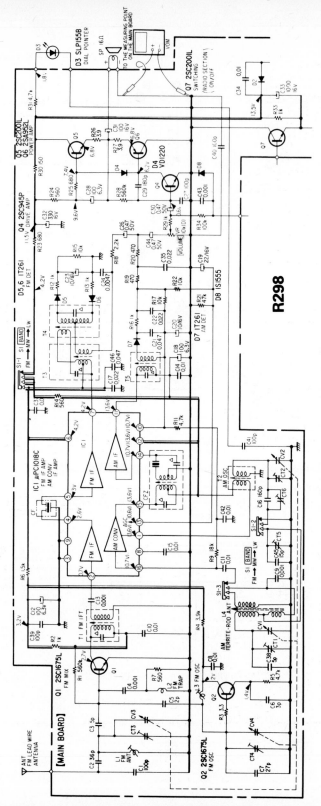

(R298) CIRCUIT DIAGRAM (RADIO SECTION)—MODEL ICF-C12L

R298

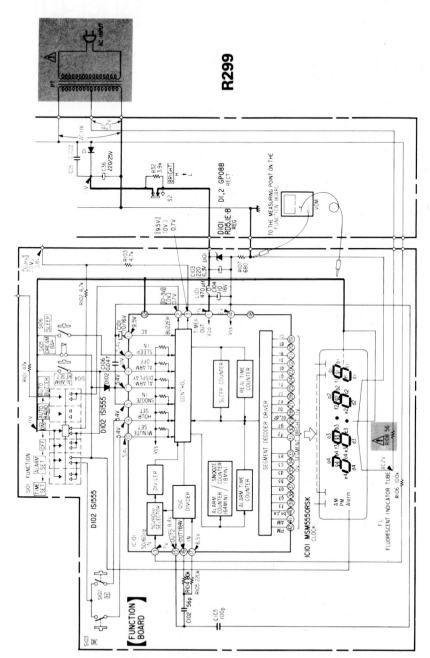

(R299) CIRCUIT DIAGRAM (CLOCK SECTION)—MODEL ICF-C12L

SONY   Model ICF-C16L

General Description: A three-waveband A.M./F.M. digital clock-radio receiver operating from mains supplies with clock supply maintained by back-up battery. An integrated circuit is used for A.M. and I.F. signal processing.

Mains Supply: 240 volts, 50Hz.

Battery: 9 volts.

Wavebands: L.W. 150–255kHz; M.W. 530–1605kHz; F.M. 87·5–108MHz.

Loudspeaker: 16 ohms impedance.

Access for Service

Remove 7 screws from the rear of the cabinet. Remove the pointer holder from the drive cord and pointer guide. Remove 2 screws from the pointer guide panel. Access to the component side of the printed panel may be gained after taking off the drive cord and dial wheel. The board is held by 2 screws.

Alignment

A.M. R.F.; M.W. 520kHz (T2), 620kHz (L4), 1680kHz (CT2), 1400kHz (CT1); L.W. 145kHz (T2), 160kHz (L4), 265kHz (CT6), 220kHz (CT5).
F.M. I.F. 10·7MHz (T1, T3, T4); 87·1MHz (L3, L1), 108·5MHz (CT4, CT3).

Stringing

Turn the dial drum fully counterclockwise.

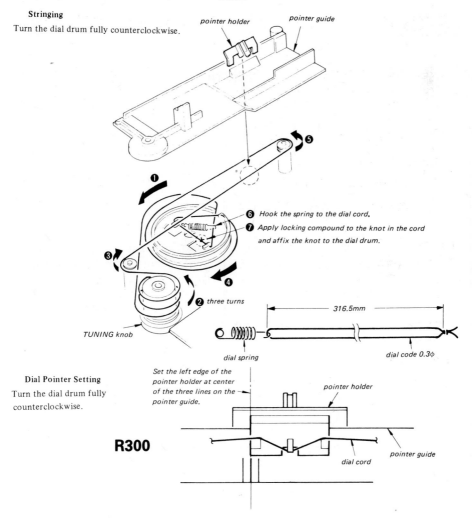

6 Hook the spring to the dial cord.

7 Apply locking compound to the knot in the cord and affix the knot to the dial drum.

pointer holder

pointer guide

❷ *three turns*

TUNING knob

dial spring

316.5mm

dial code 0.3φ

Dial Pointer Setting

Turn the dial drum fully counterclockwise.

Set the left edge of the pointer holder at center of the three lines on the pointer guide.

pointer holder

R300

pointer guide

dial cord

(R300) DRIVE CORD—MODEL ICF-C16L

667

R301a

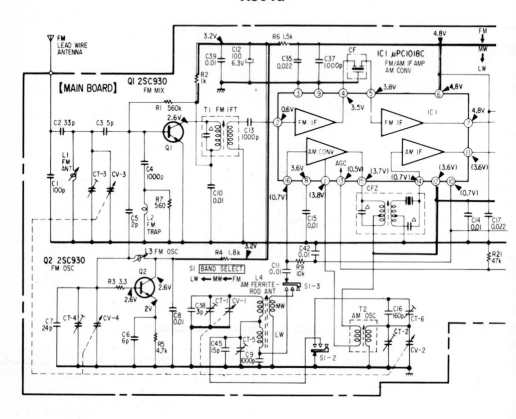

(R301a) CIRCUIT DIAGRAM (RADIO SECTION)—MODEL ICF-C16L (*PART*)

R301b

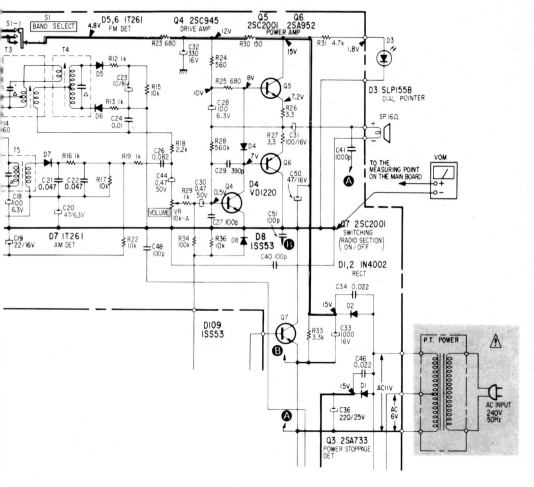

(R301b) CIRCUIT DIAGRAM (RADIO SECTION)—MODEL ICF-C16L (*CONTINUED*)

(R302) CIRCUIT DIAGRAM (CLOCK SECTION)—MODEL ICF-C16L

R302

SONY   Model ICF-400

General Description: A three-waveband A.M./F.M. portable radio receiver operating from battery supplies. An integrated circuit is used in the A.M. and I.F. stages and a socket is provided for the connection of an earphone.

Batteries: 6 volts (4×1·5 volts).

Wavebands: L.W. 150–255kHz; M.W. 530–1605kHz; F.M. 87·6MHz.

Loudspeaker: 8 ohms impedance.

Access for Service

Remove the battery cover and take out 5 screws to release the rear of the case. Remove the smaller gear wheel and the screw adjacent to it to release the tuner board.

Alignment

A.M. I.F. 455kHz (T1, T4); M.W. 1680kHz (CT5), 1400kHz (CT3), 520kHz (L5), 620kHz (L4); L.W. 145kHz (CT6), 160kHz (L4), 240kHz (CT4).
F.M. I.F. 10·7MHz (T2, T3); 87·35MHz (L3-1, 2, L1-1, 2), 108·25MHz (CT2, CT1).

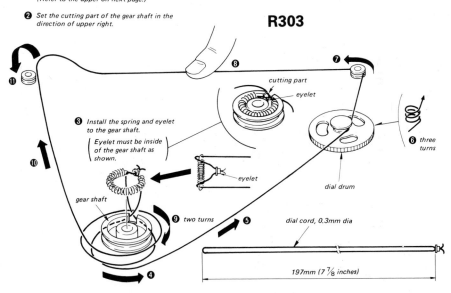

❶ Turn the dial drum gear fully counterclockwise. (Refer to the upper on next page.)

❷ Set the cutting part of the gear shaft in the direction of upper right.

R303

❽   ❼

cutting part

eyelet

❸ Install the spring and eyelet to the gear shaft.

(Eyelet must be inside of the gear shaft as shown.)

eyelet

gear shaft

❾ two turns   ❺

❻ three turns

dial drum

dial cord, 0.3mm dia

❹

197mm (7⅞ inches)

(R303) DRIVE CORD—MODEL ICF-400

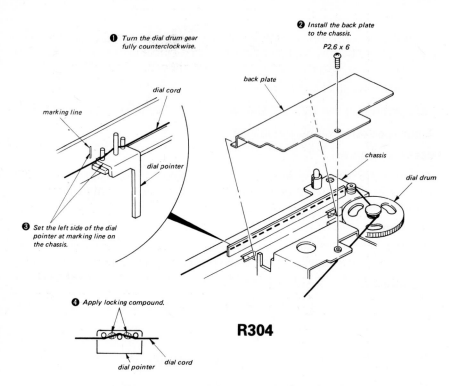

❶ Turn the dial drum gear fully counterclockwise.

❷ Install the back plate to the chassis.

P2.6 x 6

back plate

dial cord

marking line

dial pointer

chassis

dial drum

❸ Set the left side of the dial pointer at marking line on the chassis.

❹ Apply locking compound.

dial pointer   dial cord

R304

(R304) DIAL POINTER SETTING—MODEL ICF-400

672

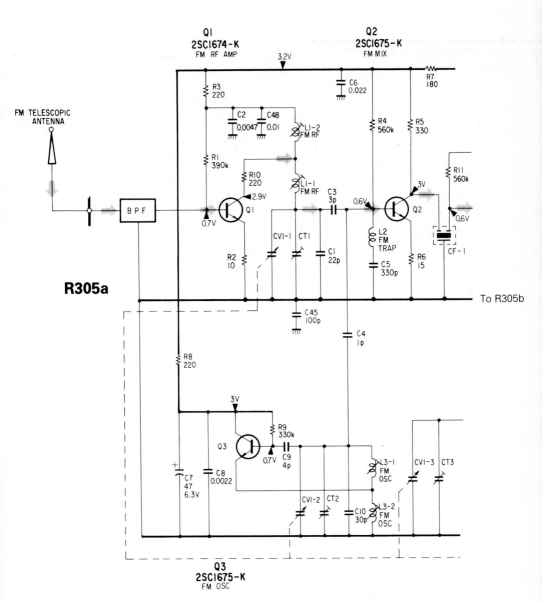

(R305a) CIRCUIT DIAGRAM—MODEL ICF-400 (*PART*)

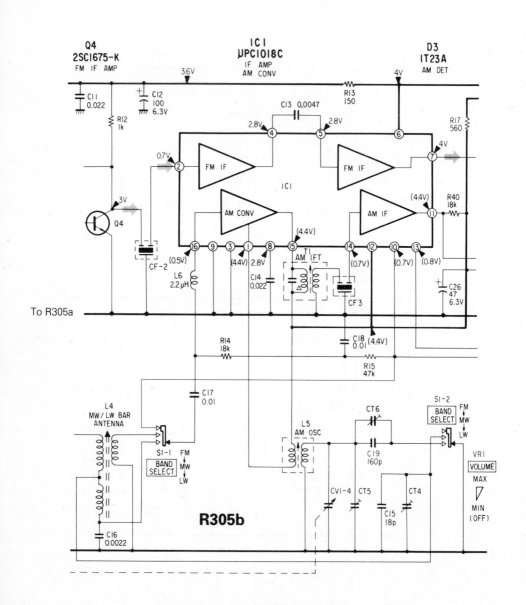

(R305b) CIRCUIT DIAGRAM—MODEL ICF-400 (*PART*)

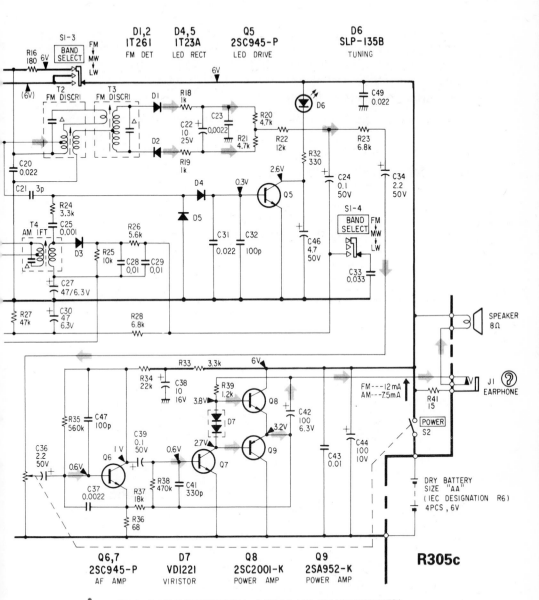

(R305c) CIRCUIT DIAGRAM—MODEL ICF-400 (*CONTINUED*)

R305c

SONY   Model ICF-7600A

General Description: A portable A.M./F.M. radio receiver operating from battery supplies and covering F.M., M.W. and 9 shortwave bands. Sockets are provided for the connection of external supplies, tape recorder and earphones. Integrated circuits are used in the I.F. and A.F. stages.

Batteries: 6 volts (4×1·5 volts).

Wavebands: F.M. 76–108MHz; S.W. 5·95–6·20MHz; S.W. 7·10–7·30MHz; S.W. 9·50–9·80MHz S.W. 11.70–12·00MHz; S.W. 15·10–15·45MHz; S.W. 17·70–17·90MHz; S.W. 21·45–21·75MHz; M.W. 530–1605kHz.

Access for Service (See Fig. R306).

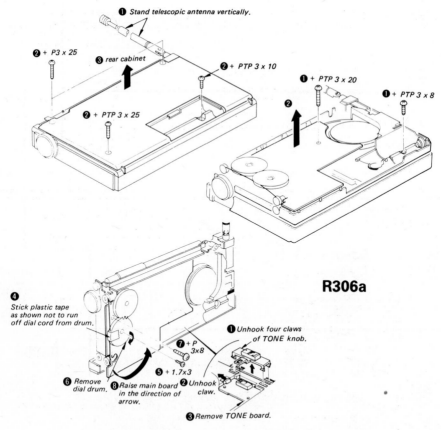

(R306a) DISMANTLING PROCEDURE—MODEL ICF-7600A

676

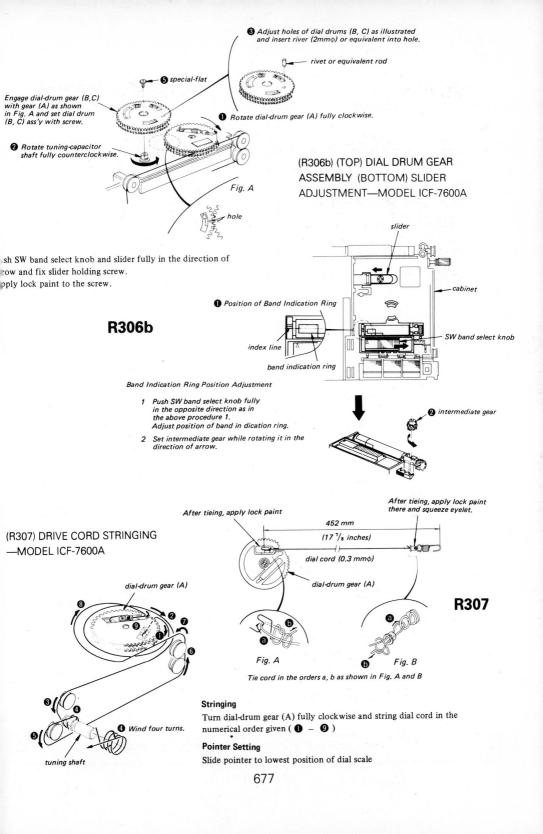

❸ Adjust holes of dial drums (B, C) as illustrated and insert river (2mmφ) or equivalent into hole.

rivet or equivalent rod

❺ special-flat

Engage dial-drum gear (B,C) with gear (A) as shown in Fig. A and set dial drum (B, C) ass'y with screw.

❶ Rotate dial-drum gear (A) fully clockwise.

❷ Rotate tuning-capacitor shaft fully counterclockwise.

Fig. A

hole

(R306b) (TOP) DIAL DRUM GEAR ASSEMBLY (BOTTOM) SLIDER ADJUSTMENT—MODEL ICF-7600A

slider

...sh SW band select knob and slider fully in the direction of ...ow and fix slider holding screw.
...pply lock paint to the screw.

cabinet

R306b

❶ Position of Band Indication Ring

index line

band indication ring

SW band select knob

Band Indication Ring Position Adjustment

1 Push SW band select knob fully in the opposite direction as in the above procedure 1.
   Adjust position of band in dication ring.

2 Set intermediate gear while rotating it in the direction of arrow.

❷ intermediate gear

After tieing, apply lock paint there and squeeze eyelet.

After tieing, apply lock paint

452 mm

(17 ⁷/₈ inches)

dial cord (0.3 mmφ)

(R307) DRIVE CORD STRINGING —MODEL ICF-7600A

dial-drum gear (A)

dial-drum gear (A)

R307

dial-drum gear (A)

❽ ❷ ❼ ❾ ❶ ❻

Fig. A

ⓑ
ⓐ

ⓐ
ⓑ

Fig. B

❸ ❹
❺

❹ Wind four turns.

tuning shaft

Tie cord in the orders a, b as shown in Fig. A and B

Stringing

Turn dial-drum gear (A) fully clockwise and string dial cord in the numerical order given (**❶** − **❾**)

Pointer Setting

Slide pointer to lowest position of dial scale

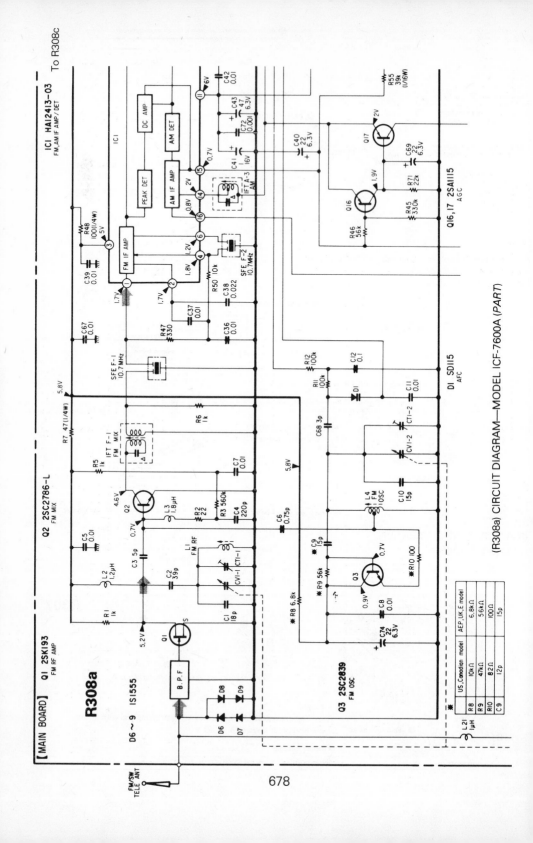

(R308a) CIRCUIT DIAGRAM—MODEL ICF-7600A (PART)

678

R308b

(R308b) CIRCUIT DIAGRAM—MODEL ICF-7600A (PART)

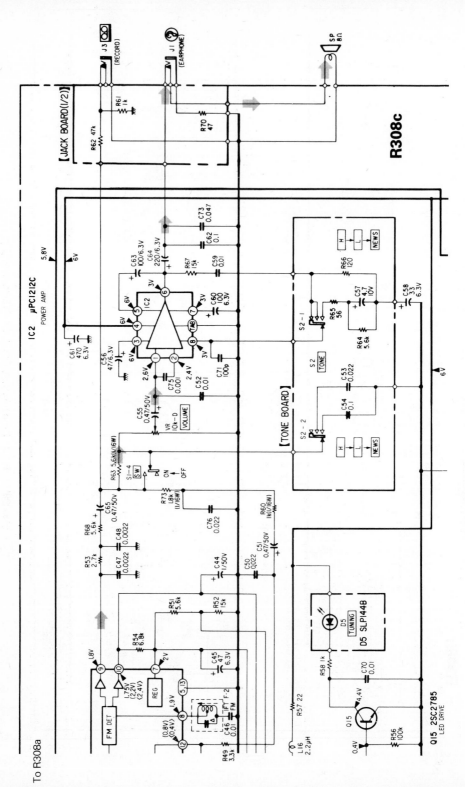

(R308c) CIRCUIT DIAGRAM—MODEL ICF-7600A (*PART*)

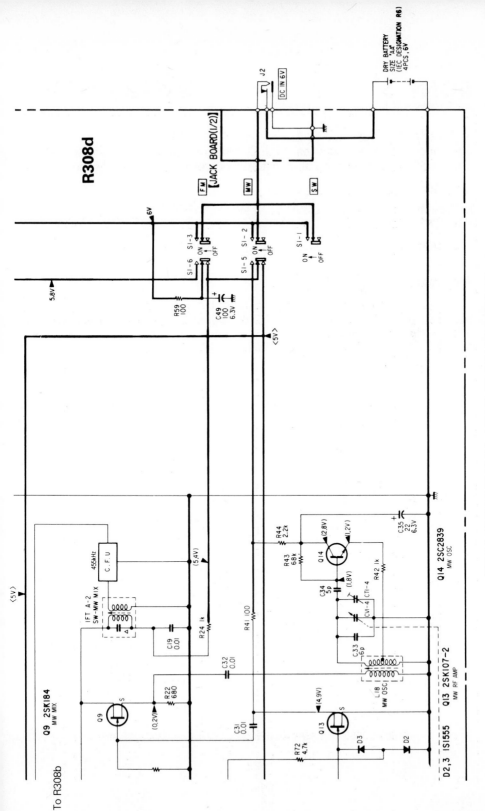

(R308d) CIRCUIT DIAGRAM—MODEL ICF-7600A (CONTINUED)

SONY   Model ICF-900

General Description: A four-waveband A.M./F.M. portable radio receiver operating from battery supplies. A L.E.D. tuning indicator is fitted and sockets provided for the connection of an external power supply and earphones.

Batteries: 4·5 volts (3×1·5 volts).

Wavebands: L.W. 150–280kHz; M.W. 530–1605kHz; S.W. 6–15·8MHz; F.M. 87·5–108MHz.

Loudspeaker: 8 ohms impedance.

Alignment

A.M. I.F. 455kHz (T3); L.W. 145kHz (L11, L7), 300kHz (CT7, CT4); M.W. 520kHz (L10), 620kHz (L6), 1680kHz (CT6), 1400kHz (CT3); S.W. 16MHz (CT8, CT5), 5·8MHz (L12, L8).
F.M. I.F. 10·7MHz (T1, T2); 87·35MHz (L3, L1), 108·25MHz (CT2, CT1).

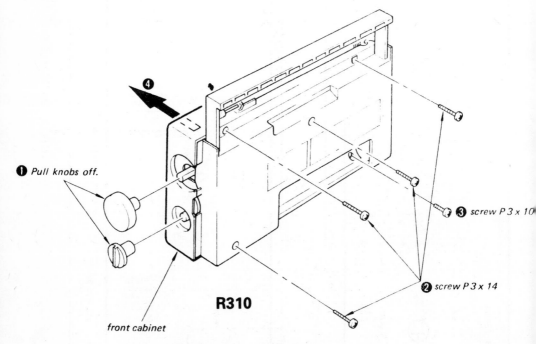

❶ *Pull knobs off.*

❸ *screw P 3 x 10*

❷ *screw P 3 x 14*

R310

front cabinet

(R310) ACCESS FOR SERVICE (FRONT CABINET)—MODEL ICF-900

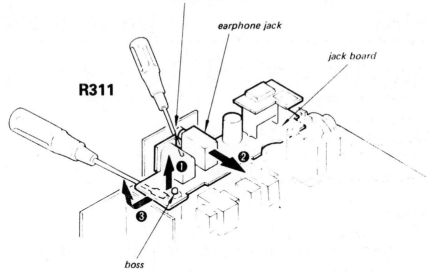

R311

Do not break case.

earphone jack

jack board

boss

(R311) JACK BOARD REMOVAL—MODEL ICF-900

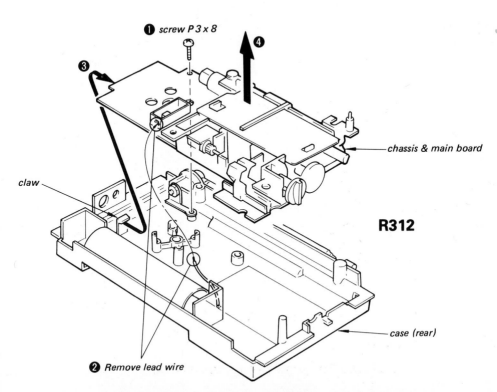

❶ screw P 3 x 8

claw

chassis & main board

R312

case (rear)

❷ Remove lead wire

(R312) CHASSIS AND MAIN BOARD REMOVAL—MODEL ICF-900

❶ *Remove the dial cord.*

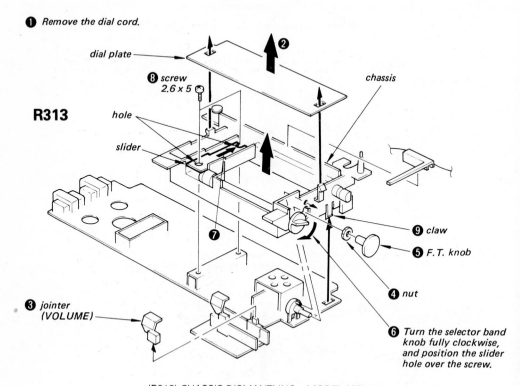

R313

dial plate

❷

chassis

❽ *screw 2.6 x 5*

hole

slider

❼

❾ claw

❺ F.T. knob

❹ nut

❸ *jointer (VOLUME)*

❻ *Turn the selector band knob fully clockwise, and position the slider hole over the screw.*

(R313) CHASSIS DISMANTLING—MODEL ICF-900

Dial Cord Stringing

Eyelet *dial cord 0.5 mm dia.*

314 mm
(12³/₈")

*Perform the dial stringing in the
numerical order given.
Retain the drum while performing
the dial stringing① ~ ⑤ .*

Turn the tuning capacitor shaft fully counterclockwise.

R314

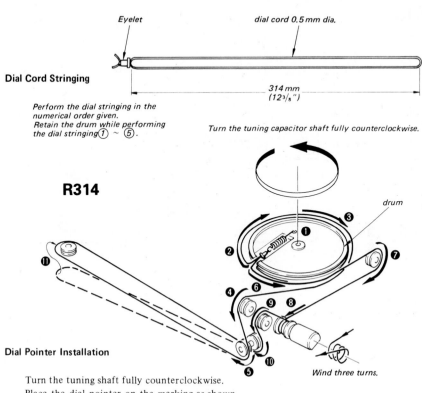

Dial Pointer Installation

Turn the tuning shaft fully counterclockwise.
Place the dial pointer on the marking as shown
below.
Secure the dial pointer to the dial cord with a
suitable locking compound.

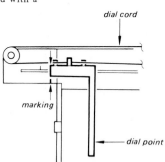

dial cord

marking

dial point

(R314) DRIVE CORD—MODEL ICF-900

685

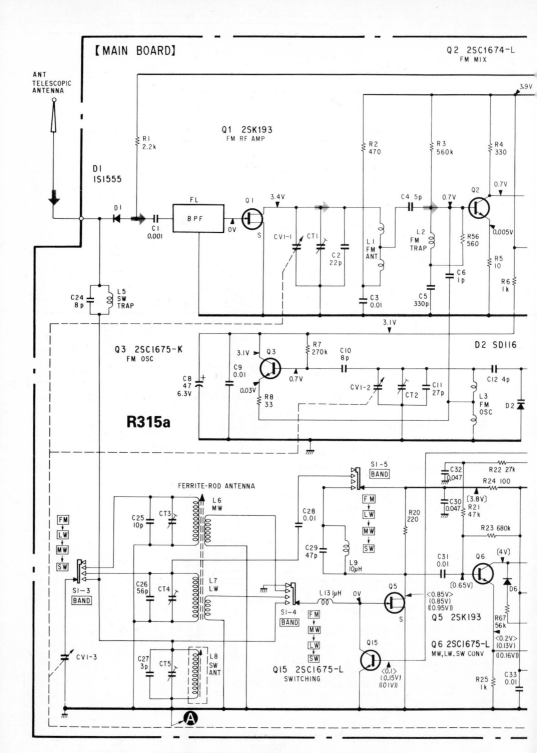

(R315a) CIRCUIT DIAGRAM—MODEL ICF-900 (*PART*)

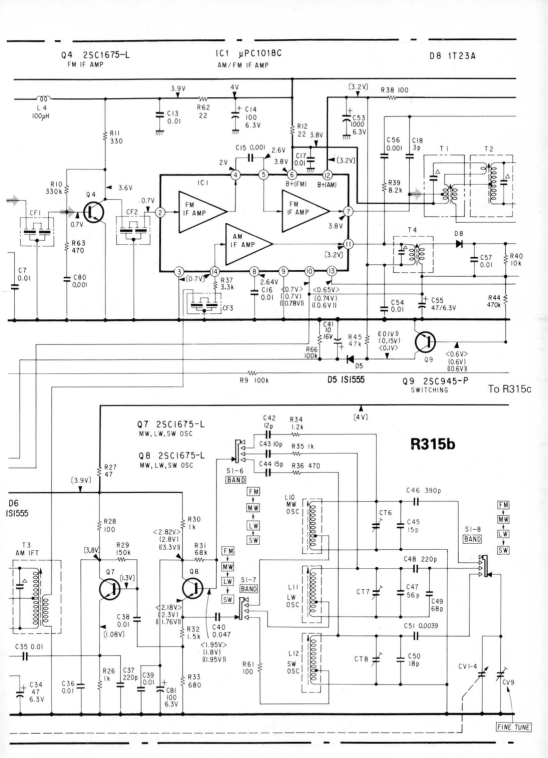

(R315b) CIRCUIT DIAGRAM—MODEL ICF-900 (*PART*)

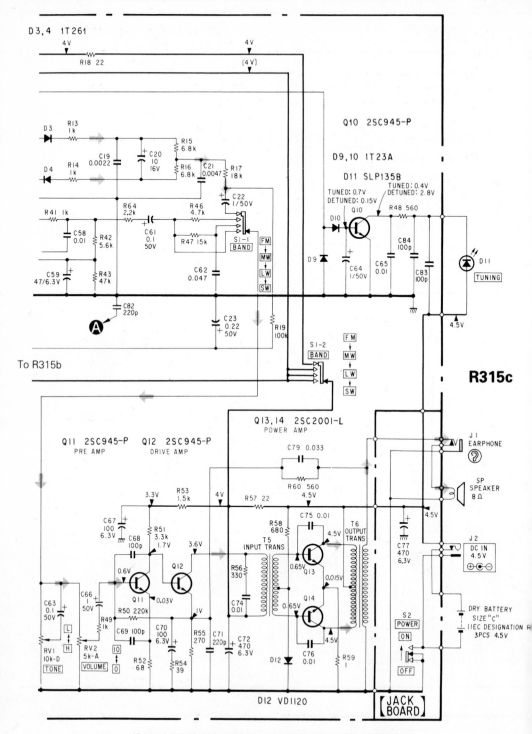

(R315c) CIRCUIT DIAGRAM—MODEL ICF-300 (*CONTINUED*)

SONY   Model ICR-4800

General Description: A six-waveband A.M. portable radio receiver covering medium and 5 shortwave bands. Battery power supplies are used and a socket is provided for the connection of an earphone.

Batteries: 3 volts (2×1·5 volts).

Wavebands: M.W. 530–1605kHz; S.W. 5·95–18MHz in five bands.

Loudspeaker: 8 ohms impedance.

Access for Service

Pull out the telescopic aerial and remove 2 back screws.
Pull off the volume control knob and remove 2 screws.

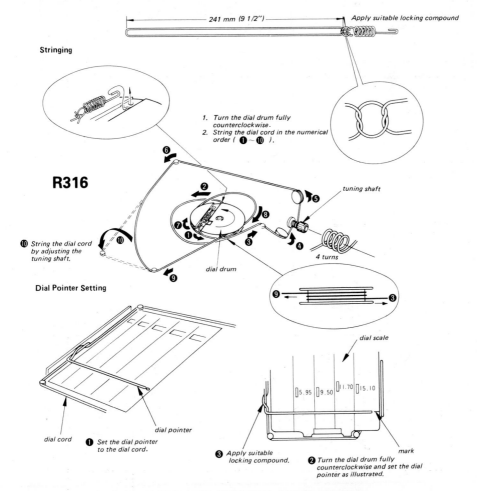

Stringing

241 mm (9 1/2")

Apply suitable locking compound

1. Turn the dial drum fully counterclockwise.
2. String the dial cord in the numerical order (❶ ~ ❿).

R316

⑩ String the dial cord by adjusting the tuning shaft.

tuning shaft

4 turns

dial drum

Dial Pointer Setting

dial scale

5.95   9.50   11.70   15.10

dial cord

dial pointer

mark

❶ Set the dial pointer to the dial cord.

❸ Apply suitable locking compound.

❷ Turn the dial drum fully counterclockwise and set the dial pointer as illustrated.

(R316) DRIVE CORD ASSEMBLY—MODEL ICR-4800

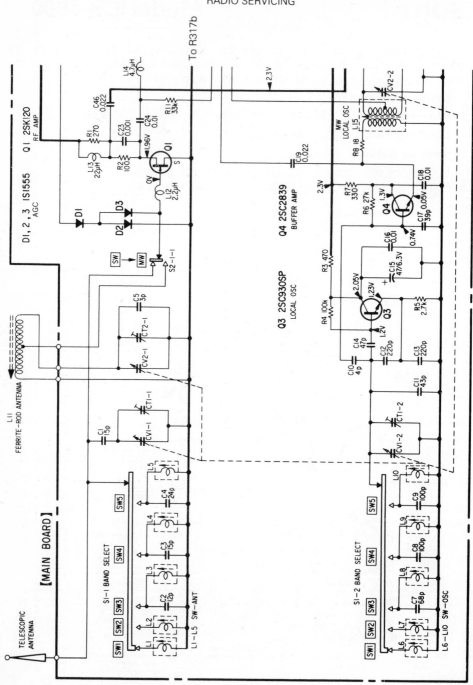

(R317a) CIRCUIT DIAGRAM—MODEL ICR-4800 (PART)

R317a

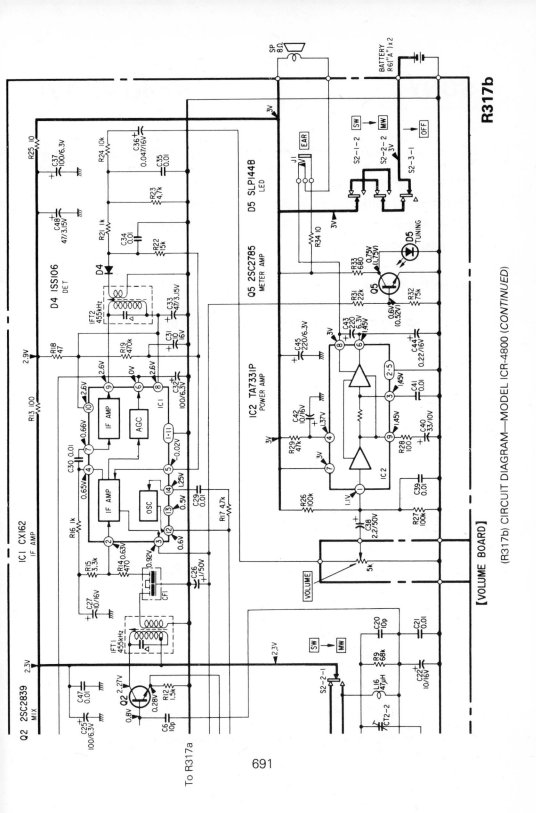

(R317b) CIRCUIT DIAGRAM—MODEL ICR-4800 (CONTINUED)

[VOLUME BOARD]

SONY   Model VX-1W

General Description: A portable two-waveband A.M./F.M. radio receiver operating from battery supplies and incorporating a clock module with voice synthesized time announcing. Sockets are provided for the connection of external supplies and earphone.

Batteries: 3 volts (Radio); 3 volts+1·5 volts (Clock).

Wavebands: M.W. 530–1605kHz; F.M. 87·5–108MHz.

Loudspeaker: 8 ohms impedance.

Access for Service

Remove 5 back screws (1 in battery compartment). The clock module and jack board are held by plastic claws and the chassis by one screw.

Alignment

A.M. R.F. 1680kHz (CT4), 1400kHz (CT3), 520kHz (L4), 620kHz (L1).
F.M. R.F. 87·1MHz (L3, L2), 108·5MHz (CT2, CT1).

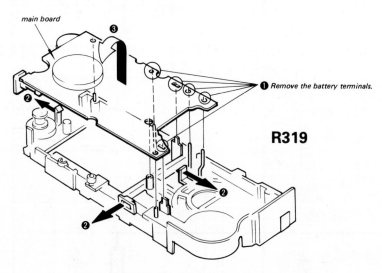

(R319) MAIN BOARD REMOVAL—MODEL VX-1W

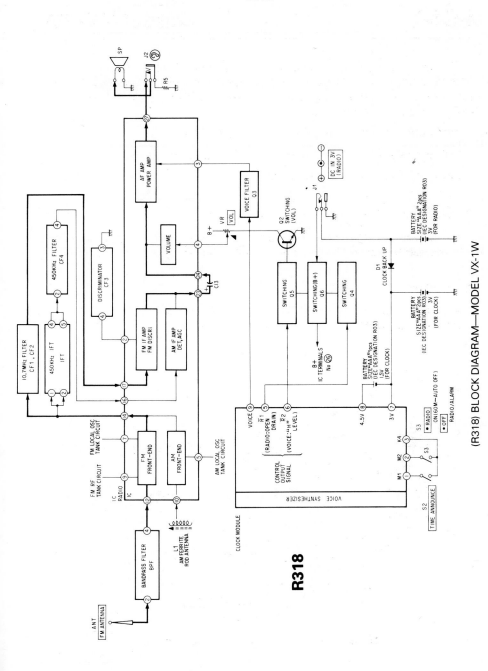

R318

(R318) BLOCK DIAGRAM—MODEL VX-1W

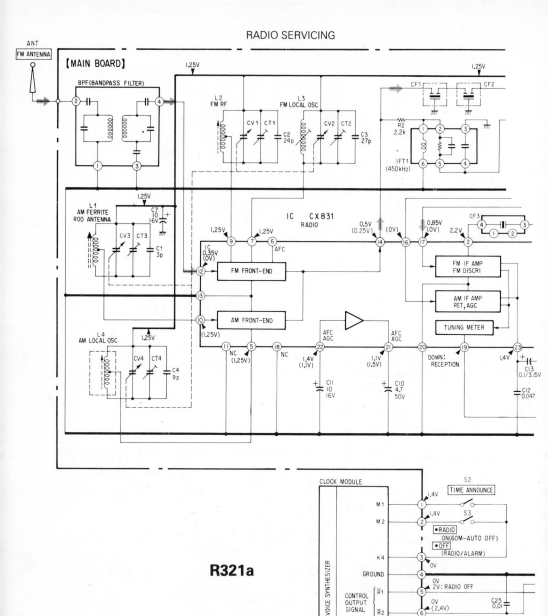

R321a

(R321a) CIRCUIT DIAGRAM—MODEL VX-1W (*PART*)

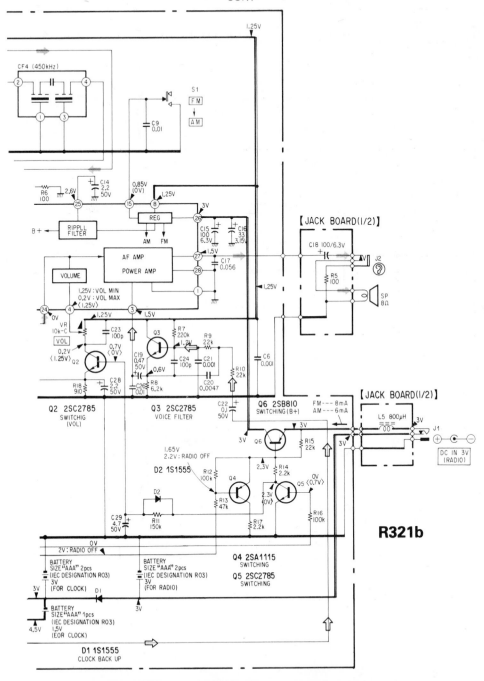

R321b

(R321b) CIRCUIT DIAGRAM—MODEL VX-1W (*CONTINUED*)

Shape the dial cord

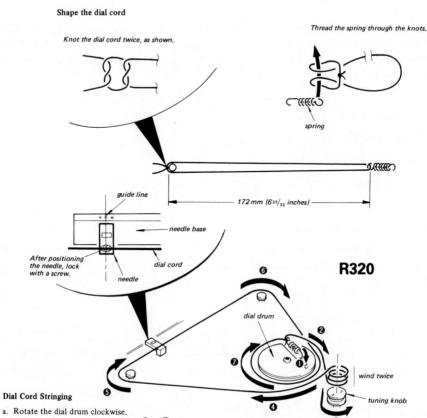

Knot the dial cord twice, as shown.

Thread the spring through the knots.

spring

guide line

172 mm (6²⁵/₃₂ inches)

needle base

After positioning the needle, lock with a screw.

dial cord

needle

dial drum

R320

wind twice

tuning knob

❻ ❷ ❼ ❶ ❺ ❹ ❸

Dial Cord Stringing

a. Rotate the dial drum clockwise.

b. String the dial cord following steps ① ～ ⑦ as shown.

c. Turn the tuning knob to rotate the dial drum fully counter clockwise.

d. Line up the center of the needle with the guide line.

(R320) DRIVE CORD—MODEL VX-1W

TOSHIBA   Model KT-R2

General Description: A personal portable F.M. stereo radio and separate cassette-recorder supplied with stereo headphones. External loudspeakers may also be connected to give 200mW per channel. Integrated circuits are used in the I.F. and decoder stages of the radio and in the equalizing, motor control and A.F. output stages of the recorder.

Microphones are fitted to the case and sockets provided for the connection of auxiliary inputs and power supply.

Batteries: 6 volts (4×1·5 volts).

Waveband: 88–108MHz.

Headphones: 32 ohms impedance per channel.

External Loudspeakers: 8 ohms per channel.

Access for Service (Recorder)

Remove battery cover and remove 6 screws (2 at the back, 2 underneath and 2 at the side). The cover of the tuner unit is held by 6 screws on the back of its case.

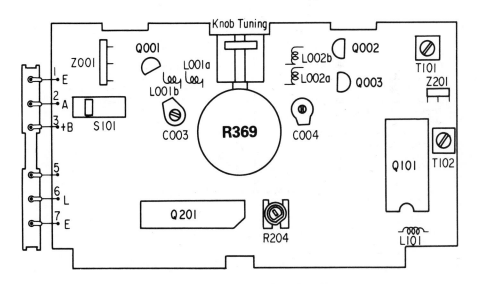

(R369) ALIGNMENT ADJUSTMENTS—MODEL KT-R2

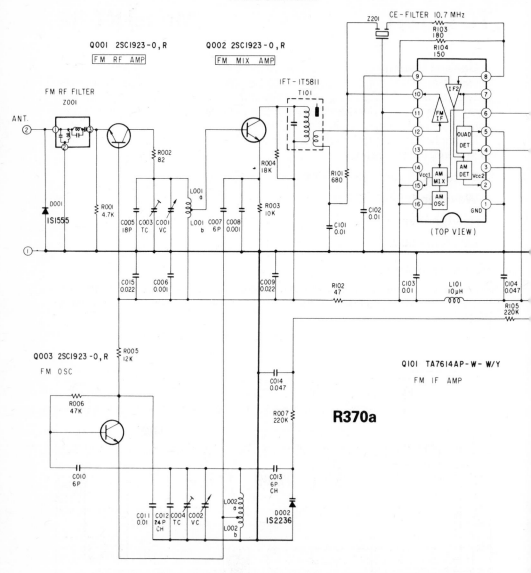

(R370a) CIRCUIT DIAGRAM (F.M. TUNER)—MODEL KT-R2 (*PART*)

F.M.–I.F. Alignment (See Fig. R369)

Connect the R.F. Sweep Signal Output from the signal generator through the loop antenna to the receiver. Connect the oscilloscope vertical input

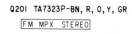

Q201 TA7323P-BN, R, O, Y, GR

FM MPX STEREO

R370b

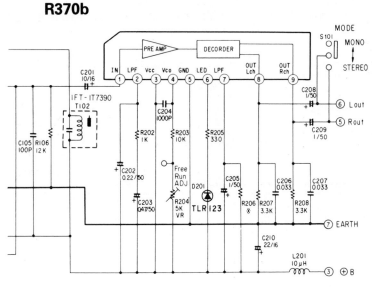

Note

S101 Slide Switch (STEREO/MONO Select) --- MONO Position.

(R370b) CIRCUIT DIAGRAM (F.M. TUNER)—MODEL KT-R2 (*CONTINUED*)

directly to the test point L. or R. and connect the shielded lead to the test point Earth.

Connect the Sweep Voltage Output of the sweep generator to the oscilloscope.

Signal coupling	Equip	Tuning	Connection	Adjust. point	Pattern
Connect sweep generator output to aqlthree-turn loop antenna of 10 cm diameter	Sweep generator of 10·7Mhz centre freq. with 10·7MHz marker	Tuning Knob fully counter-clockwise (Highest frequency)	Set scope for connecting output signal from Tuneout to vertical axis of scope 'V' and sweep generator output to horizontal axis 'H'	L004 T102	Turn the coil T102 fully counter-clockwise to obtain a single peak. Adjust coil T101 in order until the best single peak is obtained. Finally turn the coil T102 to obtain 'S' curve

F.M.–R.F. Alignment

Connect the V.T.V.M. across a 15K ohm dummy load on the R. or L. output terminals (Pins 5 or 6).

Adjust the signal generator fequency as indicated in F.M.–R.F. Alignment Chart, and maintain a sufficient signal output level to provide a measurable indication.

Step	Signal Generator	Radio Dial Setting	Adjustment	Remarks
1	87·5MHz	Tuning knob fully counter-clockwise (Lowest frequency)	Ogs. Coil L002	Adjust for maximum output indication
2	108 MHz	Tuning knob fully clockwise (Highest frequency)	OSC. Trim. C004	Adjust for maximum output indication
3	Repeat steps 1 and 2 as required			
4	90MHz	Tune to signal	R.F. coil L001	Adjust for maximum output indication
5	106MHz		Ant. trim. C003	
6	Repeat steps 4 and 5 as required			

Decoder Frequency Alignment

Adjust R204 under no signal condition so as to obtain 76kHz ±150Hz at the test point.

Semiconductor Voltages

0409 TA7658P

Pin	1	2	3	4	5	6	7 Play	7 Rec	7 Editor	8	9	10	11	12	13	14
Voltage	5·25V	1·4V	2·1V	0V	1·2	0V	0V	0·7–1·0V	1·35V	0V	0V	1·2V	0V	2·1V	1·4V	0V

Q303, 304 TA7313AP

Pin	1	2	3	4	5	6	7	8	9
Voltage	0·73V	2·9V	2·7V	3·0V	0V	3·0V	6·0V	5·7V	2·75V

Q501 TA7522S

Pin	1	2	3	4	5	6	7	8	9
Voltage	0·03V	0·63V	5·96V	0·03V	0V	0·5–1·4V	0·6–0·85V	0·59–0·88V	6·3V

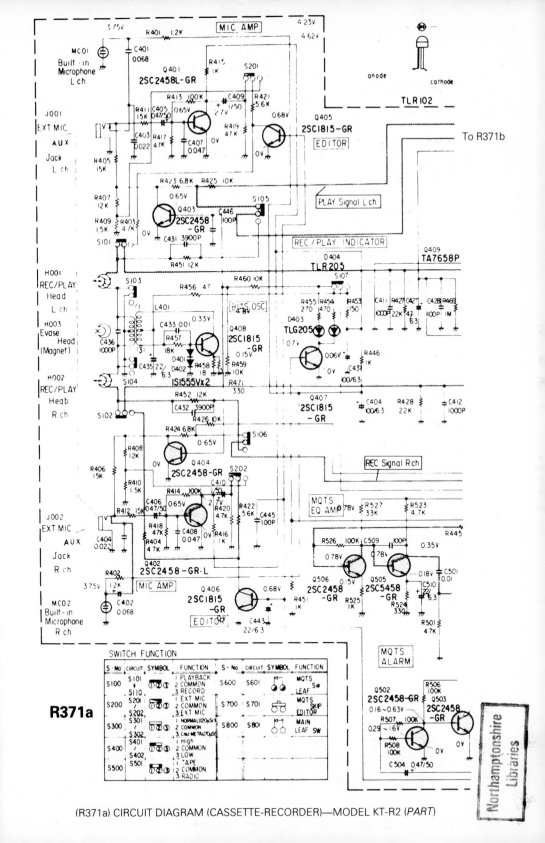

(R371a) CIRCUIT DIAGRAM (CASSETTE-RECORDER)—MODEL KT-R2 (*PART*)

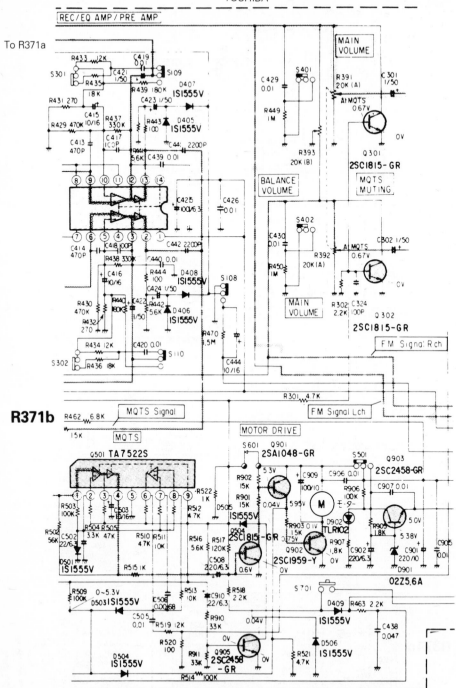

(R371b) CIRCUIT DIAGRAM (CASSETTE-RECORDER)—MODEL KT-R2 (*PART*)

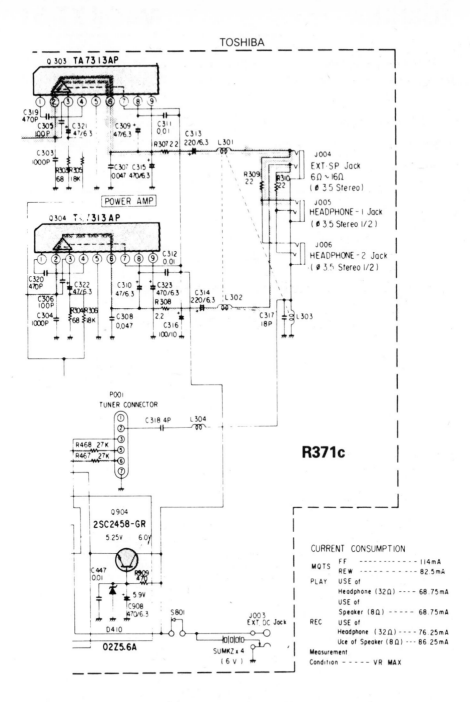

CURRENT CONSUMPTION

MQTS	FF	------------ 114mA
	REW	---------- 82.5mA
PLAY	USE of Headphone (32Ω)	---- 68.75mA
	USE of Speaker (8Ω)	----- 68.75mA
REC	USE of Headphone (32Ω)	---- 76.25mA
	Uce of Speaker (8Ω)	--- 86.25mA

Measurement
Condition ----- VR MAX

(R371c) CIRCUIT DIAGRAM (CASSETTE-RECORDER)—MODEL KT-R2 (*CONTINUED*)

TOSHIBA | Model KT-S1

General Description: A personal portable stereo F.M. radio and cassette tape-player unit operating from battery supplies with output power sufficient for headphone listening. The radio section of this model is that used in model KT-R2 and is described elsewhere in this volume. Signal processing in the tape-player is by integrated circuit.

Batteries: 6 volts (4×1·5 volts.

Headphones: 32 ohms impedance.

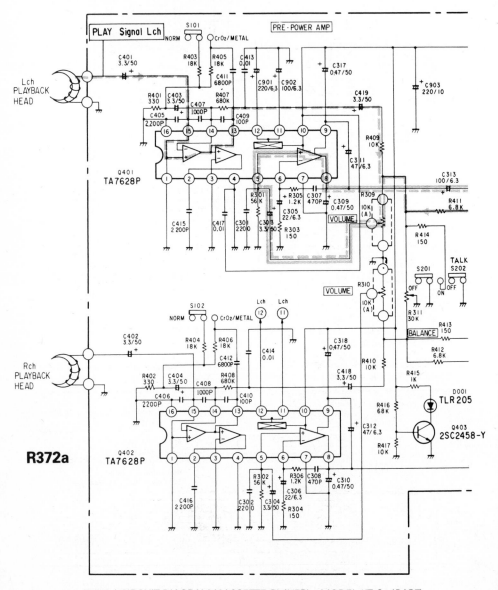

(R372a) CIRCUIT DIAGRAM (CASSETTE-PLAYER)—MODEL KT-S1 (*PART*)

Access for Service

Remove 4 screws (2 on the back, 1 on the bottom and 1 on the side) to free the cabinet back. 3 screws hold the P.C.B. and mechanism to the cabinet front. Open the cassette cover and remove 2 screws holding the cover and connector.

Integrated Circuit Voltages

Q401.402

Pins	1	2	3	4	5	6	7	8	9	10	11	12	13	14	15	16
Voltage	0V	5·0V	0V	0V	0V	0·6V	0V	30V	5·8V	6·0V	6·9V	4·8V	1·6V	1·6V	1·1V	0V

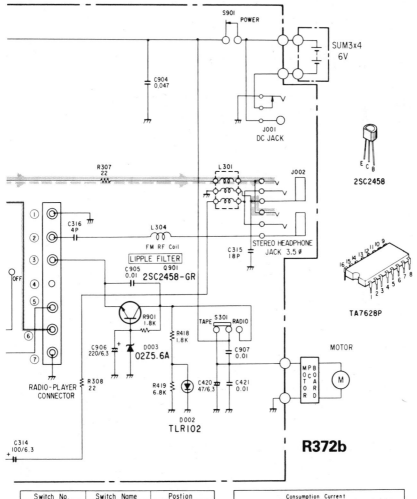

R372b

Switch No.	Switch Name	Postion		
S101 ~ S102	TAPE SELECT	NORM		CrO2/MEAL
S201 ~ S202	TALK	OFF		ON
S301	TAPE – RADIO SELECT	OFF		ON
S901	POWER			

Consumption Current		
Operation postion	Current	Measurement Condition
F . F	About 170 mA	• Used Tape C-60
R E W	About 120 mA	• Tape at near the Center
P L A Y	About 110 mA	• VR Minium
At time tuning	About 90 mA	V R Minium

(R372b) CIRCUIT DIAGRAM (CASSETTE-PLAYER)—MODEL KT-S1 (*CONTINUED*)

TOSHIBA   Model KT-S2

General Description: A personal portable stereo F.M. radio and cassette tape-player unit operating from battery power supplies with sufficient power to drive stereo headphones. The radio section of this model is that used in Model KT-R2 and is described elsewhere in this volume. Signal processing in the tape-player is by integrated circuit.

Batteries: 6 volts (4×1·5 volts).

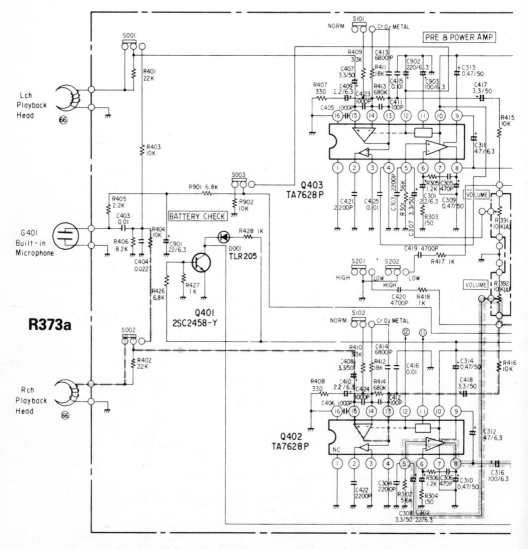

(R373a) CIRCUIT DIAGRAM (TAPE-PLAYER)—MODEL KT-S2 (*PART*)

706

Access for Service

Remove the battery cover and take out three screws (1 in battery compartment, 1 in the back and 1 in the base of the case).

Transistor Voltages

Q402, Q403

Pins	1	2	3	4	5	6	7	8	9	10	11	12	13	14	15	16
Voltage	0V	5·0V	0V	0V	0V	0·6V	0V	3·0V	5·8V	6·0V	6·9V	4·8V	1·6V	1·4V	1·1V	0V

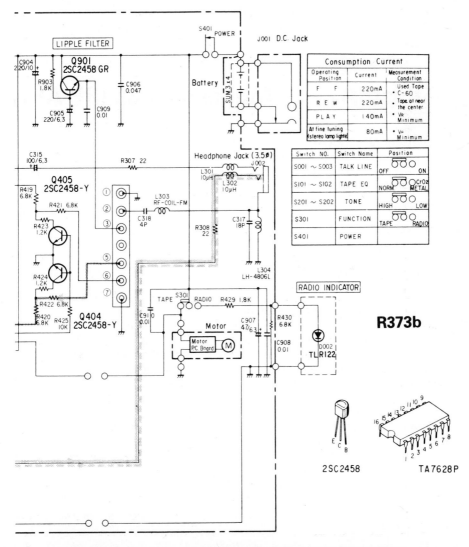

(R373b) CIRCUIT DIAGRAM (TAPE-PLAYER)—MODEL KT-S2 (*CONTINUED*)

TOSHIBA   Model KT-S3

General Description: A personal portable stereo radio and cassette-recorder. The radio section of this model is the same as that used in Model KT-R2 and the cassette unit is similar to that found in Model KT-S1. These units are described elsewhere in these pages.

TOSHIBA   Model KT-VS1 (RP-AF1)

General Description: A personal portable A.M./F.M. radio tuner and cassette-recorder for personal earphone listening. Integrated circuits are widely used in both tuner and cassette circuits. The tuner pack in this model is designated RP-AF1.

Batteries: 3 volts (2×1·5 volts).

Wavebands: M.W. 525–1605kHz; F.M. 88–108MHz.

Headphones: 32 ohms impedance.

Access for Service

Remove two screws from each side of the front case of the cassette unit and four screws retaining the cabinet back.

(Radio Section) Remove three screws from the case bottom to take off the cabinet top. When servicing the tuner, remove two hinge screws from the front of the cassette case and remove the front. Load the tuner into the cassette holder. Set the cassette player and tuner in the 'Play' mode. Pack tuner with insulated strip into holder and lift outwards.

Alignment (See Fig. R358)

A.M.–I.F. Alignment: Turn on both sweep generator and oscilloscope, and allow a fifteen-minute warm-up period.

Connect the R.F. Sweep Signal Output from the signal generator through the loop antenna to the receiver.

Connect the oscilloscope vertical input directly to the test point L or R and connect the shielded lead to the test point Earth.

Connect the Sweep Voltage Output of the sweep generator to the oscilloscope.

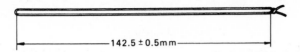

—142.5 ± 0.5mm—

Turn the tuning knob counterclockwise fully (to the direction of lower frequency).
Wind the dial cord in numerical order.
Fix the dial pointer on the cord so as to fit the pointer margin to the marking line on the mould frame.

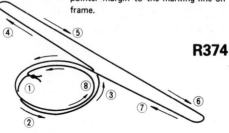

R374

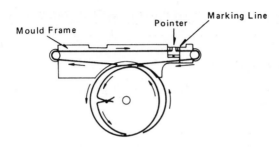

Marking Line

Pointer

Mould Frame

(R374) DRIVE CORD—MODEL KT-VS1 (RP-AF1)

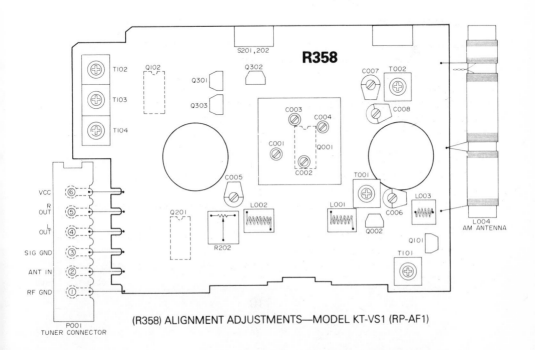

R358

(R358) ALIGNMENT ADJUSTMENTS—MODEL KT-VS1 (RP-AF1)

A.M.–I.F. Alignment Chart:

Step	Signal coupling	Equip.	Tuning	Connection	Adjust. point	Pattern
1	Connect sweep generator output to a loop antenna	Sweep generator of 455kHz centre freq. with 455kHz marker. (YY . . . 460kHz)	Tuning knob fully counter-clockwise (Highest frequency)	Set scope for connecting output signal from Tune Out to vertical axis of scope and sweep generator output to horizontal axis	T102 T104	Adjust coil T102 and T104 until the best single peak is obtained

A.M. Alignment: Connect the V.T.V.M. across a 15K ohm dummy load. Adjust the signal generator frequency as indicated in F.M.–R.F. Alignment Chart, and maintain a sufficient signal output level to provide a measurable indication.

A.M. Alignment Chart:

Step	Signal generator	Radio dial setting	Adjustment	Remarks
1	520kHz	Tuning knob fully counter-clockwise (Lowest frequency)	Osc. Coil T102	Adjust for maximum output indication
2	1650kHz	Tuning knob fully clockwise (Highest frequency)	Osc. Trim. C007	Adjust for maximum output indication
3	Repeat steps 1 and 2 as required			
4	600kHz	Tune to signal	R.F. Coil L004	Adjust. for maximum
5	1400kHz		Ant. trim C008	Output indication
6	Repeat steps 4 and 5 as required			

F.M.–I.F. Alignment Chart:

Step	Signal coupling	Equip.	Tuning	Connection	Adjust. point	Pattern
1	Connect sweep generator output to a three-turn loop antenna of 10 cm diameter	Sweep generator of 10·7MHz centre freq. with 10·7 MHz marker	Turning knob fully counter-clockwise (Highest frequency)	Set scope for connecting output signal from Tune Out to vertical axis of scope and sweep generator output to horizontal axis	T101 T103	Turn the coil T103 fully counter-clockwise to obtain a single peak. Adjust coil T101 in order until the best single peak is obtained. Finally turn the coil T103 to obtain 'S' Curve

F.M.–R.F. Alignment Chart:

Step	Signal generator	Radio dial setting	Adjustment	Remarks
1	87·5MHz	Tuning knob fully counter-clockwise (Lowest frequency)	Osc. Coil L002	Adjust for maximum output indication
2	108MHz	Tuning knob fully clockwise (Highest frequency)	Osc. Trim. C005	Adjust for maximum output indication
3	Repeat steps 1 and 2 as required			
4	90MHz	Tune to signal	R.F. Coil L001	Adjust for maximum
5	106MHz		Ant. trim. C006	output indication
6	Repeat steps 4 and 5 as required			

Stereo Decoder: Adjust R202 to obtain a frequency of 76 ±1kHz at Pin 3 of Q201.

S.C. Voltages

Q001 (TA7335F)

	1	2	3	4	5	6	7	8
F.M.	0V	2·3V	0V	3·0V	3·0V	3·0V	3·0V	3·0V
A.M.	0V	0V	0V	0·15V	0·15V	0·15V	0·15V	0·15V
	9	10	11	12	13	14	15	16
F.M.	0V	3·0V	0V	2·8V	0·4V	0V	1·4V	0V
A.M.	0V	0·15V	0V	0·15V	0·4V	0V	1·35V	0V

Q102 (TA7687F)

	1	2	3	4	5	6	7	8
F.M.	0·05V	0·05V	3·0V	1·4V	1·4V	0·1V	1·75V	0V
A.M.	0·95V	0·95V	0·2V	1·4V	1·4V	0V	2·8V	0V
	9	10	11	12	13	14	15	16
F.M.	0·95V	3·0V	3·0V	3·0V	2·8V	2·8V	2·7V	3·0V
A.M.	1·2V	3·0V	3·0V	3·0V	2·65V	2·75V	2·8V	3·0V

Q002

	F.M.	A.M.
D	3·0V	3·0V
G	0V	0V
S	0·3V	0·2V

Q101

	F.M.	A.M.
E	0·8V	0·1V
C	2·9V	0·1V
B	1·4V	0·7V

Q201 (TA7342F)

	1	2	3	4	5	6	7	8
F.M.	3·0V	3·0V	2·6V	2·6V	0V	0V	1·9V	2·2V
A.M.	3·0V	3·0V	2·4V	2·4V	0V	0V	2·25V	3·0V
	9	10	11	12	13	14	15	16
F.M.	0·9V	0·9V	0·9V	0·9V	0·9V	0·6V	0·6V	2·2V
A.M.	0·9V	0·9V	0·9V	0·9V	0·9V	0·6V	0·6V	2·2V

Q301

	F.M.	A.M.
E	0V	2·2V
C	0V	2·2V
B	0·7V	2·7V

Q302

	F.M.	A.M.
E	0V	0V
C	0·05V	2·2V
B	0·5V	0·15V

Q303

	F.M.	A.M.
E	0V	0V
C	0V	0V
B	0·3V	0·65V

Q304

1	2	3	4	5	6	7	8
1·35V	1·4V	1·4V	1·4V	—	2·2V	1·45V	0V
9	10	11	12	13	14	15	16
3·0V	1·45V	2·25V	—	3·0V	1·4V	1·4V	1·35V

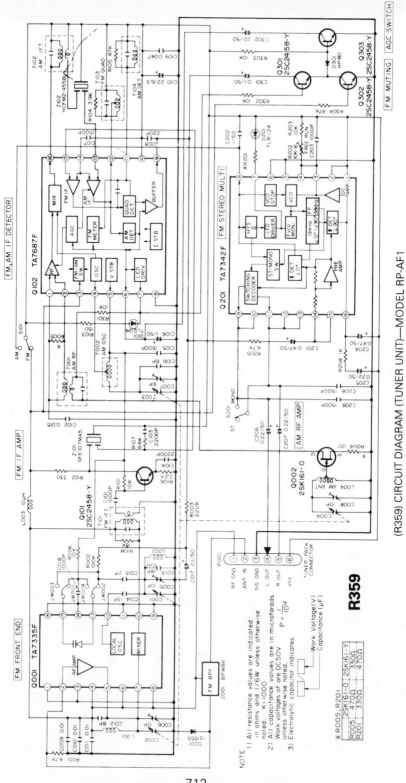

(R359) CIRCUIT DIAGRAM (TUNER UNIT)—MODEL RP-AF1

R359

NOTE
1) All resistance values are indicated in ohms and 1/6W unless otherwise noted. K=1000
2) All capacitance values are in microfarads. Work voltages of are DC50V unless otherwise noted. $P = \frac{1}{10^2}$
3) Electrolytic capacitor indicates.

Work Voltage(V)
Capacitance (µF)

※ R005, R201	25K161-O	25K161-Y
R005	470Ω	330Ω
R201	330Ω	470Ω

712

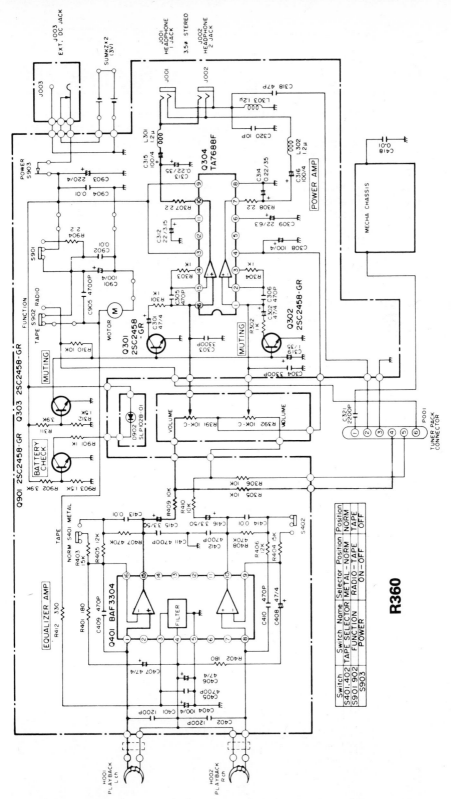

(R360) CIRCUIT DIAGRAM (CASSETTE-PLAYER)—MODEL RP-AF1

R360

Switch	Switch Name	Selector Position	Position	Position
S401.402	TAPE SELECTOR	METAL – NORM	NORM	TAPE
S901.902	FUNCTION	RADIO – TAPE	TAPE	OFF
S903	POWER	ON – OFF		

VEGA   Model 'Spidola' 250

General Description: A multiband A.M./F.M. battery-operated portable receiver covering long, medium and six shortwave bands on A.M. and the V.H.F. broadcast band on F.M.

An internal ferrite aerial is used for long and medium, and an external telescopic aerial for shortwave A.M. and V.H.F./F.M. reception. An input socket for an external A.M. aerial is provided.

Modular construction is used; the units comprise an independent V.H.F. tuner, combined A.M./F.M. I.F. and A.F. with power output, an eight-coilstrip turret for A.M. reception, the ferrite aerial, and a 4-push-button switch module. Changeover from A.M. to F.M. reception is effected by a sliding switch in the I.F./A.F. unit, cam-operated from the turret mechanism.

Two scales are fitted, one in the horizontal and one in the vertical plane. The main horizontal scale indicates wavebands and wavelengths, a vertical logging scale is visible from the cabinet front. Both use a common pointer.

Four pushbutton switches are for on/off, dial lamp switching, bass control, and A.F.C. on F.M.

Sockets at the receiver rear allow for connection of a tape recorder, headphones or external loudspeaker, external A.M. aerial and a 9V external D.C. supply.

Batteries: 9 volts (6×SP2).   **Consumption (no signal):** 25mA (approx.)

Wavebands: L.W. 150–408kHz; M.W. 525–1605kHz; S.W. 2·0–5·0MHz; 5·0–7·4MHz; 9·5–9·77MHz; 11·7–12·1MHz; 15·1–17·9MHz; 21·45–21·75MHz; F.M. 87·5–108MHz.

Loudspeaker: 8 ohms impedance.

Transistor Voltages

(Measured with 20 kilohm/volt meter, with receiver switched to V.H.F. All readings are negative with respect to chassis.)

Unit	Transistor	Base	Emitter	Collector
V.H.F. tuner	Tr1	—	1·75V	4·0V
	Tr2	—	2·55V	3·95
I.F./A.F.	Tr1	1·2V	0·92V	2·8V
	Tr2	2·6V	0·38V	1·7V
	Tr3	0·48V	0·35V	1·3V
	Tr4	1·3V	1·1V	3·7V
	Tr5	0·91V	0·67V	—
	Tr6	0·26V	0·12V	2·6V
	Tr7	2·6V	2·4V	8·5V
	Tr8	2·1V	0·4 to 1·6V*	—
	Tr9	0·14V	—	9·0V
	Tr10	0·14V	—	9·0V
	Tr11	0·75V	—	8·3V
	Tr12	8·2V	8·4V	—
	Tr13	2·7V	2·6V	8·2V

*According to signal input, for 0·7V max. A.F. output.

714

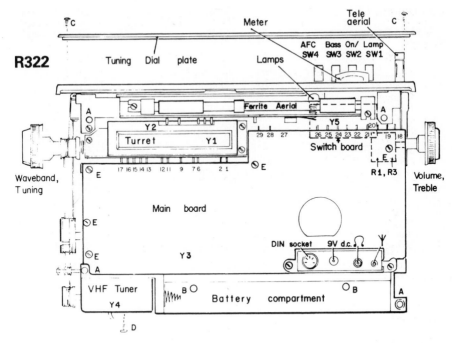

R322

Tuning Dial plate

Meter

Tele aerial

AFC Bass On/ Lamp
SW4 SW3 SW2 SW1

Lamps

Ferrite Aerial

Y2

Turret   Y1

Y5

Switch board

R1, R3

Waveband,
Tuning

17 16 15 14 13 12 11 9 7 6   2 1

E

Main   board

Volume,
Treble

Y3

DIN socket   9V d.c.

VHF Tuner
Y4

Battery   compartment

D

(R322) DISMANTLING PROCEDURE—MODEL 'SPIDOLA' 250

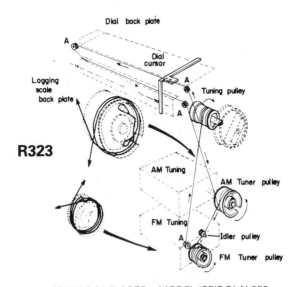

Dial back plate

Dial
cursor

Logging
scale
back plate

Tuning pulley

R323

AM Tuning

AM Tuner pulley

FM Tuning

Idler pulley

FM Tuner pulley

(R323) DRIVE CORD—MODEL 'SPIDOLA' 250

Component List

V.H.F. Tuner:

Resistors

R1	1k ohms
R2	43k ohms
R3	3·9k ohms
R4	30 ohms
R5	3k ohms
R6	15k ohms
R7	6·2k ohms
R8	180 ohms
R9	5·1k ohms
R10	56k ohms
R11	2·7k ohms
R12	12 ohms

Capacitors

C1	6·2pF
C2	30pF
C3	0·01µF
C4	10pF
C5	0·01µF
C7	2·2−16pF
C8	6·2pF
C9	18pF
C10	510pF
C11	0·047µF
C12	0·01µF
C13	4·3pF
C14	68pF
C15	0·01µF
C16	47pF
C17	2·2−16pF
C18	36pF
C19	6·2pF
C20	0·01µF

Transistors

Tr1	TT313B
Tr2	TT313A

Diodes

D1	D20
D2	D902
D3	D104

I.F./A.F. Board Y3

Resistors

R1	6·8k ohms
R2	560 ohms
R3	560 ohms
R4	100 ohms
R5	100 ohms
R6	6·8k ohms
R7	3·9k ohms
R8	180 ohms
R9	1k ohms
R10	1·2k ohms
R11	22k ohms
R12	33k ohms
R13	330 ohms
R14	18k ohms

R15	3·9k ohms
R16	10k ohms
R17	560 ohms
R18	5·1k ohms
R19	10k ohms
R20	330 ohms
R21	15k ohms
R22	3·3k ohms
R23	820 ohms
R24	1k ohms
R25	2·7k ohms
R26	390 ohms
R27	47k ohms
R28	270 ohms
R29	1·2k ohms
R31	1·5k ohms
R32	10k ohms
R33	8·2k ohms
R34	180 ohms
R35	47k ohms
R36	68 ohms
R37	47 ohms
R38	120k ohms
R39	560 ohms
R40	1·2k ohms
R41	8·2k ohms
R42	36 ohms
R43	180 ohms
R44	22k ohms
R45	1k ohms
R46	27 ohms
R47	1 ohms
R48	4·7k ohms
R49	10k ohms
R50	270 ohms
R51	27k ohms
R52	47k ohms
R53	390 ohms
R54	6·8k ohms
R55	100 ohms
R56	1k ohms
R57	5·6k ohms
R58	1·8k ohms
R59	2k ohms
R61	3·3k ohms
R62	6·8k ohms
R63	6·8k ohms
R64	15k ohms
R65	15k ohms
R66	5·1k ohms
R67	15k ohms
R69	47k ohms
VR1	47k ohms
VR3	10k ohms*

Capacitors

C1	300pF
C2	1000pF
C3	6·2pF
C4	330pF
C5	120pF
C6	10−430pF
C7	130pF

C8	10−430pF
C9	0·033µF
C10	0·033µF
C11	0·01µF
C13	68pF
C14	5µF
C15	270pF
C16	0·047µF
C17	6800pF
C18	50µF
C19	0·033µF
C20	0·05µF
C21	3300pF
C22	0·1µF
C23	100µF
C24	62pF
C26	0·01µF
C27	620pF
C28	0·033µF
C29	0·05µF
C30	1µF
C31	22pF
C32	9·1pF
C33	20µF
C34	47pF
C35	0·047µF
C36	3·9pF
C37	510pF
C38	50µF
C40	510pF
C41	20µF
C42	6·2pF
C43	560pF
C44	0·01µF
C45	100µF
C46	20µF
C47	0·033µF
C48	50µF
C49	62pF
C50	0·01µF
C51	0·01µF
C52	20µF
C53	1000pF
C54	6800pF
C55	0·033µF
C56	33pF
C57	39pF
C58	68pF
C59	10µF
C60	390pF
C61	180pF
C62	0·033µF
C63	62pF
C64	1000pF
C65	10pF
C67	47pF
C68	3300pF
C69	270pF
C70	270pF
C71	0·05µF
C72	10µF
C73	270pF
C74	5µF

C75	3300pF	Tr4	MN41	**Diodes**	
C76	20µF	Tr5	TT322A	D3	D20
C77	0·047µF	Tr6	MNA1	D4	D9B
C78	500µF	Tr7	MN41	D5	7TE2A
C80	8·2pF	Tr8	TT322B	D6	D20
		Tr9	TT402E	D7	D20
Transistors		Tr10	TT402E		
Tr1	TT322B	Tr11	TT322A	* variable	
Tr2	TT322A	Tr12	MN37		
Tr3	MN41	Tr13	MN40		

Turret Coil Unit Y1 Components:

Band	R1	C1	C2	C3	C4	C5
M.W.	1k ohms	4·15pF	5·20pF	360pF	—	—
L.W.	1·8k ohms	10pF	5·20pF	220pF	91pF	75pF
60–150m	—	15pF	5·20pF	22pF	1800pF	—
41–60m	—	240pF	62pF	5·20pF	100pF	—
31m	—	—	—	—	—	—
25m	—	—	—	—	—	—
16–19m	—	150pF	150pF	—	—	—
13m	—	75pF	—	—	—	—

Alignment

Equipment required: A.M. signal generator covering 150kHz to 22MHz, modulated 400Hz at 30 per cent. F.M. signal generator, covering 10·7MHz, 97 to 108MHz, modulated 1kHz, 30 per cent, deviation ±15kHz. V.T.V.M. or suitable output meter. Input matching components as detailed in text.

A.M. I.F. Stages:

Preliminaries: Connect output meter across loudspeaker terminals. Select 'M.W.' and tune receiver to the low frequency end (100 on the logging scale) end of the dial. On the I.F./A.F. board connect short-circuit links across the I.F. filter coil L3 and points 12, 13 (the A.M. oscillator output). Connect the A.M. signal generator, tuned to 465kHz, modulated, to transistor bases as listed below via a diode probe and a 0·05µF capacitor. Turn volume to maximum.

Progressively reduce signal input as circuits come into alignment to avoid A.G.C. action.

Inject signal to Tr5 base, adjust L19, L20, L12, filters L10, L7, L6 and then L11 for maximum output.

Transfer signal input to Tr2 base, remove short-circuit link from filter coil L3, and adjust L3 for maximum.

R.F. Stages: (See P.C.B. and turret coil strip layout diagrams for locations of coils and trimmers.)

Connect A.M. signal generator output via a 0·05µF capacitor to the external aerial socket and earth for the shortwave bands, and inject the signal via an inductive loop to the ferrite aerial for the L.W. and M.W. bands. Check travel of dial cursor in relation to the tuning gang at both ends.

717

M.W. (select 'M.W.'):

1. Tune receiver and generator to 560kHz (84 on logging scale); adjust L1, L2 (on M.W. coilstrip) for maximum.

2. Retune receiver and generator to 1500kHz (8 on logging scale); adjust trimmer C2 for maximum.

3. Repeat steps 1 and 2 for optimum result.

4. Tune receiver to 560kHz; adjust ferrite aerial coils L3, L4 (by sliding these along rod) for maximum.

5. Retune receiver to 1500kHz: adjust aerial trimmer C1 (on coilstrip) for maximum.

6. Repeat steps 4 and 5 for optimum result.

L.W. (select 'L.W.':

7. Tune receiver and generator to 160kHz (10 on logging scale). Adjust L1, L2 (on L.W. coilstrip) for maximum.

8. Retune receiver and generator to 390kHz (75 on logging scale). Adjust trimmer C2 for maximum.

9. Repeat steps 7 and 8 for optimum result.

10. Tune receiver and generator to 160kHz; adjust aerial coils L1, L2 for maximum.

11. Seal ferrite aerial and coil strip cores and trimmers.

S.W. (60m):

12. Tune receiver and generator to 2·1MHz (84 on logging scale). Adjust coils L3, L4 for maximum.

13. Retune receiver and generator to 4·7MHz (10 on logging scale). Adjust trimmer C3 for maximum.

14. Repeat steps 12 and 13 for optimum result.

15. Tune receiver and generator to 2·1MHz; adjust coils L1, L2 for maximum.

S.W. (41m):

16. Tune receiver and generator to 5·1MHz (98 on logging scale). Adjust coils L3, L4 for maximum.

17. Retune receiver and generator to 7·4MHz (8 on logging scale). Adjust trimmer C3 for maximum.

18. Repeat steps 16 and 17 for optimum result.

Tune receiver and generator to 5·1MHz. Adjust coils L1, L2 for maximum.

S.W. (31m): Tune receiver and generator to 9·7MHz (8 on logging scale), adjust first L3, L4 and then L1, L2 for maximum.

S.W. (25m): Tune receiver and generator to 12·0MHz (8 on logging scale). Adjust L3, L4 and then L1, L2 for maximum.

S.W. (16m): Tune receiver and generator to 18·0MHz (4 on logging scale). Adjust L3, L4 and then L1, L2 for maximum.

S.W. (13m): Tune receiver and generator to 21·8MHz (8 logging scale). Adjust L3, L4 and then L2 for maximum.

VEGA

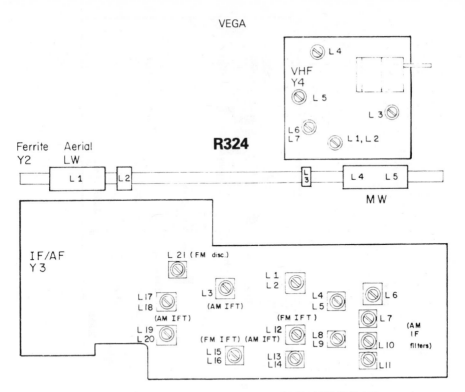

R324

(R324) ALIGNMENT ADJUSTMENTS—MODEL 'SPIDOLA' 250

R325

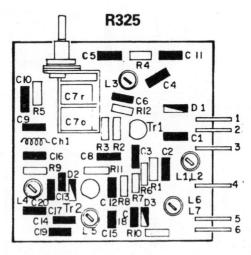

(R325) COMPONENT LAYOUT (F.M. TUNER)—MODEL 'SPIDOLA' 250

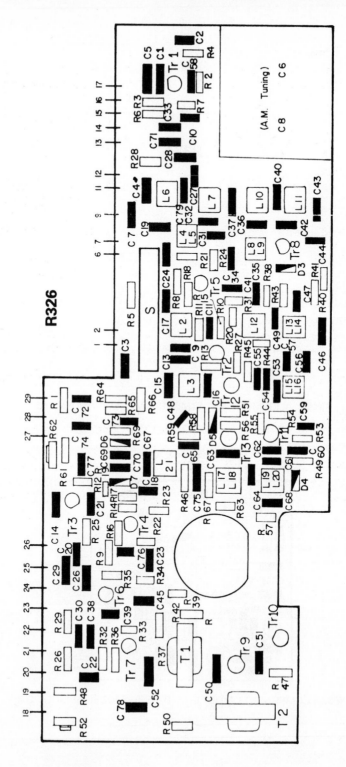

(R326) COMPONENT LAYOUT (MAIN BOARD)—MODEL 'SPIDOLA' 250

F.M. I.F. Stages:

Preliminaries: Select 'F.M.'. Connect output meter across loudspeaker terminals. Tune receiver to low frequency end of dial (100 on logging scale). Check that A.F.C. pushbutton is out (inoperative).

Inject a 10·7MHz signal, levels as given below, from the F.M. generator, modulated 1kHz, deviation 22·5kHz, to the V.H.F. tuner I.F. via a 56pF capacitor to the sleeved test input, and to the F.M. I.F.T.'s on the I.F./A.F. board via a 0·05μF capacitor. The values of signal input given are those for a 0·7V A.F. output.

1. Inject signal, level 2mV, to Tr11 base. Adjust L21, L17 for maximum.

2. Transfer signal input, level 0·4mV, to Tr8 base. Adjust L16, L15, L14 and L13 for maximum.

3. Transfer signal input, level 70μV, to Tr5 base. Adjust L9, L8, L5 and L4 for maximum.

4. Transfer signal input, level 20μV, to Tr2 base. Adjust L2, L1 for maximum.

5. Transfer signal input, level 20μV, to test point on V.H.F. tuner. Adjust L6, L5 in tuner for maximum.

6. Adjust 'S' curve in discriminator L21 output as follows:

(a) Connect a D.C. voltmeter across capacitor C73.

(b) Increase signal input to 200μV and switch off modulation.

(c) Using the meter, adjust L21 until zero point of S-curve occurs at 10·7MHz.

(d) Inject a 30 per cent A.M. signal and adjust pre-set R62 for minimum reading (A.M. rejection).

R.F. Stages: Inject signal to telescopic aerial input of V.H.F. tuner.

Tune receiver to 3·2m (75 on logging scale) and generator to 94MHz (with 22·5kHz deviation). Adjust L4, L3 in V.H.F. tuner for maximum.

A.F.C. Check: Detune receiver from 3·2m, and with 94MHz generator input increased to 150μV, until output meter reading drops from 0·7 to 0·35V.

Press down A.F.C. button; output meter reading must not increase above 0·6V.

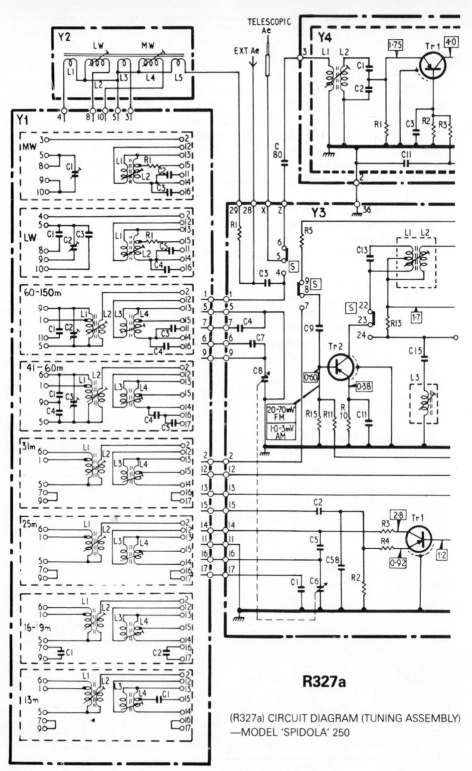

R327a

(R327a) CIRCUIT DIAGRAM (TUNING ASSEMBLY)
—MODEL 'SPIDOLA' 250

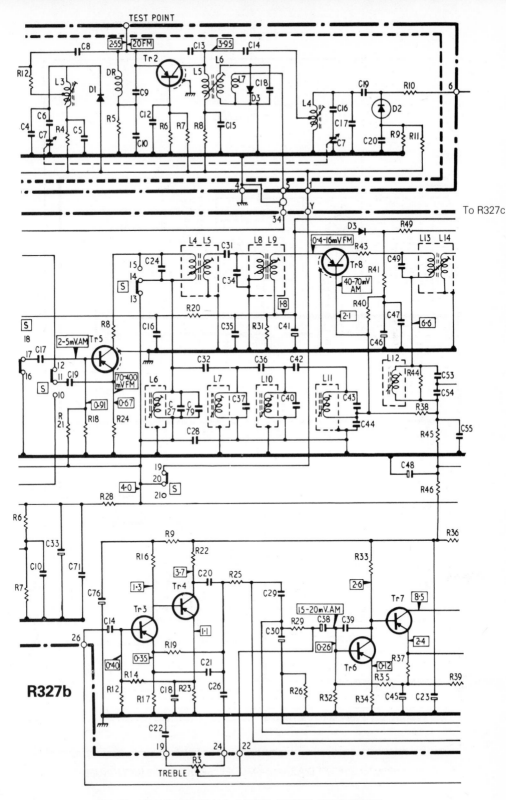

To R327c

R327b

(R327b) CIRCUIT DIAGRAM—MODEL 'SPIDOLA' 250 (*PART*)

R327c

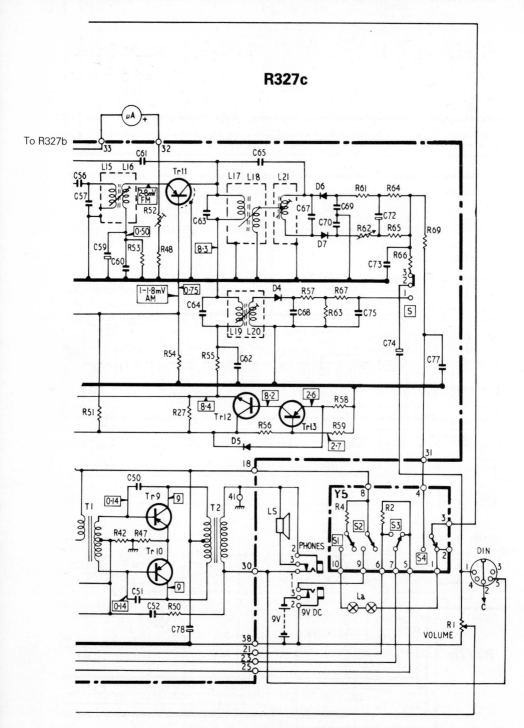

(R327c) CIRCUIT DIAGRAM—MODEL 'SPIDOLA' 250 (*CONTINUED*)

VEGA

Model 'Sapphire' 303 Mk2

General Description: A four-waveband A.M. portable radio receiver operating from battery supplies. A socket is provided for the connection of an earphone.

Battery: 6 volts (4×HB11).   **Quiescent Current:** 9mA.

Wavebands: L.W. 150–408kHz; M.W. 525–1605kHz; S.W.1 3·95–7·3MHz; S.W.2 9·5–12·1MHz.

Loudspeaker: 8 ohms impedance.

Dismantling

Place handle in upright position and push down at each corner to release.

Remove batter cover and batter holder. Remove 2 of the back retaining screws and loosen 2 of the handle retaining pegs to free cup washers beneath. Lift off back and unsolder external aerial and telescopic aerial connections. To gain access to rear of printed board:

Remove 5 retaining screws at wave change switch earthing clamp position, beneath Ferrite Rod Aerial, at bottom right hand and below tuning gang.

Test Voltages

	309B TR1	309B TR2	309B TR3	309B TR4	MN41 TR5	MN41 TR6	MN41 TR7	MN41 TR8
Collector	6·0V	6·0V	6·0V	6·0V	4·0V	6·0V	6·0V	6·0V
Base	0·4V	0·7V	1·2V	1·0V	0·6V	0·6V	0·4V	0·4V
Emitter	1·0V	0·7V	1·1V	1·0V	0·65V	0·5V	0·0V	0·0V

Alignment

Medium Wave: Using a signal of 465kHz fed into a dummy aerial placed close to set, re-align I.F. transformers 11, 10, 9 in that order for maximum gain. Set tuning cursor to 570 metres and align medium wave oscillator coil L7, then adjust medium wave aerial coil L3 on Ferrite Rod for maximum signal strength. Set tuning cursor to 200m and adjust trimmer C1e for signal of 1600kHz.

Long Wave: Tuning cursor to 2000 metres oscillator coil L8, aerial coil L4. Tuning cursor 800 metres trimmer C1f.

Short Wave 1: Tuning cursor 75m, oscillator coil L5, aerial coil L1. Tuning cursor 41m, trimmer C3.

Short Wave 2: Tuning cursor 31m, oscillator coil L6, aerial coil L2. Tuning cursor 25m, trimmer C17.

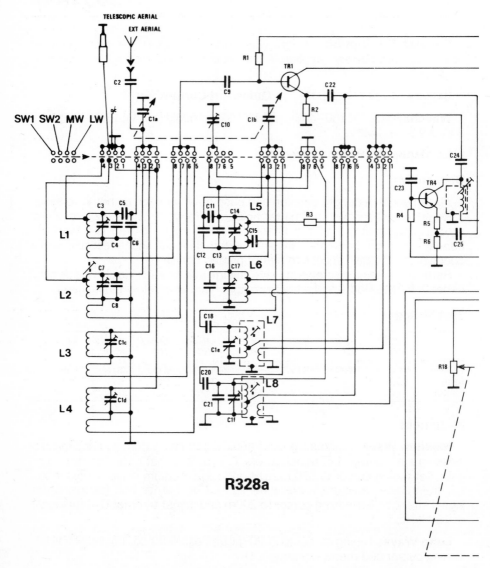

R328a

(R328a) CIRCUIT DIAGRAM—MODEL 'SAPPHIRE' 303 MK. 2 (*PART*)

Component list on p. 728

VEGA

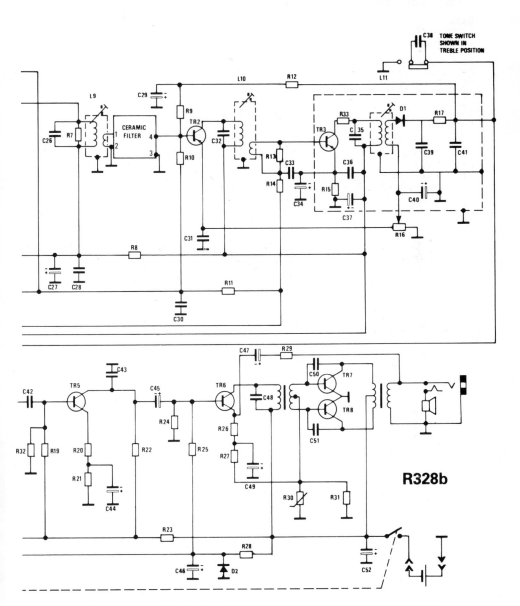

R328b

(R328b) CIRCUIT DIAGRAM—MODEL 'SAPPHIRE' 303 MK. 2 (*CONTINUED*)

Component List

Coils

L1	Aerial S.W.1.
L2	Aerial S.W.2.
L3	Aerial M.W.
L4	Aerial L.W.
L5	Osc. S.W.1.
L6	Osc. S.W.2.
L7	Osc. M.W.
L8	Osc. L.W.

Capacitors

C1a C1b	Tuning Gang
C1c	Trimmer M.W.
C1d	Trimmer L.W.
C1e	Trimmer M.W. Osc.
C1f	Trimmer L.W. Osc.
C2	5·1pF
C3	Trimmer S.W.1.
C4	75pF
C5	240pF
C6	20pF
C7	Trimmer S.W.2.
C8	43pF
C9	0·047μF
C10	Fine tuning gang
C11	240pF
C12	20pF
C13	75pF
C14	Trimmer S.W.1. Osc.
C15	510pF
C16	56pF
C17	Trimmer S.W.2. Osc.

C18	240pF
C20	120pF
C21	36pF
C22	0·047μF
C23	0·047μF
C24	0·047μF
C25	0·015μF
C26	1000pF
C27	30μF
C28	0·047μF
C29	5·0μF
C30	0·047μF
C31	0·047μF
C32	1000pF
C33	0·047μF
C34	30·0μF
C35	1000pF
C36	0·047μF
C37	30·0μF
C38	0·047μF
C29	3300pF
C40	30·0μF
C41	3300pF
C42	·10μF
C43	3300pF
C44	30·0μF
C45	10·0μF
C46	10·0μF
C47	10·0μF
C48	680pF
C49	30·0μF
C50	0·015μF
C51	0·015μF
C52	50·0μF

Resistors

R1	33K
R2	2K
R3	75 ohms
R4	6·8K
R5	13 ohms
R6	1K
R7	10K
R8	300 ohms
R9	1·2K
R10	6·8K
R11	2K
R12	10K
R13	1·2K
R14	6·8K
R15	1·0K
R16	1·0K Potentiometer
R17	2K
R18	10K Volume Control
R19	68K
R20	180 ohms
R21	1·2K
R22	3·9K
R23	100 ohms
R24	15K
R25	6·8K
R26	75 ohms
R27	300 ohms
R28	3·3K
R29	390 ohms
R30	220 ohms Thermistor
R31	220 ohms
R32	15K
R33	100 ohms

VEGA   Model 308

General Description: A compact portable battery-operated A.M./F.M. receiver, the Vega 308 covering the mediumwave band and the 31 to 49 metre S.W. bands on A.M. using an internal ferrite aerial, and the F.M. broadcast band on V.H.F., for which an external telescopic aerial is used. Logging pointers are provided on the tuning scale for quick location of preferred stations.

Waveband selection, switchable A.F.C. operating on F.M. and tone control (top cut) is by five press-button switches along the receiver top. Coarse and fine tuning (fine tuning operative on S.W.) controls are rotary.

Inputs (accessible from the receiver back) include those for external A.M. aerial and earth, and that for an external 9V D.C. supply, via a suitable plug. An earphone jack is fitted to allow private listening.

A separate V.H.F. tuner is incorporated, using an integrated circuit local oscillator/mixer; the tuning gang also houses the A.M. tuning capacitors.

Batteries: 9 volts (6×HP11).

Wavebands: M.W. 525–1605kHz; S.W. 5·9–9·8MHz; F.M. 87–108MHz.

Loudspeaker: 8 ohms impedance.

Access for Service (See Fig. R329).

Remove battery cover and batteries.

Remove two shoulder screws from cabinet back which enter hexagon pillars A.

Carefully ease off back cover and lay it to one side of cabinet front, taking care not to pull off telescopic aerial lead.

To Remove Chassis Assembly: Remove Tuning and Volume control knobs from cabinet front.

Release two hexagon pillars (with washers) A.

Remove three screws B from printed circuit board.

Ease battery compartment assembly out from moulded lugs in cabinet B.

Lift out chassis to extent of leads.

For complete removal of chassis, disconnect leads to speaker telescopic aerial, and battery compartment, having first noted connections to ensure correct reconnection.

Loudspeaker: Remove after slackening two clamps E, having first removed chassis.

Access to tuner components is by releasing two screws C and easing off screening cover.

Telescopic Aerial: Remove from cabinet back after releasing screw F. Pull aerial out from top of cabinet, leaving sleeve behind.

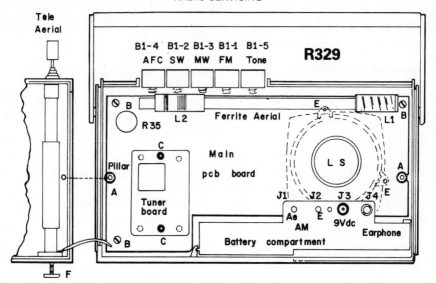

(R329) DISMANTLING PROCEDURE—MODEL 308

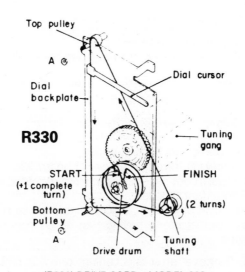

(R330) DRIVE CORD—MODEL 308

Transistor Voltages

(Check using 20 kilohm/voltmeter or electronic meter).

Transistor	Base		Emitter		Collector	
	M.W.	F.M.	M.W.	F.M.	M.W.	F.M.
Tr1	4·8V	0V	4·7V	0V	7·8V	0V
Tr2	1·1V	1·4V	1·1V	1·2V	7·8V	7·6V
Tr3	0·65V	1·1V	0·4V	1·0V	7·0V	5·6V
Tr4	1·3V	1·1V	1·2V	1·0V	7·8V	7·6V
Tr5		0·7V	0·06V		3·6V	
Tr6		5·6V	4·5V		8·3V	
Tr7		8·3V	8·5V		4·4V	
Tr8		4·4V	4·4V		9·0V	
Tr9		4·2V	4·4V		0V	
Tr101	0V	2·6V	0V	2·5V		0V

IC Y2 Pin 2: 7·2V   Pins 4, 5, 6: 1·4V
IC101 Pin 9: 2·1V

Adjustments

Output Amplifier Current: With no signal Volume control at mid-point, adjust pre-set R44 to give 5V d. on Tr6 base, and 4·4V at Tr8 and Tr9 emitters

Do NOT alter setting of pre-set R26, which is for F.M. discriminator final alignment.

Component List

Resistors

R1 150 ohms
R2 390 ohms
R3 6·8k ohms
R4 6·8k ohms
R5 100 ohms
R6 15k ohms
R7 30 ohms
R8 1k ohms
R9 68k ohms
R10 100 ohms
R11 2k ohms
R12 2k ohms
R13 6·8k ohms
R14 68k ohms
R15 1k ohms
R16 15k ohms
R17 820 ohms
R18 100 ohms
R19 15k ohms
R20 1·5k ohms
R21 6·8k ohms
R22 4·7k ohms
R23 620 ohms
R24 270 ohms
R25 200 ohms
R26 1k ohms
R27 820 ohms
R28 3k ohms
R29 4·7k ohms
R30 4·7k ohms
R31 1·5k ohms
R32 6·8k ohms
R33 2k ohms
R34 24k ohms
R35 10k ohms
R36 100 ohms
R37 56k ohms
R38 15k ohms
R39 8·2k ohms
R40 100 ohms
R41 150 ohms
R42 13 ohms
R43 15k ohms
R44 47k ohms
R45 8·2k ohms
R46 100 ohms
R47 270 ohms
R48 1·5k ohms
R49 100 ohms
R50 30 ohms
R51 820 ohms
R52 270k ohms
R101 1k ohms
R102 10k ohms
R103 15k ohms
R104 20k ohms
R105 10k ohms
R106 5·6k ohms
R107 56 ohms
R108 56 ohms
R109 150k ohms
R110 150k ohms
*Pre-set potentiometers

Capacitors

C1 5·1pF
C2 10pF
C3 0·5–4pF
C4 5–20pF
C5 5·1pF
C6 5–20pF
C7 300pF
C8 82pF
C9 0·01µF
C10 1000pF
C11 6·2F
C12 6·2pF
C13 3300pF
C14 5·1pF
C15 5–20pF
C16 0·01µF
C17 300pF
C18 5·1pF
C19 5–20pF
C20 0·047µF
C21 0·047µF
C22 0·047µF
C23 Trimmer
C24 1000pF
C25 82pF
C26 51pF
C27 0·047µF
C28 82pF
C29 0·047µF
C30 0·033µF
C31 0·033µF

C32	50μF	C48	1000pF		C64	500μF
C33	5μF	C49	1000pF		C65	100μF
C34	5μF	C50	30μF		C66	50μF
C35	0·047μF	C51	300pF		C67	3300pF
C36	0·047μF	C52	0·022μF		C68	3300pF
C37	50pF	C53	1000pF		C69	500μF
C38	82pF	C54	30μF		C101	1000pF
C39	51pF	C55	3300pF		C102	1000pF
C40	82pF	C56	3300pF		C103	10pF
C41	0·047μF	C57	10μF		C105	5−20pF
C42	1000pF	C58	500μF		C106	3·9pF
C43	0·01μF	C59	10μF		C107	150pF
C44	5·6pF	C60	0·047μF		C108	6800pF
C45	43pF	C61	3300pF		C109	6800pF
C46	0·047μF	C62	10μF		C110	5−20pF
C47	75pF	C63	200μF		C111	6·2pF

Alignment

Preliminaries: Allow test equipment to warm up before beginning alignment. Reduce signal input and volume control on receiver as circuits come into alignment to avoid A.G.C. action on A.M. Check that A.F.C. switch is Out on F.M.

A.M.

I.F. Stages (select 'M.W.'): Inject 465kHz signal from A.M. generator, modulated, to Tr2 base (junction R5/R9). Connect output meter across loudspeaker terminals. Tune receiver to a no-signal point on dial.

Adjust A.M. I.F.T.'s L15, L12, L9 and L8 (in that order) for maximum.

R.F. Stages

M.W. (select 'M.W.'): Inject A.M. signals via an inductive loop to ferrite aerial. Connect output meter across loudspeaker terminals.

1. Tune signal generator to 520kHz, receiver to low frequency of dial. Adjust M.W. oscillator coil L6 for maximum.

2. Retune signal generator to 1610kHz, receiver to high frequency end of dial. Adjust M.W. oscillator trimmer C19 for maximum.

3. Repeat steps 1 and 2 for optimum result.

4. Tune signal generator and receiver to 525kHz. Adjust M.W. aerial coil L2 (by sliding this along ferrite rod) for maximum.

5. Retune generator and receiver to 1600kHz. Adjust M.W. aerial trimmer C6 for maximum.

6. Repeat steps 4 and 5 for optimum result.

S.W. (select 'S.W.')

7. Tune signal generator to 5·6MHz, receiver to low frequency to low frequency end of dial. Adjust S.W. oscillator coil L5 for maximum.

8. Retune generator to 9·9MHz, receiver to high frequency end of dial. Adjust S.W. oscillator trimmer C15 for maximum.

9. Repeat steps 7 and 8 for optimum result.

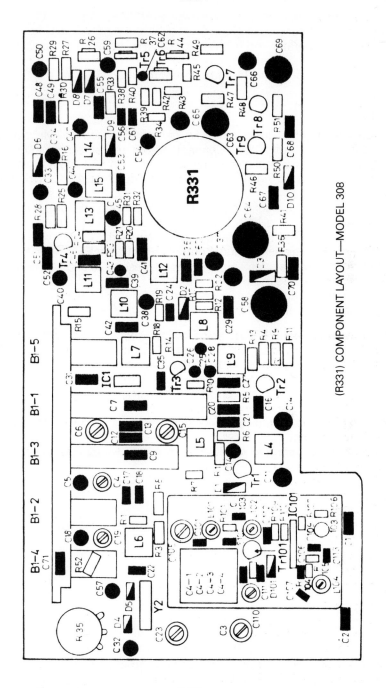

(R331) COMPONENT LAYOUT—MODEL 308

10. Tune signal generator and receiver to 5·6MHz. Adjust S.W. aerial coil L1 (on ferrite rod) for maximum.

11. Retune generator and receiver to 9·5MHz. Adjust S.W. aerial trimmer C4 for maximum.

12. Repeat steps 10 and 11 for optimum result.

F.M. (select 'F.M.')

I.F. Stages: Inject signals from F.M. generator, tuned 10·7MHz, deviated ±22·5kHz to pin 1 of IC101. Connect output meter across loudspeaker terminals, and centre-zero meter across test points KT2 and KT3 and earth. Tune receiver to a no-signal point on dial. Check that A.F.C. switch is Up (inoperative).

1. Set core of F.M. discriminator L14 just flush with coil former top.

2. Adjust F.M. I.F.T.'s L13, L11, L10 and L7 and V.H.F. tuner mixer coil L104, in that order, for maximum output meter reading.

3. Transfer F.M. signal generator output to test point KT1 (increasing signal input lever to produce a reading) and adjust preset R26 and L14 to give zero reading on centre-zero meter.

4. If necessary, repeat step 2 to obtain maximum output while maintaining zero reading on centre-zero meter.

R.F. Stages: Inject R.F. F.M. signals to telescopic aerial input.

1. Tune signal generator to 86MHz and receiver to low frequency end of dial. Adjust F.M. oscillator coil L104 for maximum.

2. Retune generator to 108MHz and receiver to high frequency end of dial. Adjust F.M. oscillator trimmer C18 for maximum.

3. Repeat steps 1 and 2 for optimum result.

4. Tune signal generator and receiver to 88MHz. Adjust F.M. aerial coil L102 for maximum.

5. Retune generator and receiver to 106MHz. Adjust F.M. aerial trimmer C5 for maximum.

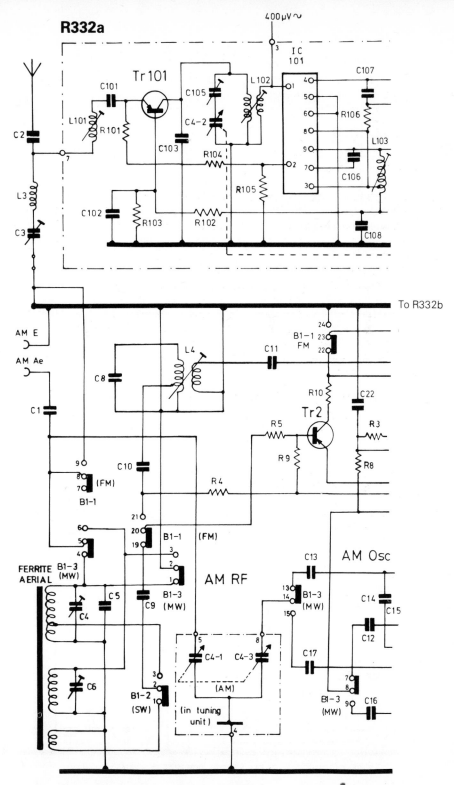

R332a

400µV ~

To R332b

(R332a) CIRCUIT DIAGRAM—MODEL 208 (*PART*)—Component list on p. 731

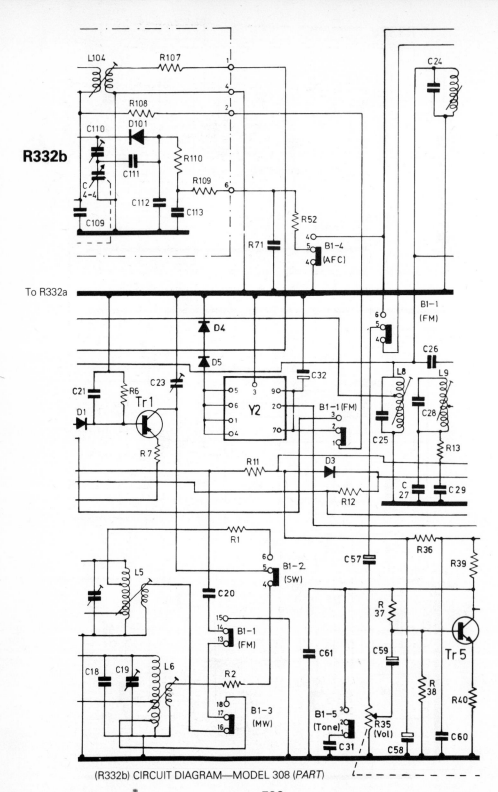

(R332b) CIRCUIT DIAGRAM—MODEL 308 (*PART*)

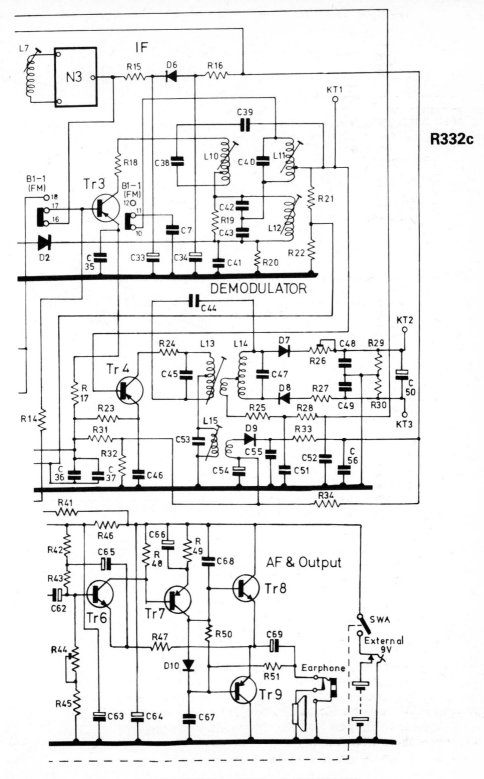

R332c

(R332c) CIRCUIT DIAGRAM—MODEL 308 (*CONTINUED*)

VEGA   Model 'Selga' 404

General Description: An eight-transistor, battery operated, portable radio receiver covering Long and Medium wavebands.

Batteries: 9 volts (6×1·5 volts SP7).

Quiescent Current: 10mA.

Wavebands: L.W. 150–408kHz; M.W. 525–1605kHz.

Loudspeaker: 8 ohms impedance.

(R333) DRIVE CORD—MODEL 'SELGA' 404

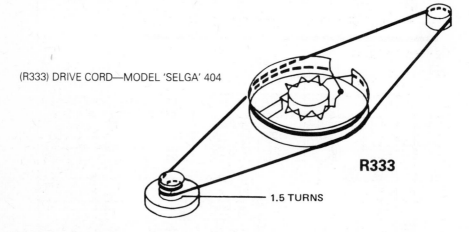

R333

1.5 TURNS

Dismantling

Case Removal: Remove the screw from the wax-filled hole on the rear cover which is situated between the Volume control and Tuning knob. Also remove the screw which is accessible after removing the battery carrier.

Chassis removal: Remove the single screw situated, again, between the Volume control and the Tuning knob, and push the chassis slightly to the left and lift.

Re-alignment

Before alignment is attempted, ensure that the receiver is functioning.

Alignment

I.F. Stages: Switch receiver to mediumwave. Connect Ma meter in series with battery supply, or output meter across speaker. Inject modulated 465kHz signal to base of TR1 transistor, via a 0·05 M.F.D. capacitor. Adjust L13/14, L11/L12, L9/L10, for maximum output, reducing generator output after each adjustment. Repeat for optimum. Remove generator lead.

R.F. Stages: Inject R.F. signal to receiver via an inductive loop (an old ferrite rod fitted with a car coupling coil is ideal) positioned parallel approximately 1 foot from receiver. Tune each stage for maximum output. Tuning gang fully closed (medium wave). Change generator frequency to 515kHz. Adjust L5. Generator signal to 540kHz tuning receiver to signal. Adjust L1/L2 on ferrite rod. Tuning gang fully open. Generator frequency to 1640kHz. Adjust trimmer C9. Generator frequency to 1500kHz. Tune receiver to signal. Adjust trimmer C1. Switch to longwave gang fully closed. Generator signal to 147kHz. Adjust L7. Generator frequency to 160kHz. Tune receiver to signal. Adjust L3/L4 on ferrite aerial. Generator frequency to 420kHz tuning gang to maximum. Adjust trimmer C10. Change generator to 390kHz tuning receiver to signal. Adjust trimmer C2. Repeat for optimum. Seal coils on ferrite rod.

Transistors: TR1–TR6 BF229 TR7 and TR8 AC132

Component List

Capacitors

C1	4–15pF Trimmer
C2	4–15pF Trimmer
C3	130pF
C4	39pF
C5	5–270pF
C6	5·1pF
C7	5–270pF
C8	330pF
C9	6–25pF Trimmer
C10	6–25pF Trimmer
C11	0·033μF
C12	220pF
C13	4700pF
C14	10μF
C15	0·022μF
C16	1000pF
C17	6·8pF
C18	1000pF
C19	0·022μF
C20	20μF
C21	1μF
C22	220pF
C23	0·01μF
C24	3300pF
C25	50μF
C26	50μF
C27	20μF
C28	0·033μF
C29	4700pF
C30	4700pF
C31	4·7pF
C32	2000pF
C33	0·033μF
C34	50μF
C35	3·300pF
C36	2200pF

Resistors

R1	100
R2	1K8
R3	10K
R4	5K6
R5	3K9
R6	15K
R7	6K8 Vol. Cont
R8	3K6
R9	6K2
R10	75K
R11	10K
R12	10K
R13	18K
R14	5K1
R15	390
R16	27
R17	1K5
R18	4K3
R19	470
R20	510
R21	120
R22	12K
R23	330 Pre-set
R24	910
R25	510
R26	6
R27	100

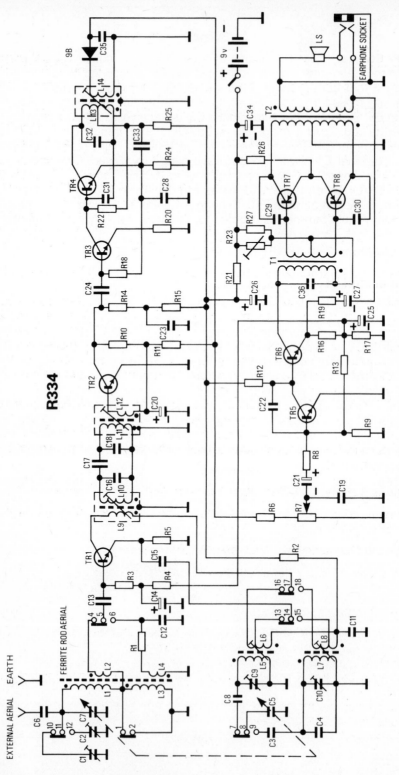

R334

(R334) CIRCUIT DIAGRAM—MODEL 'SELGA' 404

VEGA   Model 705

General Description: A stereo radio recorder for internal battery or A.C. mains operation combining a four-band A.M./F.M. radio tuner, with M.P.X. decoder for F.M. stereo broadcasts, and a stereo cassette-deck, each of which feeds a common amplifier with a 2W per channel output.

The radio tuner covers long and medium wavebands on A.M., using an internal ferrite aerial, shortwave A.M. and the F.M. broadcast band on V.H.F., reception for which is provided by an external telescopic aerial.

The cassette-deck accepts normal (ferric) tape, is fitted with a digital counter, and has automatic end-of-tape stop. Record level control is automatic, but twin VU meters indicate the levels for both recording and playback.

These meters also act as a battery condition indicator (L.H.) and a tuning meter (R.H.) when in the 'Radio' mode.

Controls are rotary for tuning, press-button for waveband selection, toggle switch for on/off or timer 'sleep' function (this allows the 705 to be switched off after a period determined by the remaining playing time of a cassette loaded and the tape transport moving). Other toggle switches are for stereo/mono selection, selection of 'Line-in' (i.e. from an external amplifier via a Din in/out socket) or from an external microphone, and for determining the 'Radio' or 'Tape' operating mode.

Twin internal capacitor microphones are fitted. Inputs and outputs provided include a 5-pin Din socket, external microphone jacks with a remote tape transport control facility, and external F.M. aerial twin-feeder terminals. Outputs include separate channel extension loudspeaker jacks, and that for a stereo headset.

Mains Supply: 230 volts, 50Hz.   **Batteries:** 9 volts (6×HP2).

Wavebands: L.W. 155–320kHz; M.W. 540–1600kHz; S.W. 6–18MHz; F.M. 88–104MHz.

Fuse: 1 Amp antisurge.   **Lamps:** 9 volts 100mA; L.E.D.

Tape Bias: 50kHz.   **Loudspeakers:** 4 ohms impedance.

Dismantling

Disconnect mains lead from supply (if in use), remove mains lead and battery compartment covers, and remove batteries.

Remove the cabinet back cover as follows:

Remove the top 3 screws (two self-tapping screws and one threaded screw).

From inside the mains lead and battery compartments, remove 2 screws and one central (longer) screw.

741

Ease off the cabinet back and detach the external aerial lead connector from the tuner P.C.B. tag.

The back can now be laid down below the front as shown in the diagram. To remove the back completely, unplug the two D.C. supply leads from tags on the amplifier P.C.B.

The internal moulded chassis, housing all the main assemblies, is removed from the cabinet front as follows:

Ease off the main tuning and the four slider control knobs.

Remove 2 screws accessible through holes in the P.C.B.s, 2 screws from the cassette deck chassis, and 1 screw from the tuner P.C.B., lower right hand corner.

Lift the lower edge of the chassis and manoeuvre the switch toggles, cassette deck keys and slider control levers clear of the cabinet top escutcheon. Then ease out the chassis complete.

To free the chassis from the cabinet front, unsolder the leads to the dial lamp (noting the connection points for refitting), unplug connectors 3, 4, and 5 (headphone jack leads) from the amplifier board, and release two pairs of screws to free the internal microphones.

The chassis can now be taken clear of the cabinet front, complete with the microphones, leaving the headphone jack, loudspeakers and dial lamp assembly in the cabinet.

This stage of dis-assembly should be sufficient for most service operations. To remove the main P.C.B.s, however:

Release 3 screws to free the radio tuner board (taking care to maintain tension on the dial drive drum).

Release 2 screws and 4 screws to free the amplifier and cassette-deck board (to gain access to the cassette-deck mechanism).

Other components and assemblies can be removed from the cabinet back and front mouldings as follows:

Mains power unit board: release screws.

Mains transformer: release 2 screws.

Mains/battery switch: release 2 screws.

Mains lead: disconnect from terminal block, release screws to slacken clamp.

Terminal block is held to cabinet back by a screw.

Loudspeakers are each held to the cabinet front by 4 screws.

Dial lamp: remove single screw to free mounting at dial glass end.

Components List

Radio Tuner:

Resistors					
R1	270 ohms	R8	680 ohms	R16	3·3k ohms
R2	220k ohms	R9	220 ohms	R18	100k ohms
R3	100 ohms	R10	220k ohms	R19	220 ohms
R4	220 ohms	R11	47 ohms	R20	100 ohms
R5	1·5k ohms	R12	1·5k ohms	R21	2·2k ohms
R6	220k ohms	R13	4·7 ohms	R22	100 ohms
R7	100 ohms	R14	33 ohms	R23	33k ohms
		R15	33 ohms	R24	12k ohms

R25	560 ohms
R26	1k ohms
R27	1k ohms
R28	100k ohms
R29	5·6k ohms
R30	5·6k ohms
R31	220 ohms
R32	330 ohms
R33	1·5k ohms
R34	10k ohms
R35	150 ohms
R36	3·3k ohms
R37	100 ohms
R38	680 ohms
R39	6·8k ohms
R40	3·3k ohms
R41	3·3k ohms
R42	220 ohms
R43	3·9k ohms
R44	3·9k ohms
R45	2·2k ohms
R46	3·9k ohms
R47	2·2k ohms
R48	3·9k ohms

Potentiometers

SR1	50k ohms
SR2	5k ohms
SR3	2k ohms

Capacitors

C1	10pF
C2	20pF
C3	0·0015μF
C4	10pF
C5	7pF
C6	15pF
C7	8pF
C8	15pF
C9	20pF
C10	470pF
C11	0·002μF
C12	0·002μF
C13	0·04μF
C14	10pF
C15	150pF
C16	0·04μF
C17	3pF
C18	30pF
C20	0·001μF
C21	0·001μF
C22	5pF
C23	5pF
C24	22pF
C25	0·04μF
C26	0·04μF
C27	0·01μF
C28	0·02μF
C29	30pF
C30	15pF
C31	82pF
C32	10pF
C33	380pF
C34	150pF

C35	0·005μF
C36	0·04μF
C37	10μF
C38	0·02μF
C39	0·001μF
C40	0·02μF
C41	10μF
C42	0·01μF
C43	0·04μF
C44	0·47μF
C45	1000μF
C46	0·04μF
C47	0·02μF
C48	3pF
C49	0·01μF
C50	0·02μF
C51	0·02μF
C52	0·04μF
C53	0·02μF
C54	4·7μF
C55	300pF
C56	300pF
C57	2·2μF
C58	1μF ·
C59	220μF
C60	1500pF
C61	470μF
C62	0·015μF
C63	0·015μF
C64	0·47μF
C65	0·47μF
C66	0·47μF
C67	4·7μF
C68	4·7μF
C69	0·001μF
C70	0·001μF
C177	0·002μF
C178	0·001μF
C179	0·001μF
C180	0·002μF
C181	0·047μF

Transistors

Tr1	2SC930E
Tr2	2SC930E
Tr3	2SC930E

Integrated circuits

IC1	LA1201
IC2	LA3350

Diodes

D1	IS188
D2	IS188
D3	IS188
D4	IS188
D5	IS553
D6	5·6V Zener
D7	IS188
D8	IS188
D9	IS188
D10	IS188
D11	IS188
D12	IS188
D13	LED

Amplifier and Recorder:

Resistors

R49	3·3k ohms
R50	10 ohms
R51	10k ohms
R52	4·7k ohms
R53	68k ohms
R54	680 ohms
R55	56k ohms
R56	12k ohms
R57	680k ohms
R58	1·5k ohms
R59	6·8k ohms
R60	12k ohms
R61	12k ohms
R62	15k ohms
R63	180k ohms
R64	5·6k ohms
R65	470 ohms
R66	100k ohms
R67	18k ohms
R68	390k ohms
R69	1·5k ohms
R71	8·2k ohms
R72	8·2k ohms
R73	1M ohms
R74	4·7k ohms
R75	270 ohms
R76	10k ohms
R77	4·7k ohms
R78	1·8k ohms
R79	12k ohms
R80	4·7k ohms
R81	2·2k ohms
R83	15k ohms
R84	15k ohms
R86	100 ohms
R87	220 ohms
R88	56k ohms
R89	5·6k ohms
R90	3·3k ohms
R91	10 ohms
R92	10k ohms
R93	4·7k ohms
R94	68k ohms
R95	680 ohms
R96	56k ohms
R97	12k ohms
R98	1M ohms
R99	680k ohms
R100	1·5k ohms
R101	6·8k ohms
R102	12k ohms
R103	12k ohms
R104	15k ohms
R105	180k ohms
R106	5·6k ohms
R107	470 ohms
R108	100k ohms
R109	18k ohms
R110	390k ohms

R111	1·5k ohms	C76	150pF	C161	150pF
R113	8·2k ohms	C77	4·7μF	C162	150pF
R114	8·2k ohms	C78	0·47μF	C163	0·0027μF
R115	1M ohms	C79	33μF	C164	47μF
R116	4·7k ohms	C80	0·47μF	C165	0·0027μF
R117	4·7k ohms	C121	150pF	C166	0·01μF
R118	10k ohms	C122	0·47μF	C167	220μF
R119	4·7k ohms	C123	33μF	C168	3000μF
R120	1·8k ohms	C124	0·47μF	C169	0·01μF
R121	12k ohms	C125	560pF	C170	0·01μF
R122	4·7k ohms	C126	0·47μF	C171	0·01μF
R123	2·2k ohms	C127	330μF	C172	0·01μF
R124	1·2k ohms	C128	0·0033μF	C173	0·04μF
R125	15k ohms	C129	0·01μF	C174	3000μF
R126	15k ohms	C130	0·0033μF	C175	220μF
R128	100 ohms	C131	0·47μF		
R129	220 ohms	C133	150pF		
R130	3·3k ohms	C135	0·01μF		
R131	10 ohms	C136	4·7μF		
R132	220 ohms	C137	0·33μF		
R133	33k ohms	C138	150pF		
R134	33 ohms	C139	4·7μF		
R135	470 ohms	C140	220μF		
R137	150 ohms	C141	150pF		
R138	150 ohms	C142	0·47μF		
R139	47k ohms	C143	0·47μF		
R140	47k ohms	C144	0·0015μF		

Transistors

Tr4	2SC536G
Tr5	2SC536G
Tr6	2SC536G
Tr7	2SC536G
Tr8	2SC536G
Tr9	2SC536G
Tr10	2SC536G

Integrated circuits

IC3	LA3210
IC4	LA4102
IC5	LA3210
IC6	LA4102

Potentiometers

SR4	500 ohms
Volume L	50k ohms
Volume R	50k ohms
Treble	50k ohms
Bass	50k ohms

C145	0·0027μF		
C146	0·0015μF		
C147	0·0015μF		
C148	1000pF		
C149	4·7μF		
C150	33μF		
C151	10μF		
C152	220μF		
C153	820pF		
C154	820pF		
C155	0·1μF		
C156	220μF		
C157	1000μF		
C158	0·1μF		
C160	33μF		

Diodes

D14	Silicon
D15	Silicon
D16	Silicon
D17	Silicon
D18	Silicon
D19	Silicon
D20	Silicon
D21	1N4002
D22	1N4002
D23	1N4002
D24	1N4002

Capacitors

C71	0·47μF
C72	4·7μF
C73	0·0015μF
C74	1000pF
C75	33μF

Radio Tuner Alignment (See Fig. R335)

Preliminaries: Allow receiver and test equipment to warm up thoroughly before starting alignment. Progressively reduce signal input as circuits come into alignment, so as to avoid A.G.C. action on A.M., 'limiting' on F.M. See radio tuner P.C.B. and alignment diagrams for locations of coils and trimmers.

Procedure

A.M. (select 'M.W.'):

I.F. Stages: Connect A.M. signal generator output, modulated, via a 0·04μF capacitor, to test point TP5 (junction C8, C9 and L1), or via an inductive loop to the ferrite aerial. Connect output meter across either the extension loudspeaker jack for one channel, or across the terminals of one loudspeaker. Select 'Mono' and tune receiver to a no signal point on the dial scale.

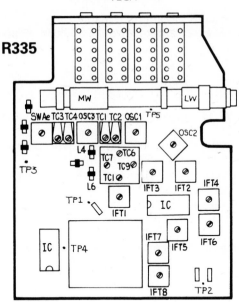

VEGA

R335

(R335) ALIGNMENT ADJUSTMENTS—MODEL 705

Tune signal generator to 455kHz, modulated, and adjust A.M. I.F. transformers IFT6, IFT3 and IFT2 for maximum output on meter.

R.F. Stages: Connect signal generator output, modulated, via an inductive loop to the ferrite aerial for long and medium wavebands. Connect output meter across one loudspeaker terminals or an extension speaker jack. Select 'Mono'.

L.W. (select 'L.W.'):
1. Tune signal generator to 145kHz and receiver to low frequency end of scale. Adjust L.W. oscillator coil OSC1 for maximum.
2. Retune generator to 355kHz and receiver to high frequency end of scale. Adjust L.W. oscillator trimmer TC2 for maximum.
3. Repeat steps 1 and 2 for optimum result.
4. Tune signal generator and receiver to 170kHa. Adjust L.W. aerial coil (by sliding this along ferrite rod) for maximum.
5. Retune generator and receiver to 320kHz. Adjust L.W. aerial trimmer TC1 for maximum.
6. Repeat steps 4 and 5 for optimum result.

M.W. (select 'M.W.'):
7. Tune signal generator to 510kHz and receiver to low frequency end of scale. Adjust M.W. oscillator coil OSC2 for maximum.
8. Retune generator to 1650kHz, receiver to high frequency end of scale. Adjust M.W. oscillator trimmer TC5 for maximum.

745

9. Repeat steps 7 and 8 for optimum result.

10. Tune signal generator and receiver to 600kHz. Adjust M.W. aerial coil (on ferrite rod) for maximum.

11. Retune generator and receiver to 1400kHz. Adjust M.W. aerial trimmer TC6 for maximum.

12. Repeat steps 10 and 11 for optimum result.

S.W. (select 'S.W.'): Connect signal generator output to telescopic aerial input point on P.C.B.

13. Tune signal generator to 5·85MHz, receiver to low frequency end of scale. Adjust S.W. oscillator coil OSC3 for maximum.

14. Retune generator to 18·7MHz and receiver to high frequency end of scale. Adjust S.W. oscillator trimmer TC4 for maximum.

15. Repeat steps 14 and 15 for optimum result.

16. Tune signal generator and receiver to 7·0MHz. Adjust S.W. aerial coil for maximum.

17. Retune generator and receiver to 16·0MHz. Adjust S.W. aerial trimmer TC3 for maximum.

18. Repeat steps 16 and 17 for optimum result.

F.M. (select 'F.M.'):

I.F. Stages: Connect F.M. signal generator via sweep marker generator to test point TP-1 (Tr2 base) via a 0·04μF capacitor. Connect oscilloscope probe between testpoint TP-2 (junction R29/R30) and earth via a 0·01μF capacitor. Set marker generator to sweep 10·7MHz ±100kHz, marker at 10·7MHz.

1. Adjust F.M. I.F. transformers IFT8, IFT7, IFT5, IFT4 and IFT1 for maximum amplitude of response curve at 10·7MHz on display.

2. Readjust discriminator IFT8 to produce a symmetrical 'S' curve with the straight part passing through zero at 10·7MHz.

R.F. Stages: Connect F.M. signal generator output, modulated, to the telescopic aerial input point on the tuner P.C.B. Connect output meter across the terminals of one loudspeaker or to the corresponding speaker jack.

1. Tune signal generator to 87MHz and receiver to the low frequency end of the dial scale. Adjust F.M. oscillator coil L6 (by varying turns spacing) for maximum.

2. Retune generator to 105MHz, receiver to high frequency end of scale. Adjust F.M. oscillator trimmer TC8 for maximum.

3. Repeat steps 1 and 2 for optimum result.

4. Tune signal generator and receiver to 90MHz. Adjust F.M. aerial coil L4 for maximum.

5. Retune generator and receiver to 102MHz. Adjust F.M. aerial trimmer TC7 for maximum.

6. Repeat steps 4 and 5 for optimum result.

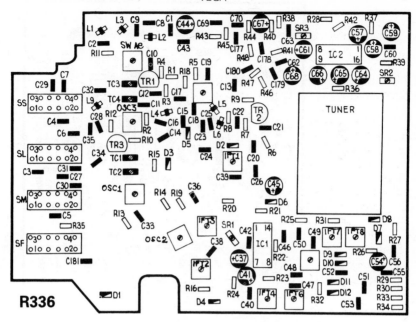

(R336) COMPONENT LAYOUT (RADIO TUNER)—MODEL 705

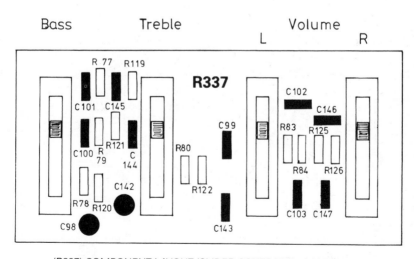

(R337) COMPONENT LAYOUT (SLIDER CONTROLS)—MODEL 705

R338

(R338) COMPONENT LAYOUT (RECORDER BOARD)—MODEL 705

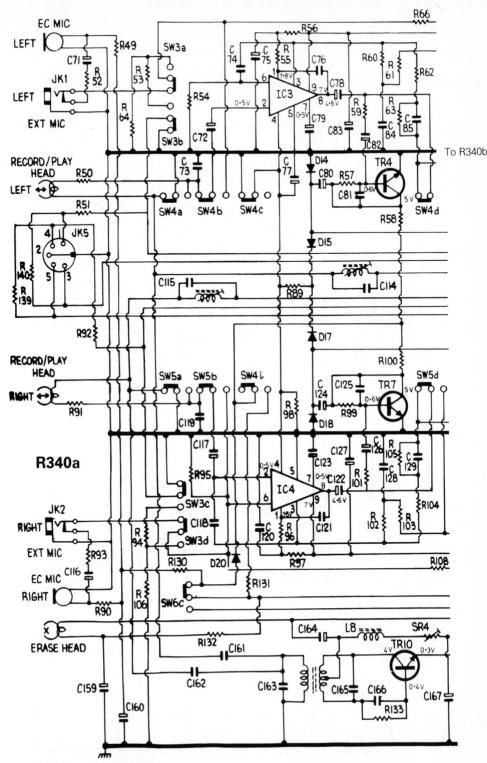

(R340a) CIRCUIT DIAGRAM (RECORDER AND AMPLIFIER)—MODEL 705 (*PART*)—
Component list on p. 742

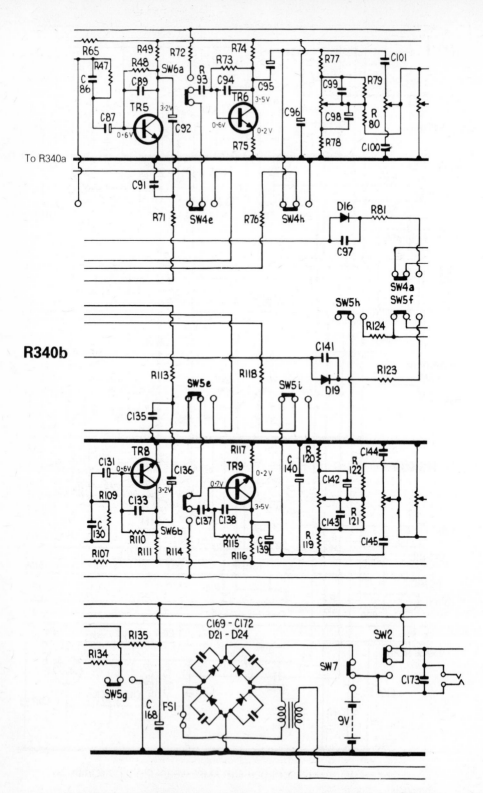

(R340b) CIRCUIT DIAGRAM (RECORDER AND AMPLIFIER)—MODEL 705 (*PART*)

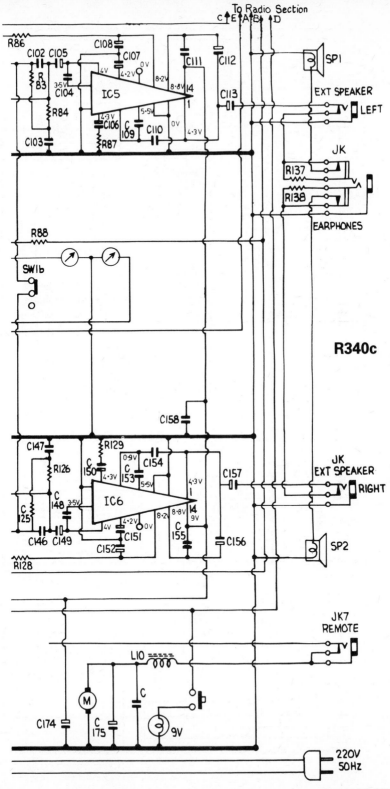

(R340c) CIRCUIT DIAGRAM (RECORDER AND AMPLIFIER)—MODEL 705 (*CONTINUED*)

(R339a) CIRCUIT DIAGRAM (RADIO TUNER)—MODEL 705 (PART)

R339a

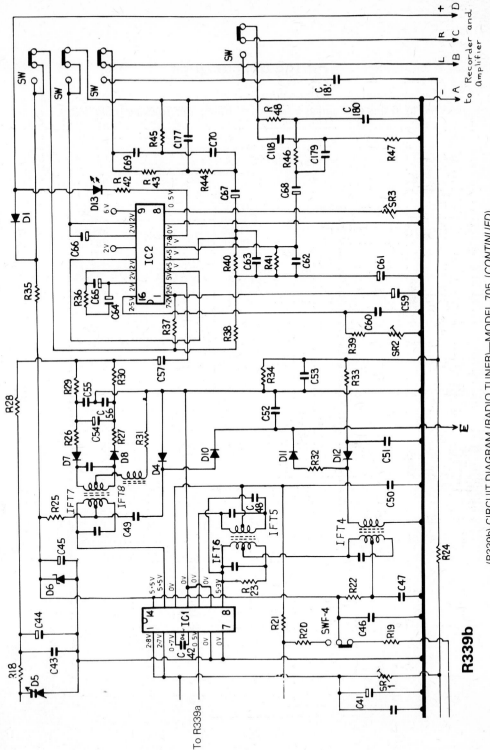

(R339b) CIRCUIT DIAGRAM (RADIO TUNER)—MODEL 705 *(CONTINUED)*

R339b

CROSS REFERENCES

The models listed here will be found to be generally similar to the models referred to in this and earlier volumes of *Radio and Television Servicing*.

Make	Model	Refer to	Volume
Ferguson	3781, 3795, 3796, 37003	T.C.E. TX9	1981–82
	3870, 3871,	T.C.E. 1697 Series	This volume
	38000, 38010	T.C.E. 1696 Series	This volume
G.E.C.	C1615-H	G.E.C. C1650	1981–82
Murphy	MTV 100	Sanyo 12T280	1980–81
Philips/Pye	23BC2001	Philips TC2	1981–82
	12TX1501, 12TX2502, 31BX1016, 31BX1026, 31BX1124, 31BX1125, 35BX2045	Philips TX12	1981–82
	3060, 3062, 3157, 3237, 3715, 3726, 3761, 20CT3227, 20CT3723, 51KT3260, 51KT3262, 51KT3267,51KT3272	Philips KT3	This volume and 1980–81
	22CS1234, 26CS1250, 56KS1062,66KS1152	Philips K30	This volume and 1981–82
	44T7220	Philips E2	1980–81

SUPPLEMENTARY SERVICING INFORMATION

The following pages contain abstracts from service bulletins issued by major manufacturers during the past year. This information, which is of a random nature, lists modifications to circuit designs already published.

The information given here is arranged in manufacturer order. (See also 'Supplementary Servicing' and 'Recent Developments' sections in volumes 1971–72 onwards).

1982–83 Supplement

Philips Service (Philips/Pye etc.)
Colour Television: G11, K30, KT3 Chassis
Monochrome Television: TX, TX14 Chassis

Sony (U.K.) Ltd.
Colour Television: KV-1612UB, KV-2212UB

Thorn Consumer Electronics (Ferguson, Ultra etc.)—Now Thorn-EMI-Ferguson
Colour Television: 9000 Series, TX9 Chassis
Audio: 3T12

PHILIPS SERVICE (Philips/Pye)

Colour Television

G11 Chassis

Random Fuse Blowing: In cases where random blowing of the 3·15A fuse FS1301 (and/or FS1302 where fitted) occurs a 1 Ω 4 watt resistor can be connected in series with the anode of each thyristor SCR18/20.

To fit the additional components, cut the leads of each resistor to 19mm (¾") then connect and position each resistor, using the spare tag for terminating one resistor. It is important to keep the resistors well clear of heat sinks, other components, and existing wiring.

Note: This change is most effective when fitted to receivers with R1310 (1.2 Ω) already fitted on the mains input panel. However, it may be applied with benefit to those receivers with or without R1310 fitted.

For certain technical reasons, no attempt should be made to add R1310 to early-type mains input panels.

Substitute Type for Diode BYX55-600: The diode type BYX55-600 which is used in positions 3132 and 3138 on the Line Scan Panel is no longer available. Diode type BYV95C is now supplied as a substitute.

Engineers should ensure that when replacing the substitute version, it is positioned 12mm (½") away from the printed panel, and well clear of surrounding components.

Failure of BU208A and TBS2600: Experience has indicated that the electrolytic capacitor C4029 can develop a high resistance at its terminations, causing intermittent surges in H.T. voltage, resulting in possible failure of the BU208A and/or TDA2600 devices.

Capacitors of PYE/CCL manufacture should be replaced with DALY/I.T.T. types, which are available from Spare Sales Department at Waddon under code number 124 47056.

I.T.T. capacitors type KV3253T should also be replaced with DALY/I.T.T. type.

(a) On some chassis C1422 may be changed in value from 2.2μF to 0.47μF, R1424 changed from 47K Ω to 100k Ω and a 15μF electrolytic capacitor added between the junction of D1562/D1567 and pin 15 of the line output transformer. If incorporated, this change may be left in circuit, but there is no need for it to be fitted otherwise.

(b) To overcome an H.T. variation occurring at switch-on change of channel or sometimes when retuning the receiver, a 100nF capacitor (**C337**) was fitted between the base of TS336 and pin 12 of IC322 on the supply control module. This addition should be retained—on chassis where this change has not been incorporated, a new capacitor should be fitted.

(c) On some chassis, a 2.7 Ω resistor may be found to be fitted across R1461 on the monocarrier board—this change must be removed from a chassis when the modifications stated below in (d) are carried out.

(d) The current factory production changes are as follows:

(i) R7329 and C7329 are deleted, and a wire link fitted in place of R7329.

(ii) The value of R7354 is changed from 270 Ω to 560 Ω (or 2×270 Ω resistors wired in series).

Notes: (i) The factory coding of the supply module incorporating the changes in (d) is BY02, and the panel is available under code number 212 21114—this panel will eventually replace existing panels with code numbers 212 20614 (KT3 chassis) and 212 20795 (K30 chassis).

(ii) Change (d) (i) must not be applied to existing panels.

(iii) Change (d) (ii) can be applied to early panels and will cure most problems, but to provide a complete cure to the problem a BY02 version panel should be fitted.

(iv) Engineers must ensure that if a 2.7 Ω resistor is fitted across R1461, it must be removed before fitting a BY02 panel, or carrying out the change detailed in (d) (ii) above.

KT3 chassis (Edition II)

Production Change: C583 on the monocarrier panel has been changed in production from 47μF to a value of 4μF.

KT3 Chassis

'Plop' Effect on Sound: In order to eliminate a 'plop' effect on sound when operating the speech/music switch, a 47k Ω resistor has been added between pins 10 and 11 of the sound module socket, the resistor being fitted to existing holes in the mono-carrier panel.

Philips/Pye KT3 and K30 chassis

Returning Tuner Drawers for Replacement: The tuner drawer used in these chassis is fitted with an A.F.C. switch, but the length of the leads and terminations vary, depending on the model in which the assembly is fitted.

As a consequence of factory standardisation, one type of drawer only is supplied suitable for various models, but less the A.F.C. switch. Engineers should note that before tuner drawers are returned for replacement, the A.F.C. switch should be removed and retained for fitting to the replacement drawer.

Change of Fuses on Teletext Models: To increase the operating current margins, fuses VL001 and VL002 which are used on the Teletext Power Supply Panel have been changed in value and type from 500mA QA to 630mA TL.

Note: Failure of this fuse will produce symptoms of no text/picture, but with a visible raster.

Luminance—Chrominance/Interface Change: To standardise production panels, the following change has been introduced on the luminance-chrominance/interface panel used on teletext models:

R3055 changed in value from 1k Ω to 220 Ω

C2055 changed in value from $2\mu2$ to $6\mu8$

Philips/Pye K30 chassis

Incorrect Transistor Marking (TS490) on Motherboard: On certain motherboard panels using this chassis, the emitter of TS490 is incorrectly shown as being connected to earth print—the correct marking should be collector to earth as per the circuit diagram.

Philips/Pye K30 chassis

To Cure Non-operation of Position 1 on the Selector Keyboard: The following production change has been made to the encoder/CCAM panel, to cure non-operation of position 1 on the selector keyboard:

Two ceramic plate capacitors of value 10 or 12pF respectively have been fitted, one between pin 21 and chassis, and the other between pin 22 and chassis, on the integrated circuit SAA1082—mounted on the print side of the panel.

Philips and Pye KT3/K30 chassis (Edition II)

Poor Playback from V.H.S. Machines: Some cases of vision noise in the form of random horizontal bars near black level, have been reported where chassis fitted with a 'single chip' luminance/chrominance module (Edition II) are used in conjunction with V.H.S. video recorders.

The problem is caused by noise on the video signal from the recorder, and may be alleviated by increasing the value of capacitors C44-45-46 from 22nF to $10\mu F$ (observe correct polarity when fitting replacement capacitors).

KT3/KT30 Chassis

Spurious H.T. Shut-down ('Hiccups' in the Safety Circuit): Over the period of time since the start of production of these chassis, a number of cases have been reported where spurious shut-down of the H.T. supply occurs.

In the event of either excess output voltage or current, the switch-mode power supply unit is designed to automatically shut-down to protect the rest of the receiver. This protection is instantaneous, and it will react by 'hiccuping' to overload faults, particularly in the line output stages, as well as C.R.T. flashover and severe spiky pulses on the mains input.

If repeated random 'hiccuping' is experienced, engineers are advised firstly to make the following checks:

(i) Check mains supply plug and socket and fuse for possible intermittent contact.

(ii) Check for correct value H.T. voltage on KT3 (129V) or K30 (140V).

(iii) Check earth connections to C.R.T. Aquadag and 'P' band. On K30 chassis, check E.H.T. lead connections at both ends.

(iv) Check for dry joints which may cause arcing, and/or poor edge connections to modules.

Philips 22CS1234/05T—26CS1250/05T (K30 chassis)
Philips 20CT3723/05T and Pye 46KT3157—51KT3237(KT3)

The following information relates to factory production changes which have been introduced at various times, and concludes with the very latest changes incorporated to overcome the problem.

Luminance/Chrominance Interface Panel Change: Following the deletion of the 'MIX-MODE' contrast reduction circuit on the teletext panel, the pre-set potentiometer on the luminance/chrominance interface panel has also been deleted and replaced by a wire link.

Panels incorporating this change will have the BA/HU code raised by one to BA/HU-02.

Improved Chroma Rejection: The following production change has been introduced to the luminance/chrominance interface panel used in teletext receivers employing the K30 chassis.

The TDK filter KT44 has been deleted and replaced by a subpanel assembly code number 212 27541.

Note: Teletext receivers using the KT3 chassis will continue to be fitted with the TDK filter. For replacement purposes, only luminance/chrominance interface panels fitted with the sub-panel assembly will be supplied under code number 212 27535.

Beam Current Limiter Change: To ensure operation of the Beam Current Limiter on Teletext and Picture in 'MIX-MODE', a change has now been made to the luminance/chrominance interface panel which involves moving the cathode of D6010 (type 1N4248) to pin 14 of IC6023 (type HEF4053BP).

Service engineers should check the position of D6010 on all receivers serviced which have a later-version decoder fitted. If D6010 is *not* connected to pin 14 of IC6023, pins 12 and 13 of the device should be short-circuited to achieve the desired result.

Note: When fitting a later-version decoder to early models, the following options are available:

(*a*) *To retain MIX-MODE contrast reduction facility:* Add diodes D6054 and D6055 (type 1N4148) Delete R40.

(*b*) *To delete MIX-MODE contrast reduction:* Convert interface panel to later type by shorting pins 12 and 13 of IC6023.

Deletion of Contrast Reduction Circuit: The following production change to delete the contrast reduction circuit in 'MIX-MODE' has been made to the teletext panel-code number 212 27508.

Diode D6054 deleted

Diode D6055 deleted and replaced by a wire link

A 15k Ω resistor (R6040) added between pin 2 of IC7043 (type SAA5050) and the 5 volt supply line.

Monochrome Television

Philips/Pye TX Chassis

Uncontrollable Volume: In cases where symptoms of uncontrollable volume are experienced, the cause may automatically be attributed to a faulty volume control. However, service experience has shown that in many cases the real cause has been a faulty IC310 (type TBA120AS).

Since the control of volume is by D.C. level, a fault in the I.C. can override the effect of the volume control, consequently, engineers are advised to support their diagnosis by means of the usual resistance checks on the control.

Intermittent Line Collapse: To prevent the possibility of an intermittent line collapse, resulting in a bright vertical line appearing on the C.R.T. screen, a production change has been introduced increasing the value of C393 from 560pF to 1.5nF.

Note: Engineers are advised to check the value of C393 on all receivers handled which bear Factory code HU on the chassis or serial plate, and ensure that a 1·5nF capacitor is fitted.

Philips/Pye TX 14 Chassis

Black Patch Effect on Picture: To eliminate the possibility of a black patch appearing at the corner of the C.R.T. screen, the following production changes have been introduced.

R511 changed in value from 820k Ω to 1M Ω

R513 changed in value from 120k Ω to 150k Ω

Distortion at Low Volume: To improve the sound quality at low volume, R311 (⅛W) has been changed in value from 16 Ω to 27 Ω, and R322 (2W) changed in value from 1k2 Ω to 1k5 Ω.

Note: Check also that the value of R300 is 27k Ω.

SONY (U.K.) LTD

Colour Television

KV-1612UB

Should it become necessary to replace the tuner in the above model, to existing tuner, Type U321, has now been substituted by the Type U341.

In order that the replacement tuner is fully compatible, the following circuit changes are necessary:

(i) Change R249 from 1K8 to 3K9

(ii) Remove R285 and replace with a shorting link

(iii) Remove R256

(iv) Remove Q214 and fit shorting link between the base and emitter connections

(v) Change C246 from 22uF 16v to 10uF 16v

(vi) Remove R250 and replace with shorting link

(vii) Remove C265

KV-2212UB

The following circuit change has been introduced to increase the contrast range.

R311 new value 680 ohm, ¼ w

R807 new value 150k ohm, ½ w

Q302 voltages to read: e. 1.3v, b. 1.8v, c. 4.8v

THORN-EMI-FERGUSON (formerly T.C.E.)

Colour Television

9000 Series

W704—Switched Mode Power Supply and Line Output Module (PC752 or PC558): Currently being supplied for the component in this position is diode type BYW96E.

Two alternative types have previously been used. They were:

Type F249—this device was bolted through the VT701 heat sink, being cylindrical in shape with a threaded stud on the cathode side.

Type SKE5F: A rectangular device with two off-set legs, one leg was attached to the heat sink with a small nut and bolt.

The latest device is co-axial and requires no attachment to a heat sink. In PC752 it can be fitted directly to the P.C.B. and should be stood off from the board. In PC558 it will be more convenient to fit the component vertically – leaving the axial leads as long as possible.

The cathode of the BYW96E is marked with a black circle round the body of the component.

Since it is fitted with substantial lead-outs it may be necessary to enlarge the holes in the P.C.B. to fit it. Care should be taken not to damage the copper print when drilling out these holes.

TX9 Series—Mains Fuse: Random fuse-blowing with no fault condition observed after replacement of fuse: It must be emphasised that due to the over-voltage and over-current protection circuits incorporated in these chassis, almost any fault condition can rupture the fuse and if a genuine fault condition does exist, the modifications listed will obviously not clear the problem.

It should be noted, however, that if the fault report indicates that a receiver which was working normally, failed whilst actually operating, the problem could be due to a mains-borne transient and the 002 version of L65 has been specifically engineered to alleviate problems attributable to this cause. If, on the other hand, the failure mode indicates a switch-on/switch-off failure, the Zener device now recommended in W85 position has been selected for improved characteristics in the mode which could be responsible for a random fuse-blowing report.

After ensuring that an intermittent fault condition does not exist, the following action is recommended.

PC1001 only. Alter the value of: C146 to 220µF, 16 volt.
R223 to 470 ohm 5% 0·5W fusible*.
R216 to 1k ohm, 0·5W.
TR67 should be replaced by the later 'in-house' device T6059V*.

PC1001 and PC1040: L65 should be changed to the later device held under P/No. 90D3-028-002* (Part Number printed on choke).

D85 (called up as W85 on PC1001) should be changed to the currently approved Zener Diode held under P/No. 02V4-718* (The device can be recognised by the 3-segmented circle symbol printed on the body).

C134 and C135 should be replaced by 10kpF+80/−20% 1kV capacitors held under P/No. C5100-EW410-CBC1.

Components listed above marked with an asterisk are either 'Safety' devices or 'in-house' types and as such should only be obtained from TEF Service Division sources.

Model 3T12

Failure of S202 (Tape/Radio/Sleep Switch): Deterioration of the switch contacts of S202 on early receivers may be experienced due to C231 discharging directly into C227 via the switch.

In later production, a 1 ohm, 1 watt resistor (R107) was added as shown in the circuit Fig. 1. It is recommended that this modification is carried out as a routine procedure on any unit returned for service where the resistor is not already fitted.

Mechanically, the resistor should be mounted vertically on the Power Supply P.C.B. The existing Red lead to S202B should be unsoldered from the + ve tag on the P.S.U., the resistor soldered to this tag and the sleeving slipped over the red lead before reconnecting it to the free end of the resistor. The sleeve should then be slipped down over the resistor and joint.

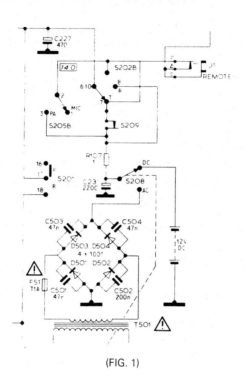

(FIG. 1)

INDEX

* indicates a cross-reference to information given in earlier volumes of
Radio and Television Servicing.

AMSTRAD

Television Receivers:
CTV1400, 9

Radio Receivers etc.:
6011, 416
8060, 425

BLAUPUNKT

Television Receivers:
FM120 Chassis, 15

BUSH

Radio Receivers etc.:
6090, 430
9450, 434

DECCA

see TATUNG

DORIC

Television Receivers:
Mk. 4 Series
CD51402D, 26
CD51404DRC, 26
CD56402D, 26
CD56404DRC, 26
CD67402D, 26
CD67404DRC, 26
CU51402D, 26
CU51404DRC, 26
CU56402D, 26
CU56404DRC, 26
CU67402D, 26
CU67404DRC, 26
CV51402D, 26
CV51404DRC, 26
CV56402D, 26
CV56404DRC, 26

DORIC *(continued)*

CV67402D, 26
CV67404DRC, 26

EVER READY

Radio Receivers etc.:
'Skytime', 440

FERGUSON

Television Receivers:
3781, 754*
3795, 754*
3796, 754*
3870, 754*
3871, 754*
37003, 754*
38000, 754*
38010, 754*
3T14, 46

Radio Receivers etc.:
3R05, 442
3R06, 446
3R07 'Roadstar', 450
3R11, 454
3T12, 460
3T16, 468
3T17, 476
3T18, 484
3T19, 494
3T20, 502
3T21, 510
3967, 517
3967B, 517
3967C, 517
3967D, 517
3968A, 534
3968B, 534
3977, 542
Music Centre 90, 534
Music Centre 100D, 517
Music Centre 101, 542

FIDELITY

Radio Receivers etc.:
1S System, 543

G.E.C.

Television Receivers:
C1615H, 754*
'Starline', 60
C2065H, 60
C2067H, 60
C2069H, 60
C2265H, 60
C2267H, 60
C2269H, 60
C2273H, 60

GRUNDIG

Television Receivers:
CUC120 Chassis, 83
CUC220 Chassis, 93
CUC720 Chassis, 102
A3102/5, 83
A3402/5, 93
A3412/5, 93
A4202/5, 83
A4402/5, 93
A5102/5, 83
A5402/5, 93
A6100/2/3/4, 83
A6200/2/3/4, 83
A6400/2/3/4/7, 93
A6410/2/3/4, 93
A7100/2/3/4, 83
A7110/2/3, 83
A7200/2/3, 83
A7210/2/3, 83
A7400/2/3/7, 93
A7401/2/3, 93
A7410/2/3, 93
A8400/2/3/6/7, 102
A8410/2/3, 102
A8420/2/3, 102